Roma International Conference on Astroparticle Physics 2014 (RICAP-14)

EPJ Web of Conferences Volume 121 (2016)

Noto, Italy
30 September - 3 October 2014

Editors:

P. Piattelli
R. Coniglione
M. De Vincenzi
A. Morselli
A. Capone
G. De Bonis
C. Distefano
P. Sapienza

ISBN: 978-1-5108-2604-5

Printed from e-media with permission by:

Curran Associates, Inc.
57 Morehouse Lane
Red Hook, NY 12571

proceedings.com

Some format issues inherent in the e-media version may also appear in this print version.

Printed by Curran Associates, Inc. (2016)

For additional information, please contact EDP Sciences – Web of Conferences
at the address below.

EDP Sciences – Web of Conferences
17, Avenue du Hoggar
Parc d'Activité de Courtabœuf
BP 112
F-91944 Les Ulis Cedex A
France

Phone: +33 (0) 1 69 18 75 75
Fax: +33 (0) 1 69 28 84 91

contact-edps@webofconferences.org

Additional copies of this publication are available from:

Curran Associates, Inc.
57 Morehouse Lane
Red Hook, NY 12571 USA
Phone: 845-758-0400
Fax: 845-758-2633
Email: curran@proceedings.com
Web: www.proceedings.com

TABLE OF CONTENTS

Local Organizing Committee

- Antonio Capone (Università di Roma Sapienza & INFN, Roma)
- Marco Ciuchini (INFN, Roma3)
- Rosa Coniglione (INFN, LNS)
- Giacomo Cuttone (INFN, LNS)
- Giulia De Bonis (INFN, Roma)
- Mario De Vincenzi (Università Roma Tre & INFN, Roma)
- Marcella Diemoz (INFN, Roma)
- Carla Distefano (INFN, LNS)
- Francesco Fucito (Università Roma Tor Vergata & INFN, Roma)
- Aldo Morselli (Università Roma Tor Vergata & INFN, Roma)
- Paolo Piattelli (INFN, LNS) – Chair
- Piera Sapienza (INFN, LNS)

International Organizing Committee

- Felix Aharonian (Dublin Institute for Advanced Studies, Ireland)
- Roberto Battiston (University & INFN, Perugia, Italy)
- Tom Gaisser (University of Delaware, USA)
- Francis Halzen (University of Wisconsin, USA)
- Juan Jose Hernandez (University of Valencia, Spain)
- Paolo Lipari (INFN, Roma, Italy)
- Szabolcs Marka (Columbia University, Ney York, USA)
- Antonio Masiero (University & INFN, Padova, Italy)
- Giorgio Matthiae (University "Tor Vergata" & INFN, Roma, Italy)
- Carlos Muñoz (Universidad Autonoma de Madrid, Spain)
- Christian Spiering (DESY Zeuthen, Germany)
- Eli Waxman (The Weizmann Inst. of Science, Rehovot, Israel)
- Juande Zornoza (Instituto de Física Corpuscular, Valencia, Spain)

Sponsors

Università di Roma "Sapienza"

Università degli Studi di Roma Tor Vergata

Università degli Studi Roma Tre

Istituto Nazionale di Fisica Nucleare

Associazione Romana per le Astro-Particelle

Hamamatsu Photonics

C.A.E.N.

edp sciences EPJ Web of Conferences

Statement of Peer review

In submitting conference proceedings to EPJ Web of Conferences, the editors of the proceedings certify to the Publisher that they adhere to the **Policy on Publishing Integrity** of the journal in order to safeguard good scientific practice in publishing.

1. All articles have been subjected to peer review administered by the proceedings editors.
2. Reviews have been conducted by expert referees, who have been requested to provide unbiased and constructive comments aimed, whenever possible, at improving the work.
3. Proceedings editors have taken all reasonable steps to ensure the quality of the materials they publish and their decision to accept or reject a paper for publication has been based only on the merits of the work and the relevance to the journal.

Title, date and place of the conference

ROMA INTERNATIONAL CONFERENCE ON ASTROPARTICLE PHYSICS 2014 (RICAP-14)
NOTO (ITALY)
SEPTEMBER 30 - OCTOBER 3, 2014

Proceedings editor(s):

PAOLO PIATTELLI (CHAIR)
ANTONIO CAPONE
ROSA CONIGLIONE
GIULIA DE BONIS
MARIO DE VINCENZI
CARLA DISTEFANO
ALDO MORSELLI
PIERA SAPIENZA

Date and editor's signature 20/2/2016

17, avenue du Hoggar - PA de Courtabœuf – BP 112 - 91944 Les Ulis Cedex A (France)
Tél. : 33 (0)1 69 18 75 75 - Fax : 33(0)1 69 07 45 17 – www.edpsciences.org

Preface

RICAP-14 was the fifth edition of the RICAP Conference held for the first time outside Rome.

Organized by the Laboratori Nazionali del Sud of the Istituto Nazionale di Fisica Nucleare, the Conference was hosted in the beautiful town of Noto by the Consorzio Universitario Mediterraneo Orientale (CUMO).

The conference was an excellent opportunity to discuss the most recent results in astroparticle physics and debate some of its outstanding open questions. The most recent results from cosmic-rays, high-energy gamma rays and high-energy neutrino experiments were presented with a particular focus on multi-messenger searches. An outlook towards the future experiments was also the aim of a dedicated session. Dark Matter searches were also addressed.

Our thanks go to the Consorzio Universitario Mediterraneo Orientale and to its Director Prof. Salvatore Cavallo for the warm hospitality in their beautiful building.

We would also like to thank the Mayors and the municipalities of Noto and Portopalo di Capo Passero for the support to the conference and in particular for having granted the patronage to the event.

Finally our thanks go to the members of the International Advisory Committee, who helped in setting up an outstanding programme, the chairpersons and all the participants.

Paolo Piattelli (for the organizing committee)

EPJ Web of Conferences **121**, 01001 (2016) DOI: 10.1051/epjconf/201612101001
RICAP-2014

First real–time detection of solar pp neutrinos by Borexino

M. Pallavicini[1,2,a], G. Bellini[3], J. Benziger[4], D. Bick[5], G. Bonfini[6], D. Bravo[7], B. Caccianiga[3], F. Calaprice[8], A. Caminata[1,2], P. Cavalcante[6], A. Chavarria[8], A. Chepurnov[9], D. D'Angelo[3], S. Davini[10], A. Derbin[11], A. Empl[10], A. Etenko[12], K. Fomenko[13,6], D. Franco[14], F. Gabriele[6], C. Galbiati[8], S. Gazzana[6], C. Ghiano[1,2], M. Giammarchi[3], M. Göger-Neff[15], A. Goretti[8], M. Gromov[9], C. Hagner[5], E. Hungerford[10], Al. Ianni[6], An. Ianni[8], M. Kayser[5], V. Kobychev[25], D. Korablëv[13], G. Korga[11], D. Kryn[14], M. Laubenstein[6], B. Lehnert[17], T. Lewke[15], E. Litvinovich[16], F. Lombardi[6], P. Lombardi[3], L. Ludhova[3], G. Lukyanchenko[12], I. Machulin[16], S. Manecki[7], W. Maneschg[19], S. Marcocci[24], Q. Meindl[5], E. Meroni[3], M. Meyer[5], L. Miramonti[3], M. Misiaszek[18], M. Montuschi[6], P. Mosteiro[8], V. Muratova[11], L. Oberauer[15], M. Obolensky[14], F. Ortica[19], K. Otis[21], L. Papp[7], L. Perasso[2], A. Pocar[21], G. Ranucci[3], A. Razeto[6], A. Re[3], A. Romani[19], N. Rossi[6], R. Saldanha[8], C. Salvo[2], S. Schönert[15], H. Simgen[19], M. Skorokhvatov[12,16], O. Smirnov[13], A. Sotnikov[13], S. Sukhotin[12], Y. Suvorov[12,22], R. Tartaglia[6], G. Testera[2], D. Vignaud[14], R.B. Vogelaar[7], F. von Feilitzsch[15], H. Wang[22], J. Winter[15], M. Wojcik[18], M. Wurm[23], O. Zaimidoroga[13], S. Zavatarelli[2], K. Zuber[17], and G. Zuzel[18]

[1] Università di Genova, Dipartimento di Fisica, via Dodecaneso 33, 16146 Genova, Italy
[2] INFN Sezione di Genova, via Dodecaneso 33, 16146 Genova, Italy
[3] Dipartimento di Fisica, Università degli Studi and INFN, Milano 20133, Italy
[4] Chemical Engineering Department, Princeton University, Princeton, NJ 08544, USA
[5] Institut für Experimentalphysik, Universität Hamburg, Germany
[6] INFN Laboratori Nazionali del Gran Sasso, Assergi 67010, Italy
[7] Physics Department, Virginia Polytechnic Institute and State University, Blacksburg, VA 24061, USA
[8] Physics Department, Princeton University, Princeton, NJ 08544, USA
[9] Lomonosov Moscow State University Skobeltsyn Institute of Nuclear Physics, Moscow 119234, Russia
[10] Department of Physics, University of Houston, Houston, TX 77204, USA
[11] St. Petersburg Nuclear Physics Institute, Gatchina 188350, Russia
[12] NRC Kurchatov Institute, Moscow 123182, Russia
[13] Joint Institute for Nuclear Research, Dubna 141980, Russia
[14] APC, Univ. Paris Diderot, CNRS/IN2P3, CEA/Irfu, Obs. de Paris, Sorbonne Paris Cité, France
[15] Physik Department, Technische Universität München, Garching 85747, Germany
[16] National Nuclear Research University "MEPhI", 31 Kashirskoe shosse, Moscow, Russia
[17] Technische Universität Dresden, Institut für Kern und Teilchenphysik, Zellescher Weg 19, 01068 Dresden, Germany
[18] M. Smoluchowski Institute of Physics, Jagiellonian University, Crakow 30059, Poland
[19] Dipartimento di Chimica, Biologia e Biotecnologie, Università e INFN, Perugia 06123, Italy
[20] Max-Plank-Institut für Kernphysik, 69117 Heidelberg, Germany
[21] Physics Department, University of Massachusetts, Amherst, MA 01003, USA
[22] Physics ans Astronomy Department, University of California Los Angeles (UCLA), Los Angeles, CA 90095, USA

[a]e-mail: marco.pallavicini@ge.infn.it

[23] Institut für Physik, Johannes Gutenberg Universität Mainz, Germany
[24] Gran Sasso Science Institute (INFN), 67100 L'Aquila, Italy
[25] Kiev Institute for Nuclear Research, Kiev 06380, Ukraine

Abstract. Solar neutrinos have been pivotal to the discovery of neutrino flavour oscillations and are a unique tool to probe the reactions that keep the Sun shine. Although most of solar neutrino components have been directly measured, the neutrinos emitted by the keystone pp reaction, in which two protons fuse to make a deuteron, have so far eluded direct detection. The Borexino experiment, an ultra–pure liquid scintillator detector running at the Laboratori Nazionali del Gran Sasso in Italy, has now filled the gap, providing the first direct real time measurement of pp neutrinos and of the solar neutrino luminosity.

1. Introduction

The study of low energy solar neutrinos (i.e. below 1–2 MeV, see Fig. 1) is relevant for two main reasons: on one hand, they offer a unique opportunity to investigate the behaviour of the Sun's interior and test the predictions of the Standard Solar Model; on the other, they allow to probe the MSW-LMA neutrino oscillation scenario in an energy range that is not accessible to water Cherenkov detectors [1, 2], which can detect solar neutrinos only above an energy threshold of about 3.5 MeV.

The original main goal of Borexino [3, 4] was the precise measurement of the rate induced by the monochromatic electron neutrinos (0.862 keV) produced by the electron capture decay of ^{7}Be in the Sun [5, 6]. However, the very high radio purity of the scintillator offered new unexpected results, such as a clear evidence of the pep solar neutrinos [7], a low energy threshold (3 MeV) detection of ^{8}B neutrinos [8], an unambiguous detection of geo-neutrinos [9], and several additional results on the search of rare events [10–13].

In this paper we report about an even less anticipated result, namely the real time detection of the solar neutrinos of lowest energy, the so called pp neutrinos, which were detected in the early 90's only by means of geochemical techniques [14], and its flux extracted indirectly. The Borexino direct measurement was made possible by the very low radioactive background, particularly in ^{85}Kr, achieved after an extensive purification campaign performed in 2010-2011 and thanks to the extremely good performance of the detector as a whole.

The pp reaction $pp \longrightarrow d + e^+ + \nu_e$ in the core of the Sun is the keystone process for energy production and is the source of the largest component of the neutrino flux. Its measurement is a major experimental milestone in solar neutrino physics, and paves the way to a deeper understanding of the Sun dynamics through a high precision comparison between photon and neutrino luminosity that future Borexino inspired detectors will be able to achieve.

Borexino is a liquid scintillator calorimeter [16–18] designed for the real–time observation of low energy solar neutrinos and anti–neutrinos. Neutrinos are detected through their elastic scattering on electrons and by means of the consequent emission of scintillation light, which is produced with a yield of about 500 p.e./MeV; anti-neutrinos are detected through inverse beta decay on protons. The total cross section of electron neutrinos depends on both charged and neutral currents weak interactions, while that of other neutrino flavours is induced by neutral currents only. The interaction rate depends therefore on the neutrino type at the target, a fact that makes the experiment sensitive to neutrino oscillations occurring along the path from the production site in the Sun's up to the detector.

The inner part of the Borexino detector [16] is an unsegmented stainless steel sphere (SSS) that is both the container of the scintillator and the mechanical support of the photomultipliers. Within this sphere, two nylon vessels separate the scintillator volume in

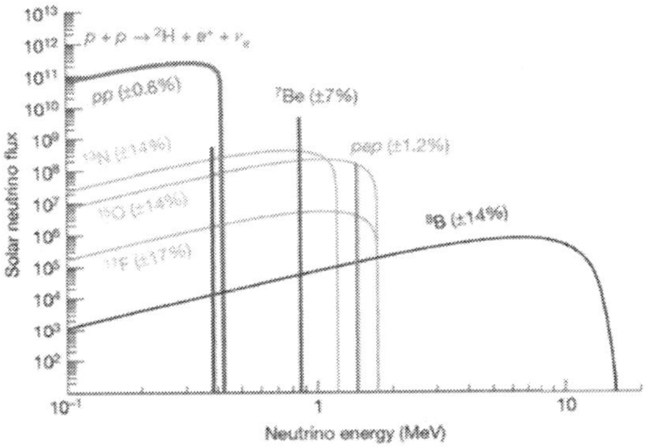

Figure 1. The spectrum of solar neutrinos as predicted by the Standard Solar Model [15].

three shells of radii 4.25 m, 5.50 m and 6.85 m, the latter being the radius of the SSS itself. The inner nylon vessel (IV) contains the liquid scintillator solution (PC+PPO), while the second and the third shell contain PC with a small amount of light quencher in order to further reduce the scintillation yield of pure PC. The PC/PPO solution that we adopted as liquid scintillator guarantees high scintillation yield ($\approx 10^4$ photons/MeV), high light transparency, and fast decay time (≈ 3 ns), all essential for good energy resolution, precise spatial reconstruction, and good discrimination between β-like events and events due α particles [19].

The scintillation light is collected by 2212 photomutipliers (PMTs) that are uniformly attached to the inner surface of the SSS. All but 384 PMTs are equipped with light concentrators that are designed to reject photons not coming from the active scintillator volume, thus reducing the background due to radioactive decays originating in the buffer liquid or γ's from the PMTs. The 384 PMTs without concentrators are used to study this background, and to help identify muons that cross the buffer, but not the Inner Vessel. Due to ageing, the actual number of operational PMTs during data taking was normally smaller, typically about 1800 in total.

The SSS is supported by 20 steel legs and enclosed within a large tank (WT) that is filled with ultra-pure water. The WT has a cylindrical base with a diameter of 18 m and a hemispherical top with a maximum height of 16.9 m. The WT is a powerful shielding against external background (γ rays and neutrons from the rock) and is also used as a Cherenkov muon counter and muon tracker. The muon flux, although reduced by a factor of 10^6 by the 3800 m w.e. depth of the Gran Sasso Laboratory, is of the order of 1 m^{-2} h^{-1}, corresponding to about 4000 muons per day crossing the detector. A strong additional reduction factor (about 10^4) is necessary to meet the Borexino requirements. Therefore the WT is equipped with 208 photomultipliers that collect the Cherenkov light emitted by muons in water. More details about the detector and its design principles can be found in [16–18].

The very high purity of the scintillator and the low radioactive background obtained in the innermost volume of the detector are the unique feature that has allowed all Borexino measurements and the one described in this paper in particular. The achievement of this good performance is the result of a long R&D effort that is described elsewhere [3, 4, 20, 21].

Borexino data taking is split into two main periods: the so-called Phase-I, which lasted from 2007 until 2011, where we measured ^{7}Be, ^{8}B, pep and geo-neutrinos; the Phase-II, which has begun at the end of 2011 and which is still in progress. The pp-neutrinos analysis was done with Phase-II data only, being Phase-I data affected by a non–negligible ^{85}Kr background. In the next sections we describe the pp analysis and discuss the result.

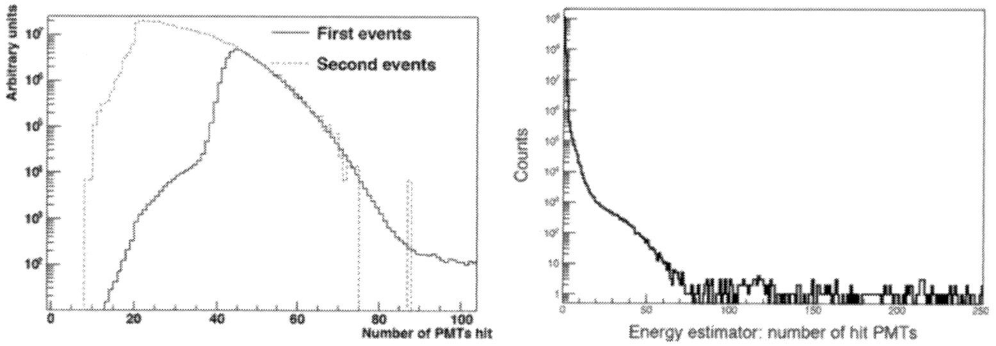

Figure 2. Left: spectra of first and second events in trigger windows acquired during the live time considered in the pp analysis. Second clusters are not subject to the effects of trigger threshold, and are thus useful for the independent estimate of the ^{14}C rate. Right: spectrum obtained by slicing random triggers into smaller time windows with the same duration as those used in regular data for the computation of our energy estimator.

2. pp neutrino analysis

The data used in this analysis was collected between Jan. 2012 and Jun. 2013, containing the first 408 live days of data of Borexino Phase-II. These data are characterised by reduced levels of some of the most relevant radiogenic backgrounds, particularly ^{85}Kr, ^{210}Bi and ^{210}Po, thanks to a series of successful purification campaigns based on water extractions and low Argon-Krypton nitrogen sparging [23]. Basic selection criteria of this analysis are the same as previous Borexino solar neutrino papers and have been described in detail in [23], in the context of Borexino Phase-I. In particular, the criteria used in the pp analysis are:

1. No coincidence with muon events, identified both the the Cherenkov detector and with pulse–shape analysis in the scintillator;
2. Event triggered by a scintillation event in the inner detector;
3. No electronic noise from PMTs or electronics racks;
4. No coincidence with isotopic decays in the radon branch of the uranium series;
5. Position reconstruction within the fiducial volume; for this analysis, the fiducial volume is a truncated sphere with radius R = 3.021 m, $|z|$ = 1.67 m; the fiducial volume is 86 m^3, corresponding to 75.5 ton of scintillator target.

Most of the remaining background is ^{14}C β-decays, whose end–point is 156 keV, very close to the pp region of interest. To estimate its rate independently from the main analysis, we looked at a sample of data in which the event causing the trigger is followed by a second event within the same trigger window (16 μs long, while physical events last 100–500 ns). These second events are not subject to the trigger threshold (the typical threshold was 25 hits during data taking, corresponding to approximately 40 keV in β-equivalent energy deposit), so they register down to much lower energies. The spectra of first and second events are shown in Fig. 2–left. By fitting this spectrum against the known ^{14}C spectral shape, we obtain a ^{14}C rate in Borexino of 40 $\pm$ 1 Bq per 100 t. The error accounts for statistical and systematical uncertainties, most notably due to uncertainty in the ^{14}C spectral shape and fit conditions. Due to its high rate, ^{14}C naturally generates a considerable amount of pile-up: two physical events occur so close to each other in time that the detector fails to resolve them. With the above measurement of the ^{14}C rate, the expected rate of ^{14}C-^{14}C pile-up is 100 per day per 100 t. This is comparable to the expected pp neutrino interaction rate. In addition, the end–point of the ^{14}C-^{14}C pile-up spectrum is 312 keV (twice the ^{14}C spectrum end-point), also close to the endpoint of the pp-induced recoil spectrum, E_R^{max}=264 keV. This underlines the importance of a careful assessment of pileup events, both ^{14}C-^{14}C and other pile-up.

The pile-up component can be determined using an independent, data-driven method which we called synthetic pileup. Real triggered events with no cuts are artificially overlapped with random data obtained from the ends of real trigger windows, uncorrelated with the triggering event. By overlapping four random data samples with each physical trigger, we obtain a sample of synthetic events that has four times the exposure of the real data set used. The synthetic events are then reconstructed using the same software used for real events, and selected with the same criteria. Using this method, we obtained the true rate and spectral shape of pile-up in our detector. This method naturally accounts for the effect of fiducialization on pile-up events.

With the ^{14}C and pile-up rates fixed, we obtained the pp rate by fitting the data spectrum in a window of 165–590 keV against the known spectral shapes. The fit was performed with a tool named spectral-fitter, previously used in ^{7}Be [5] and pep [7] analyses. The tool was improved to work on any Borexino analysis. In particular, we included a treatment of the scintillator energy response to mono-energetic electrons at high statistics; a modified treatment of the energy resolution at low energy; and the introduction of synthetic pile-up. In addition to pp, ^{14}C and pile-up, other species included in the fit are other solar neutrinos (^{7}Be, CNO and pep), ^{85}Kr, ^{210}Bi, ^{210}Po and ^{214}Pb. The ^{7}Be rate was fixed to the rate previously obtained by Borexino [5]; CNO and pep rates were fixed to the values predicted by the high-metallicity Standard Solar Model; remaining background rates are left free in the fit, except for ^{214}Pb, which is fixed to the value measured by looking at ^{214}Bi-^{214}Po coincidences. The scintillator light yield and two energy resolution parameters are free to vary in the fit. The energy spectrum after cuts, together with the best fit, is shown on Fig. 3. The uncertainties shown in the figure are statistical only.

We considered also an alternative method to account for pile-up, which we called the convolution method. Random trigger with 16 μs-long samples of data are regularly collected by Borexino. These data are sliced into time windows of the same duration as the window used for our energy estimator. By computing the energy estimator inside each of these small windows, we generate the distribution shown in Fig. 2–right. This distribution is the spectrum collected by Borexino without a trigger threshold, and it includes electronic noise, as well as all physical events generated by signal and background species. Pile-up can be generated by convolving each spectral component (neutrinos and backgrounds) with this spectrum. Fit conditions were varied to estimate the systematic uncertainty. Effects considered include variations in: 1) The pile-up evaluation method (synthetic and convolution); 2) The length of the time window in which we count PMTs hit; 3) The definition of the fiducial volume; 4) The expected value of the CNO neutrino detection rate; 5) The fit start and end points.

Fits were performed using many different combinations of all these. The histogram of the rates obtained in all these conditions is shown in Fig. 4. The root mean square of the distribution is 9.8 $(d \cdot 100\,t)^{-1}$. In addition, the resolution of the position reconstruction algorithm (20 cm; Sect. 2) induces a 2% uncertainty in the determination of the fiducial volume. These uncertainties are added in quadrature to obtain the total systematic uncertainty of our measurement. The final pp neutrino interaction rate is:

$$R_{pp} = 144 \pm 13(stat) \pm 10(syst)\,(d \cdot 100\,t)^{-1}.$$

We also studied possible variations in the result caused by: 1) uncertainties in the spectral shapes of ^{214}Po and ^{210}Bi; 2) variations of the expected pep rate within the measured and theoretical uncertainties; 3) leaving the ^{214}Po rate free in the fit; 4) varying the veto time after muons cross the detector. No significant variations were observed. We also studied as a potential source of background the β-emitter ^{87}Rb, with a Q-value of 283.3 keV, very

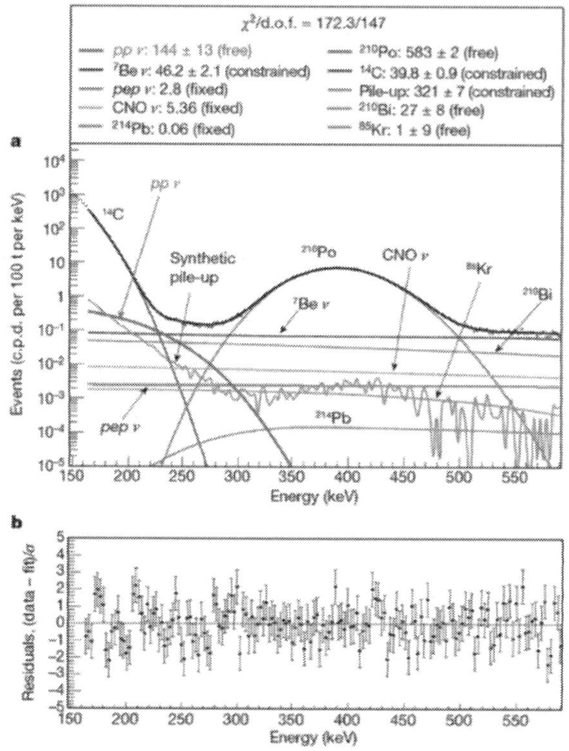

Figure 3. a) The fit result with all components explained within the figure. The fitted pp spectrum is shown in red; b) fit residuals as a function of the energy in the region of interest of pp neutrinos.

similar to the end-point of the pp-induced electron recoil spectrum E_R^{max}. Using the known crustal abundances, isotopic abundances and half–lives of ^{87}Rb and ^{40}K, we can calculate the ratio between their natural activities to be 1 to 4, respectively. Under the assumption that the relative abundances of ^{87}Rb and ^{40}K in Borexino are equal to their relative crustal abundances, and knowing that the rate of ^{40}K in Borexino is < 0.4 (d · 100 t)$^{-1}$, we can constrain the ^{87}Rb rate to <0.1 (d· 100 t)$^{-1}$.

3. Interpretation and conclusions

The electron neutrino survival probability, P_{ee}, is the probability for a solar neutrino produced as electron type to arrive at Earth as an electron-neutrino. It can be computed from the most up-to-date oscillation parameters and interaction cross sections, and the electron number density in Borexino, $(3.307 \pm 0.003) \times 10^{31}$ (100 t)$^{-1}$ [6] (see [22] for details). Using that number and our measured pp neutrino interaction rate, we can calculate the pp neutrino flux to be $(6.6 \pm 0.7) \times 10^{10}$ cm^{-2} s^{-1}, consistent with the SSM prediction. The uncertainty of the present measurement does not allow us to distinguish between the high and low–metallicity solar models.

Conversely, using pp neutrino flux predicted by the SSM as input, we calculate the survival probability $P_{ee} = 0.64 \pm 0.12$, consistent with expectations from the LMA-MSW solution [24], as shown in Fig. 5. Finally, the fact that the present measurement of the pp neutrino flux is consistent with expectation justifies the assumption made by the SSM that the photon and neutrino luminosities from the Sun can be correlated; i.e., that the Sun has

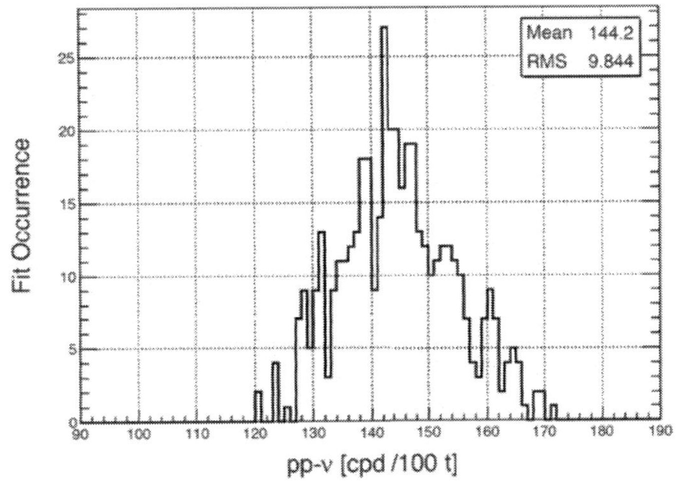

Figure 4. Distribution of the fit central value obtained by varying several fit conditions and parameters; the central value was taken as measurement central value; the sigma was used to estimate the systematic error to the measurement.

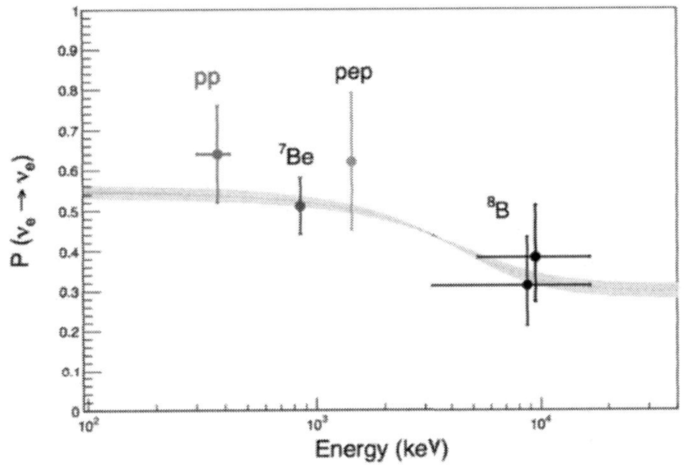

Figure 5. The contribution of Borexino to the experimental knowledge of the solar neutrino electron survival probability. In red, the measurement described in this work.

been stable for the past 10^5 years. This invalidates a previous hypothesis that changes in the solar core over these time scales could be responsible for terrestrial global epochs [25].

In conclusion, the pp reaction in the Sun is the keystone process for energy production and is the source of the largest component of neutrino flux. The observation of the low-energy (0–420 keV) pp neutrinos made by Borexino was possible because of the unprecedentedly low level of radioactivity reached inside the detector. The measured value is in good agreement with the predictions of both the high-metallicity and the low-metallicity SSMs. Although the experimental uncertainty does not yet allow the details of these models to be distinguished, this measurement confirms our understanding of the Sun. Future Borexino-inspired experiments might be able to measure solar pp neutrinos with the level of precision ($\approx$1% or better) needed to cross-compare photon and neutrino solar luminosities, while providing insight into solar dynamics over 10^5-yr timescales. At the same time, such a precise measurement of pp neutrinos would yield the ultimate test for the MSW-LMA neutrino oscillation model and allow precision tests for exotic neutrino properties.

The Borexino experiment has measured all neutrino components related to the pp cycle. An attempt to measure CNO neutrinos, and therefore contribute to the solution of the metallicity problem, will be done in the next years, possibly with the help of additional purification cycles. The Borexino detector will also be used to search for sterile neutrinos by means of artificial neutrino and anti-neutrinos sources in the framework of the SOX project [26].

The Borexino program is made possible by fundings from INFN (Italy); NSF (USA); BMBF, DFG and MPG (Germany); JINR, RFBR, RSC and NRC Kurchatov Institute (Russia); and NCN (Poland). We acknowledge the generous support of the Laboratori Nazionali del Gran Sasso (Italy) and that of the Gran Sasso Science Institute. SOX is funded by ERC with grant ERC-AdG-2012 N. 320873.

References

[1] K. Abe et al. (Super-Kamiokande Collaboration), Phys. Rev. **D 83** (2011) 052010
[2] B. Aharmin et al. (SNO collaboration), Phys. Rev. **C 81** (2010) 055504
[3] G. Alimonti et al. (Borexino Collaboration), Astropart. Phys. **16** (2002) 205
[4] C. Arpesella et al. (Borexino Collaboration), Astropart. Phys. **18** (2002) 1
[5] G. Bellini et al. (Borexino Collaboration), Phys. Rev. Lett. **107** (2011) 141302
[6] G. Bellini et al. (Borexino Collaboration), Phys. Lett. **B 707** (2012) 22–26
[7] G. Bellini et al. (Borexino collaboration), Phys. Rev. Lett. **108** (2012) 051302
[8] G. Bellini et al. (Borexino collaboration), Phys. Rev. **D 82** (2012) 033006
[9] G. Bellini et al. (Borexino collaboration), Phys. Lett. **B 722** (2013) 295–300
[10] G. Bellini et al. (Borexino collaboration), JCAP **1308** (2013) 049
[11] G. Bellini et al. (Borexino collaboration), Phys. Rev. **D85** (2012) 092003
[12] G. Bellini et al. (Borexino collaboration), JINST **6** (2011) P05005
[13] G. Bellini et al. (Borexino collaboration), Phys. Lett. **B696** (2011) 191–196
[14] W. Hampel et al. (Gallex Collaboration), Phys. Lett. **B447** (1999) 127–133
[15] A.M. Serenelli, W.C. Haxton, W. C and C. Penã-Garay, Astrophys. J. **743** (2011) 24
[16] G. Alimonti et al. (Borexino collaboration), Nucl. Instr. and Meth. **A 600** (2009) 568
[17] G. Bellini et al. (Borexino collaboration), JINST **6** (2011) P05005
[18] G. Bellini et al. (Borexino collaboration), Nucl. Instr. and Meth. **A 609** (2009) 58–78
[19] H.O. Back et al. (Borexino collaboration), Nucl. Instr. and Meth. **A 584** (2008) 98–113
[20] H.O. Back et al. (Borexino collaboration), Phys. Lett. **B 563** (2003) 23–34
[21] H.O. Back et al. (Borexino collaboration), Phys. Lett. **B 525** (2002) 29–40
[22] G. Bellini et al. (Borexino Collaboration), Nature, **512** (2014) 383
[23] G. Bellini, et al. (Borexino Collaboration), Phys. Rev. **D 89** (2014) 112007
[24] J. N. Bahcall, P. I. Krastev, A. Y., Smirnov, Phys. Rev. **D 58** (1998) 096016
[25] G. A. Cowan, W. C. Haxton, Los Alamos Science **3** (2) (1982) 46
[26] G. Bellini et al. (Borexino Collaboration), JHEP **08** (2013) 038

EPJ Web of Conferences **121**, 01002 (2016) DOI: 10.1051/epjconf/201612101002
RICAP-2014

Short distance neutrino oscillations with Borexino

A. Caminata[1,a], M. Agostini[2], K. Altenmüller[2], S. Appel[2], G. Bellini[3], J. Benziger[4], N. Berton[5], D. Bick[6], G. Bonfini[7], D. Bravo[8], B. Caccianiga[3], F. Calaprice[9,10], P. Cavalcante[7], A. Chepurnov[11], M. Cribier[5], D. D'Angelo[3], S. Davini[10], A. Derbin[12], L. di Noto[1], M. Durero[5], A. Empl[13], A. Etenko[14,15], S. Farinon[1], V. Fischer[5], K. Fomenko[16], D. Franco[17], F. Gabriele[7], J. Gaffiot[5], C. Galbiati[9], C. Ghiano[7], M. Giammarchi[3], M. Göger-Neff[2], A. Goretti[9], M. Gromov[11], C. Hagner[6], Th. Houdy[5], E. Hungerford[19], Aldo Ianni[7], Andrea Ianni[9], N. Jonquères[20], M. Kaiser[6], V. Kobychev[21], D. Korablev[16], G. Korga[19], D. Kryn[17], T. Lachenmaier[22], T. Lasserre[5], M. Laubenstein[7], B. Lehnert[23], J. Link[8], E. Litvinovich[14,15], F. Lombardi[7], P. Lombardi[3], L. Ludhova[3], G. Lukyanchenko[14], I. Machulin[14,15], W. Maneschg[24], S. Marcocci[1,10], J. Maricic[25], G. Mention[5], E. Meroni[3], M. Meyer[6], L. Miramonti[3], M. Misiaszek[7,26], M. Montuschi[7], V. Muratova[12], R. Musenich[1], B. Neumair[2], L. Oberauer[2], M. Obolensky[17], F. Ortica[27], M. Pallavicini[1], L. Papp[2], L. Perasso[1], A. Pocar[28], G. Ranucci[3], A. Razeto[7], A. Re[3], A. Romani[27], N. Rossi[7], S. Schönert[2], L. Scola[5], H. Simgen[24], M. Skorokhvatov[14,15], O. Smirnov[16], A. Sotnikov[16], S. Sukhotin[14], Y. Suvorov[7,14,29], R. Tartaglia[7], G. Testera[1], C. Veyssière[5], M. Vivier[5], R.B. Vogelaar[8], F. von Feilitzsch[2], H. Wang[29], J. Winter[30], M. Wojcik[26], M. Wurm[30], O. Zaimidoroga[16], S. Zavatarelli[1], K. Zuber[23], and G. Zuzel[26]

[1] Dipartimento di Fisica, Università degli Studi e INFN, 16146 Genova, Italy
[2] Physik-Department and Excellence Cluster Universe, Technische Universität München, 85748 Garching, Germany
[3] Dipartimento di Fisica, Università degli Studi e INFN, 20133 Milano, Italy
[4] Chemical Engineering Department, Princeton University, Princeton, New Jersey 08544, USA
[5] Commissariat à l'Energie Atomique et aux Energies Alternatives, Centre de Saclay, IRFU, 91191 Gif-sur-Yvette, France
[6] Institut für Experimentalphysik, Universität Hamburg, 22761 Hamburg, Germany
[7] INFN Laboratori Nazionali del Gran Sasso, 67100 Assergi, Italy
[8] Physics Department, Virginia Polytechnic Institute and State University, Blacksburg, Virginia 24061, USA
[9] Physics Department, Princeton University, Princeton, New Jersey 08544, USA
[10] Gran Sasso Science Institute (INFN), 67100 L'Aquila, Italy
[11] Lomonosov Moscow State University Skobeltsyn Institute of Nuclear Physics, 119234 Moscow, Russia
[12] St Petersburg Nuclear Physics Institute, 188350 Gatchina, Russia
[13] Department of Physics, University of Houston, Houston, Texas 77204, USA
[14] NRC Kurchatov Institute, 123182 Moscow, Russia
[15] National Research Nuclear University MEPhI (Moscow Engineering Physics Institute), 115409 Moscow, Russia
[16] Joint Institute for Nuclear Research, 141980 Dubna, Russia
[17] APC, Université Paris Diderot, CNRS/IN2P3,CEA/Irfu, Observatoire de Paris, Sorbonne Paris Cité, 75205 Paris Cedex 13, France

[a] e-mail: alessio.caminata@ge.infn.it

[18] Institut für Experimentalphysik, Universität Hamburg, 22761 Hamburg, Germany
[19] Department of Physics, University of Houston, Houston, Texas 77204, USA
[20] Commissariat à l'Energie Atomique et aux Energies Alternatives, Centre de Saclay, DEN/DM2S/SEMT/BCCR, 91191 Gif-sur-Yvette, France
[21] Kiev Institute for Nuclear Research, 03680 Kiev, Ukraine
[22] Kepler Center for Astro and Particle Physics, Universitt Tübingen, 72076 Tübingen, Germany
[23] Department of Physics, Technische Universität Dresden, 01062 Dresden, Germany
[24] Max-Planck-Institut für Kernphysik, 69117 Heidelberg, Germany
[25] Department of Physics and Astronomy, University of Hawai'i, Honolulu, HI 96822, USA
[26] M. Smoluchowski Institute of Physics, Jagiellonian University, 30059 Krakow, Poland
[27] Dipartimento di Chimica, Biologia e Biotecnologie, Universita' degli Studi e INFN, 06123 Perugia, Italy
[28] Physics Department, University of Massachusetts, Amherst, Massachusetts 01003, USA
[29] Physics and Astronomy Department, University of California Los Angeles (UCLA), Los Angeles, California 90095, USA
[30] Institut für Physik, Johannes Gutenberg Universität Mainz, 55122 Mainz, Germany

Abstract. The Borexino detector has convincingly shown its outstanding performances in the low energy, sub-MeV regime through its unprecedented accomplishments in the solar and geo-neutrinos detection. These performances make it the ideal tool to accomplish a state-of-the-art experiment able to test unambiguously the long-standing issue of the existence of a sterile neutrino, as suggested by the several anomalous results accumulated over the past two decades, i.e. the outputs of the LSND and Miniboone experiments, the results of the source calibration of the two Gallium solar neutrino experiments, and the recently hinted reactor anomaly. The SOX project will exploit two sources, based on Chromium and Cerium, respectively, which deployed under the experiment, in a location foreseen on purpose at the time of the construction of the detector, will emit two intense beams of neutrinos (Cr) and anti-neutrinos (Ce). Interacting in the active volume of the liquid scintillator, each beam would create an unmistakable spatial wave pattern in case of oscillation of the ν_e (or $\bar{\nu}_e$) into the sterile state: such a pattern would be the smoking gun proving the existence of the new sterile member of the neutrino family. Otherwise, its absence will allow setting a very stringent limit on its existence.

1. Introduction

Neutrino flux measurements by real-time detectors have confirmed the neutrino oscillations between its three known flavor states. Borexino, a world-class example of this kind of detector, has measured different components of the solar neutrinos [1, 2], including the *pp* neutrinos, which are produced in the reaction that releases 99.77% of the Sun energy [3].

In the near future, Borexino aims to investigate the existence of neutrino oscillations in sterile states (beyond the known ν_e, ν_μ and ν_τ states) which are detectable as a deficit in a neutrino flux. The *Short distance neutrino Oscillations with BoreXino* project will explore the Δm^2_{14} region of $\sim 1\ eV^2$ [4].

2. The Borexino detector

Borexino is currently installed in Hall C of the Laboratori Nazionali del Gran Sasso (LNGS) in Italy. Its design is based on the principle of graded shielding with the inner scintillating core surrounded by concentric vessels of decreasing radio purity from inside to outside (Fig. 1). The active scintillator is a solution of PPO (2,5-diphenyloxazole), which

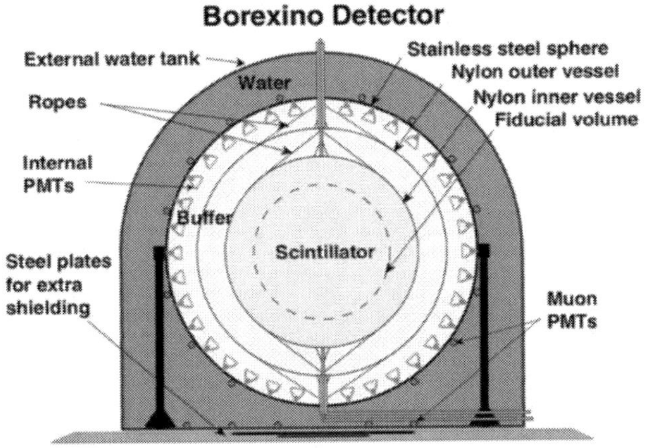

Figure 1. A schematic view of the Borexino detector.

is a wavelength-shifter, at a concentration of 1.5 g/l in pseudocumene (PC, 1,2,4-trimethylbenzene). The scintillator is contained in a 125 μm thick spherical inner nylon vessel (IV) with 4.25 m radius. The IV is surrounded by a passive shield made by PC in which a small quantity of DMP (dimethylphthalate) is added to inhibit the scintillation process. The scintillation light is collected by 2212 8 inch PMTs. The scintillator and the PMTs are located inside a 18-meter-diameter stainless steel sphere which is contained in a water Čerenkov tank acting as active muon[1] veto (efficiency greater than 99.99%).

The light yield is about 500 photoelectrons/MeV with a threshold of about 60 keV. Since scintillation light coming from nuclear decays inside the inner vessel is indistinguishable from the neutrino elastic-scattering off electrons in PC, an extreme radiopurity level must be maintained. Thanks to specific scintillator purification activities, an unprecedented level of several 10^{-19} grams of ^{238}U and ^{232}Th per gram of scintillator is reached. From the beginning of data taking, four major calibration campaigns with different kinds of radiation sources (α, β, γ and neutrons) have been taken place. The spatial resolution of the detector is about 10 cm at the energy of 1 MeV while the energy resolution is about 5% [1].

3. The SOX project

The SOX experiment [4] aims at the complete confirmation or at a clear disproof of the so-called neutrino anomalies, a set of circumstantial evidences of electron neutrino disappearance observed at LSND [5], MiniBoone [6], with nuclear reactors [7] and with solar neutrino Gallium detectors [8]. If successful, SOX will demonstrate the existence of sterile neutrino components and will open a brand new era in fundamental particle physics and cosmology. A solid signal would mean the discovery of the first particles beyond the Standard Electroweak Model and would have profound implications in our understanding of the Universe and of fundamental particle physics. In case of a negative result, it is able to close a long standing debate about the reality of the neutrino anomalies, would probe the existence of new physics in low energy neutrino interactions, would provide a measurement of neutrino

[1] The Gran Sasso d'Italia mountains, under which the LNGS are located, provide a 3600 meters of equivalent water shielding against cosmic rays.

magnetic moment, and would yield a superb energy calibration for Borexino which will be very beneficial for future high-precision solar neutrino measurements.

The Borexino detector is placed over a small tunnel designed to deploy in it neutrino sources to calibrate the experiment. The first stage of the experiment (SOX-A) will consist in the deployment of high-intensity ν_e (^{51}Cr) and $\bar{\nu}_e$ (^{144}Ce) sources in the pit, at a distance of 8.25 m from the detector's center.

3.1 ^{51}Cr neutrino source

A first option to test the existence of sterile neutrinos will be placing a 10 MCi ^{51}Cr source in the pit. The ^{51}Cr source will be produced by irradiating a large sample of 38%-^{50}Cr-enriched material in a nuclear reactor. Since a high flux nuclear reactor is needed, the enrichment could be done in High-Flux Isotope Reactor (HFIR) in Oak Ridge National Laboratory USA or in the Mayak reactor in Russia. ^{51}Cr decays via electron capture into ^{51}V emitting two neutrino lines of 750 keV (90%) and 430 keV (10%). Since the energy is very similar to the energy of ^{7}Be neutrinos (or close to the pp cut-off for the less dominant component), Borexino has already proven its capability to detect neutrinos of these energies. Due to ^{51}Cr short half-life ($t_{1/2} = 27.7$ days), a very quick transportation to LNGS is needed.

3.2 ^{144}Ce anti-neutrino source

A recently developed plan is to deploy a $\approx$ 150 kCi ^{144}Ce antineutrino source in the pit. Antineutrinos are detected by means on inverse beta decay (IBD, threshold 1.8 MeV) on protons. The low radioactivity and the clean tag offered by the space-time coincidence between the prompt e^+ and the subsequent neutron capture ($\tau = 254\ \mu s$ [9]) make accidental background essentially zero. The ^{144}Ce source β-decays to ^{144}Pr ($t_{1/2} = 296$ days) which rapidly ($\tau =$ 17 min) β-decays to ^{144}Nd emitting $\bar{\nu}_e$s. The detection threshold for IBD is 1.8 MeV, matching adequately the energy of the $\bar{\nu}_e$ emitted by the source (the end point of the ^{144}Pr is about 3 MeV).

3.3 Calorimetry and data analysis

The sources, with the proper shielding, will be placed in water-flow calorimeters to measure their activities. For the ^{51}Cr source, due to the shorter lifetime, calorimetry will be performed continuously during the data-taking while the source is deployed under the detector. For the ^{144}Ce source, the measurement will be performed before inserting the source under Borexino and at the end of data-taking, when the source will be removed from the pit. Calorimeters design is mostly finalized and a prototype with resistive heaters will soon be starting acceptance tests in the first part of 2015.

Borexino can study short distance neutrino oscillations in two ways. The first way is the standard disappearance technique (rate technique): if oscillations occur, the total count rate is lower than that expected without oscillations. The second way is based on an oscillometry measurement within the detector volume (rate+shape technique). If the values of Δm^2_{41} are of the order of 1 eV^2 and the neutrino energy is of the order of 1 MeV, the typical oscillation length is about few meters and, being the spatial resolution about 10 cm (Sect. 2), the resulting oscillations waves can be directly seen in Borexino [4]. This technique will allow a direct measurement of Δm^2_{41} and θ_{41}, investigating most of the anomaly phase space (Fig. 2). The best fit values for the current allowed phase space will be investigated to generate a discovery/exclusion signal at 5 σ [10]. In parallel with the sterile neutrino search, some precision measurements will be done. For example a much lower energy ($\approx$ 1 MeV)

DOI: 10.1051/epjconf/201612101002

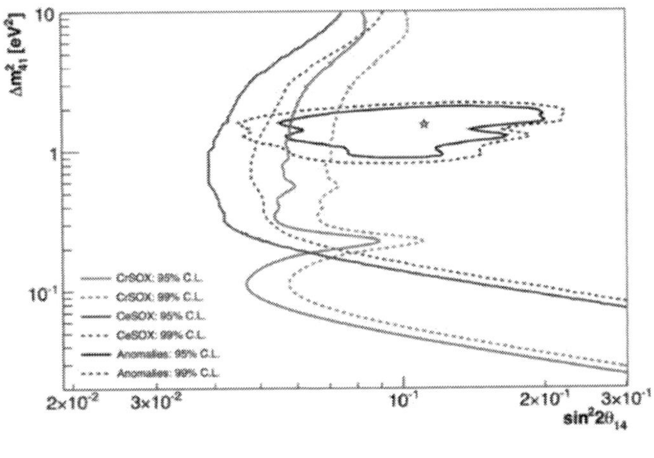

Figure 2. Allowed parameter space accordingly to one of the latest review of global fits [10] and region of space covered by SOX-A experiment.

measurement of the Weinberg angle, g_A and g_V parameters at low energies and studies of the neutrino magnetic moment.

The plan for the mid-term future is to insert a 75 kCi ^{144}Ce source inside the water tank (SOX-B, 2017) and in the center of Borexino (SOX-C, later). This projects require a refurbishment of the detector and a more challenging active cooling of the source and will be performed after the completion of the Borexino solar physics program. With SOX-B and SOX-C it will be possible to investigate the gallium+reactor anomaly 99%-allowed phase space to 95% and 99% confidence level respectively.

4. Conclusions

The Borexino detector is an ideal detector to search for sterile neutrinos. Thanks to the high radio-purity and the low energy threshold, it is possible to test the existence of sterile neutrinos using both a neutrino and an anti-neutrino source. The data taking will start at the end of 2015 deploying a ^{144}Ce anti-neutrino generator in the pit under the Borexino detector and the first results will be expected in 2016.

References

[1] G. Bellini et al., Physical Review D **89**, 112007 (2014)
[2] G. Bellini et al., Physical Review D **82**, 033006 (2010)
[3] G. Bellini et al., Nature **512**, 383 (2014)
[4] G. Bellini et al., Journal of High Energy Physics **38**, 1 (2013)
[5] A. Aguilar et al., Physical Review D **64**, 112007 (2001)
[6] A. Aguilar et al., Physical Review Letters **110**, 161801 (2013)
[7] G. Mention et al., Physical Review D **83**, 073006 (2011)
[8] C. Giunti et al., Physical Review C **83**, 065504 (2011)
[9] G. Bellini et al., Physics Letters B **722**, 295–300 (2013)
[10] C. Giunti et al., Physical Review D **88**, 073008 (2013)

EPJ Web of Conferences **121**, 02001 (2016) DOI: 10.1051/epjconf/201612102001
RICAP-2014

Gravitational waves and multimessenger astronomy

Fulvio Ricci[1,2,a]

[1] Dipartimento di Fisica, Università di Roma La Sapienza, Italy
[2] INFN Sezione di Roma, Rome, Italy

Abstract. It is widely expected that in the coming quinquennium the first gravitational wave signal will be directly detected. The ground-based advanced LIGO and Virgo detectors are being upgraded to a sensitivity level such that we expect to be measure a significant binary merger rate.
Gravitational waves events are likely to be accompanied by electromagnetic counterparts and neutrino emission carrying complementary information to those associated to the gravitational signals. If it becomes possible to measure all these forms of radiation in concert, we will end up an impressive increase in the comprehension of the whole phenomenon.
In the following we summarize the scientific outcome of the interferometric detectors in the past configuration. Then we focus on some of the potentialities of the advanced detectors once used in the new context of the multimessenger astronomy.

1. From the first to the second generation of GW detectors

Since 1960 various laboratory experiments have been carried on to detect gravitational waves (GW). The detector technique evolved from the simpler room resonant detectors of J. Weber to the cryogenic ones and at present the field is dominated by the operation of the giant interferometric devices installed in Europe, Japan and USA. The device sensitivity and its bandwidth is increased progressively expanding the search domain on a larger variety of GW signals. However, since the Weber's epoch it was clear that this experimental challenge requires the coordinate effort of more than one detector installed in different Earth locations.

The first attempt to observe a gravitational wave signal using a world wide network was in 1986 the coincidence experiment of the three cryogenic resonant-mass detectors set up by the universities of the Louisiana state, Rome and Stanford. [1]. In more recent years the International Gravitational Event Collaboration, derived upper limit on the event rate of GW bursts using observational data of the resonant detectors in operation taken during almost three years of run [2]. Nowadays, the network of the GW advanced detectors is based on all the data produced by the interferometers of the GEO [3], LIGO [4] and Virgo [5] allowing for maximal sky coverage and permitting the transient source location by triangulation. In the near future we expect that the KAGRA instrument [6], installed in Japan, will join the network improving significantly the capability of the sky localisation of the GW event.

GEO600 in Ruthe, Germany is a 600 m long arm interferometer in a dual-recycled configuration, i.e. it includes both power and signal recycling. It is equipped with a homodyne

[a] e-mail: `fulvio.ricci@roma1.infn.it`

detection system and quantum squeezing. The instruments results from an international collaboration of Germany and United Kingdom. It does not include Fabry-Perot cavities in its arms, as it is the case for LIGO and Virgo, and its best sensitivity corresponds to an amplitude spectral density $\sim 3 \times 10^{-21}\,1/\sqrt{\text{Hz}}$ in a 2 kHz bandwidth around 1 kHz.

Advanced LIGO are two interferometers, now in the commissioning phase, located in Hanford (Washington state, USA), and in Livingston (Louisiana state, USA). Groups at Caltech and MIT are operating these 4 km long arm instruments. In the past years LIGO achieved a sensitivity of about $2 \times 10^{-23}\,1/\sqrt{\text{Hz}}$ at 150 Hz.

The Virgo detector is installed at Cascina (Italy) and the European Gravitational Wave consortium of the CNRS-France and INFN-Italy is managing the site. The instrument is operated by an international collaboration collaboration of five european nation, which includes France, Italy, Nederland, Hungary and Poland. The facility has perpendicular 3 km long arms and took data during several scientific runs achieving a sensitivity of $6 \times 10^{-23}\,1/\sqrt{\text{Hz}}$ in a broad bandwidth around 200 Hz.

In 2008, the installation of the advanced LIGO (aLIGO) detectors was started and three years later the advanced Virgo project followed. The new instruments will use a higher power laser to reduce the shot noise at high frequencies, the signal recycling and fused silica monolithic suspensions for the last stage of mirror suspension, to reduce the thermal noise, which limits the sensitivity at low and mid frequencies. The advanced detectors have a designed sensitivity about ten times better than the initial detectors and thus will be capable of observing sources from a spatial volume a thousand times larger.

Both the advanced LIGO detectors at Hanford and Livingston are in the commissioning phase and we will begin operating in 2015, gradually achieving the final design sensitivity with many plausible detections.

The Virgo detector will join the network one year later providing localization.

Deep under the mountain of the Kamioka mine in Kamioka-cho, Gifu-prefecture of Japan, the Kamioka Gravitational Wave detector (KAGRA) is under construction. It is now likely to enter operation in 2018/2019. To achieve the required sensitivity, several advanced techniques will be employed such as a low-frequency vibration-isolation system, cryogenic mirrors, a resonant side band extraction method.

Since the detection probability increases with the number of detectors, it is very likely that gravitational wave detections will happen in the next quinquennium, starting the new era of gravitational wave astronomy. These signals produced by the coherent motion of large masses will permit to probe the properties of the gravitational interaction. The GW detection will open the study of gravitational field propagation, which is described by the Einstein General Relativity, a theory based on the principle of local invariance with respect to arbitrary transformations of the coordinate systems. This invariance is extended to the description of the force fields as it happen for example in the case of the electroweak interactions, providing a conceptual basis for the unification of gravity with the other elementary forces.

In addition to the impact on fundamental physics, the first direct detection of gravitational wave signals will open a new era for Astronomy, as we will try to demonstrate in a following section.

2. The legacy of the first generation

Here we will try to summarise the scientific results obtained using the km-size interferometers of the first generation. For convenience we will follow the classic division between the searches for continuous signals and those related to the transients.

2.1 Continuous signals

The way to search for continuous GW signals depends on how much about the source is known. At present, about 2000 neutron stars in our Galaxy have been identified by their radio, X-ray or γ ray emissions. The effort to detect a GW signal emitted by one of these sources is called *target search*. However, our Galaxy is estimated to contain roughly 100 million neutron stars, so the *all-sky search* for continuous GWs might detect new ones and tell us about their structure.

Aiming at signals potentially emitted by isolated radio pulsars and incorporating phase information deduced by radio observations, the target search was able to set limits on the GW metric perturbation h_c as low as a few 10^{-24}, limits which could be translated into star ellipticities $\epsilon < 10^{-5}$ for the four closest pulsars, thus starting to probe plausible astrophysical values.

Using the data of the last run of Virgo, thanks to the good low-frequency sensitivity, an upper limit on the emission by the Vela pulsar [7] was obtained on the assumption that the phase of the GW signal can be modeled on the radio emission, and that spin axis and wave polarization angles are known. The limit lyes in the interval $h_c < (1.9 - 2.2) \times 10^{-24}$, thus beating the spin down limit $h_{c,max} = 3.3 \times 10^{-24}$ derived from the pulsar period evolution measured using the electromagnetic emission of the pulsar.

The targeted search was extended to 116 known millisecond and young pulsars using LIGO and Virgo data allowing to set a limit on the Crab GW emission, a factor seven below the spin-down limit, which means less than 2 % of the available spin-down power. For several of the other young pulsars, the limits are only slightly above the spin-down limits. The best results were obtained for J1603-720, with a limit $h_c < 2.3 \times 10^{-26}$, and for J2124-3358, with a limit $\epsilon < 7 \times 10^{-8}$ [8].

The all-sky search for periodic signals based on LIGO and Virgo data [9], has been carried on in the frequency band of 50–800 Hz assuming a first derivative of the emission frequency lying in the interval $6 \times 10^{-9} - 0\,\mathrm{Hz/s}$ and we set a limit of $h_c < 3.8 \times 10^{-24}$ for all polarizations and sky locations.

A different category of continuos signals is the GW stochastic background of both astrophysical and cosmological origins. In this case, the detection technique is based on the cross-correlation of the outputs of different instruments. During the LIGO run S5 and the first Virgo Science run, an upper limit on the normalized energy density of GW stochastic background $h_o^2\Omega_{GW}$ at 150 Hz was set under the hypothesis of an isotropic signal distribution [11],

$$h_o^2\Omega_{GW} < 6.9 \times 10^{-6}\ at\ 150\,\mathrm{Hz}$$

and at higher frequencies [12]

$$h_o^2\Omega_{GW} < 0.16\left(\frac{f}{900\,\mathrm{Hz}}\right)^3\ for\ \ 600\,\mathrm{Hz} < f < 1000\,\mathrm{Hz}.$$

2.2 The transient signals

In this category we distinguish between generally unmodeled signals (GW burst) and those of the binary system coalescence with defined GW waveform.

In the latter case the analysis of LIGO S6 data and Virgo second and third science run data provides for the low-mass binary systems the best results so far [13]. Assuming that the total mass of the binary system M_{tot} is in the interval 2–25 solar masses the search had a distance limit for events associated to a neutron star-neutron star system (NS-NS) as far as 40 Mpc, a distance averaged over the observation time and over source sky location and orientation. This corresponds to set a limit for the event rate per unit volume of the Universe

$$R_{NS-NS} < 1.3 \times 10^{-4}\,\mathrm{yr}^{-1}\,\mathrm{Mpc}^{-3}$$

In the case of higher mass system including one or two black holes (BH) we end up with

$$R_{NS-BH} < 3.1 \times 10^{-5}\,\mathrm{yr}^{-1}\,\mathrm{Mpc}^{-3}$$

and

$$R_{BH-BH} < 6.4 \times 10^{-6}\,\mathrm{yr}^{-1}\,\mathrm{Mpc}^{-3}.$$

The same data allowed to improve limits on binary black hole events, with an analysis which exploited all phases (inspiral, merger, ringdown) of the signal [14]: for M_{tot} ranging in the interval of 25–100 solar masses, the search achieved the best distance value of 300 Mpc setting the limit

$$R_{BH-BH} < 3.3 \times 10^{-7}\,\mathrm{yr}^{-1}\,\mathrm{Mpc}^{-3}$$

for systems with non-spinning components in the interval 19–28 solar masses.

For signals of unknown shape, it is generally convenient to define the strength of the detected signal in terms of the root sum square (*rss*) amplitude h_{rss} of the signal $h(t)$.

The joint LIGO S6 run, and the second and third Virgo runs allowed to achieve the best results [15]: over a total observation time of 207 days, for signals shorter than 1 s lying in the 64–5000 Hz frequency band, we set an upper limit on the rate of 1.3 events/year in the range of amplitude spectral density sensitivity of $5 \times 10^{-22} - 1 \times 10^{-20}\,\mathrm{Hz}^{-1/2}$. For a burst emission in the center of our Galaxy (~10 kpc), these values allow a detect with 50% efficiency, if the sources emit an energy of about 2.2×10^{-8} solar masses, a value not far from the prediction of the modern models of a supernova explosion [16, 17].

3. Advanced detectors and multimessenger astronomy

The technological progress achieved in the construction of the new generation of advanced detectors, and the knowledge gained in astrophysical sources and in data analysis techniques promises that the next years will generate exciting progress in the comprehension of several astrophysical phenomena.

Compact binary coalescence is one of the most important sources for ground-based interferometers and the waveform of the related GW signal at the inspiral phase is fairly well understood theoretically. The expected detectable distance in the advanced detector era for NS-NS binaries is around 150–200 Mpc and the most probable event rate is around 40 yr^{-1} for advanced LIGO and Virgo.

The signal is dominated by the inspiral phase, well described by the post-newtonian approximation, so that standard matched filtering allows to search for events in the detector noise. However the most interesting information is hidden in the final phase near the merger when the stars are colliding, and the corresponding part of the GW signal brings the information on the equation of state of the nuclear matter. In fact, during the last stages of inspiral, the Newtonian tidal field of one component induce a quadrupole moment in the other. The effect depends on the tidal deformability of the massive star in a way that is determined by its equation of state. The neutron stars deformation has an influence on the orbital motion, in particular the phase, which up to a factor of two is also the phase of the emitted gravitational wave signal. In particular, it has been shown [18] that, in a realistic data analysis setting, quantitative information about the size of the tidal deformability at a given reference mass can be obtained with a $2 - \sigma$ statistical uncertainty of $\sim$10% after a few tens of detections, and hypothesis ranking will be able to distinguish between a hard, moderate, and soft equation of state.

Direct GW measurement of a coalescing binary provides a distance-ladder-independent measure of the luminosity distance D_L to a source. While D_L may be well measured, the source redshift cannot be inferred directly from the GW signal. An independent measure of the event redshift is therefore required to measure cosmological parameters with GW events. If an electromagnetic (EM) signature is detected, spectroscopic observations of the event or the galaxy hosting the EM event should be obtainable. Thus, we can use inspirals as cosmological probes when we associate the gravitational event with an electromagnetic counterpart.

This is the first example of potential outcome of a multimessenger approach providing a new tool for precision cosmology, but there are several other benefits of joint GW and EM observations. EM localization of a GW transient event reduces significantly the parameter space of search templates, and correspondingly increasing the observed signal-to-noise ratio (SNR). Likewise, an identification drastically cuts down the search range in time, reducing the threshold signal-to-noise required for confident detection.

Many astrophysical sources, the majority of which originate from cataclysmic events, are expected to produce both GWs and high energy neutrinos (HEN). While GWs are linked to the dynamics of the bulk motion of the source progenitor, HENs trace the interactions of accelerated protons (and possibly heavier nuclei) with matter and radiation in and around the source. Stellar core collapse during a supernovae (SN) explosion releases enough gravitational binding energy in less than one second, involving mass asymmetric motion to generate strong gravitational-wave signals.

Long-soft γ ray bursts, which represent the majority of γ ray events (GRB) events, have been related to massive stellar explosions [19]. However, neither all SNe, nor even all SNe of Type Ibc produce GRBs. The degree of differential rotation in the collapsing iron core of massive stars when they die may be what makes the difference. The detection of GWs from a SN strongly constrains the rotation of the collapsed core. A SN close enough to be a strong GW source, which is also a strong neutrino source, with the information associated to the photon emission become a powerful tool to study the physics of the matter at nuclear densities.

The study of the origin of short-hard γ ray bursts (SHB) has shown an indirect evidence that they are associated to an older stellar population, driving the hypothesis that binary mergers ($NS - NS$ or $NS - BH$) are the short γ ray progenitors. However, a number of other origins remain viable: a concurrent GW inspiral event in the same place and time will solve the enigma. In the standard picture, the mechanism responsible for the enormous EM flux in the prompt emission and in the afterglow of a SHB is the dissipation (via internal and external shocks) of bulk kinetic into highly relativistic particles; the particles are accelerated to a nonthermal energy distribution in a relativistically expanding fireball ejected by the GRB central engine. The internal or external shocks accelerate protons that interact with the gamma rays and/or other protons inside the fireball, producing charged pions and kaons that subsequently decay into high energy neutrinos (HEN). While γ ray and HEN emissions from GRBs are related to the mechanisms driving the relativistic outflow, GW emission is closely connected to the central engine and hence to the progenitor of the GRB. In addition to the information extracted from the GW signal shape, the difference between the times of arrival of GW and HEN signals provides an important information on the emission mechanism: the detection of a HEN signal prior to a GW one may indicate that the strongest GW emission from the source is not connected to the onset of the activity of the central engine that one might expect from core-collapse models.

To complete the picture, we note also that the emission of short GRBs and the related HEMs is beamed and, hence we can expect also cases in which GWs will be observed from such sources without an observed GRB or HEN signal. However, an orphan afterglow, macronovae or radio flares might be observed in these cases. In conclusion, combining information from all these independent detectors can provide original ways of constraining the physical processes driving the source.

Finally we have to cite that on the order of 20 highly magnetized neutron stars are observed in our Galaxy as soft γ repeaters (SGR) and anomalous X-ray pulsars (AXP). They emit short bursts of X-rays and soft γ rays and sometimes giant flares or storms of bursts. The energy release may excite neutron star crustal f-modes in the kHz region emitting GW signals with the typical ring down signature. Torsional (shear) s-mode *core-quakes* may release significantly more GW energy [20].

A upper limit on the GW emission from six magnetars was set already using LIGO and Virgo data [21] and we expect a factor 100 better sensitivity (in emitted energy) with advanced detectors.

4. Conclusion

The global advanced detector network, which includes the two LIGO detectors, GEO-HF, Virgo, KAGRA will be operative in the next quinquennium and possibly LIGO-India will join in the future. The first science run of LIGO is approaching and next year Virgo will be put in operation. The network enables more confident detection by requiring the same or similar signal in widely separated instruments. More important, a network locate the sources in the sky, since individual detectors have essentially no ability determine the source direction: the network acts as a gravitational-wave interferometer, determining source direction through timing and amplitude information of a wave detected coherently across a global baseline. Good source localisation allows one to look for other forms of radiation coming from that location, nearly simultaneously with the GWs: optical, radio, X-ray or gamma electromagnetic radiation, or neutrinos. Such multi-messenger signals may be expected from sources containing ordinary atomic or nuclear matter.

Thus, we expect that the searches undertaken by advanced detectors will be indeed likely capable of detecting, constraining, or excluding several existing astrophysical models, contributing in the future to a significant progress in Astronomy.

References

[1] E. Amaldi et al., Astron. and Astrophys. **216**, 325 (1989)
[2] P. Astone et al., Phys. Rev. **D68**, 022001 (2003)
[3] `http://www.geo600.org`
[4] `http://www.ligo.caltech.edu`
[5] F. Acernese et al., Class. Quantum Grav. **32**, 024001 (2015)
[6] `http://gwcenter.icrr.u-tokyo.ac.jp/en`
[7] J. Abadie et al., Astrophys. J. **737**, 93 (2011)
[8] B. P. Abbott, et al., Astrophys. J. **713**, 671 (2010)
[9] J. Abadie et al., Phys. Rev. **D85**, 022001(2012)
[10] J. Abadie et al., Phys. Rev. **D85**, 022001 (2012)
[11] B. Abbott et al., Nature **460**, 990 (2009)
[12] J. Abadie et al., Phys. Rev. **D85**, 122001 (2012)
[13] J. Abadie et al., Phys. Rev. **D85**, 082002 (2012)

[14] J. Aasi, et al., Phys. Rev. **D87**, 022002 (2013)
[15] J. Abadie et al., Phys. Rev. **D85**, 122007 (2012)
[16] C. D. Ott, Class. Quantum Grav. **26**, 063001 (2009)
[17] K. N. Yakunin, P. Marronetti, A. Mezzacappa, S. W. Bruenn, C.-T. Lee, M. A. Chertkow, W. R. Hix, J. M. Blondin, E. J. Lentz, O. E. Bronson Messer, and S. Yoshida, Class. Quantum Grav. **27**, 194005 (2010)
[18] W. Del Pozzo, T. G.F. Li, M. Agathos, C. Van Den Broeck, Phys. Rev. Lett. **111**, 071101 (2013)
[19] S. E. Woosley, J.S. Bloom, Ann. Rev. Astron. & Astrophys. **44**, 507 (2006)
[20] R. C. Duncan, C. Thompson, Astrophys. J. **392**, L9 (1992)
[21] J. Abadie et al., Astrophys. J. **734**, L35 (2011)

EPJ Web of Conferences **121**, 03001 (2016) DOI: 10.1051/epjconf/201612103001
RICAP-2014

Precision Cosmic Ray physics with space-born experiment

Marco Incagli[a]

Istituto Nazionale di Fisica Nucleare (INFN), Pisa, Italy

Abstract. More than 100 years after their discoveries, cosmic rays have been extensively studied, both with balloon experiments and with ground observatories. More recently, the possibility of mounting detectors on satellites or on the International Space Station has allowed for a long duration (several years) continuous observation of primary cosmic rays, i.e. before their interaction with the earth atmosphere, thus opening a new regime of precision measurements. In this review, recent results from major space experiments, as Pamela, AMS02 and Fermi, as well as next generation experiments proposed for the International Space Station, for standalone satellites or for the yet to come Chinese Space Station, will be presented. The impact of these experiment on the knowledge of Cosmic Ray propagation will also be discussed.

1. Introduction

Charged Cosmic Rays cover many orders of magnitude, both in flux and in energy range. Therefore many different techniques have been established to study them over the whole spectrum. In particular, large areas have to be covered in order to observe cosmic rays with energy above 10^{15} eV, where $\simeq$1 particle/m^2/year is expected[1].

With current techniques, observations at such energies and above can only be performed on ground[2].

On the other hand, particle fluxes at energies 1 GeV–1 TeV are of order 1 particle/m^2/second and experiments having dimensions $\simeq 1\,\text{m}^3$ can be used. This has allowed the deployment of some detectors in space, either as free-flyers – PAMELA, FERMI-GLAST – or as payloads mounted on the International Space Station (ISS) – AMS02 and, in the near future, CALET, ISS-CREAM.

The clear advantage of such detectors, with respect to ground based ones, is that they are sensitive to the *primary* CR component, where by *primary* I mean "before interacting with the earth atmosphere", they can reach a much higher precision on the energy determination, in particular for electromagnetic particles, and on the chemical composition. Moreover, if equipped with a magnet, they are sensitive to anti-particles. On the other hand, due to the

[a] e-mail: marco.incagli@pi.infn.it

[1] Here I am referring to the integral flux. The flux of particular species (nuclei, leptons, ...) can vary by orders of magnitude, as we will see in the next sections.

[2] A noticable exeption to this paradigma is the possibility of observing the fluorescence light, emitted by very energetic air showers (> 10^{20} eV), from a telescope orbiting at an altitude of $\simeq$ 400 km. Several such experiments have been proposed in the past – EUSO, JEM-EUSO, KLYPVE, OWL; as of now, only a small prototype (MINI-EUSO) has a good chance of being launched on the International Space Station in the near future.

large cost of space missions, such detectors are limited both in mass and in geometrical dimensions, which results in a severe limitation in the energy range due to the steeply falling CR flux.

As an intermediate step between ground and space detectors, balloon experiments have to be mentioned. Flying at an altitude of 37 km, with only 5 gr/cm^2 of residual atmosphere above, these experiments share similar advantages as space ones, at a much lower cost. However balloon flights last at most 3 weeks, so the integration time is quite limited.

In this review I will focus on space experiments discussing the general characteristics (Sect. 2) with a specific reference to geometrical acceptance and background rejection (Sect. 3). I will then discuss the physics case and the recent results from the two major experiments focused on charged cosmic rays: PAMELA and AMS02 (Sects. 4 and 5). As there are specific presentations on photon physics, in these proceedings, I will not discuss here results from the FERMI experiment, although it will be mentioned while describing the general characteristics of space experiments. Finally I will briefly mention the detectors foreseen for the near or the more far future (Sect. 6).

2. Space detectors

Although several techniques can be (and have been) used for building or designing space detectors, three general classes of experiments can be identified:

1. magnetic spectrometers (*à la* AMS02);
2. pair-conversion telescopes (*à la* FERMI);
3. calorimeters (*à la* ATIC or CREAM, but also CALET, DAMPE or ISS-CREAM) with a possible enphasis on hadrons (nuclei) or leptons ($e^{\pm}, \gamma$) which can, on their turn, be more specialized on hadrons or on electromagnetic showers.

Of course, different combinations of the different techniques are possible.

2.1 Magnetic spectrometers

Spectrometers allow for the measurement of the particle momentum and of the sign of its charge. As a consequence, it is possible to have access to anti-particles, in particular to positrons and anti-protons, but also, at least in principle, to anti-deuteron or anti-helium nuclei.

Moreover, by combining the momentum with the measurement of the velocity β, performed either through Cherenkov or through Time Of Flight (TOF) techniques, it is possible to determine the nuclear mass number A via the relation:

$$A = \frac{RZe}{m_N c\beta\gamma} \tag{1}$$

where m_N is the nucleon mass[3], thus giving access to isotopes (^{3}He,^{4} He,^{9} Be,^{10} Be, ...). A review of the techniques used for isotope identification requires a specific discussion of TOF or Cherenkov detectors, which goes beyond the scope of this generic introduction.

With current technologies, two solutions are possible for magnets: permanent magnets and superconducting magnets working at the superfluid helium temperature. Several interesting *R&D* are in progress to develop superconducting magnets operating at higher

[3] It is easy to write an expression in which the neutron and the proton masses appear explicitely instead of the average m_N. However, the error introduced by using this approximation is much smaller than the one due to the momentum or velocity measurements.

temperatures ([1]). However these techniques have still to be fully developed, and then it will be necessary to test them for space. So this possibility will not be discussed any further.

Permanent magnets are heavy and their magnetic field is limited to values of $\simeq$1 kGauss; superconducting magnets are hard to operate in space and require a reservoir of Helium to be carried on board. For example, in the first project of the AMS02 experiment a dewar with 3500 l of Helium was foreseen. With this amount of helium, the lifetime of the experiment was planned to be $\simeq$28 months ([2]).

2.2 Pair-conversion telescopes

A *pair-conversion telescope* is a detector with a dedicated tracking stage having a thickness of 1-2 radiation lengths, normally obtained by interleaving silicon active material with tungsten layers, in which photons interact converting in an e^+e^- pair. The two leptons are precisely traced and their energy is measured by the calorimeter located below. So these detectors are mostly dedicated to precision γ physics at energies which can be as low as $\simeq$1 MeV, which is the threshold energy for electron pair production. As a matter of fact, electrons produced with a momentum of few MeV undergo multiple scattering in the passive layers and their momentum cannot be effectively measured. So the lower energy limit is really $\simeq$100 MeV. An example of such an experiment is FERMI ([3]). To reach the 1 MeV limit, a modified structure in which the passive layers are replaced by additional active silicon has been proposed in ([4]).

These type of detectors have an excellent angular resolution, or Point Spread Function (PSF). However the presence of the *heavy tracker* introduces some complexity in the system which has some impact in the Field Of View (FOV) and in the energy resolution.

2.3 Calorimeters

Calorimeter-based detectors are specialized in measuring the nuclear charge and the energy of electromagnetic showers. The limit on the total weight and dimensions does not allow to build hadronic calorimeters as on ground. However it is nevertheless possible to measure the energy of hadrons by using some special techniques, as described later for the ISS-CREAM experiment.

The main characteristic of calorimeters, however, is clearly to maximize the geometrical acceptance, by dedicating all the mass budget to the detector.

Comparing calorimeters to gamma-converters and to magnetic spectrometers, it is possible to note that the pointing capability for photons is $\simeq 0.5°$, more than enough for anisotropy studies, and that, although not sensitive to anti-particles, calorimeters can accurately study the total $e^+ + e^-$ flux, thus being able to detect sharp variations in the spectrum induced by a potential DM source.

3. Statistics vs. Acceptance and the issue of background

The CR flux decreases very rapidsly with the energy. As an example, Fig. 1 shows the number of electrons ($e^+ + e^-$) and of positrons expected above a given energy for several geometrical acceptances, indicated in the plot by different lines.

The two colored dashed and dotted lines approximately indicate the geometrical acceptances of FERMI and AMS02 experiments. The numbers represent the effective geometrical acceptance integrated over one year, where *effective* means that all inefficiencies, including dead time, trigger efficiency and so on, are accounted for. For an apparatus having the size of Fermi, which is currently the high energy space detector with the largest acceptance, at most 100 electron events per year are expected at energies E = 2–3 TeV. This number rapidly

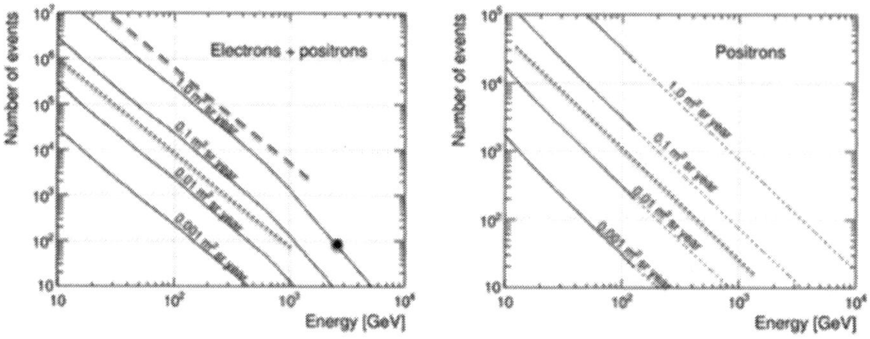

Figure 1. Number of events above a given energy (efficiency cuts not considered) for different geometrical exposures. The red dotted and the green dashed lines appreoximately represent the geometrical acceptance of the AMS02 and the FERMI experiment, respectively.

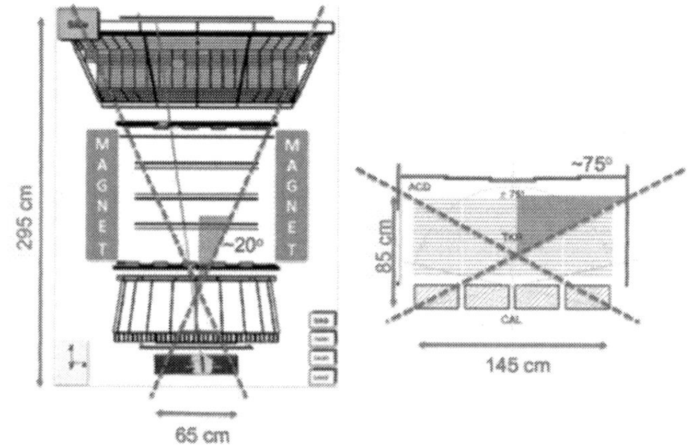

Figure 2. AMS02 and FERMI detectors in scale. Different dimensions and different geometrical acceptances are evident.

decreases for a detector whose FOV is limited by the presence of the magnet. Figure 2 shows the two detectors on the same scale, putting in evidence the different choices made.

However statistics is not the whole story. Different channels, in particular anti-particles, have fluxes which can be suppressed by several orders of magnitude with respect to the most common ones. Figure 3 shows the flux, as a function of energy, for some particle types with respect to protons. In this plot, three regions can be identified: ep-rejection, charge confusion and photons.

- ep-rejection: electrons suffer from a proton background of 10^2-10^3, depending on the energy, while for positrons this background is $\simeq 10^4$; as a consequence, to measure the electron spectrum in the TeV region with a precision better than 10%, an electron-proton (ep) rejection of $10^{-4}-10^{-5}$ must be achieved, while still keeping a high signal efficiency to keep under control the statistical error.
- charge confusion: protons are approximately 4 orders of magnitude more abundant than anti-protons. Therefore the sign of the charge has to be determined with such a level of precision in order to avoid the *charge confusion* effect, i.e. the wrong sign reconstruction.

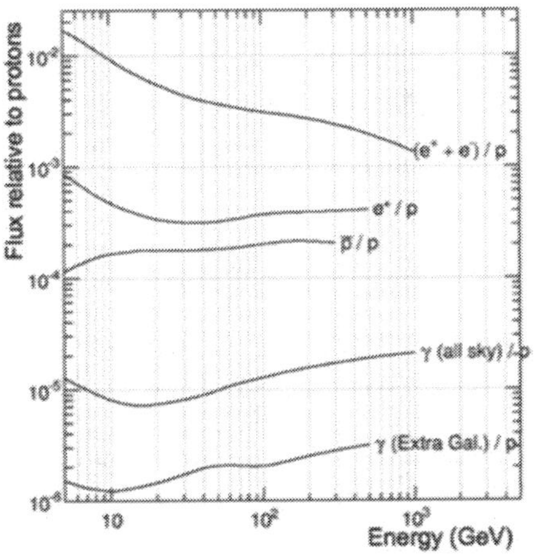

Figure 3. Cosmic ray flux as a function of energy relative to protons for several species.

- photons: to measure photons, a proton suppression factor of order ppm is required.

All these constraints must be considered while designing experiments for space.

4. The physics case: Cosmic Rays and Dark Matter search

Many topics may be studied in space experiments. The two main ones are: Cosmic Rays (CR) origin and propagation and Dark Matter (DM) searches.

4.1 Cosmic Rays origin and propagation

The reference theoretical framework which is widely used to interpret CR nuclear data is based on the following main assumptions/approximations:

- CR diffusion is treated as isotropic: the interstellar diffusion tensor is assumed to be a scalar independent from rigidity, the so called diffusion coefficient D;
- the diffusion coefficient is assumed to be spatially uniform and (in most cases) to have a simple power-law dependence on the particle rigidity: $D(\rho) \propto \rho^{\delta}$;
- all primary nuclear species share the same single (or broken) power-law acceleration shape up to $E \simeq 10^{15}$ eV, the so called *knee*;
- CR sources are continuosly distributed in the Galactic disk.

With these approximations, the main experimental features of primary and secondary CRs are reasonably well reproduced (fluxes, $\bar{p}/p$ ratio, primaries/secondaries spectra, ...). However new and more precise experimental data are in tension with the predictions of this standard scenario and call for a more detailed and complete theoretical framework.

Among the observables which are in tension with the standard CR description, some of the ones which are relevant for space experiments are:

- *p/He ratio*: ATIC, CREAM and PAMELA found different spectral indexes for proton and Helium CRs. Preliminary AMS02 results seem to confirm this difference. This is hardly explainable in terms of Fermi acceleration theory, which does not differentiate between elements at ultra-relativistic rigidities.

- *CR spectra hardening*: PAMELA found a sharp spectral breaking of p and He spectra at the same rigidity $p_{break} \simeq 240\,\text{GeV}$. Although AMS02 preliminary data do not confirm this result, it surely allows for a better connection with CREAM data in the 10−500 TeV energy range. Since a spectral breaking is not predicted by Fermi acceleration theory, this hardening may either require to abandon the continuos source distribution limit and/or to introduce major changes in the way CR propagation is treated.
- *anisotropy and secondary to primary ratio*: an anisotropy in CR arrival direction is expected, with a larger contribution from the galactic center with respect to the external part of the galaxy. Models which predict a diffusion coefficient $\delta > 0.5$, favored by recent B/C ratio data, severely overpredict this anisotropy.

These observational issues call, on one hand, for an improved description of the CR production and propagation and, on the other hand, for more experimental data with improved resolution to further constraint the theory. The main role of space experiments, in this respect, is there capability of discriminating the different components by measuring the Z of nuclei by dE/dx in different materials, tipically scintillators or tracker silicon layers, or by other techniques (e.g. the amount of photons observed in the rings produced by particle crossing radiators in RICH counters).

4.2 Dark Matter identification

There is now a overwhelming experimental evidence of the presence of a large component of a new type of particles, generally called *Dark Matter* (DM), whose mass makes up $\simeq 80\%$ of the mass budget of the universe. Although there is no evidence of the energy scale of this new form of matter, which could range from invisible to macroscopic objects, there is a particular solution which has several attractive features: the Weakly Interacting Massive Particles (WIMP). There is a singular coincidence between the parameters of the Standar Model of particle physics and of the Cosmological Model to provide valid DM candidates at the electroweak scale ($\simeq$1 TeV) with a cross section typical of weak processes.

These particles are being studied in different processes:

- the annihilation, or indirect, channel, in which to DM particles annihilate to produce two standard ones. This channel is the relevant one for space experiments;
- the interaction, or direct channel, is pursued in underground laboratories which look for the interaction of the WIMP wind, caused by the earth motion in the solar system, and in the galaxy in general, with ordinary ultra-pure matter;
- the production in collisions of Standard Model particles, as studied in colliders.

The annihilation channel can be detected by an excess of particles with respect to the ones produced in standard Cosmic Rays. It is normally expected that the DM annihilation provides a particle-antiparticle pair, therefore, being antiparticles much rarer than particles in CRs, cosmic rays, the eventual excess would be more relevant in the positron or the antiproton channel than in the correspondants particles.

For this reason, it was proposed in the 90s that monoenergetic positrons from halo annihilations could be a significant and distinctive signal for massive dark-matter particles [7]. The signature would be an increase in the *positron fraction* ($e^+/(e^+ + e^-)$) with respect to the one expected from the interaction of primary cosmic rays with the interstellar medium. To detect positrons a magnetic detector is required, which led to the construction and set in orbit first of the PAMELA experiment, on a dedicated satellite, and then of AMS02, installed on the ISS.

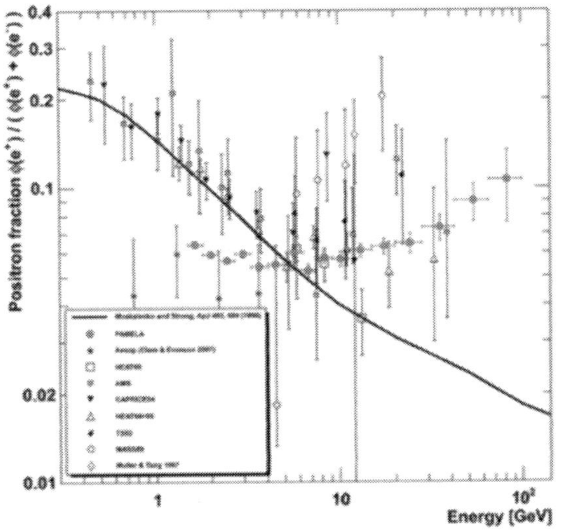

Figure 4. Positron fraction measured by PAMELA experiment.

5. The magnetic spectrometers: PAMELA and AMS02

The two detectors are described in detail in several publications (e.g. [5, 6]). The main structure is a permanent magnet with a dipole field, in both cases with intensity $\simeq 1$ kG, equipped with a set of silicon tracking layers, followed by an electromagnetic calorimeter having a thickness of $\simeq 17\,X_0$ (X_0 = *radiation length*). Both detectors are also equipped with a Time Of Flight trigger system and a set of Anti-coincidence counters to reject tracks entering the apparatus from the sides. AMS02, in addition, has a Transition Radiation Detector, for electron-proton (or positron-proton) rejection, and a Ring Image Cherenkov Counter, which can measure the particle velocity β and can provide information on isotope composition. However the main difference between the two detectors are the dimensions and, as a consequence, the Geometrical Acceptance: PAMELA weighs $\simeq 500$ kg and has an acceptance $A \simeq 20\,\mathrm{cm}^2\mathrm{sr}$, while AMS02 payload weighs $\simeq 7$ tons[4] and its effective geometrical acceptance (i.e. after selection) is of $\simeq 450\,\mathrm{m}^2$ sr. PAMELA was launched in June 2006; AMS02 in May 2011.

5.1 Positron excess and anti-proton flux

The first clear evidence of an excess of positrons in the flux of Cosmic Rays was published by PAMELA in April 2009 on Nature ([5]) and it is represented in Fig. 4. The observed spectrum is incompatible with a secondary production, which is expected to smoothly decrease as in the solid line of Fig. 4. Therefore, either a significant modification in the acceleration and propagation models for cosmic rays is needed, or a primary component is present.

A modification in the propagation of secondary cosmic rays has been proposed by several authors ([8, 9]). The proposed effect, which could explain the observed "excess", is that a fraction of the secondary positrons are produced inside the same astrophisical source which accelerate the primaries, such as supernova remnants. The flux of positrons produced at the shock is naturally harder than the standard one and could explain the rise in the observed

---

[4] Almost half of this weight is due to the support system which connects the detector to the ISS.

positron fraction. The same mechanism, however, predicts a similar effect also on the ratio $\bar{p}/p$ and B/C (Boron over Carbon), excess which, at the moment, has not been observed[5].

Alternatively, the excess could be due to the existence of a primary source which, tipically, produces e^+, e^- pairs. A very intriguing possibility is that these pairs are produced by the annihilation of DM particles, as suggested in [7]. However there are also more standard scenarios in which positrons are produced by a nearby astrophysical source; for example a pulsar could emit energetic photons that, interacting with the strong magnetic field surrounding the neutron star, decay in an e^+e^- pair [12].

Although this excess generated a lot of excitement in the community, in view of the potential DM explanation, the measurement of the anti-proton flux by the same collaboration showed no excess with respect to standard secondary production [13]. While this result is well compatible with the pulsar hypothesis, a *well behaving* DM particle would instead produce a sizeable proton-antiproton flux, unless some *ad hoc* mechanism is invoked. An additional difficulty in the DM explanation is the magnitude of the effect, as it exceeds what is expected by the thermal cross section requiring sizeable *boost effects*. Several ways out have been proposed (see, e.g., [14] or [15]), but to discriminate between the different possible explanations, more precise measurements extending at higher energies are necessary.

5.2 The AMS02 experiment

A new measurement of the positron and electron fluxes, as well as of the positron fraction, has been recently performed by the AMS02 experiment with a much higher statistics. Details are given in [16]. Here we note that positron events suffer from two different backgrounds:

1. *ep-rejection*: protons which are wrongly reconstructed as electromagnetic showers (protons are 4 orders of magnitude more abundant than positrons);
2. *charge confusion*: electrons whose charge has been wrongly identified in the tracker and are reconstructed as positrons.

In AMS02 electrons are distinguished from protons by three different, and in large part independent, detectors:

- the Transition Radiation Detector (TRD) located on top of AMS02 measures the δ- electrons emitted by relativistic particles when crossing the surface between materials with different dielectric constant. This effect is summarized in a likelihood estimator, as shown in Fig. 5(a);
- the ECAL measures the 3D shower shape and, by using a multivariate approach, distinguishes electrons from protons as shown in Fig. 5(b);
- finally, by comparing the total energy deposited in ECAL with the tracker momentum, leptons peak at $E/p = 1$, while hadrons are shifted to zero, having a sizable leakage in the ECAL (Fig. 5(c).

Details of the analysis are in [16]. The resulting positron fraction is shown in Fig. 6, which extends the previous measurements up to 500 GeV (upper edge of last bin). The excess is confirmed with high precision; moreover, a clear flattening in the fraction occurs starting from $\simeq$200 GeV. This behaviour is compatible both with the pulsar and with the DM explanation, as in both cases a cutoff is expected. In the first case the cutoff is due to the energy of the emitted photons, while in the second one the upper limit is determined by the DM mass. A

[5] See, however, [11] for a recent review which includes recent preliminary AMS02 data on B/C.

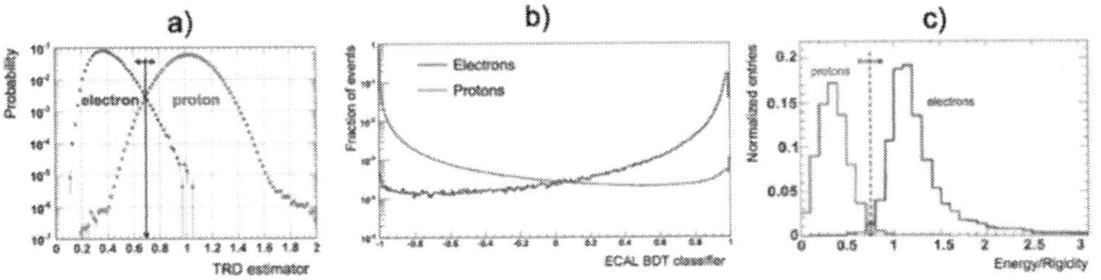

Figure 5. Electron-proton rejection in AMS02 through a) Transition Radiation Detector b) Electromagnetic Calorimeter c) Energy/Rigidity ratio.

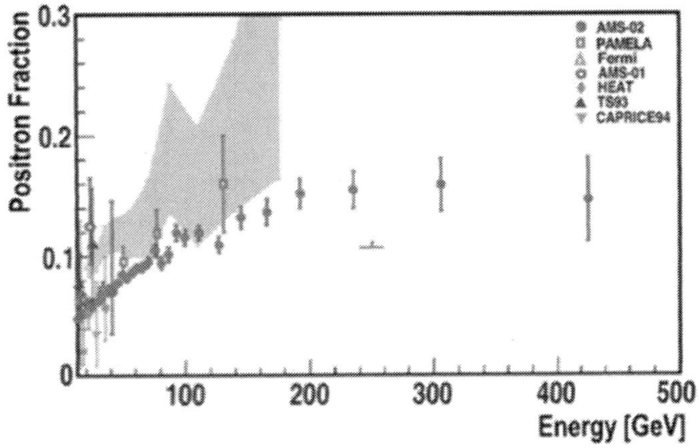

Figure 6. Positron fraction measured by AMS02 for energies above 10 GeV, where the fraction starts to increase. AMS02 shows, for the first time, a clear flattening of the fraction starting at 200 GeV.

possibility to discriminate between the two scenarios is shown in Fig. 7 in which the AMS02 results are extrapolated under two different models: typical DM-like models show a sharper cutoff with respect to pulsar-like ones. However this conclusion is not so firmly established (see, for example, [17]).

The separate fluxes have also been measured [18] and are shown in Fig. 8, where the positrons have been scaled up by a factor 10 in order to be compared with the electrons (positron scale on the right axis). A comparison between the two histograms shows how the effect of this eventual primary source has a different impact on the two components. Due to the higher statistics, the electron flux has been extended up to 700 GeV.

Although a more detailed information on the electron and positron spectrum is certainly useful, only a multi-messenger approach may provide sufficient information to disantangle among the different explanations. On this respect, several other results are expected from the AMS02 collaboration:

- proton and helium flux
- B/C, C/O and B, C, O fluxes
- anti-protons
- light nuclei and Nitrogen (Li, Be, N)
- photons
- heavier nuclei
- isotopes ($^{3}He/^{4}He$ and $^{10}Be/^{9}Be$)

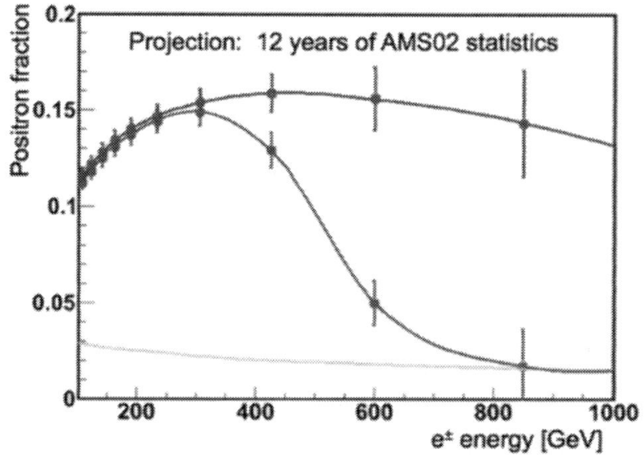

Figure 7. Projected statistical sensitivity of AMS02 experiment after 12 years of data taking. The lower, green line represents the expected background from secondaries. The brown middle line includes a DM contribution with mass 700 GeV; the upper blue line represents a Pulsar like model. These models are only representative.

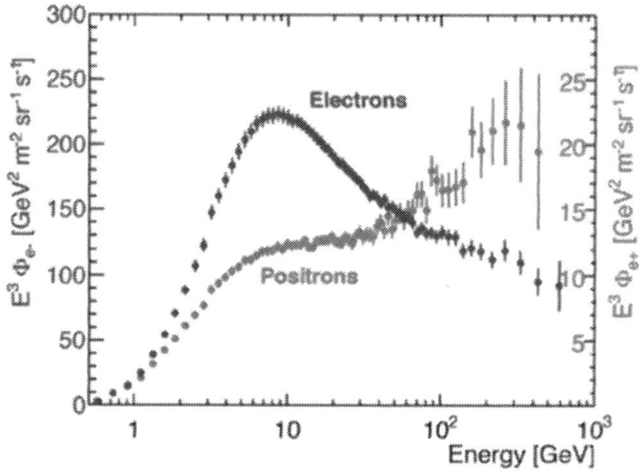

Figure 8. AMS02 flux for electrons (left axis) and positron (right axis).

- anti-D
- anti-He.

All these measurements are useful either as a direct probe of DM, as anti-protons and, possibly, anti-deuteron, or as a constraint to the propagation of CR in the insterstellar medium, as B/C or as the isotope ratio.

6. Future experiments

6.1 Near future

Within 2017, three more space experiments are foreseen, all of calorimetric type: CALET [19], ISS-CREAM [20], DAMPE [21].

The first two will be mounted in 2015/2016 on the Japanese module of the International Space Station, while DAMPE is supposed to be launched at the end of 2016 on board of a chinese satellite.

CALET and DAMPE have very similar characteristics: the main detector is a thick electromagnetic calorimeter of $\simeq 30\,X_0$, with very good energy resolution ($\simeq$1% at 1 TeV), pointing capability (0.5 degrees) and an excellent ep-rejection (at least 10^{-4}), with a *heavy* tracker in front in which photons can convert. In CALET the tracker active material is made of 1 mm scintillating fibers, so the main focus is on the identification of electromagnetic and hadronic showers, while DAMPE uses silicon wafers, with more emphasis on photon pair production. Both detectors have very good Z discrimination capabilities. The geometrical acceptance is 0.1–0.2 m^2sr, thus a factor of 4 larger than AMS02, but still an order of magnitude smaller than FERMI.

ISS-CREAM, instead, is a modified version of the CREAM experiment, who had six succesful antartic flights on aerostatic balloons between the years 2004 and 2010, for a cumulative exposure of 162 days. In spite of the success of the balloon missions, the experiment integrated less than half a year of equivalent data for a space mission, therefore the possibility of installing the apparatus on the ISS would boost the physics potential, at least in terms of accumulated statistics. The CREAM experiment extends the direct measurement of cosmic-ray composition to energies of hundreds of TeV, at which CRs are capable of generating gigantic air showers which have been observed on the ground, thereby providing calibration for indirect measurements. A specific technique, which uses a preshower carbon detector half interaction length thick to induce hadronic cascades, is used to reach such energies with a *thin* calorimeter (*thin* in terms of hadronic showers). So, differently from previous experiments, this one is more focused on the precise measurements of the energy dependance of elemental spectra at the highest energies.

6.2 Medium term experiments

6.2.1 GAMMA-400

GAMMA-400 [22] is a russian-italian experiment which will fly on board of the russian *Navigator* spacecraft. The experiment will be initially installed on a highly elliptical orbit (with apogee 300.000 km, perigee 500 km and an inclination of 51.4°), with 7 days orbital period; the orbit will then be gradually changed, in 5 months, to a circular one with a radius of 200.000 km radius. At such an orbit the Earth will not cover a significant fraction of the sky, as is usually the case for the Low Earth Orbit (LEO) of the experiments mentioned so far. Also the geomagnetic effect on charged particles will become negligible, which allows for a clean study of the low energy part ($E < 10\,\text{GeV}$) of the spectrum.

The launch is foreseen for 2020. The project started with the main goal of improving FERMI results on photon physics. Therefore the apparatus has a converting tracker and a calorimeter having an energy resolution of $\simeq$ 1% at 1 TeV. To reach this energy resolution, the calorimeter has to be thicker than the FERMI one, with the consequence that the geometrical acceptance is almost a factor of 2 lower.

However, the project has been further developed and now the proposed calorimeter uses the CaloCube technique [23], which allows to use not only the upper face but also the lateral ones to detect CRs, with the consequence that the geometrical acceptance is increased by a factor 5, at least for charged particles.

The introduction of a highly segmented homogeneous calorimeter made with CsI(Tl) cubes, with improved energy resolution and extended geometrical factor, coupled to the improvement of the converter-tracker structure make GAMMA-400 an excellent dual instrument, optimized both for photons (in the 100 MeV–1 TeV energy range) and charged cosmic rays (electrons up to 10 TeV and high energy nuclei up to the knee region).

6.2.2 The chinese space station

China has recently announced an ambitious space program timetable with the goal of establishing its first space station around 2022. As part of this program, a major space experiment for Cosmic Ray detection is foreseen. Two possible designs are currently under study: a calorimetric detector surrounded by a pair-conversion tracker, similar in many aspects to GAMMA-400 but with a wider geometrical coverage, and a magnetic spectrometer.

The calorimetric apparatus option is more advanced, with a collaboration already building up around the design of HERD (High Energy cosmic Radiation Detection), while the magnetic spectrometer option, possibly an AMS03 mission, is still very preliminar.

Although at the moment it is too preliminary to make detailed comments about these possible detectors, it is evident the potentiality of such an observatory.

7. Conclusions

Many important and useful informations can be acquired by studying cosmic rays, both on their origin and propagation mechanism but also, possibly, on new physics, in particular Dark Matter.

Due to the large spread in energy and in flux, many different techniques have been developed. Among these, in the recent years space experiments have been characterized by the unique possibility of studying with great care the energy range from few hundred MeV up to hundreds of TeV, at least for protons.

Space experiments have the unique opportunity of measuring anti-particles and isotopes, and have a greatly improved capability of discriminating different nuclear charges Z. Several important experiments are still collecting data: PAMELA, which is almost at the end of its life after 8 years of glorious mission, AMS02 and FERMI. Many new experiments have been proposed or are being built for the near and for the far future.

Only through a comparison of many different channels, the so-called "multi-messenger" approach, it will be possible to constraint the propagation of standard CRs and thus put severe limits on, or exclude some models of, Dark Matter.

References

[1] A. Ballarino 2010, J. Phys.: Conf. Ser. 234 032003 ; "World-record current in the MgB2 superconductor", CERN Bulletin 16–17, April 2014
[2] AMS collaboration, "Upgrade of the Alpha Magnetic Spectrometer (AMS-02) for long term operation on the International Space Station (ISS)", Nucl. Instrum. Meth. A **654** (2011) 639-648
[3] W. B. Atwood et al., 2009 ApJ **697**, 1071
[4] A. Morselli et al., Nuclear Physics B Proc. Supp. **239–240** (2013) 193-198
[5] Pamela collaboration, Nature **45B** (2009) 607
[6] M. Aguilar et al., Phys. Rev. Lett. **110**, 141102 (2013)
[7] M. S. Turner, F. Wilczek, Phys. Rev. D **42** (1990) 1001
[8] P. Blasi, Phys. Rev. Lett. **103**, 051104 (2009)
[9] M. Ahlers, P. Mertsch, S. Sarkar, Phys. Rev. D **80**, 123017 (2009)
[10] P. Blasi, P. Serpico, Phys. Rev. Lett. **103**, 081103 (2009)

[11] P. Mertsch, S. Sarkar, Phys. Rev. D **90** (2009) 061301
[12] I. Buesching et al., Astrophys. J. **78** (2008) L39
[13] Pamela Collaboration, PRL **105** (2010) 121101
[14] M. Cirelli et al., Nucl. Phys. B **813**, 1 (2008)
[15] I. Cholis et al., Phys. Rev. D **80**, 123518 (2009)
[16] AMS collaboration, Phys. Rev. Lett. **113**, 121101 (2014)
[17] G. Bertone, M. Lattanzi, M. Pato, JCAP **1012** (2010) 020
[18] AMS collaboration, Phys. Rev. Lett. **113**, 121102 (2014)
[19] Journal of Physics: Conference Series **409** (2013) 012026
[20] E. S. Seo et al., "Cosmic Ray Energetics And Mass (CREAM) for the ISS JEM-EF," the 2013 conference on TeV Particle Astrophysics (TeVPA) 2013
[21] J. Chang, in The 7th international workshop "Dark Side of the Universe (DSU 2011)" (2011)
[22] Nuclear Physics B (Proc. Suppl.) **239–240** (2013) 204
[23] O. Adriani et al., Development of a 3-D cubic crystal calorimeter for space: CaloCube, Proceedings of the International Conference on Calorimetry for the High Energy Frontier (CHEF 2013)

EPJ Web of Conferences **121**, 03002 (2016) DOI: 10.1051/epjconf/201612103002
RICAP-2014

Cosmic ray measurements around the knee

Andrea Chiavassa[a]

Dipartimento di Fisica, Università degli Studi di Torino, via P. Giuria 1, 10125 Torino, Italy

Abstract. Primary cosmic rays of energy greater than $\sim 10^{14}$ eV must be studied by indirect experiments measuring the particles generated in the EAS (Extensive Air Shower) development in atmosphere. These experiments are mainly limited by the systematic errors due to their energy calibration. I will discuss the main sources of these errors: the choice of the hadronic interaction model and of the mass of the primary particle (that cannot be measured on a event by event basis).

I will then summarize some recent measurements of the all particle spectrum, and I will show that, keeping into account the differences due to the energy calibration, they all agree on the spectral shape. Then I will describe the measurements of the light and heavy primaries mass groups spectra, discussing the claimed features. Using a simple calculation of the elemental spectra (based on the hypothesis that the knee energies follow a Peter's cycle) I will try to discuss if all these results can be interpreted in a common picture.

1. Introduction

Primary cosmic rays with energy greater than 10^{14} eV cannot be studied by direct experiments operating on balloon or on satellites and their detection is only possible by means of indirect EAS experiments. All the main characteristics of the primary particle (i.e. mass, energy and arrival direction) must therefore be derived measuring the Extensive Air Showers (EAS) generated by the interaction of primary cosmic rays with atmospheric nuclei.

The experiments operating in the 10^{14}–10^{18} eV energy range can be divided in two groups:

- Surface arrays: sampling the EAS at fixed atmospheric depth. Almost all of these arrays are able to simultaneously detect more than one EAS components: usually the electromagnetic and the muonic ones. Detecting the particle density and arrival times at different distances from the shower core these arrays derive the arrival direction of the primary cosmic ray, the number of charged particles (N_{ch}) and the number of muons (N_μ) in the EAS at observation level. Both N_{ch} and N_μ are derived as normalization of the lateral distribution of the particle density and can be defined either as the total number of particles at observation level or as the number of particles at a fixed distance from the shower core (distance that has to be fixed by every single experiment and depends both on the detector layout and on the primary energy range studied by the array). These detectors operate with a 100% duty cycle.
- Cherenkov arrays: detecting the cherenkov light emitted by particles during EAS development. The big advantage of these arrays is that they perform an almost

[a] e-mail: achiavas@to.infn.it

calorimetric measurement of the primary energy. The atmospheric depth of the shower maximum (a parameter indicating the mass of the primary particle) can be estimated comparing the cherenkov photon densities measured at two different distances from the shower core. These arrays operate only during clear moonless nights, thus their duty cycle is $\sim$10–15%.

The main problem of these experiments is that energy and mass calibrations have to be performed by means of EAS simulations that are based on hadronic interaction models extrapolated from lower energy collider measurements. Recently the results coming from the LHC experiments are filling this gap, nevertheless collider experiments do not cover the forward region that is the relevant one to describe the EAS development.

In spite of all the mentioned calibration problems the results of last generation experiments allowed an understanding of details of the knee of the primary cosmic ray spectrum. The scenario resulting from such measurements favors an astrophysical interpretation of the knee, even if the resulting picture is not complete and, to further improve our knowledge, high precision experiments are needed. I will show that recent results can, at least qualitatively, be described by elemental spectra with change of slope at energies scaling with Z, even if the recent ARGO-YBJ results [1] cannot be easily included in a simple framework and an additional galactic component dominating the spectrum above 10^{16} eV must be introduced [2, 3].

2. Energy and mass calibration of indirect experiments

2.1 Energy calibration

Surface arrays derive the energy of the primary particle either from the measured value of the number of charged particles (N_{ch}) either from the measured value of the number of muons (N_μ) or from a combination of both.

These shower parameters depend not only on the energy of the primary cosmic ray but also on the atmospheric depth crossed by the EAS, thus the zenith angle θ has to be taken into account. The shower evolution in atmosphere can be treated in two different ways: a first one is based on the experimental data supposing that the arrival directions of cosmic rays are isotropically distributed, the so called "constant intensity cut" [4], while a second one describes the EAS atmospheric evolution by means of a complete simulation.

The main sources of systematic errors in the energy calibration of any indirect experiment are related to the use of a complete EAS simulation and are therefore connected to the two main hypotheses that must be done when running these codes: the mass of the primary particle generating the shower and the hadronic interaction model describing high energy interactions.

Considering the choice of the mass of the primary particle three different strategies are possible:

- the primary energy is calculated for the two extreme values of the primary mass (Hydrogen and Iron). The spectra that are obtained for these two cases represent the upper and lower limits bracketing the "true" spectrum.
- Starting from the primary chemical composition measured at lower energy by direct experiments and assuming its evolution with energy the mean value of the primary mass is calculated as a function of the primary energy. This mean value of the primary mass is then used to convert the experimental observable to primary energy.
- Combining the N_{ch} and N_μ values measured for each event a parameter correlated with the primary mass is evaluated. Then this parameter is used in the conversion from the

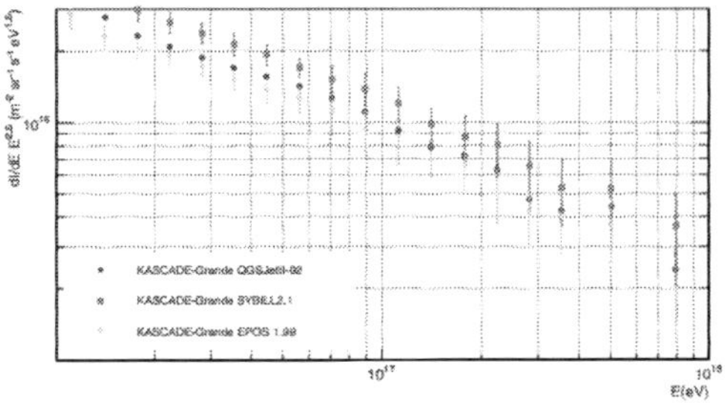

Figure 1. KASCADE-Grande measurements of the all particle energy spectra calibrated by mean of three different hadronic interaction models: QGSJetII-02, Sibyll2.1 and EPOS1.99.

experimental observable to the primary energy [5]. In this way no assumption is made about the primary chemical composition, but a further dependence on the hadronic interaction model is added (in the definition of the parameter correlated with the primary mass).

Cherenkov experiments derive the primary energy from the photon density measured at a fixed distance from the shower core, 200 m in the case of the TUNKA-133 experiment [6]. Considering TUNKA-133 as reference experiment the correlation between $Q(200)$ and the primary energy is given by: $E_0 = C\,Q(200)^g$. The parameter g is evaluated by mean of the CORSIKA [7] simulation code, $g = 0.94 \pm 0.01$. The value of the constant C is obtained normalizing the integral energy spectrum of each single night of data taking to a reference one measured by the QUEST experiment [8].

The second main source of systematic error in the energy calibration of indirect experiments is the choice of the hadronic interaction model used in the EAS simulation. The center of mass energy of the interaction between an high energy cosmic ray and air nuclei is well above the ones studied at collider experiments. The more often used hadronic interaction models used by current EAS experiments are: QGSJetII-02 [9], Sibyll2.1 [10] and EPOS1.99 [11]. The KASCADE-Grande experiment has evaluated the differences in the energy assignment due to the hadronic interaction model choice [12], finding that they are smaller than 20%. The all particle spectra derived from the KASCADE-Grande data calibrated by means of these three different hadronic interaction models are shown in Fig. 1.

2.2 Mass calibration

Due to the intrinsic fluctuations in the EAS development indirect experiments cannot access the mass of every single event, even at shower maximum (where the fluctuations are minimized) the differences between EAS generated by primaries of similar mass (i.e. H and He, He and CNO, Si and Fe) are smaller than one standard deviation. Studies of the primary chemical composition can be performed using statistical methods like, for instance, the unfolding analysis introduced by the KASCADE experiment [13], or separating events in two mass groups, i.e. light and heavy primaries.

The shower observables sizable to such purpose are:

- the measurement of the shower maximum development. With nowadays techniques this information can be directly obtained only by fluorescence light telescopes, while cherenkov detectors can infer it from the ratio of the photon densities at two different distances from the shower core.

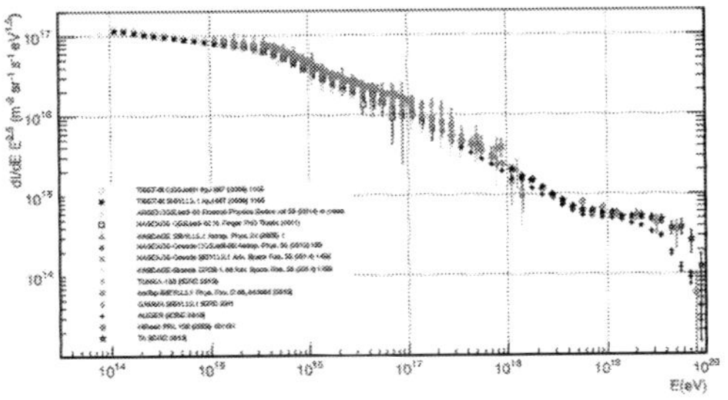

Figure 2. Compilation of all particle energy spectra measurements [1, 5, 6, 12–20].

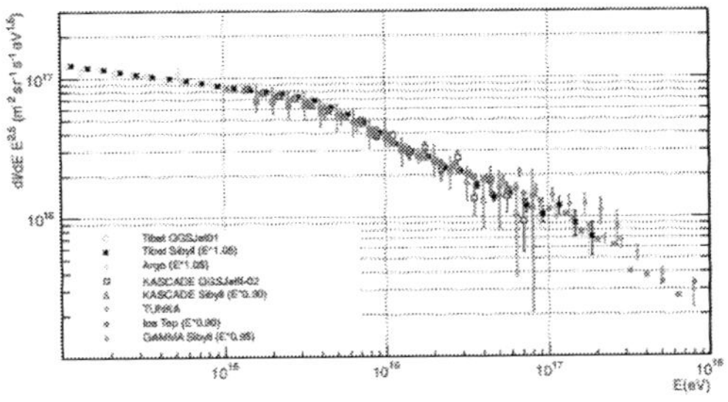

Figure 3. All particle energy spectrum measured in the 10^{14}–10^{18} eV energy range. The energy calibration of the experiments are shifted by an arbitrary amount smaller than the calibration systematic errors. References can be found in the caption of Fig. 2, the energy shifts applied are reported in the legend.

- The ratio between observables representing the charged particle and the muon numbers at observation level.
- A correlation between a parameter sensible to the number of particles at observation level and another one reflecting the shape of the particle lateral distribution.

In order to assign a value of the primary mass to any of the previously indicated experimental observables a complete EAS simulation is needed, therefore all these analyses depend on the high energy hadronic interaction model used.

3. Experimental results

Having discussed the main sources of systematic errors affecting the energy and mass measurements performed by indirect experiments, in this section I will discuss some of the more recent results obtained in the 10^{14}–10^{18} eV energy range.

3.1 All particle energy spectrum measurements

Figure 2 shows a compilation of the all particle energy spectrum measured by different experiments using different measurement techniques, operating at different height above see level and calibrated by different hadronic interaction models. Looking at this plot we can have the impression of a confused situation, with measurements showing big differences between each other, but simply shifting, as already proposed in [3], the single experiment energies by an amount lower than the previously discussed systematic errors (i.e. $\Delta E/E \leq 20\%$) all the results agree much better, as can seen in Fig. 3. The differences shown in Fig. 2 can thus be

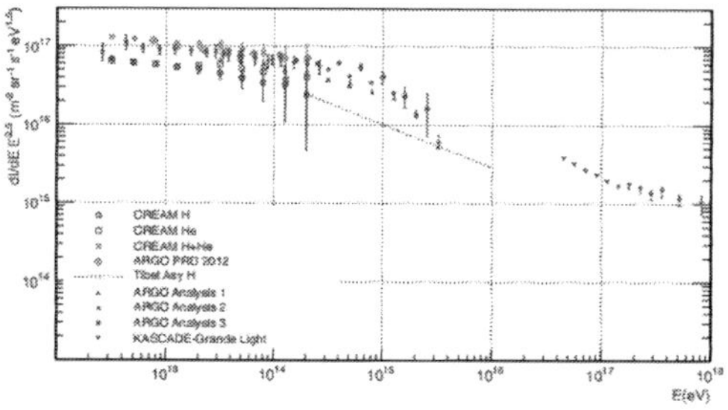

Figure 4. Spectra of the light primaries measured by ARGO-YBJ (the different analysis shown by the collaboration are described in [1]) and KASCADE-Grande. Lower energy direct measurements of the CREAM experiment are shown as a reference. The former measurement obtained by the KASCADE collaboration is not shown in this plot as it was displayed in a non energy calibrated way [21].

attributed to the energy calibration of the experiments and there is a general agreement about the structure of the spectrum showing, in addition to the well known feature known as the knee, two faint but significant structures at $\sim 10^{16}$ eV and at $\sim 10^{17}$ eV.

3.2 Mass groups energy spectra

Indirect experiments can measure the spectra of mass groups of primary cosmic rays by two different approaches that will be described in the following.

A first analysis technique separates the events in two samples according to the measured ratio between the muon and charged particles numbers, converted, by means of the constant intensity cut, to a reference zenith angle. The events sample with a high value of this parameter (called Y by the KASCADE collaboration who introduced this analysis technique [21]) is generated by heavy primaries and the one with low Y values by the light ones. The KASCADE-Grande experiment has then modified this analysis [22, 23] simply calculating the ratio between muon and charged particles numbers in intervals of zenith angles (i.e. atmospheric depth).

A similar approach has been used by the ARGO-YBJ collaboration [1] that, being a full coverage detector operating at higher altitude (i.e. almost near to shower maximum) and not having a muon detector, selected the events on the basis of the ratio between the number of particles measured within a distance of 8 m from the shower core (Np_8) and the slope (s') of the lateral distribution function.

In both techniques a selection level must be defined to divide the two samples, this cut has to be defined according to the results of a complete EAS simulation. In the case of the KASCADE and KASCADE-Grande experiments it was shown that the level used to define the two samples affects only the absolute fluxes of the spectra, while eventual spectral features are visible independently from this choice, only enhancing them. For a specified hadronic interaction model the choice of the cut value corresponds to a primary mass, while for a specified primary mass its choice means selecting simulations performed with a particular hadronic interaction model. Thus a spectral feature measured independently from the cut value does not depend on the hadronic interaction model used in the EAS simulation.

Figure 4 shows a summary of the, energy calibrated, KASCADE-Grande and ARGO-YBJ measurements of the light primaries spectra. In the plot also the direct measurements of the H and He spectra performed by the CREAM collaboration [24] are shown, their agreement with the low energies measurement of the ARGO-YBJ experiments confirms the validity of the energy calibration of this experiment and suggests also the good reliability (at least around 10^{13} eV) of the hadronic interaction models included in the CORSIKA code.

The KASCADE experiment published [21] the spectra of the "electron rich" (i.e. light primaries) and the "electron poor" (i.e. heavy primaries) events. These spectra are not calibrated in energy (they are shown as a function of the muon density at fixed core distance), but using the integral flux above the change of slope we can identify this feature with the knee of the all particle energy spectrum.

The ARGO-YBJ collaboration has shown [1] the results of three different analysis that differs between each other by an amount lower than the declared systematic errors, all the three spectra show a break well below 10^{15} eV.

The KASCADE-Grande experiment published the spectra of the light and heavy components in two slightly different energy ranges with selection cuts defined in order to enhance the spectral features of these two components:

- an hardening of the light component spectrum [23]; at $10^{17.08\pm0.08}$ eV the spectral slope changes from $\gamma = -3.25 \pm 0.05$ to $\gamma = -2.79 \pm 0.08$.
- A steepening of the heavy primaries spectrum [22]; at $\log(E/\text{eV}) = 16.92 \pm 0.04$ the spectral slope γ changes from $\gamma = -2.76 \pm 0.02$ to $\gamma = -3.24 \pm 0.05$.

4. Discussion

In this section I present the results of a calculation, based on the work of T. Gaisser et al. [3], of elemental spectra following a Peter's cycle (i.e. with knee energies at constant rigidity). My goal is neither reaching a detailed interpretation of the data nor making any fit to obtain the elemental spectra describing them, I will only search for a qualitative agreement between the data reported in Sect. 3 and the calculated spectra.

The elemental spectra are calculated with an expression [25] describing a power law with index changing from γ_1 to γ_2 at energy E_{knee}.

$$\Phi(E) = K E^{\gamma_1} \left[1 + \left(\frac{E}{E_{knee}}\right)^{\epsilon}\right]^{\frac{(\gamma_2-\gamma_1)}{\epsilon}} . \tag{1}$$

The parameter ϵ represent the sharpness of the change of slope, and in these calculations has been kept constant ($\epsilon = 10$) corresponding to a smooth knee.

The elemental fluxes have been calculated with the following hypotheses:

1) the absolute normalization is given by the intensities measured by CREAM [24] at 10^{13} eV.
2) The slopes of the H and He spectra before the knee are the ones measured by CREAM, the slope of the He spectrum is used also for heavier elements.
3) All the spectra suffer the same change of slope, values from $\Delta\gamma = 0.5$ to $\Delta\gamma = 0.7$ have been used.
4) The knee energy has been calculated following a Peter's cycle: $E_{knee}(Z) = Z \times E_{knee}(H)$. The Hydrogen knee is fixed at two different energies: $E_{knee}(H) = 6.5 \times 10^{14}$ eV following the ARGO-YBJ results and $E_{knee}(H) = 1.5 \times 10^{15}$ eV attributing the knee of the all particle spectrum to a light element, i.e. either H or He.
5) To account for the hardening of the light elements spectrum a flatter H component ($\gamma = -2.66$) becoming dominant at $\sim 10^{17}$ eV is added.

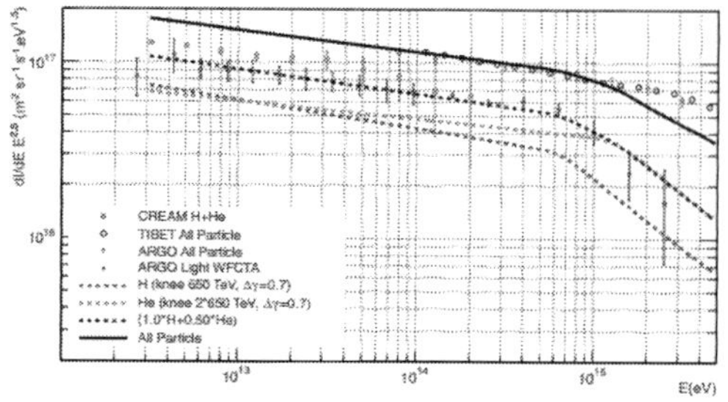

Figure 5. All particle and light elements spectra measured by the ARGO-YBJ experiment [1] compared with the calculations described in Sect. 4.

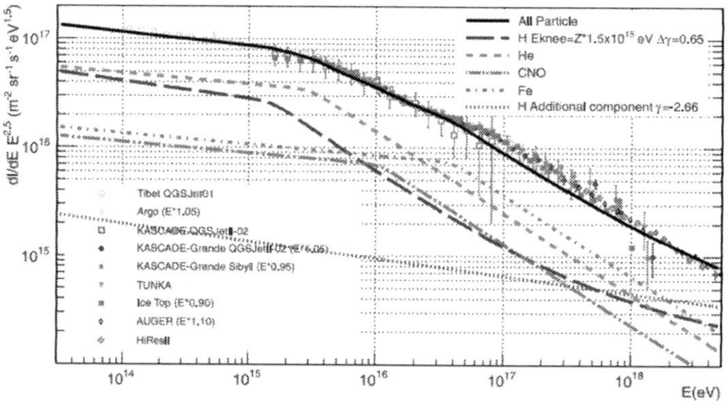

Figure 6. Summary of all particle spectrum measurements, shifted by an arbitrary amount (reported in the legend), and compared with the calculations described in Sect. 4.

Comparing the results of these calculations with the mass group spectra we must always keep in mind that the intensities measured with these algorithms aim to enhance spectral feature and not to determine the absolute fluxes.

In Fig. 5 the ARGO-YBJ light and all particle spectra [1] are shown together with the results of the previously described calculations. We can see as the sharp knee of the light elements spectrum can be described fixing a huge value for the change of slope ($\Delta\gamma = 0.7$), but summing all elements to calculate the all particle spectrum we cannot describe the Tibet-III [15] and ARGO-YBJ measurements. This disagreement is smaller assuming that the energy of the knee scales with the mass of the primaries (A) or if we use harder spectra for elements heavier than He, but it is in any case difficult to have a good description of both the light and the all particle spectra. If the ARGO-YBJ results will be confirmed an additional component must be invoked.

Extending the elemental spectra calculations from 10^{14} eV to 5×10^{18} eV we can compare the expectations with the measurements in a wide energy range, in Fig. 6 I show the calculated fluxes and the experimental data shifted by an arbitrary scale. The shown fluxes correspond to: H (blue long-dash line), He (magenta short-dash line), C (green dash-dotted line), Fe (red dotted line) and their sum (black thick solid line). We can see that the knee structure is well reproduced by elemental spectra following Peter's cycle, but further details must be clarified by future experiments: 1) which element is dominating the spectrum at the knee? 2) If all the claimed results are confirmed, is there a contradiction between measurements performed at high altitudes and at sea level? 3) Which is the exact energy of the knee? 4) The knees of different elements scale with the charge (Z) or with the mass (A) of the primaries? 5) At energies smaller than the knee are the spectral indexes of all elements (except H) equal or are they different? 6) Is the change of slope at the knee the same for all elements?

Thus new measurements are required, these can be performed either by direct, high acceptance measurements or by indirect high resolutions arrays. Both techniques have their advantages and limitations:

- direct measurements. Long duration satellite experiments, like the foreseen iss-cream [26], will be able to collect enough statistics to cover the knee energy and will have the charge resolution to separate all elements. Their main limitation will be the energy resolution ($\sim \Delta E/E \sim 40\%$) as they will have strong constrain on their mass, therefore calorimetric detectors will contain only a small fraction of the shower generated the primary particle interaction. Furthermore no man made sources will be available at these energies, thus the absolute energy calibration will be performed indirectly by simulations.
- Indirect measurements. The main limitation of EAS experiments will always be the EAS development fluctuations. The best way to minimize them is to locate the array at high altitudes detecting showers at their maximum development. This is the goal of the LHAASO experiment [27] that will be constructed in China, in the Sichuan province at ~4300 m *a.s.l.*. This experiment will sample the electromagnetic and muonic shower components with an unprecedented active surface reaching therefore a huge resolution.

References

[1] G. Di Sciascio on behalf of the ARGO collaboration, Frascati Physics Series **58** (2014) in press
[2] A.M. Hillas, J. Phys. G: Nucl. Part. Phys. **31**, R95 (2005)
[3] T.K. Gaisser et al., Front. Phys. **8**(6), 748 (2013)
[4] J. Hersil et al., Phys. Rev. Lett. **6**, 22 (1961); D.M. Edge et al., J. Phys. A **6**, 1612 (1973)
[5] W.D. Apel et al., Astropart. Phys. **36**, 183 (2012)
[6] V.V. Prosin et al., Nucl. Instr. Meth. A **756**, 94 (2014)
[7] D. Heck et al., Report FZKA 6019, Forschungszentrum Karlsruhe (1998)
[8] E.E. Korosteleva et al., Nucl. Phys. B (Proc. Suppl.) **165**, 6837 (2007)
[9] S. Ostapchenko, Phys. Rev. D **74**, 014026 (2006)
[10] E.-J. Ahn et al., Phys. Rev. D **80**, 094003 (2009)
[11] K. Werner, F.M. Liu, T. Pierog, Phys. Rev. C **74**, 044902 (2006)
[12] W.D. Apel et al., Advances in Space Research **53**, 1456 (2014)
[13] T. Antoni et al., Astropart. Phys. **24**, 1 (2005)
[14] M.G. Aartsen et al., Phys. Rev. D **88**, 042004 (2013)
[15] M. Amenomori et al., ApJ **678**, 1165 (2008)
[16] M. Finger, PhD Thesis, Karlsruhe University (2011)
[17] A.P. Garkaya et al., J. Phys. G: Nucl. Part. Phys. **35**, 115201 (2008)
[18] A. Aab et al., arXiv:13075059v1 (2013)
[19] R.U. Abbasi et al., Phys. Rev. Lett. **100**, 101101 (2008)
[20] T. Abu-Zayyad et al., arXiv:1305:7273 (2013)

[21] T. Antoni et al., Astropart. Phys. **16**, 373 (2002)
[22] W.D. Apel et al., Phys. Rev. Lett. **107**, 171104 (2011)
[23] W.D. Apel et al., Phys. Rev. D **87**, 081101(R) (2013)
[24] Y.S. Yoon et al., ApJ **728**, 122 (2011)
[25] S.V. Ter-Antonyan and L. Haroyan, arXiv:hep-ex/0003006 (2000)
[26] E.S. Seo et al., Proc. of the 33rd ICRC (Rio de Janeiro, Brasil), #0629 (2013)
[27] Cao Z., Chinese Physics C **34**, 249 (2010)

EPJ Web of Conferences **121**, 03003 (2016) DOI: 10.1051/epjconf/201612103003
RICAP-2014

Latest results from the Pierre Auger Observatory

Isabelle Lhenry-Yvon[1,a] for the Pierre Auger Collaboration[2,b]

[1] Institut de Physique Nucléaire d'Orsay (IPNO), Université Paris 11, CNRS-IN2P3, Orsay, France
[2] Observatorio Pierre Auger, Av. San Martiín Norte, 304, 5613 Malargüe, Argentina

Abstract. The Pierre Auger Observatory has been designed to investigate the origin and nature of Ultra High Energy Cosmic Rays (UHECR) with energies from 10^{17} to 10^{20} eV. In this paper we will review some of the most recent results obtained from data of the Pierre Auger Observatory, namely the spectrum of cosmic rays, the anisotropies in arrival directions and the studies related to mass composition and to the number of muons measured at the ground. We will also discuss the implication of these results for assembling a consistent description of the composition, origin and propagation of cosmic rays.

1. Introduction

The origin of the UHECR (above 10^{18} eV) is still a challenge today. Even though cosmic rays have been discovered more than 100 years ago, their study remains of great interest, especially at the highest energies where their existence is still a puzzle for astrophysicists. At these energies, they can only be observed indirectly through air showers initiated in the Earth's atmosphere. The Pierre Auger Observatory [1], conceived to characterize the properties of the UHECRs, is a hybrid instrument for air shower detection. It combines an array of particle detectors, the Surface Detector array (SD) [2], to sample the air shower front as it reaches the ground, and Fluorescence Detector (FD) telescopes to collect the ultraviolet light emitted by nitrogen as showers develop in the atmosphere [3]. The SD consists of 1660 water-Cherenkov stations, spread over a total area of $\approx$3000 km^2 on a 1.5 km grid spaced close to Malargüe, province of Mendoza, Argentina; it measures the lateral density distribution of charged particles at the ground and their temporal distribution in each station. The FD overlooks this region from four peripheral sites, each with six FD telescopes and records shower longitudinal profiles. A small area of $\sim$30 km^2, the Infill, is instrumented with 71 surface detectors on a 750 m grid and with three FD with elevated fields of view to lower the energy threshold of the Observatory down to 10^{17} eV. A set of high quality instruments is installed in the field to monitor the atmospheric conditions; on-line and long-term performances of the detectors and data quality are monitored continuously [4]. The overall uptime of the SD is above 98% and it is 100% efficient in detection showers above 3×10^{18} eV, while the FD, which can operate only in clear moonless nights, is operational only 13% of the time. Thanks to the possibility of combining the information from the two types of detectors, the reconstruction capabilities are enhanced with respect to the individual detectors.

---

[a] e-mail: `lhenry@ipno.in2p3.fr`
[b] Full author-list: `http://www.auger.org/archive/authors_2014_09.html`

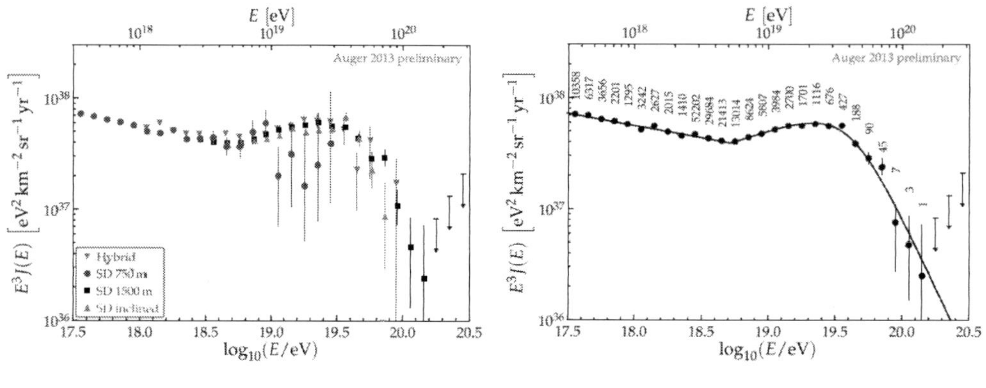

Figure 1. Left: the energy spectra obtained from hybrid events, and from data of the 750 m and 1500 m SD arrays for vertical and inclined showers measured at the Auger Observatory. Right: the combined spectrum, with the number of events shown in each energy bin. The arrows represent upper limits at an 84% confidence level [5].

2. Spectrum

Four different data sets are used to compose the Pierre Auger Observatory spectrum: events with the 750 m SD subarray and events with the 1500 m full SD array, for events with zenith angle below 55° and 60 ° respectively (hereafter referred as vertical events), hybrid events measured with both the FD and the SD, events with zenith angle above 60° (called inclined events). Those spectra are shown in Fig. 1, left. The flux $J(E)$ is multiplied by E^3 to favor an easy identification of features in the power-law spectrum $E^{-\gamma}$.

These data sets are combined to compose the spectrum shown in Fig. 1, right, using a maximum-likelihood method providing the relative normalization. Smearing corrections are applied to take into account possible bin-to-bin migrations. About 130,000 events compose this spectrum. The systematic uncertainty in the energy scale is 14% [6] and the energy resolution for events measured only with the SD is better than 12% above 10^{19} eV [5]. Two features can be clearly identified in the resulting spectrum: a sharp hardening of the spectrum at the so-called ankle, and a suppression of the flux at the highest energies. At the energy of 5×10^{18} eV, corresponding to the ankle, the power-law index changes from 3.23 ± 0.07 to 2.63 ± 0.04. Above 2×10^{19} eV the spectrum starts to deviate from this power law, being suppressed with respect to what would be expected by extrapolating the spectrum above the ankle. At 4×10^{19} eV the flux is half of that expected from that extrapolation. The significance of the suppression is 20 σ.

Details of the Auger spectrum were compared with theoretical predictions assuming different extragalactic astrophysical scenarios, pure-light or pure-heavy compositions, and two maximal acceleration energies at the sources, 10^{20} and $10^{20.5}$ eV. Auger flux measurements over the whole energy range from below the ankle up to the suppression region are in better agreement with predictions assuming a maximal acceleration energy of 10^{20} eV, and exclude a pure-iron composition. The energy spectrum of UHECRs turns out to be most likely the result of multiple processes involving the spatial distribution of their sources, the acceleration process and the interactions along their propagation paths. To distinguish between the different scenarios mentioned above and thus to understand the origin of the highest energy cosmic rays, complementary measurements of the mass composition are necessary as well as studies of the anisotropy in the arrival directions.

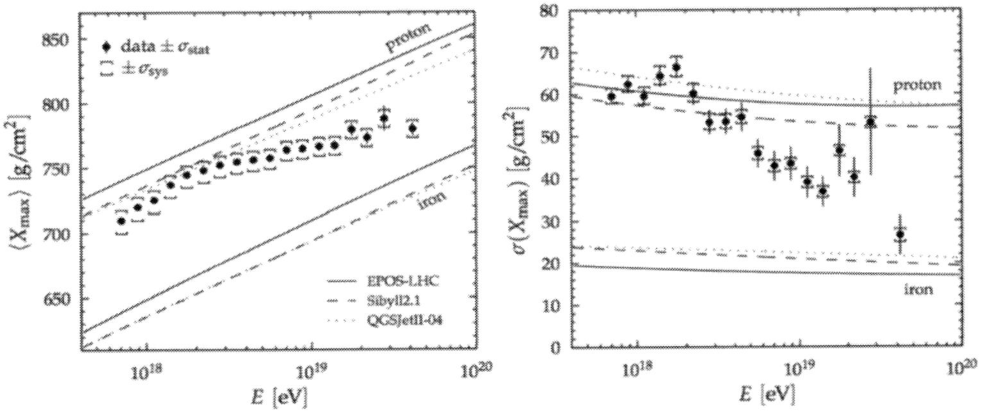

Figure 2. $\langle X_{\text{max}} \rangle$ and its RMS as a function of energy, compared to the predictions of the most recent, post-LHC, air shower simulations using different hadronic interaction models (see text) [10].

3. Composition studies

The Pierre Auger Observatory hybrid detectors measure air shower observables containing information about the mass of the primary particle. The fluorescence telescopes allow us to follow the longitudinal development of the electromagnetic component of the air showers in the atmosphere and thus to measure the depth of maximum development, $\langle X_{\text{max}} \rangle$, and its fluctuations, $\sigma(X_{\text{max}})$. A difference of $\sim$100 g/cm^2 is expected between the $\langle X_{\text{max}} \rangle$ of showers generated by protons and iron nuclei, the latter developing higher in the atmosphere. A narrower distribution (smaller $\sigma(\text{X}_{\text{max}})$) is also expected from heavy nuclei as compared to light ones. $\langle X_{max} \rangle$ is related to the depth of the first interaction of the primary and to the subsequent development of the shower; for this reason, the interpretation in terms of composition is complicated by the large uncertainties in the hadronic interaction models used in the simulations.

Once corrected for the detector resolution, $\langle X_{\text{max}} \rangle$ and $\sigma(\text{X}_{\text{max}})$ can be directly compared to the predictions of the latest models, Sybill 2.1 [7], QGSJetII-04 [8] and EPOS-LHC [9]. As shown in Fig. 2, all models predict a constant elongation rate as a function of energy for pure compositions. This is not compatible with our measurements, which clearly show a change of slope at a few EeV. If we assume that no change in hadronic interactions comes into play at these energies, our data show evidence for a composition getting heavier above the ankle region.

The analysis of $\langle X_{\text{max}} \rangle$ and $\sigma(\text{X}_{\text{max}})$ relies on measurements of hybrid events, thus suffering from the lower number of events measured during the shorter duty cycle of the FD.

The nature of the primaries can be studied with more statistics by exploiting the SD data. The time structure of particles reaching the surface detector array on the ground allows us to measure the distribution of muon production distances and from that distribution infer the slant depth at which the production of muons reaches its maximum, $\langle X^{\mu}_{\text{max}} \rangle$. The depths of both maxima are, of course, strongly correlated, mainly through the depth of the first interaction. At the same time, these measurements provide a means of testing and constraining models of hadronic interactions at the energies involved in the showers. The analysis of $\langle X^{\mu}_{\text{max}} \rangle$ exploits the higher number of showers measured with the SD, and therefore also reaches higher energies. Results from simulations indicate that this quantity depends on the mass of the primary particle in a similar way to $\langle X_{\text{max}} \rangle$, its average and dispersion being

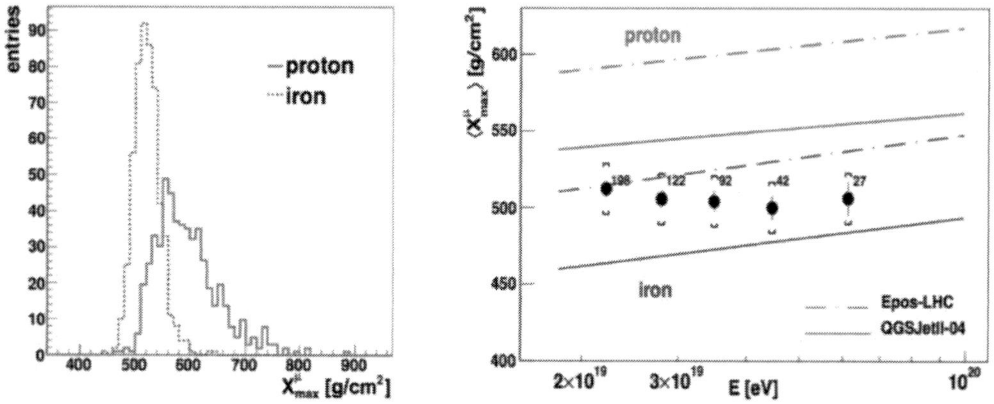

Figure 3. Left: $\langle X^{\mu}_{max} \rangle$ distributions in showers initiated by protons and iron nuclei of 3×10^{19} eV and angles between 55° and 65°, simulated with EPOS-LHC. Right: evolution of $\langle X^{\mu}_{max} \rangle$ measured versus energy, compared to the predictions of simulations using EPOS-LHC and QGSJetII-04 for proton and iron showers. The brackets represent the systematic uncertainty. The number of showers is shown for each energy bin [11].

smaller for heavier primaries compared to those obtained for lighter ones. An example of the $\langle X^{\mu}_{max} \rangle$ distribution for showers initiated by protons and iron nuclei of 3×10^{19} eV is shown in Fig. 3, left, obtained with simulations using EPOS-LHC as the interaction model.

The analysis of muon production depths was performed on showers of energies above 2×10^{19} eV detected at the Pierre Auger Observatory from January 2004 up to December 2013. In the construction of the distributions of muon production depths only muons from inclined showers with zenith angles between 55° and 65° and within a distance from the shower core between 1700 m and 4000 m were included, thus allowing a clear selection of the signal due to the muonic component, without any bias arising from the electromagnetic component. In Fig. 3, right, the resulting $\langle X^{\mu}_{max} \rangle$ are shown for five energy bins, and compared to predictions from simulations adopting QGSJetII-04 or EPOS-LHC. Both models predict a linear evolution of $\langle X^{\mu}_{max} \rangle$ with the logarithm of the energy, although with a systematic difference of about 50 g/cm^2. The data points are bracketed by the lines following predictions of the simulations for proton and iron primaries with the QGSJetII-04 model, and fall below those predicted from simulations with EPOS-LHC. Furthermore, the data show a flatter trend than predicted for a pure-proton or a pure-iron composition. From these results it is also possible to use the muon production depth as a tool to constrain hadronic interaction models.

In the frame of the same interaction models, both $\langle X^{\mu}_{max} \rangle$ and $\langle X_{max} \rangle$ can be converted to $\langle \ln A \rangle$, the mean logarithmic mass, hopefully converging to a coherent result [11]. As shown in Fig. 4, there is a discrepancy in the EPOS-LHC case, which could be most probably explained as a collective effect of nuclei in the atmosphere (as compared to the proton-proton interactions at LHC) [5].

As for QGSJetII-04, the good agreement is only apparent, as this model does not accurately reproduce the rapidity gap distribution of LHC data. From these considerations, although insufficient to exclude specific models, it appears that the UHECR data have the power to constrain high energy interaction models in unexplored energies and kinematic regions.

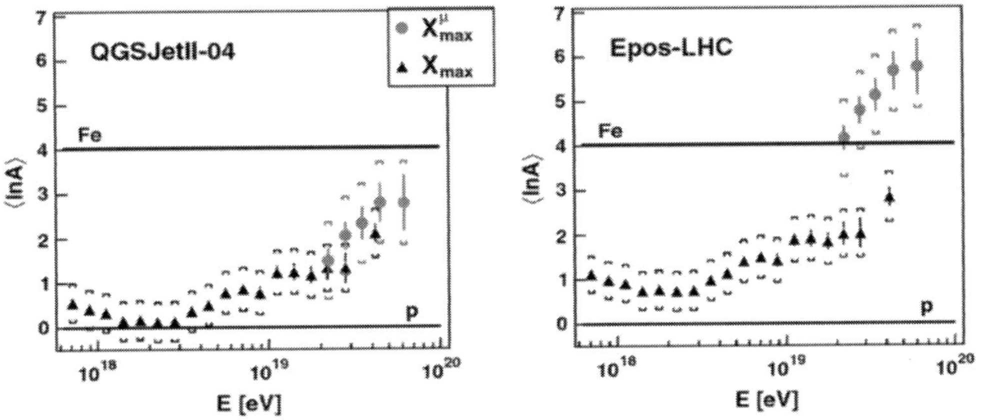

Figure 4. Conversion of $\langle X_{\mathrm{max}}\rangle$ (black triangles) and $\langle X^{\mu}_{\mathrm{max}}\rangle$ (red circles) to ⟨lnA⟩, as a function of energy, for the hadronic interaction models QGSJetII-04 (left) and EPOS-LHC (right).

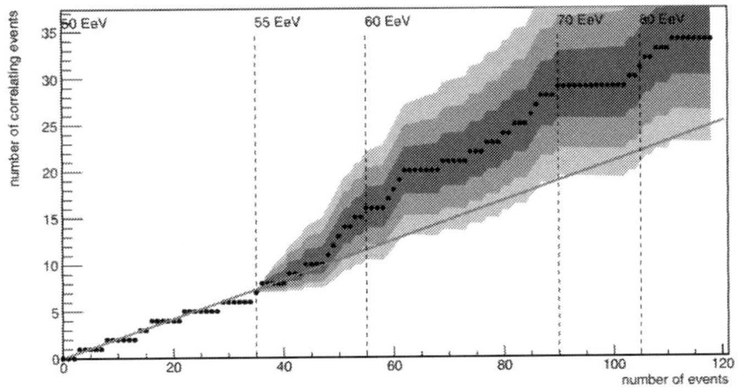

Figure 5. Auger events correlating with the positions of AGN from the VCV catalogue, in an energy ordered plot [5].

4. Anisotropies in arrival directions

A key observable to obtain information about the sources and nature of UHE cosmic rays is the distribution of their arrival directions. At these energies, the particles are most probably extragalactic; if the observed flux suppression can be attributed to propagation effects, they would be limited by a GZK horizon of about 100 Mpc. A correlation between the arrival directions of the highest energy cosmic rays (above 50 EeV) and the positions of the nearby AGN from the Véron-Cetty-Véron (VCV) catalogue was reported in 2007 [12]. The most recent update of this search is shown in Fig. 5 where the current estimate of the fraction of correlation is $(33 \pm 5)\%$, against a 21% expected under the hypothesis of isotropy.

The observation of large scale anisotropies in the arrival directions could give a hint on the galactic or extragalactic origin of the cosmic rays in different energy ranges. Recently, the Pierre Auger Collaboration and the Telescope Array Collaboration published the results for the dipole and quadrupole moments, combining their data sets of arrival directions above $\sim 10^{19}$ eV in a full-sky coverage. The region of the sky for which the fields of view of both observatories overlap provides the cross-calibration factor for the different exposures and the compensation of systematic uncertainties due to different energy scales. No significant deviation from isotropy at any scale is found throughout the various analyses. Upper limits on the amplitude of the dipole and quadrupole moments are derived as a function of the direction in the sky, resulting between 7% and 13% for the dipole, and between 7% and 10% for a symmetric quadrupole [13].

5. Conclusion

Results from the Pierre Auger Observatory clearly confirmed the called ankle transition in the spectrum index at 4×10^{18} eV as well as the flux suppression above 5×10^{19} eV. These results are not enough to understand whether the suppression is caused by the energy loss of cosmic rays during their propagation, or is rather a signal of the maximum acceleration at their sources. The measurement of the composition of the detected cosmic rays is a crucial key to answer this question and to understand their origin. For this reason the Pierre Auger Collaboration is currently proposing an upgrade of the Observatory, that consists in an addition of scintillator detectors aiming at a better identification of the chemical composition of the primary particles, mostly increasing the mass sensitivity of the Observatory in the highest-energy region. It will also allow an improved study of anisotropies by selecting those primaries with a small charge and of hadronic interactions at center-of-mass energies an order of magnitude above those presently attainable at the LHC.

References

[1] J. Abraham et al. (Pierre Auger Collaboration) Nucl. Instrum. Meth. A **523**, 50–95 (2004)
[2] J. Abraham et al. (Pierre Auger Collaboration) Nucl. Instrum. Meth. A **613**, 29–39 (2010)
[3] J. Abraham et al. (Pierre Auger Collaboration) Nucl. Instrum. Meth. A **620**, 227–251 (2010)
[4] P. Abreu et al. (The Pierre Auger Collaboration) J. Instrum. **7**, P09001 (2012)
[5] A. Letessier-Selvon for the Pierre Auger Collaboration, Proc. 33rd ICRC, arXiv:1310.4620 (2013), Braz. J. Phys. **44**, 560–570 (2014)
[6] V. Verzi for the Pierre Auger Collaboration, Proc. 33rd ICRC, arXiv:1307.5059 (2013)
[7] E.-J. Ahn et al., Phys. Rev. D **80**, 09400 (2009)
[8] S. Ostapchenko, Phys. Rev. D **83**, 014018 (2011)
[9] T. Pierog et al., arXiv:1306.0121 [hep-ph](2013)
[10] A. Aab et al. (The Pierre Auger Collaboration) Phys. Rev. D **90**, 122005 (2014)
[11] A. Aab et al. (The Pierre Auger Collaboration) Phys. Rev. D **90**, 012012 (2014)
[12] The Pierre Auger Collaboration, Science **318**, 938 (2007)
[13] A. Aab et al. (The Pierre Auger and Telescope Array Collaborations) Astrophys. J. **794**, 172 (2014)

EPJ Web of Conferences **121**, 03004 (2016) DOI: 10.1051/epjconf/201612103004
RICAP-2014

Results from Tunka-133 (5 years observation) and from the Tunka-HiSCORE prototype

V.V. Prosin[1,a], S.F. Berezhnev[1], N.M. Budnev[2], M. Brückner[3], A. Chiavassa[4], O.A. Chvalaev[2], A.V. Dyachok[2], S.N. Epimakhov[5], A.V. Gafarov[2], O.A. Gress[2], T.I. Gress[2], D. Horns[5], N.N. Kalmykov[1], N.I. Karpov[1], S.N. Kiryuhin[2], E.N. Konstantinov[2], A.V. Korobchenko[2], E.E. Korosteleva[1], V.A. Kozhin[1], M. Kunnas[5], L.A. Kuzmichev[1,2], B.K. Lubsandorzhiev[6], N.B. Lubsandorzhiev[1], R.R. Mirgazov[2], R.D. Monkhoev[2], R. Nachtigall[5], E.A. Osipova[1], A.L. Pakhorukov[2], M.I. Panasyuk[1], L.V. Pankov[2], V.A. Poleschuk[2], E.G. Popova[1], A. Porelli[7], V.S. Ptuskin[8], G.I. Rubtsov[6], M. Rüger[7], V.S. Samoliga[2], Y.A. Semeney[2], A.A. Silaev[1], A.A. Silaev (jun.)[1], A.V. Skurikhin[1], C. Spiering[7], L.G. Sveshnikova[1], M. Tluczykont[5], R. Wischnewski[7], A.V. Zagorodnikov[2], and A.V. Zurbanov[2]

[1] Skobeltsyn Institute of Nuclear Physics, Lomonosov Moscow State University, Moscow, Russian Federation
[2] Institute of Applied Physics, Irkutsk State University, Irkutsk, Russian Federation
[3] Institute für Informatik, Humboldt University, Berlin, Germany
[4] University of Turin, Turin, Italy
[5] Institute fur Experimentalphysik, University of Hamburg, Hamburg, Germany
[6] Institute of Nuclear Research, Russian Academy of Sciences, Moscow, Russian Federation
[7] DESY Zeuthen, Germany
[8] Pushkov Institute of Terrestrial Magnetism, Ionosphere and Radio Waves Distribution, Russian Academy of Sciences, Moscow, Russian Federation

Abstract. Data obtained with two detectors located at the Tunka Cosmic Ray facility are presented. The Cherenkov light array for registration of extensive air showers (EAS) Tunka-133 collected data during 5 winter seasons since 2009 to 2014. The differential energy spectrum of all particles and the dependence of the average maximum depth on the energy in the range of $6 \cdot 10^{15}-10^{18}$ eV measured for 1540 hours of observation are presented.

The preliminary all particle energy spectrum by the data of Tunka-HiSCORE prototype array, installed in 2013, is presented. Some additional experiments in the Tunka Valley are briefly described.

1. Introduction

EAS Cherenkov light array Tunka-133 with $\sim 3\,\text{km}^2$ geometric area operated since 2009 [1]. Five winter seasons of data acquisition ($\sim 10^7$ triggers) and high quality of information permitted us to reconstruct primary energy spectrum and mass composition in the energy range $6 \cdot 10^{15}-10^{18}$ eV. This energy range is the most important for understanding of transition from Galactic to extragalactic cosmic rays (CR).

[a] e-mail: `v-prosin@yandex.ru`

To expand the measurement range to lower energies the new array Tunka-HiSCORE [2] data has been used. The HiSCORE concept has been described in [3] and [4]. The preliminary primary energy spectrum in the range 200 TeV – 20 PeV by the data of the first season of 9 HiSCORE stations operation is presented. Tunka-HiSCORE stations are located at the same area as Tunka-133 detectors. In 2013 the 9 station array, arranged on a rectangular grid, covered an area of $0.09\,\mathrm{km}^2$, which will be increased to about $0.25\,\mathrm{km}^2$ in 2014.

2. The experiments development In Tunka Valley

Some additional experiments are carried out in the Tunka Valley astrophysical site.

The scintillation detectors used for some years at KASCADE-Grande array [5] has been transported to the Tunka Valley in 2013. These detectors were originally designed for the EAS-TOP array [6] in Italy. The area of each detector is $0.64\,\mathrm{m}^2$. They are planned to be grouped into 19 stations located near the centers of Tunka-133 clusters. In each station 12 detectors will be located on the ground surface and 8 detectors – underground for registration of muons.

The Tunka radio extension (Tunka-REX) [7] experiment is operating together with Tunka-133.

The main purpose of Tunka-HiSCORE is to select high energy gamma quanta from point like sources. To solve this task it is planned to reach an accuracy of EAS arrival direction measurement better than $0.1°$.

It is planned to install several IACT telescopes [2] for common operation with Tunka-HiSCORE.

3. Data processing

The Tunka-133 data were processed using the original codes, in which all fitting and conversion functions were obtained from the analysis of events simulated by CORSIKA for the energy range of $10^{15}–10^{18}$ eV. The details were described in [8].

Shower arrival direction, characterized by zenith and azimuth angles is reconstructed by fitting the measured delays with curve shower front: $T_i - T_f = R \cdot (R + 500)/(c \cdot F)$, where T_f – estimated delay for a plane front, R – the perpendicular distance from the shower axis in meters c- the speed of light, F is a parameter. This approximate formula is derived from the analysis of CORSIKA simulated showers. The formula, on the one hand, has a non-zero value of the derivative at $R = 0$ (conical shape, typical for Cherenkov radiation at short distances from the axis) and on the other hand, has only one shape parameter which is essential for processing of relatively small showers.

EAS maximum depth is derived from two parameters – the steepness of the amplitude-distance function (ADF) and the pulse width at a core distance of 400 m $\tau(400)$ [8].

Data for 262 clear moonless nights of observation was collected during 5 winter seasons of 2009–2014. Trigger condition was a coincidence of pulses in 3 detectors of a cluster inside the time gate of 0.5 μs. It was accumulated about 10^7 trigger events for 1540 hours during these nights.

Tunka-HiSCORE data presented here have been collected during 80 h for 13 clean moonless nights from February 21st to April, 3th, 2014. All the 9 stations recorded about $1.5 \cdot 10^7$ Cherenkov light pulses. The pulses treated to belong to the same shower if the time difference among them was less than 2 μs. It was selected about $1.7 \cdot 10^6$ showers with 4 and more hitted stations. The showers selected were processed with algorithms similar to those used for Tunka-133 [8].

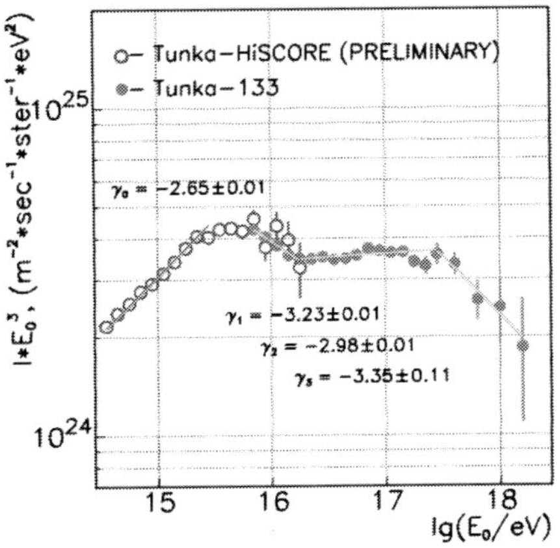

Figure 1. Differential primary cosmic rays (CR) energy spectrum.

4. Energy spectrum

To construct the Tunka-133 spectrum the events were selected with zenith angles $\theta \leq 45°$ and the core position inside a circle of radius $R_c \leq 450$ m for energy $E_0 < 5 \cdot 10^{16}$ eV and a circle of radius $R_c < 800$ m for showers with energy $E_0 \geq 5 \cdot 10^{16}$ eV. Comparison of the spectra for these two effective areas showed that starting from the above mentioned energy spectra within the error bars are the same, but the event statistics in the second round is 3 times more which is essential for energies $E_0 \geq 10^{17}$ eV.

Efficiency of shower selection inside the circle with $R_c \leq 450$ m reaches 100% for energies $E_0 \geq 6 \cdot 10^{15}$ eV. Total number of events above this energy is 270.000. About 3.000 events, selected in the circle with $R_c < 800$ m have $E_0 \geq 10^{17}$ eV.

The events for construction of the preliminary Tunka-HiSCORE spectrum were selected from events recorded inside the zenith angle $\theta \leq 25°$ and the effective area consisting of four circles of radius 75 m around the centers of 4 squares arranged by 9 stations. The number of events with energy $E_0 \geq 3 \cdot 10^{14}$ eV (for which the efficiency is close to 100%) is about 140.000.

The resulting differential Tunka-133 energy spectrum is shown in Fig. 1 together with the preliminary spectrum of Tunka-HiSCORE. One can see that Tunka-HiSCORE can provide information on the detailed shape of the spectrum at and before the knee. But more statistics is needed for this purpose. The spectrum of Tunka-133 shows a number of features – deviations from the power law. At the energy of about $2 \cdot 10^{16}$ eV the power law index changes from $\gamma = 3.23 \pm 0.01$ to $\gamma = 2.98 \pm 0.01$. Points of the spectrum are consistent with such an index till the energy $E_0 = 3 \cdot 10^{17}$ eV. Above this energy spectrum becomes steeper with $\gamma = 3.35 \pm 0.11$ (the second "knee"). There is a possibility of more complex description of the second "knee" with a small intermediate increase of the absolute value of the index to $\gamma = 3.06 \pm 0.03$ at the energy of $8 \cdot 10^{16}$ eV.

In Fig. 2 the spectrum is compared with a number of other experimental data. The spectra of all the experiments shown in Fig. 2: KASCADE [10], EAS-TOP [6], Tibet [11], HEGRA [12] – are practically indistinguishable at the energy of the first (classical) knee.

At energies below the knee there is an agreement between the preliminary result of Tunka-HiSCORE on the one hand and results of HEGRA [12] and Tibet [11] on the other hand.

There is an agreement between the result of Tunka-133 and the spectra of GAMMA [9], KASCADE-Grande [5] and Ice-TOP [13] in the intermediate energy range 10^{16}–10^{17} eV. Noticeable in Fig. 2 difference among the spectra at E_0 about 10^{17} eV can be eliminated by

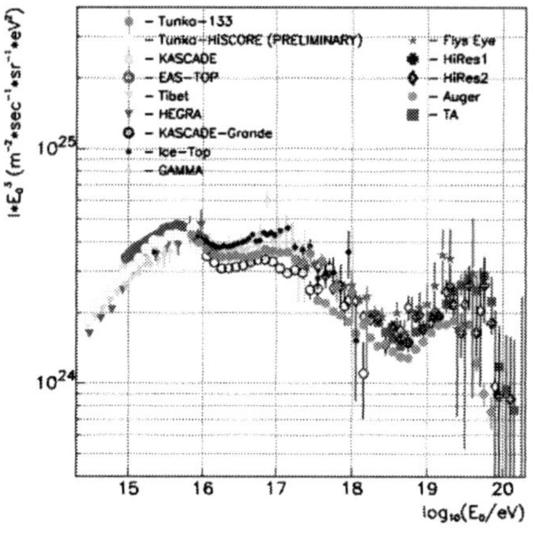

Figure 2. Comparison of energy spectra obtained at Tunka with some other experimental results.

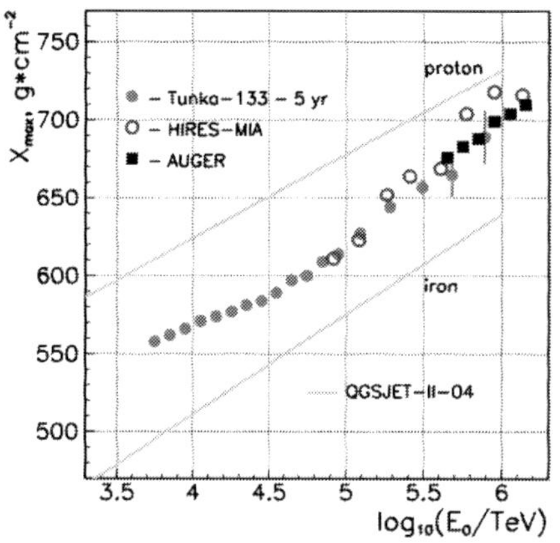

Figure 3. Mean Depth of Maximum vs. Energy.

correcting energy by 3% only. Such energy shift is much smaller than the absolute accuracy of these experiments.

For highest energies Tunka-133 data are in agreement with Fly's Eye [14], HiRes [15] and Telescope Array (TA) [16] data, and also (to less extend) with Auger [17].

5. The depth of maximum

The methods of EAS maximum depth $X_{\max}$ measurement are described in [8]. Events for this analysis are selected in a circle of radius 450 m. To get a uniform estimation for b_A over a wide range of energies, we remove from the analysis detectors at distances larger than 250 m from the core during the last step of parameters reconstruction. The experimental dependence of mean $\langle X_{\max} \rangle$ vs. primary energy E_0 in the energy range of $6 \cdot 10^{15} - 10^{18}$ eV is presented in Fig. 3. The experimental points are compared with the points of HiRes-MIA experiment [15] and fluorescent light detectors of Pierre Auger Observatory [17]. One can see an agreement of Cerenkov light Tunka experiment results with direct fluorescent light observations. The new measurements are compared with the theoretical curves simulated with the QGSJET-II-04 model for primary protons and iron nuclei.

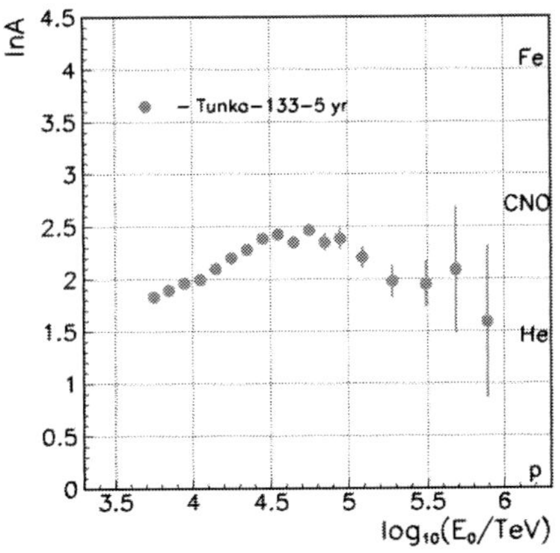

Figure 4. Mean Logarithmic Mass vs. Energy.

The mean values of $\langle X_{max} \rangle$ can be recalculated to the mean values of $\langle \ln A \rangle$ by a simple method of interpolation. The result of such an approach for the points derived from the ADF steepness analysis are shown in Fig. 4. Primary mass composition becomes heavier in the energy range 10^{16}–$3 \cdot 10^{16}$ eV and lighter again in the range 10^{17}–10^{18} eV.

6. Conclusion

1. Primary CR energy spectrum in the range of $6 \cdot 10^{15}$–10^{18} eV has a number of features: the spectrum becomes flatter (the index changes from $\gamma = 3.23 \pm 0.01$ to $\gamma = 2.98 \pm 0.01$) at $E_0 = 2 \cdot 10^{16}$ eV and steeper ($\gamma = 3.35 \pm 0.11$) at $E_0 = 3 \cdot 10^{17}$ eV. There is a possibility of more complex description of the second "knee" with a small intermediate increase of the absolute value of the index to $\gamma = 3.06 \pm 0.03$ at $E_0 = 8 \cdot 10^{16}$ eV.
2. Tunka-HiSCORE will provide information on the detailed shape of the spectrum at and before the knee after getting more statistics. A preliminary spectrum has been presented.
3. In the energy range of 10^{16}–10^{17} eV observed Tunka spectrum is consistent with the spectra of KASCADE-Grande [5] and Ice-TOP [13].
4. Beyond the energy 10^{17} eV Tunka-133 spectrum is consistent with fluorescent light experiments: Fly's Eye [14], HIRES [15] and hybrid experiment TA [16].
5. The depth of maximum X_{max} does not contradict to the results obtained by fluorescent light experiments HiRes-MIA [15] and Auger [17].
6. Primary mass composition becomes heavier in the energy range 10^{16}–$3 \cdot 10^{16}$ eV and lighter again in the range 10^{17}–10^{18} eV.

This work was supported by the Ministry of Education and Science of the Russian Federation (State Contract 14.V25.31.0010, project 1366, zadanie N 3.889.2014/K), the Russian Foundation for Basic Research (grants 13-02-12095, 13-02-00214, 15-02-05769, 15-02-10005), the Helmholtz Association (grant HRJRG-303) and the Deutsche Forschungsgemeinschaft (grant TL 51-3).

References

[1] Berezhnev S.F., et al. (Tunka-133 Collaboration), Nucl. Instr. and Methods in Physics Research A **692**, 98 (2012)
[2] Budnev N.M., et al., Journal of Instrumentation **9**, C09021 (2014)
[3] Tluczykont M., et al., Astropart. Phys. **56**, 42 (2014)

[4] Tluczykont M., et al., 24th ECRS, Kiel (in press) (2014)
[5] Apel W.D., et al. (KASCADE-Gr. Collaboration), Astropart. Phys. **36**, 183 (2012)
[6] Aglietta M., et al., EAS-TOP Collaboration, Astropart. Phys. **10**, 1 (1999)
[7] Kostunin D., et al., Nucl. Instr. and Methods in Physics Research A **742**, 89 (2014)
[8] Prosin V.V., et al., Nucl. Instr. and Methods in Physics Research A **756**, 94 (2014)
[9] Garyaka A.P., et al., Journal of Physics G: Nuclear and Particle Physics **ID=35**, 115201 (2008)
[10] Antony T., et al., Nucl. Instr. and Methods in Physics Research A **513**, 490 (2003)
[11] Amenomori M., et al., (Tibet Col.), Astrophys. J. **678**, 1165 (2008)
[12] Arqueros F., et al., The HEGRA Collaboration, Astron. Astrophys. **359**, 682 (2000)
[13] Aartsen M.G., et al., Phys. Rev. Lett. D **88**, 042004 (2013)
[14] Abu-Zayyad T., et al., Phys. Rev. Lett. **84**, 4276 (2000)
[15] Sokolsky P. for the HiRes Collaboration, Nuclear Physics B (Proc. Suppl.) **212–213**, 74 (2011)
[16] Abu-Zayyad T., et al., Astropart. Phys. **48**, 16 (2013)
[17] Schulz A. for the Pierre Auger Collaboration, Proc. 33rd ICRC Rio De Janeiro **ID=769**, (2013)

EPJ Web of Conferences **121**, 03005 (2016) DOI: 10.1051/epjconf/201612103005
RICAP-2014

AMS-02: Cosmic electron and positron (e^- + e^+) spectrum up to 1 TeV

J. Bazo[1,2,a] on behalf of the AMS-02 Collaboration

[1] INFN, Sezione di Perugia, 06123 Perugia, Italy
[2] INFN, Sezione di Tor Vergata, 00133 Roma, Italy

Abstract. The AMS-02 spectrometer, on the ISS since 2011, performs highly accurate measurements of cosmic rays up to the TeV region. We review the analysis of the cosmic ($e^+ + e^-$) flux in the energy range between 0.5 GeV and 1 TeV, based on 10.6 million ($e^+ + e^-$) events. The high statistics and detector energy resolution allow for a study of the spectral shape of unprecedented quality, thus improving our understanding of the production, acceleration and propagation of cosmic rays. The resulting energy spectrum does not show prominent features.

1. Introduction

A large experimental effort has been undertaken in the last 50 years by balloon, space born and ground based detectors in order to study the cosmic ray (CR) electron component in an extended energy range. Electrons can probe the origin and propagation of cosmic rays in the the local interstellar medium in a complementary way with respect to the hadronic CR component.

Production of CR that is solely due to super nova remnants (SNRs) is not enough to explain the observed spectral features of CR electrons under the standard propagation model. These features (e.g. the increase of the positron component in the positron fraction) could have different explanations, for example exotic sources (e.g. neutralino annihilation), or pulsar production. To understand the processes behind the observed spectral features high accuracy and high statistics measurements are needed.

The Alpha Magnetic Spectrometer (AMS) [1, 2] has precisely measured the positron fraction and the positron flux $\Phi(e^+)$ up to 500 GeV and the electron flux $\Phi(e^-)$ up to 700 GeV. They provide information on the combined flux $\Phi(e^+ + e^-)$ up to 500 GeV. In this proceeding we summarize the measurement of $\Phi(e^+ + e^-$ up to 1 TeV [3] with reduced statistical and systematic errors.

2. The AMS-02 detector

AMS is a general purpose high-energy particle physics detector installed on the International Space Station (ISS) to conduct a long-duration (~ 20-year) mission of fundamental physics research in space [4]. It consists of a tracker, a magnet, a time of flight (TOF), anti-coincidence counters, a ring imaging Cherenkov detector, an electromagnetic calorimeter (ECAL) and a transition radiation detector (TRD).

[a] e-mail: JoseLuis.Bazo@pg.infn.it

The 9-layer double-sided silicon microstrip tracker determines the trajectory and absolute charge |Z| of CR using multiple measurements of the coordinates and energy loss. Together with the 0.14 T permanent magnet, the tracker measures the particle rigidity.

The four TOF planes trigger the readout of all the detectors and measure the particle velocity and direction. The anti-coincidence counters inside the magnet bore are used to reject particles outside the geometric acceptance. The tracker, TOF, and TRD measure |Z| independently. The curvature measured with the tracker and the magnet and the direction of the particle measured with the TOF give the sign of the charge.

The 3-dimensional imaging capability of the 17 radiation length ECAL allows for an accurate measurement of the $(e^+ + e^-)$ energy E scaled to the top of AMS and of the shower shape. An ECAL estimator, based on a boosted decision tree algorithm, is used to differentiate $(e^+ + e^-)$ from protons by exploiting their different shower shapes. To further differentiate between $(e^+ + e^-)$ and protons, signals from the 20 layers of proportional tubes in the TRD are combined into a TRD classifier formed from the product of the probabilities of the $(e^+ + e^-)$ hypothesis.

AMS operates continuously on the ISS and is monitored and controlled from the ground. The detector performance is steady over time. A Monte Carlo program based on the GEANT 4.9.4 package is used to simulate physics processes and detector signals.

3. Analysis

A major experimental advantage of the combined flux analysis compared to the measurement of the individual positron and electron fluxes, particularly at high energies, is that the selection does not depend on the charge sign. Another advantage is that it has a higher overall efficiency. Consequently, this measurement is extended to 1 TeV with less overall uncertainty over the entire energy range.

Data collected from May 19th, 2011 to November 26th, 2013 have been analyzed. The isotropic $(e^+ + e^-)$ flux measured in energy bins of width ΔE is given by:

$$\Phi(e^+ + e^-) = \frac{N(E)}{A_{\mathrm{eff}}(E)\epsilon_{\mathrm{trig}}(E)\epsilon_{\mathrm{ECAL}}(E)T(E)\Delta E} \tag{1}$$

where N is the number of $(e^+ + e^-)$ events, A_{eff} is the effective detector acceptance, ϵ_{trig} is the trigger efficiency, ϵ_{ECAL} is the signal selection efficiency of the ECAL estimator and T is the exposure time.

Equation (1) is evaluated independently in 74 energy bins from 0.5 GeV to 1 TeV. The bin width is chosen to be at least twice the energy resolution. The bin-to-bin migration error is 1% at 1GeV decreasing to 0.2% above 10 GeV. The absolute energy scale uncertainty is 2% in the test beam energy range (10–290 GeV). Below 10 GeV it increases to 5% at 0.5 GeV and above 290 GeV to 5% at 1 TeV.

The event selection is done requiring a downward-going relativistic particle which hits in at least 8 out of 20 TRD layers and a single track passing through the tracker and ECAL. Events with an energy deposition compatible with a minimum ionizing particle in the first 5 layers of ECAL are rejected. Events with |Z| = 1 are selected using dE/dx in the tracker and TRD. Primary particles (i.e. not from atmospheric origin) are selected with the geomagnetic cutoff discussed below.

TRD classifier reference spectra of the $(e^+ + e^-)$ signal and the proton background are used as templates. Templates are built from data using pure samples of e^- and protons and are evaluated separately in each bin. However since the signal templates show no energy dependence above ~10 GeV, all the e^- selected in the range 15.1–83.4 GeV are taken as a

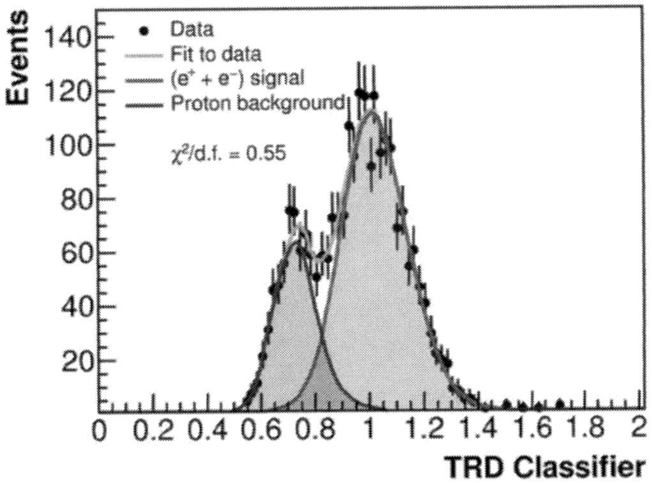

Figure 1. Result of the template fit in the 149–170 GeV energy bin showing the ($e^+ + e^-$) signal and proton background. The fit has a χ^2/d.f. = 0.55 [3].

unique signal template. The sum of the signal and background templates is fit to the data by varying their normalizations. This gives the number of signal ($e^+ + e^-$) events N and the number of background events. Figure 1 shows the data, the fit and the signal and background templates for one example bin.

A total of 10.6×10^6 ($e^+ + e^-$) events have been identified with energies from 0.5 GeV to 1 TeV.

The effective detector acceptance is defined as:

$$A_{\text{eff}} = A_{\text{geom}} \epsilon_{\text{sel}} (1 + \delta) \tag{2}$$

where A_{geom} is the geometric acceptance, ϵ_{sel} is the event selection efficiency and δ is a data driven correction. The acceptance for a particle passing through the active volumes of the tracker, TRD, TOF, and ECAL is $A_{geom} \simeq 550\,\text{cm}^2$ sr and ϵ_{sel} has typical values of 90% at 10 GeV and 70% at 1 TeV. Both A_{geom} and ϵ_{sel} are evaluated from Monte Carlo simulation.

The small acceptance correction δ is obtained comparing data and simulation efficiencies for each selection cut. This correction is a smooth, slowly varying function of energy (-0.04 at 10 GeV and -0.03 at 1 TeV).

The trigger efficiency is determined from data. The DAQ is triggered by the coincidence of all four TOF planes. AMS also records unbiased triggers which require a coincidence of any three out of four TOF planes used to measure ϵ_{trig}. The trigger efficiency is 100% above 3 GeV decreasing to 75% at 1 GeV.

The ECAL estimator efficiency ϵ_{ECAL} is measured from data using negative rigidity samples and the selection cuts. Optimal ϵ_{ECAL} values range from 75% to 95% for different energy bins, depending on the number of signal and background events.

The orbital parameters and detector status are recorded every second of data-taking. Livetime-weighted seconds are added to obtain the exposure time in a given energy bin when the minimum bin energy exceeds 1.2 times the maximum Strömer cutoff for $|Z| = 1$ particles in the AMS geometric acceptance. The exposure time excludes time spent in the South Atlantic Anomaly, during TRD gas refills, and when the AMS z axis was more than 40° from the local zenith. For the energy bins above ~30 GeV, where the effect of the geomagnetic cutoff is negligible, the exposure time is 6.2×10^7 s, decreasing to 1.5×10^7 s at 5 GeV.

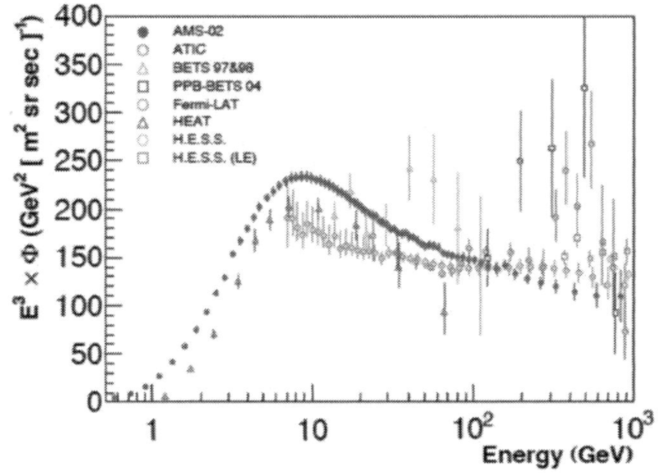

Figure 2. Electron plus positron flux $\Phi(e^+ + e^-)$ measured by AMS multiplied by $\tilde{E}^3$ versus energy. AMS total errors are the quadratic sum of the statistical and systematic errors [3]. Also shown are the results from earlier experiments [6–12].

3.1 Systematic errors

Systematic uncertainties come from the event selection, the acceptance and bin-to-bin migration discussed above.

The event selection systematic uncertainty, which includes the uncertainty from the construction of the templates, is evaluated from 2000 trials performed in each energy bin. Each trial consisted in reproducing the complete analysis. The trials were performed with different values of the ECAL estimator cut and different values of selection cuts used to construct the templates.

The number of signal events corrected by the ECAL estimator selection efficiency $N_E = N/\epsilon_{ECAL}$ is stable within the optimal range of ϵ_{ECAL}. The median value of the N_E distribution from the 2000 trials determines the flux. The RMS spread of the distribution provides an evaluation of the stability of the measurement. The difference between the width of this distribution in data and the expected statistical fluctuations gives the systematic uncertainty as <1% below ~200 GeV increasing to 4% in the 500–700 GeV bin. This is the main source of systematic uncertainty above ~500 GeV.

The acceptance systematic error is given by the uncertainty on δ. It is estimated from data to simulation comparisons. Above 3 GeV a systematic of 2% on $(1 + \delta)$ is obtained from the contributions of all cuts, increasing to 6% at 1 GeV. This is the major contribution to the systematic error below ~500 GeV. The acceptance systematic error includes a bin-to-bin correlation of 1.4% over the whole energy range.

4. Results

The measured $(e^+ + e^-)$ flux multiplied by $\tilde{E}^3$ [5] including its statistical and systematic errors is shown in Fig. 2 as a function of the energy at the top of the instrument together with previous measurements [6–12].

Below ~10 GeV solar modulation affects the behavior of $\Phi(e^+ + e^-)$. However, above 20 GeV these effects are negligible within the current experimental accuracy. The flux is smooth and shows no structures from 10 GeV to 1 TeV, revealing new information.

The flux cannot be described by a single power law ($\Phi \propto E^{\gamma}$) over the entire range. The lower energy limit above which a single power law (consistent spectral indices at the 90% C.L.) can describe the flux is estimated to be 30.2 GeV.

Fitting a single power law over the range 30.2 GeV to 1 TeV yields $\gamma = -3.170 \pm 0.008 \pm 0.008$ where the first error is the combined statistical and systematic uncertainty and the second error is due to the energy scale uncertainty.

5. Conclusions

As mentioned in [2], a single power law can describe the electron flux above 52.3 GeV and one with a different spectral index describes the positron flux above 27.2 GeV. However, the simultaneous single power law behavior of $\Phi(e^+)$, $\Phi(e^-)$, and $\Phi(e^+ + e^-)$ is unexpected.

An apparent tension between these results and some previous measurements can be avoided if the energy scale error from the different experiments is included in the figures. The energy scale error is smaller in the case of AMS-02.

The combined $(e^+ + e^-)$ flux can be described by a minimal model where to the standard primary (e-) particles accelerated by SNR and secondaries (e+) produced in the interaction of nucleons with the ISM (both described by simple power laws), there is the need to add an additional production of primary e^-/e^+ by a common source (of diverse proposed origin, as pulsars or DM annihilation) parameterized by a power law with a cutoff energy.

This work is supported at the ASI Science Data Center (ASDC) under ASI-INFN agreements C/011/11/1 and No. 2014-037-R.0.

References

[1] L. Accardo et al., Phys. Rev. Lett. **113** (2014) 121101
[2] M. Aguilar et al., Phys. Rev. Lett. **113** (2014) 121102
[3] M. Aguilar et al., Phys. Rev. Lett. **113** (2014) 221102
[4] A. Kounine, Int. J. Mod. Phys. E **21** (2012) 1230005
[5] G. D. Lafferty and T. R. Wyatt, Nucl. Instrum. Methods Phys. Res., Sect. A **355** (1995) 541 (1995). Eq. (6) was used with $\tilde{E} \equiv x_{lw}$
[6] S. Torii et al., Astrophys. J. **559** (2001) 973
[7] M. A. DuVernois et al., Astrophys. J. **559** (2001) 296
[8] J. Chang et al., Nature (London) **456** (2008) 362
[9] K. Yoshida et al., Adv. in Space Res. **42** (2008) 1670
[10] F. Aharonian et al., Phys. Rev. Lett. **101** (2008) 261104
[11] F. Aharonian et al., Astron. Astrophys. **508** (2009) 561
[12] M. Ackermann et al., Phys. Rev. D **82** (2010) 092004

EPJ Web of Conferences **121**, 03006 (2016) DOI: 10.1051/epjconf/201612103006
RICAP-2014

Positron fraction, electron and positron spectra measured by AMS-02

Cecilia Pizzolotto[1,2,a] for the AMS-02 Collaboration

[1] INFN, Sezione di Perugia, 06123 Perugia, Italy
[2] INFN, Sezione di Tor Vergata, 00133 Roma, Italy

Abstract. A precise measurement by AMS-02 of the electron spectrum up to 700 GeV and of the positron spectrum and positron fraction in primary cosmic rays up 500 GeV are presented. The combined measurement of the cosmic-ray electron and positron energy spectra and fraction provide a unique tool to improve our understanding of the production, acceleration and propagation mechanism of cosmic rays.

1. Introduction

The first results from AMS-02 on the positron fraction generated widespread interest and discussions on the origin of high-energy positrons [1]. Here we present the new results on the positron fraction and the individual electron and positron fluxes measurements in an extended energy range, all together these measurements will provide insight into the origin of cosmic rays.

2. AMS-02 detector

The AMS-02 detector [2] consists of a permanent magnet, nine planes of precision silicon tracker, a transition radiation detector (TRD), four planes of time-of-flight (TOF) counters, an array of anticoincidence counters, a ring imaging Cherenkov detector, and an electromagnetic calorimeter (ECAL). AMS-02 operates continuously on the ISS and is monitored and controlled around the clock from the ground. The timing, location, and attitude are determined by a combination of GPS units affixed to AMS-02 and to the ISS.

The TRD (above the magnet), the ECAL (below the magnet) and the tracker provide clean and redundant identification of positrons and electrons with independent suppression of the proton background. The matching of the ECAL energy, E, and the momentum measured with the tracker, p, greatly improves the proton rejection.

3. Data sample and event selection

In the positron fraction analysis, to differentiate between $e^{\pm}$ and protons in the TRD, signals from the 20 layers of proportional tubes are combined in a TRD estimator formed from the ratio of the log-likelihood probability of the $e^{\pm}$ hypothesis to that of the proton hypothesis in each layer. A similar approach is used in the individual flux measurements, where only the

[a] e-mail: cecilia.pizzolotto@pg.infn.it

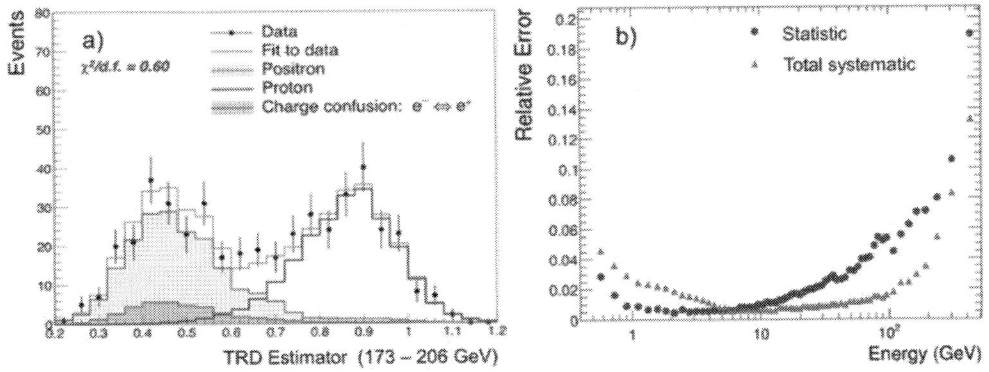

Figure 1. Panel a) shows the results of a 2D fit on the TRD estimator for the positive sample: data points (black) are fitted with the sum of proton (blue), positron (red) and charged confused electrons (magenta) contributions. Panel b) shows the relative statistic and total systematic error as a function of the energy for the positron fraction measurement. The statistical error dominates above 5 GeV.

log-likelihood probability of the $e^{\pm}$ is used as a discriminator. The proton rejection power of the TRD estimator at 90% $e^{\pm}$ efficiency measured on orbit is 10^3 to 10^4 [3].

To cleanly identify electrons and positrons in the ECAL, an estimator, based on a boosted decision tree algorithm [4], is constructed using the 3D shower shape in the ECAL. The proton rejection power of the ECAL estimator reaches 10^4 when combined with the energy-momentum matching requirement $E/p > 0.75$ [3].

Similar selection cuts are applied for the positron fraction and for the separate flux measurements: events are selected by requiring a track in the TRD and in the tracker, a cluster of hits in the ECAL, and a measured velocity $\beta \sim 1$ in the TOF consistent with a downward-going |Z|=1 particle. The tracker measures the sign of the charge. To reject the bulk of the remaining protons, an energy-dependent cut on the ECAL estimator is applied. To reject secondary positrons and electrons produced by the interaction of primary cosmic rays with the atmosphere, the energy measured with the ECAL is required to exceed, by a factor of 1.2, the maximum Størmer cutoff [5] for either a positron or electron at the geomagnetic location where the particle was detected and at any angle within the acceptance.

For the positron fraction a total of 10.9 million positron and electron events are presented. In the separate fluxes a total of 9.23×10^6 events are identified as electrons and 0.58×10^6 as positrons. These numbers are slightly less than the numbers in the positron fraction due to tighter selection criteria, such as on the exposure time, used to minimize the uncertainty of the separate flux measurements.

4. Positron fraction analysis

In each energy bin, the two-dimensional reference spectra for $e^{\pm}$ and the background are fit to data in the [TRD estimator–log(E/p)] plane by varying the normalizations of the signal and the background. The fit is performed simultaneously for the positive and negative rigidity data samples in each energy bin yielding the number of positrons, the number of electrons, the number of protons, and the amount of charge confusion. The charge confusion is defined as the fraction of electrons or positrons reconstructed with a wrong charge sign. Figure 1a shows the results of the fit on the TRD Estimator in the energy range 173–206 GeV.

Different sources of systematic errors have been carefully investigated. Compared to the previous positron fraction publication [3], systematic errors decreased with increasing statistics in the high energy region. Above 5 GeV the total systematic errors are negligible compared to the statistical ones as illustrated in Fig. 1b.

Above 100 GeV charge confusion is the dominant systematic error. There are two main sources of charge confusion: the first source is related to the finite resolution of the tracker and multiple scattering, the second to the production of secondary tracks along the path of the primary $e^{\pm}$ in the tracker. Both sources of charge confusion are found to be well reproduced by the Monte Carlo simulation and their reference spectra are derived from the Monte Carlo simulation. The systematic uncertainties due to these two effects were examined in each energy bin. To evaluate the systematic uncertainty related to event selection, the complete analysis is repeated in every energy bin over 1000 times with different cut values, such that the selection efficiency varies up to 30%. The reference spectra are determined from high statistics electron and proton data samples selected using tracker and ECAL information including charge sign, track-shower axis matching, and the ECAL estimator. The purity of each reference spectrum is verified using Monte Carlo simulation. The systematic error associated with the uncertainty of the data derived reference spectra arises from their finite statistics. The error is measured by varying the shape of the reference spectra within the statistical uncertainties. These contributions to the systematic uncertainty are mainly negligible. Below 5 GeV, the energy scale is the dominant contribution to the systematic uncertainty of the measurement. The energy scale is verified by using minimum ionizing particles and comparing data to the test beam where the beam energy is known with precision and this results in a negligible contribution to the total systematic error at energies above 5 GeV.

AMS-02 has measured the positron fraction as a function of energy from 0.5 to 500 GeV [3, 6]. Compared to previous measurements, AMS-02 extends the energy range and provides higher precision. Below $\sim$ 8 GeV, the positron fraction decreases rapidly with energy as expected from the diffuse production of positrons [7]. The fraction increases steadily from 10 to $\sim$250 GeV, but from 20 to 250 GeV the slope decreases by an order of magnitude. Above $\sim$200 GeV the positron fraction no longer increases with energy.

5. Positron and electron fluxes analysis

The isotropic fluxes of cosmic-ray electrons (Φ_{e^-}) and positrons (Φ_{e^+}) in the energy bin E of width ΔE are given by

$$\Phi_{e^{\pm}} = \frac{N_{e^{\pm}}(E)}{A_{eff}\epsilon_{trig}T(E)\Delta E} \tag{1}$$

where N_{e^-} is the number of electrons, N_{e^+} is the number of positrons, A_{eff} is the effective acceptance, ϵ_{trig} is the trigger efficiency, and T is the exposure time.

The effective acceptance is defined as $A_{eff} = A_{geom}\epsilon_{sel}\epsilon_{id}(1+\delta)$ where A_{geom} is the geometric acceptance corresponding to $\simeq 550\,\text{cm}^2\text{sr}$, ϵ_{sel} is the selection efficiency, ϵ_{id} is the identification efficiency for electrons or positrons, and δ is a minor correction obtained from the comparison of selection efficiencies for data and Monte Carlo. This correction δ is found to be a smooth, slowly falling function of energy, with a value $\sim$2% at 10 GeV and $\sim$6% at 700 GeV. The selection efficiency ϵ_{sel} is determined from the Monte Carlo simulation and found to be a smooth function of energy with a value of $\sim$70% at 100 GeV.

The trigger efficiency ϵ_{trig} is determined from data. It is 100% above 3 GeV decreasing to 75% at 1 GeV.

The exposure time as a function of energy $T(E)$ is determined by counting the live time weighted number of seconds at each location where the geomagnetic cutoff requirement is satisfied. The exposure time for this analysis at 5 GeV is 1.4×10^7 s, at 10 GeV it is 3.2×10^7 s, and above 30 GeV it is constant at 6.1×10^7 s.

In each energy bin, a two-step fit procedure is performed to determine N_{e^+} and N_{e^-}. First, a template fit is used to find the number of electrons plus positrons reconstructed with a positive charge sign N^+ and the number of electrons plus positrons reconstructed with a negative charge sign N^-. The TRD templates, that is, the shapes of the TRD likelihood distributions for signal and background, are determined from data by selecting clean samples of electrons and protons using the ECAL estimator and the charge sign measured by the tracker. The TRD templates for electrons and positrons are identical. They are found to be independent of energy above 10 GeV. A maximum-likelihood fit of the resulting TRD templates to the data yields N^+, N^-, and the number of protons in the bin. Second, N^+ and N^- are corrected for charge confusion: in each energy bin, two tracker templates are defined, one for particles with correctly reconstructed charge sign and another for particles with wrongly reconstructed charge sign. The former is defined using the negative rigidity data sample. The latter is based on the Monte Carlo simulation. These tracker templates are fit to data, bin by bin, to obtain the amount of charge confusion resulting in the determination of N_{e^+} and N_{e^-}.

An accurate study of the sources of systematic error has been carried on, in a way similar to what already descrided for the positron fraction. The systematic error associated with the uncertainty of the template shapes, of effects of charge confusion, the effective acceptance, energy scale and event selection have been considered. The total systematic error is taken as the quadratic sum of these contributions and the minute bin-to-bin migration systematic. As an example, in the energy bin from 59.1 to 63.0 GeV, the statistical error on the positron flux is 4.9% and the total systematic error is 2.9% with 0.8% from the TRD templates, 0.4% from charge confusion, 2.8% from the effective acceptance, and 0.2% from bin-to-bin migration.

The electron flux has been measured from 0.5 to 700 GeV and the positron flux from 0.5 to 500 GeV extending the measurement to higher energies than previous experiments and with better precision [8]. Below ~10 GeV, the behavior for both electrons and positrons is affected by solar modulation as seen in our data by variations of the fluxes over this data-taking interval. However, above ~20 GeV the effects of solar modulation are insignificant within the current experimental accuracy. The data show that above ~20 GeV and up to 200 GeV the electron flux decreases more rapidly with energy than the positron flux, that is, the electron flux is softer than the positron flux. This is not consistent with only the secondary production of positrons [7]. Neither the electron flux nor the positron flux can be described by single power laws ($\propto E^{\gamma}$) over the entire range.

6. Minimal fit model

We use a minimal model to fit the data of the positron fraction and of the total $(e^+ + e^-)$ flux presented in [9]. In this model the e^- and e^+ fluxes are parametrized as the sum of its individual diffuse power law spectrum and a common source term with an exponential cutoff parameter E_S:

$$\Phi_{e^+} = C_{e^+} E^{-\gamma_{e^+}} + C_S E^{-\gamma_S} e^{-\frac{E}{E_S}} \tag{2}$$

$$\Phi_{e^-} = C_{e^-} E^{-\gamma_{e^-}} + C_S E^{-\gamma_S} e^{-\frac{E}{E_S}}. \tag{3}$$

A fit of this model to the data of the positron fraction and to the $(e^+ + e^-)$ flux allows the determination of all the parameters. The measurement and the resulting fit are shown in Figs. 2a and 2b. Figures 2c and 2d show the curve predicted from the fit together with

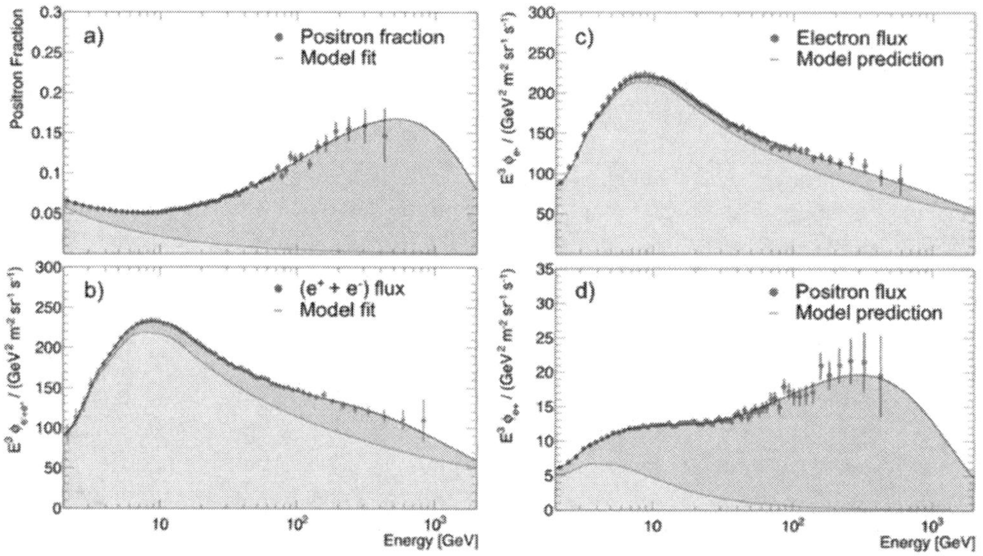

Figure 2. The positron fraction (panel a) and the total $(e^+ + e^-)$ flux (panel b) are fitted with a simple model described in Sect. 6. Once the parameter are determined, the model prediction is drawn in panel c and d where the electron flux and positron flux are shown. In all panels the filled green area represents the diffuse flux component and the filled magenta area is the source component.

the measured electron and positron spectra. There is a nice agreement between the fit and the measured points showing that cosmic rays positrons and electrons are not consistent with only diffuse power law component: an additional source is needed to explain the measured positron fraction and the positron and electron fluxes.

This work is supported at the ASI Science Data Center under ASI-INFN agreements C/011/11/1 and No. 2014-037-R.0

References

[1] L. Feng et al., Phys. Lett. B **728** 250 (2014); K. Blum et al., Phys. Rev. Lett. **111**, 211101 (2013); L. Bergströ et al., Phys. Rev. Lett. **111**, 171101 (2013); I. Cholis et al., Phys. Rev. D **88**, 023013 (2013); T. Linden et al., Astrophys. J. **772**, 18 (2013)
[2] A. Kounine, Int. J. Mod. Phys. E **21**, 1230005 (2012); B. Bertucci, Proc. Sci., EPS-HEP (2011) 67; M. Incagli, AIP Conf. Proc. **1223**, 43 (2009); R. Battiston, Nucl. Instrum. Methods Phys. Res., Sect. A **588**, 227 (2008)
[3] M. Aguilar et al., Phys. Rev. Lett. **110**, 141102 (2013)
[4] B. P. Roe et al., Nucl. Instrum. Methods Phys. Res., Sect. A **543**, 577 (2005)
[5] C. Størmer, The Polar Aurora (Oxford University Press, London, 1950)
[6] L. Accardo et al., Phys. Rev. Lett. **113**, 121101 (2014)
[7] P. D. Serpico, Astropart. Phys. **39–40**, 2 (2012); T. Delahaye et al., Astron. Astrophys. **501**, 821 (2009); I. V. Moskalenko et al., Astrophys. J. **493**, 694 (1998)
[8] M. Aguilar et al., Phys. Rev. Lett. **113**, 121102 (2014)
[9] J. L. Bazo, this conference proceedings; M. Aguilar et al., Phys. Rev. Lett. **113**, 221102 (2014)

EPJ Web of Conferences **121**, 03007 (2016) DOI: 10.1051/epjconf/201612103007
RICAP-2014

Experimental results on the atmospheric muon charge ratio

N. Mauri[a]

Dipartimento di Fisica e Astronomia dell'Università di Bologna and INFN, Sezione di Bologna, Viale Berti Pichat 6/2, 40127 Bologna, Italy

Abstract. The atmospheric muon charge ratio, defined as the number of positive over negative charged muons, is a highly informative observable both for cosmic rays and particle physics. It allows studying the features of high-energy hadronic interactions in the forward region and the composition of primary cosmic rays. In this review results from underground experiments measuring the charge ratio around 1 TeV are discussed. The measurements in the TeV energy region constrain the associated kaon production, which is particularly important e.g. for the calculation of the atmospheric neutrino flux.

1. Introduction

The atmospheric muon charge ratio is an important observable to shed light on the physics of cosmic ray interactions in atmosphere. Its measurement allows to better understand the features of high energy hadronic interactions in the forward region and to improve Monte Carlo predictions of the atmospheric neutrino flux.

Atmospheric muons constitute the penetrating charged remnants of cosmic rays under the Earth's surface. They are produced when primary cosmic rays, mainly protons, impinge on the Earth's atmosphere producing a particle cascade, in which secondary particles decay into muons. In the energy range up to $\sim$100 GeV atmospheric muons come mostly from the decay of secondary pions. At higher energies, the kaon contribution to the muon flux become significant, reaching the asymptotic value of 27% at about 10 TeV [1]. At even higher energies, at $O(100)$ TeV, also charmed hadrons are expected to contribute (*prompt* muons). In the pion dominated energy region ($E_\mu \lesssim 100$ GeV), the muon charge ratio R_μ has been measured by several experiments on surface, by balloon-borne detectors and at shallow depths [2]. It results to be constant, $R_\mu \simeq 1.27$, as expected assuming the validity of Feynman scaling in the fragmentation region [1]. Since the associated production ΛK^+ has no analog for K^-, the K^+/K^- ratio is larger than the π^+/π^- ratio [1, 3]. Thus, the increasing kaon contribution from a few hundred GeV to a few TeV causes a smooth transition to a higher value of the muon charge ratio.

In this work experimental results on the muon charge ratio in the TeV energy range, fundamental to constrain the associated kaon production and test the Feynman scaling in the forward region, are reviewed. Furthermore, this measurement allows to study the chemical composition of cosmic rays near the *knee* in the primary spectrum.

[a]e-mail: `mauri@bo.infn.it`

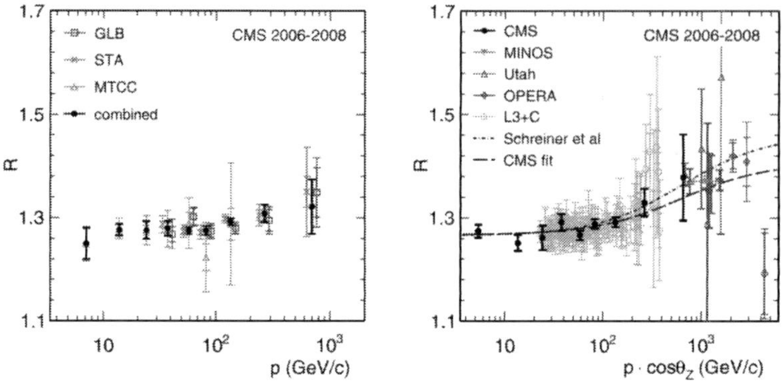

Figure 1. The atmospheric muon charge ratio measured by CMS (*black points*) as a function of (left) the muon momentum and (right) the vertical component of the muon momentum [4].

2. Underground experiments

Underground experiments naturally select high energy down-going muons, the minimum energy threshold being fixed by the rock overburden surrounding the detector. Key elements for the accurate determination of the muon charge and momentum in the O(TeV) energy range are magnetized detectors and large depth.

Recent results were obtained by the CMS [4], MINOS [5, 6] and OPERA [7, 8] experiments that measured the muon charge ratio R_μ around vertical surface energy $E_\mu \cos\theta \sim 1$ TeV. MINOS and OPERA are long-baseline neutrino experiments located deep underground, while CMS is a collider experiment located at shallow depth but provided with an intense magnetic field.

In order to disentangle the different parent components to the muon flux (π, K, charm) the correct variable to describe the evolution of R_μ is the vertical surface energy $E_\mu \cos\theta$ [9]. Taking into account the possible variation of the primary cosmic ray composition, we should consider separately the dependency on the surface muon energy E_μ and the zenith projection $\cos\theta$, see Sect. 2.3 [3].

2.1 CMS results

The CMS detector is installed in an underground cavern at the Large Hadron Collider (LHC). The experiment is located at shallow depth, with the center of the detector 89 m below the Earth's surface, corresponding to 6 to 175 m.w.e. traversed by close-to-vertical muons depending on the impact point on CMS. At the core of the apparatus, a superconducting solenoid generates a magnetic field of 3.8 T directed along the LHC beam direction (z axis). This configuration favours the reconstruction of atmospheric muons, since the bending is stronger for vertical tracks. Muon tracking in CMS is performed with the inner silicon tracker and with either three or four stations of muon chambers installed outside the solenoid. Depending on the track topology, the maximum detectable momentum ranges from 200 GeV/c up to 20 TeV/c [4].

The muon charge ratio was determined using data collected during 2006 and 2008 cosmic ray runs. The results on R_μ in the 5 GeV/c to 1 TeV/c momentum range are shown in Fig. 1. CMS provided the most precise measurement of the charge ratio below 100 GeV/c to date, $R_\mu = 1.2766 \pm 0.0032\,(stat.) \pm 0.0032\,(syst.)$. The measurements lie in the transition region between the pion dominated muon flux ($p\cos\theta < 100$ GeV/c), where R_μ is consistent

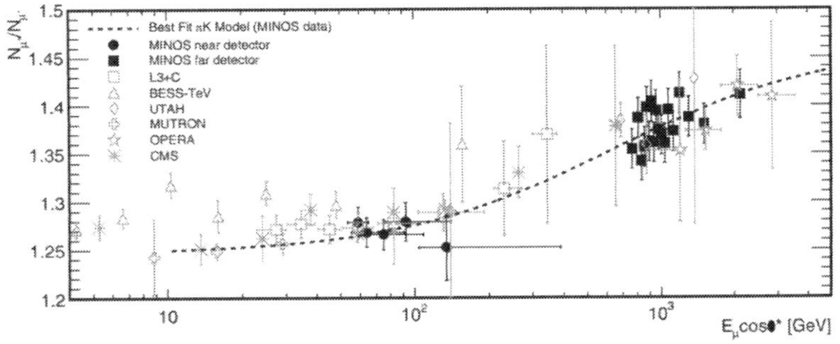

Figure 2. The atmospheric muon charge ratio measured by MINOS (*black points* for the Near Detector, *black squares* for the Far Detector) as a function of $E_\mu \cos\theta^*$ [6].

with a flat charge ratio, and the plateau corresponding to the asymptotic kaon contribution, above some TeV. CMS data are consistent with an increase of the charge ratio. Using the parameterization described in [9], the fit on stand-alone CMS data yields the fractions of charged pions and kaons decaying into positive muons, $f_{\pi^+} = 0.553 \pm 0.005$ and $f_{K^+} = 0.66 \pm 0.06$.

2.2 MINOS results

The MINOS Near and Far Detectors (ND and FD, respectively) are two functionally identical magnetic calorimeters placed in the NuMI neutrino beam. They are composed by magnetized steel planes instrumented with scintillator strips and located under a flat overburden of 94 m (ND) and 710 m (FD), corresponding to 225 m.w.e. and 2070 m.w.e., respectively. The toroidal magnetic field is stronger near the coil hole along the longitudinal z-axis (1.8 T for the FD, 2.1 T for the ND) and weaker near the edges (1 T at FD, 0.2 T at ND), with an average value of 1.3 T. This configuration focus toward the center of the detector positive muons coming from one direction and negative muons coming from the opposite direction, giving rise to a different acceptance for μ^+ and μ^-. During the data taking period, the coil current was periodically reversed in order to invert the direction of the magnetic field. Data recorded with the two magnetic field configurations, referred to as forward and reverse, were combined to cancel the acceptance bias [5, 6].

Due to the approximately flat overburden at both sites, the highest energy muons are at large zenith angles. Since the median surface energy increases as $\cos\theta$ decreases, the $E_\mu \cos\theta$ distribution is much narrower than the E_μ distribution. Due to this compensation muon events with the highest $E_\mu \cos\theta$ are those with the highest reconstructed momentum at the detector. The maximum detectable momentum is around ~ 300 GeV/c.

The results on the muon charge ratio were obtained with the FD, using data collected between 2003 and 2006 [5], and with the ND, using data collected between 2006 and 2009 [6]. R_μ is shown as a function of $E_\mu \cos\theta^*$ in Fig. 2, where θ^* is the zenith angle at the production point taking into account the Earth's curvature. The ND charge ratio is given in the 50–130 GeV vertical energy range, while the FD data are in the 800 GeV–2 TeV range. Using only MINOS Near and Far Detector data the best fit to the parameterization of R_μ [9] over (f_{π^+}, f_{K^+}) gives the result $f_{\pi^+} = 0.55$ and $f_{K^+} = 0.70$ and it is also plotted in Fig. 2.

MINOS observed an increase in the charge ratio, from $R_\mu = 1.266 \pm 0.001\,(stat.)^{+0.015}_{-0.014}\,(syst.)$ at the ND shallow depth to $R_\mu = 1.374 \pm 0.004\,(stat.)^{+0.012}_{-0.010}\,(syst.)$ at the FD large depth, consistent with an increase in the fraction of observed muons arising from kaon decays.

2.3 OPERA results

The OPERA experiment is a hybrid electronic detector/emulsion apparatus exposed to the CNGS neutrino beam from 2007 up to 2012. It is located in the underground Gran Sasso laboratory, at an average depth of 1400 m of rock corresponding to 3800 m.w.e. OPERA is the deepest experiment able to measure charge-separated atmospheric muons. The minimum surface muon energy threshold is $\sim$1 TeV (1.4 TeV averaged over all the directions and rock depths). The detector is composed of two identical parts, called supermodules, each consisting of a target section and a magnetic spectrometer. The target consists of scintillator strips interleaved with Emulsion Cloud Chambers, the spectrometer is a large dipole iron magnet instrumented with Resistive Plate Chambers. The magnetic field is almost uniform, with intensity 1.53 T, and directed along the vertical axis with opposite orientations in the two magnet arms. A muon crossing the spectrometer is deflected in the horizontal plane. The deflection is measured by six stations of vertical drift tubes, the Precision Trackers (PT), grouped in 3 pairs placed upstream of the first arm, in between the two arms and downstream of the second arm. The charge and momentum reconstruction is performed for tracks crossing at least one magnet arm using the angle $\Delta\phi$ in the bending plane, i.e. the difference between the track directions reconstructed by two PT stations before and after each magnet arm. For nearly horizontal muons up to four bending angles can be measured in the two dipole magnets.

Different muon reconstruction procedures are used for CNGS neutrino and cosmic ray induced events. Atmospheric events with single and multiple muon tracks (*muon bundles*) were reconstructed. Due to the charge-symmetry of the dipole magnet, the acceptance for μ^+ and μ^- is the same and, depending on the track topology, the maximum detectable momentum varies from a few hundred GeV/c up to 1 TeV/c [8, 10]. Data were collected with both magnetic field polarities in order to minimize systematic errors due to misalignment.

OPERA reported a first measurement of the atmospheric muon charge ratio using the 2008 Run data [7] and the final results using the complete statistics from 2008 up to 2012 [8]. In the latter analysis the two data sets collected with opposite magnet polarities were combined reaching the most accurate measurement to date of R_μ in the TeV energy region. The muon charge ratio was computed separately for single muons, $R_\mu(n_\mu = 1) = 1.377 \pm 0.006\,(stat.)^{+0.007}_{-0.001}\,(syst.)$, and for muon bundles, $R_\mu(n_\mu > 1) = 1.098 \pm 0.023\,(stat.)^{+0.015}_{-0.013}\,(syst.)$. This is the first observation of a decrease in the charge ratio of high multiplicity events with respect to single muon events. The dilution effect in R_μ is expected since the multiple muon sample selects events generated by heavier primary cosmic rays and secondaries with a low value of Feynman-x, coming from the central region [10].

Because of the Gran Sasso orography, the amount of rock crossed by muons is not directly related to the zenith angle, and the high energy tail is not completely suppressed by the $\cos\theta$ factor. OPERA presented results on R_μ (for single muons) as a function of the vertical surface energy $E_\mu \cos\theta^*$, where θ^* is the zenith angle at the muon production point, fitting data to the parameterized model described in Ref. [9]. The atmospheric muon charge ratio was measured in a large $E_\mu \cos\theta^*$ range, from 500 GeV up to $\sim$ 10 TeV, and plotted in Fig. 3. With an average value $\langle E_\mu \cos\theta^* \rangle \simeq 2$ TeV, OPERA is the magnetized experiment measuring the charge ratio at the largest vertical surface energy. The fit of OPERA and L3+C data, shown in Fig. 3, yields the fractions of charged mesons decaying into positive muons $f_{\pi^+} = 0.5512 \pm 0.0014$ and $f_{K^+} = 0.705 \pm 0.014$. The prompt muon component does not significantly contribute to R_μ up to $E_\mu \cos\theta^* \lesssim 10$ TeV.

Taking into account the primary cosmic ray composition requires to disentangle the energy and zenith angle dependencies [3]. The fit in two dimensions $(E_\mu, \cos\theta^*)$ yields

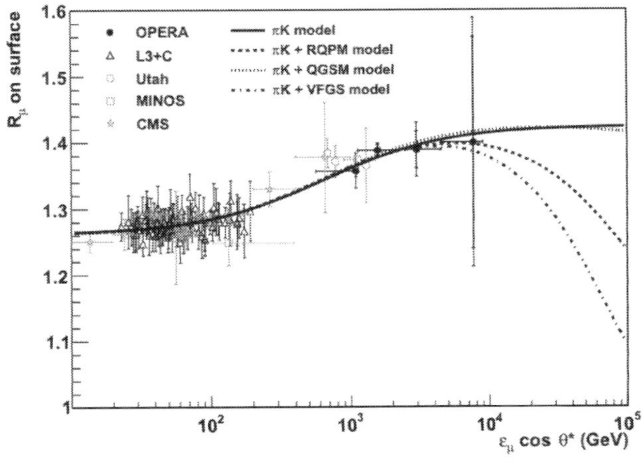

Figure 3. The atmospheric muon charge ratio measured by OPERA (*black points*) as a function of $E_\mu \cos\theta^*$.

the proton excess in primary cosmic rays $\delta_0 = 0.61 \pm 0.02$ at primary energy $\langle E_N \rangle \approx$ 20 TeV/nucleon, and the factor related to the associated kaon production, $Z_{pK^+} = 0.0086 \pm 0.0004$, determined for the first time.

3. Conclusions

The results on the atmospheric muon charge ratio from the CMS, MINOS and OPERA experiments show an increase of R_μ as a function of the vertical surface energy $E_\mu \cos\theta^*$ in the range between a few hundred GeV and a few TeV. The measurements are compatible with a simple parametric model where the rise is due to the increasing kaon contribution to the muon flux. No significant contribution from charm decay is observed up to energies ~ 10 TeV. A future experimental measurement of R_μ in the region $E_\mu \cos\theta^* > 10$ TeV, with a new detector at very large depths, could shed light on the prompt component even before the crossover between conventional and prompt muons.

The OPERA measurement of R_μ in the TeV energy range determined for the first time the spectrum weighted moment Z_{pK^+}, fundamental to accurately predict the atmospheric $\bar{\nu}/\nu$ ratio. No similar results on the associated kaon production cross section are available to date from accelerator experiments. The Z_{pK^+} factor was extracted together with the primary proton excess δ_0, which is consistent with direct measurements of the primary composition [3]. The energy behaviour of R_μ measured by OPERA above a few TeV, compatible with the plateau corresponding to the asymptotic kaon contribution, supports the validity of Feynman scaling in the fragmentation region up to primary energies/nucleon around 200 TeV.

References

[1] T.K. Gaisser, *Cosmic Rays and Particle Physics* (Cambridge University Press, Cambridge, 1990)
[2] T. Hebbeker and C. Timmermans, Astropart. Phys. **18**, 107 (2002)
[3] T.K. Gaisser, Astropart. Phys. **35**, 801 (2012)
[4] CMS Collaboration, Phys. Lett. B **692**, 83 (2010)

[5] P. Adamson et al. (MINOS Collaboration), Phys. Rev. D **76**, 052003 (2007)
[6] P. Adamson et al. (MINOS Collaboration), Phys. Rev. D **83**, 032011 (2011)
[7] N. Agafonova et al. (OPERA Collaboration), Eur. Phys. J. C **67**, 25 (2010)
[8] N. Agafonova et al. (OPERA Collaboration), Eur. Phys. J. C **74**, 2933 (2014)
[9] P.A. Schreiner, J. Reichenbacher, M.C. Goodman, Astropart. Phys. **32** (1), 61 (2009)
[10] N. Mauri, Ph.D. Thesis, Università di Bologna (2011) `http://inspirehep.net/record/1393857/files/phd_thesis-BO-2011_05_20-mauri.pdf`

EPJ Web of Conferences **121**, 03008 (2016) DOI: 10.1051/epjconf/201612103008
RICAP-2014

The Cosmic Ray p+He energy spectrum in the 3–3000 TeV energy range measured by ARGO-YBJ

S.M. Mari[1,2] and P. Montini[2,a] on behalf of the ARGO–YBJ Collaboration

[1] Dipartimento di Matematica e Fisica, Università degli Studi Roma Tre, Italy
[2] Istituto Nazionale di Fisica Nucleare (INFN), Sezione di Roma Tre, Italy

Abstract. The ARGO–YBJ experiment is a full coverage air shower detector operated at the Yangbajing International Cosmic Ray Observatory. The detector has been in stable data taking in its full configuration since November 2007 to February 2013. The high altitude and the high segmentation and spacetime resolution offer the possibility to explore the cosmic ray energy spectrum in a very wide range, from a few TeV up to the PeV region. The high segmentation allows a detailed measurement of the lateral distribution, which can be used in order to discriminate showers produced by light and heavy elements. In this work we present the measurement of the cosmic ray light component spectrum in the energy range 3–3000 TeV. The analysis has been carried out by using a two-dimensional unfolding method based on the Bayes' theorem.

1. Introduction

The Earth is permanently exposed to a flux of cosmic rays coming from outside the solar system. Their energy spectrum spans a huge energy range (from MeV up to $\sim 10^{20}$ eV), corresponding to a variation of the intensity of several orders of magnitude. The all-particle energy spectrum of cosmic rays can be roughly described by a power–law, showing a "knee" at energies around 4 PeV. Despite great experimental efforts in the last decades, the origin of the knee is still an unresolved question. In the standard picture the knee in the all particle spectrum is due to the steepening of the proton and helium spectra. Due to the rapid decrease of the intensity as the energy increases, all information about cosmic rays above 100 TeV comes from ground–based air shower experiments and the determination of the composition is essentially limited to large mass groups. A precise measurement of the proton and helium spectra in the energy region from few TeV to 10 PeV is of fundamental importance in the understanding of the origin of the knee. The ARGO–YBJ experiment is a full–coverage extensive air shower array which was in stable data taking from November 2007 to February 2013. The detector is equipped with two independent readout systems designed in order to detect showers in a very wide range of particle density. The full–coverage technique, combined with the high segmentation and spacetime resolution, allows a detailed measurement of the lateral distribution of the shower that can be used in order to discriminate events produced by light primaries.

[a]e-mail: `paolo.montini@roma3.infn.it`

2. The ARGO–YBJ experiment

The ARGO–YBJ experiment [1, 2] is a full coverage air shower detector operated in the Yangbajing International Cosmic Ray Observatory (Tibet, P.R. China, 4300 m a.s.l.). The detector has been designed and developed in order to explore the cosmic ray energy spectrum and composition in a wide energy range from few TeV up to the PeV region. The detector is made of a single layer of 1836 resistive plate chambers (RPCs) with ~93% active area and equipped with two independent readout systems. The carpet is made of 1836 RPCs, arranged in 153 clusters each made of 12 chambers. The digital readout consists of 146880 copper strips ($6.65 \times 61.80\,cm^2$) logically arranged in 18360 pads. The high segmentation and spacetime resolution of the digital readout allow the detection of showers produced by low energy primaries down down to few TeV and provide a detailed reconstruction of showers up to $\sim 23\,particles/m^2$ corresponding to a primary energy up to a few hundreds of TeV. In order to extend the operating range of the detector and fully investigate the PeV energy region, each RPC has been equipped with two large electrodes called Big Pads ($139 \times 123\,cm^2$), each providing a signal that is proportional to the number of charged particles impinging the detector surface. The analog readout system uses eight different gain scales (G0, ..., G7) giving an overlap between the analog and digital readout linearity range. Data from the highest gain scale (G7) have been used for calibration purposes. The intermediate gain scale (G4) overlaps with the digital readout data in a wide energy range between 10 and 100 TeV, providing a cross–calibration of the two techniques. Data from the lowest gain scales (G1 and G0) allow the detection of showers with about 10^4 particles/m^2 in the core region. A dedicated calibration procedure has been implemented for each gain scale [3].

3. Data analysis

As it is well known, since the development of a shower presents intrinsic fluctuations, the primary energy cannot be determined on an event–by–event basis. As widely described in [4, 5], the determination of the cosmic ray energy spectrum from the spacetime distribution of charged particles at ground level is a classical unfolding problem that can be dealt with the bayesian technique [6]. In order to evaluate the conditioned probabilities needed in the unfolding procedure a detailed simulation of the development of the shower in the Earth's atmosphere and of the detector response has been performed. A sample of simulated events was generated by using the CORSIKA code [7], including FLUKA and QGSJETII.03 hadronic interaction models. Showers produced by protons, helium, CNO and iron nuclei have been generated in the energy range (0.316–31600) TeV and in the zenith angle range 0–45 degrees according to the energy spectra described in [8]. Both the digital and analog output of the detector have been simulated by using a GEANT3 based code, including effects of time resolution, RPC efficiency, noise, etc.

4. The cosmic ray p+He spectrum

4.1 3–300 TeV energy range

The digital readout of the ARGO–YBJ experiment has been operated in its full configuration for more than five years and more than 5×10^{11} events have been recorded and reconstructed. A first selection has been based on the working condition of the detector and on the quality of the reconstruction procedure in order to obtain a sample of high–quality runs. About 3×10^{11} events have been selected. In order to obtain a better estimation of the bayesian probabilities

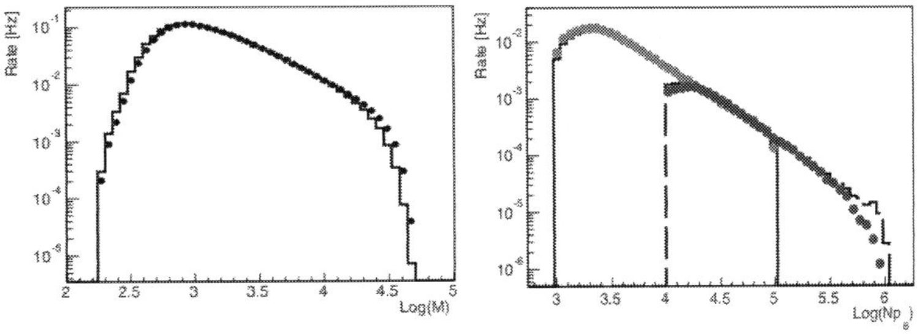

Figure 1. Left: multiplicity distribution of events selected according to the criteria described in Sect. 4.1. Values for Monte Carlo (line) and experimental data (dots) are reported. Right: distribution of Np_8 of events selected according to the criteria described in Sect. 4.2. Values for G4 Monte Carlo (solid line), G4 data (red dots), G1 Monte Carlo (dashed line) and G1 data (blue dots) are reported.

needed in the unfolding procedure the following selection criteria have been applied to both Monte Carlo and experimental data samples.

- Events with reconstructed zenith angle in the range $0° \leqslant \vartheta_R \leqslant 35°$ have been considered.
- The measured strip multiplicity M had to be in the range $150 \leqslant M \leqslant 50000$.
- The position of the cluster with the highest multiplicity must be located inside an area of $40 \times 40\,\mathrm{m}^2$ measured from the detector center
- The particle density measured by the innermost 20 clusters ρ_{in} must be higher than the one measured by the outermost 42 ones (ρ_{out}).

These selection criteria reduces bias effects in the estimation of the probabilities needed in the unfolding procedure. The particle density cut, moreover, allows the selection of showers manly produced by light primaries. In Fig. 1 the strip multiplicity distribution is reported for both data and Monte Carlo samples. The analysis has been performed on the full sample collected during the period 2008–2012 and the resulting spectrum is reported in Fig. 3. Special care has been dedicated to the evaluation of the possible systematic effects. The main sources of systematic uncertainties rely on the reliability of the detector simulation, effects related to the values of the cuts used in the event selection procedure and the contamination of heavier elements. The total systematic uncertainty has been estimated as of the order of $\pm 5\%$. The ARGO–YBJ measurement spans a wide energy range, from a few TeV up to hundreds of TeV, so far accessible only by balloon–borne or satellite experiments.

4.2 10–3000 TeV energy range

The measurement of the spectrum up to 3 PeV has been carried out by exploiting the RPC analog readout system of the ARGO–YBJ experiment. As a first step, several studies have been performed on the simulated data sample in order to identify a suitable energy estimator and a set of parameters that allow the discrimination between showers produced by light and heavy primaries respectively. The number of particles within 8 m from the core position (Np_8) seems to be well correlated with primary energy and is not affected by bias effects related to the finite detector size. The lateral distribution of charged particles have been studied at several distances from the core position with high precision. Showers produced by light primaries have a well–shaped core and the largest fraction of particles is localized around the core region. Showers generated by heavy nuclei, otherwise, show larger particle density at distances far from the core. The ratio β between the particle density at several

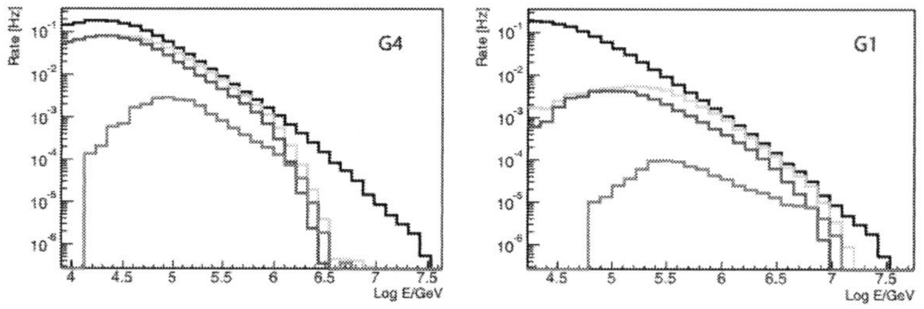

Figure 2. Energy distribution of Monte Carlo events surviving the fiducial cuts described in Sect. 4.2 (green). Energy spectra of the light (blue) and heavy (red) component after the composition cut are also reported. The simulated spectrum is reported as a reference pattern (black).

distances from the core position and the particle density measured in the core region can therefore be exploited in order to discriminate light primaries. In this work we present the analysis of the first data collected during 2010. The analysis has been performed on the data collected by using the G4 and G1 gain scales. The following fiducial cuts have been applied to both Monte Carlo and experimental data samples:

- showers with reconstructed zenith angle in the range $0° \leqslant \vartheta_R \leqslant 35°$ have been considered,
- reconstructed core position located inside an area of $40 \times 40\,\mathrm{m}^2$ around the detector center,
- Np_8 must be in the range (10^3–10^5) for G4 and (10^4–10^6) for G1 data samples.

In Fig. 1 the distribution of Np_8 is reported for both Monte Carlo and data samples, demonstrating the reliability of the simulation of the analog readout output. An additional selection criterion based on β has been applied to the events surviving the fiducial cuts. In Fig. 2 the fraction of events surviving the fiducial cuts and the discrimination cut is presented for both G1 and G4 data samples. The plot shows clearly that the selected samples are essentially made of showers produced by light primaries. The G4 and G1 data sets have been analyzed in order to measure the $N(Np_8)$ distribution, while the Monte Carlo data sample has been used in order to evaluate the probabilities needed in the unfolding procedure. The resulting spectra are shown in Fig. 3. The energy spectrum obtained by analyzing the G4 data sample spans the energy range between 10 and 150 TeV, overlapping the spectrum measured by using the digital readout data as described in Sect. 4.1. These results are fairly consistent between each other, both concerning the spectral index and flux intensity, demonstrating the reliability of the response of the analog readout system. The analysis of the G1 dataset extend the energy range up to the PeV region, clearly showing a deviation from a single power law at energies of about 700 TeV. The same sources of systematic effects mentioned in Sect. 4.1 have been considered.

5. Conclusions

The ARGO–YBJ experiment has been taking data in its full configuration for more than five years. The detector has been equipped with two independent readout systems. The high segmentation and spacetime resolution of the digital readout system allows the detection of showers produced by primaries in a wide energy range, from a few TeV up to a few hundreds of TeV. The analog readout system has been designed and implemented in order to sample showers with about 10^4 particles/m^2 around the core region, corresponding to a primary energy up to the PeV region. Three independent datasets have been analyzed in order to

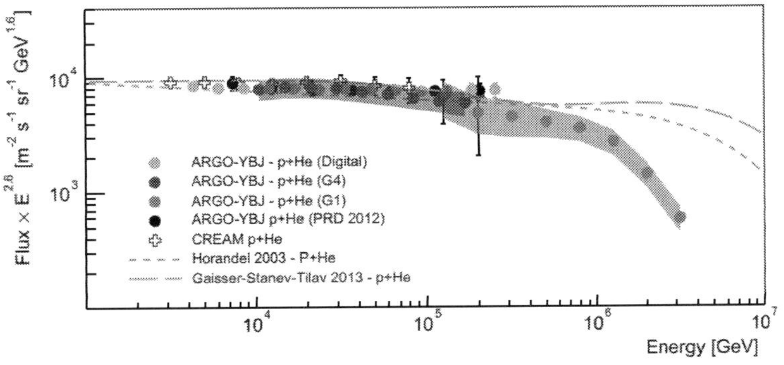

Figure 3. The p+He spectrum measured by the ARGO–YBJ experiment by analyzing the full digital data sample (green dots). Preliminary results of G4 (blue dots) and G1 (red dots) analysis are reported. The shaded area around G4 and G1 values represents the systematic uncertainty. The measurement obtained by analyzing the first data of ARGO–YBJ is also reported [5]. The sum of the proton and helium spectra measured by CREAM [9] is shown (crosses). The curves of the p+He spectrum according to the Hörandel (solid line) [8] and GST (dashed line) [10] models are also shown.

investigate the cosmic ray energy spectrum in the region (3–3000) TeV. The relation between the shower size and the primary energy has been established by using a bayesian unfolding procedure. The discrimination between showers produced by light and heavy primaries has been obtained by using a selection criterion based on the lateral distribution of charged particles in the shower front. The analysis of the digital data sample collected during the ~ 5 years lifespan of the experiment has been performed. The resulting spectrum spans the energy range (3–300) TeV, giving a spectral index $\gamma = -2.62 \pm 0.01$, in excellent agreement with the one obtained by analyzing the first data taken with the detector in its full configuration [5]. The analysis of the analog readout data sample has been performed in order to measure the p+He spectrum in the energy region (10–3000) TeV energy range. The resulting spectrum remarkably agrees with the one obtained by analyzing the digital readout full data sample in a wide energy range from 10 TeV up to 150 TeV. A deviation from a single power law is clearly evident at energies of about 700 TeV, showing a rapid decrease of the light component flux at higher energies. These results demonstrate the possibility of exploring the cosmic ray properties in a wide energy range with a single ground–based experiment and open new scenarios about the evolution of the light component energy spectrum towards the highest energies and about the origin of the knee.

References

[1] G. Aielli et al., Nucl. Instrum. Meth. **A562** (2006) 92
[2] G. Aielli et al., Nucl. Instrum. Meth. **A608** (2009) 246
[3] M. Iacovacci et al., Proc. 33th ICRC (2013) ID:1035
[4] S. Bussino, E. De Marinis, S. M. Mari, Astropart. Phys. **22** (2004) 81
[5] B. Bartoli et al., Phys. Rev. D **85** (2012) 092005
[6] G. D'Agostini, Nucl. Instr. Meth. Sect A. **362** (1995) 487
[7] D. Heck et al., Report FZKA **6019** (1998)
[8] J. R. Hörandel, Astropart. Phys. **19** (2003) 193
[9] Y. S. Yoon et al., ApJ **728** (2011) 122
[10] T.K. Gaisser, T. Stanev and S. Tilav, Front. Phys. **8**(6) (2013) 748

EPJ Web of Conferences **121**, 03009 (2016) DOI: 10.1051/epjconf/201612103009
RICAP-2014

Measurement of the cosmic ray all-particle and light-component energy spectra with the ARGO-YBJ experiment

Ivan De Mitri[1,2,a] on behalf of the ARGO-YBJ Collaboration

[1] Dipartimento di Matematica e Fisica "Ennio De Giorgi", Università del Salento, Lecce, Italy
[2] Istituto Nazionale di Fisica Nucleare, INFN, Sezione di Lecce, Italy

Abstract. ARGO-YBJ preliminary results of the measurements of the all-particle and light-component (i.e. protons and helium) energy spectra between approximately 5 TeV and 5 PeV are reported and discussed.

1. Introduction

There is a general consensus that galactic cosmic rays (hereafter CRs) up to the "knee" of the all-particle spectrum (~ 4 PeV) mainly originate in Supernova Remnants (SNRs). Recent measurements carried out by the balloon-borne CREAM experiment [1, 2] show that the proton and helium spectra from 2.5 to 250 TeV are harder compared to lower energy measurements. As pointed out by several authors, the evolution of the proton and helium spectra and their subtle differences could be indications of the presence of different populations of CR sources contributing to the overall flux and operating in environments with different chemical compositions [3, 4]. Diffusion effects during CR propagation in the Galaxy might also play an important role.

In the knee region (and above) the measurements of the CR primary spectrum are carried out by EAS arrays. In this case mass composition studies are extremely difficult and often affected by large systematic uncertainties. In the standard picture the average composition at the knee is dominated by light elements, and the knee itself is interpreted as the steepening of the p and He spectra [5]. However, several experimental results suggest an heavier composition at knee energies [6–11].

A measurement of the CR primary energy spectrum (all-particle and light-component) in the energy range few TeV–10 PeV is under way with the ARGO-YBJ experiment (for a description of the detector and a report of the latest physics results see [12]). In order to cover this wide energy range, different approaches have been followed:

- *"digital readout"*, based on the RPC readout strip multiplicity, in the 5 TeV–200 TeV range [13];
- *"analog readout"*, based on the particle density near the shower core, in the 100 TeV–10 PeV range;
- *"hybrid measurement"*, carried out by ARGO-YBJ and a wide field of view Cherenkov telescope, in the 100 TeV–PeV region [14].

[a] e-mail: `ivan.demitri@le.infn.it`

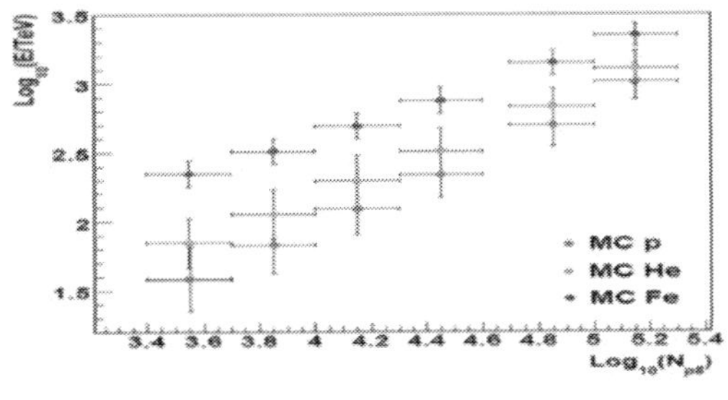

Figure 1. Primary energy as a function of the observed truncated size N_{p8} for simulated showers due to different primary nuclei.

Preliminary results concerning the all-particle and the light-component (i.e. p+He) spectra obtained with the analog readout are summarized in the following. More details on the results obtained with the 'hybrid measurement' are also given in [14, 15].

2. Measurement of the all-particle spectrum

The measurement of the CR energy spectrum up to 10 PeV is under way exploiting the RPC charge readout of the ARGO-YBJ detector which allows studying the structure of the particle density distribution in the shower core region up to particle densities of about 10^4/m^2 [16]. The study of the charged particle lateral density function (LDF) at ground is expected to provide information on the longitudinal profile of the showers in the atmosphere, that is to estimate their development stage, or the so-called *age*, which is related to X_{max}, the atmospheric depth at which the cascade reaches its maximum size. This implies the possibility of selecting showers within given intervals of X_{max} or, equivalently, of X_{dm}, the grammage between the depth of the shower maximum and the detection level. For this reason, the combined use of the shower energy and age estimations can ensure a sensitivity to the primary mass, thus giving the possibility of selecting a light (p+He) event sample with high efficiency.

Various observables were considered and analyzed in order to find a suitable estimator of the primary CR energy. Among them, according to MC simulations, N_{p8}, the number of particles detected within a distance of 8 m from the shower axis, resulted well correlated with energy, not biased by the finite detector size and not much affected by shower to shower fluctuations [17]. Nevertheless, as shown in Fig. 1, this truncated size is a mass-dependent energy estimator parameter. In order to have a mass-independent parameter we fitted the LDFs of individual showers (up to 10 m from the core) event-by-event, for different N_{p8} intervals and different shower initiating primaries, with a suitable function to get the shape parameter s' (see [17, 18] for details). From these studies we find that, for a given primary, the s' value decreases when N_{p8} (i.e. the energy) increases, this being due to the observation of younger (deeper) showers at larger energies. Moreover, for a given range of N_{p8}, s' increases going from proton to iron, as a consequence of older (shallower) showers. Both dependencies are in agreement with the expectations, the slope s' being correlated with the shower age, thus reflecting its development stage. This outcome has two important implications, since the measurements of s' and N_{p8} can both (i) help constraining the shower age and (ii) give information on the primary particle nature.

Concerning the first point, we show in Fig. 2 the s' values as obtained from the fit of the average LDFs, for each simulated primary type and N_{p8} interval, as a function of the corresponding X_{max} average value. As can be seen, the shape parameter s' depends only on the development stage of the shower, independently from the nature of the primary particle.

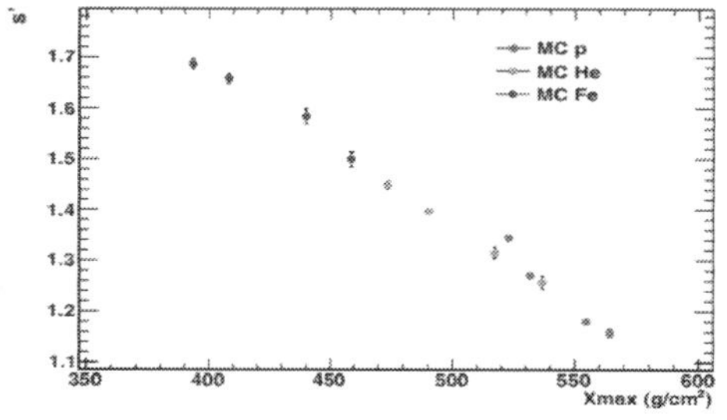

Figure 2. The age parameter s' resulting from the fit of the average LDF for simulated showers vs. the corresponding X_{max} average values.

That plot expresses an important universality of the LDF of detected EAS particles in terms of the lateral shower age. The LDF slope s' is a mass-independent estimator of the average X_{max}. Obviously shower-to-shower fluctuations introduce unavoidable systematics, whose effects can be anyway quantified and taken into account. Another implication is that s' from the LDF fit close to the shower axis, together with the measurement of the truncated size N_{p8}, can give information on the primary particle nature, thus making possible the study of mass composition and the selection of a light-component data sample.

Assuming an exponential absorption after the shower maximum, we get N_{p8}^{max}, a variable linearly correlated to the size at the shower maximum, by using N_{p8} and s' measurements for each event and simply correcting with an exponential attenuation: $N_{p8}^{max} \approx N_{p8} \cdot exp[(h_0 sec\theta - X_{max}(s'))/\lambda_{abs}]$. A suitable choice of the absorption lenght λ_{abs} (=120 g/cm^2) allows to get N_{p8}^{max}, a parameter correlated with primary energy in an almost linear and mass independent way, providing an energy estimator with a Log(E/TeV) resolution of 0.10–0.15 (getting better with energy).

As described in [16], the RPC charge readout system has 8 different and overlapping gain scale settings (G0,....,G7 from smaller to larger gains) in order to explore the particle density range $\approx 20 - 10^4$ particles/m^2. In this preliminary analysis the results obtained with two gain scales (so-called G1 and G4) are presented. Selecting quasi-vertical events ($\theta < 15°$) in terms of the truncated size N_{p8} with the described procedure we reconstructed the CR all-particle energy spectrum shown in the Fig. 3 in the energy range 100–3000 TeV. In the plot a $\pm 14\%$ systematic uncertainty, due to hadronic interaction models, selection criteria, unfolding algorithms, and aperture calculation, is shown by the shaded area. The statistical uncertainty is shown by the error bars. As can be seen from the figure, the two gain scales overlap making us confident about the event selection and the analysis procedure. The ARGO-YBJ all-particle spectrum is in fair agreement with the parametrizations provided by [19] and [3], showing evidence of a spectral index change at an energy consistent with the position of the knee. As shown in Fig. 6 this result is also consistent with previous measurements made by both direct and indirect experiments. This is also an important check on the absolute energy scale set for this analysis, whose systematic uncertainty has been anyhow conservatively estimated at the level of 10%.

3. Measurement of the light-component energy spectrum

The CR light-component energy spectrum has been measured by ARGO-YBJ from about 5 TeV to 700 TeV [12–14]. The energy range is now being extended up to the few PeV region by using the RPC charge readout information and three different approaches.

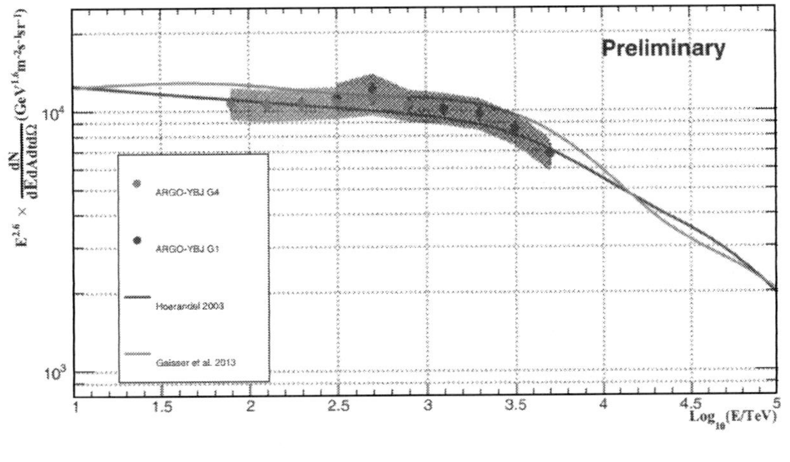

Figure 3. All-particle energy spectrum of primary CRs measured by ARGO-YBJ (see text). The parametrizations provided by [19] and [3] are shown for comparison.

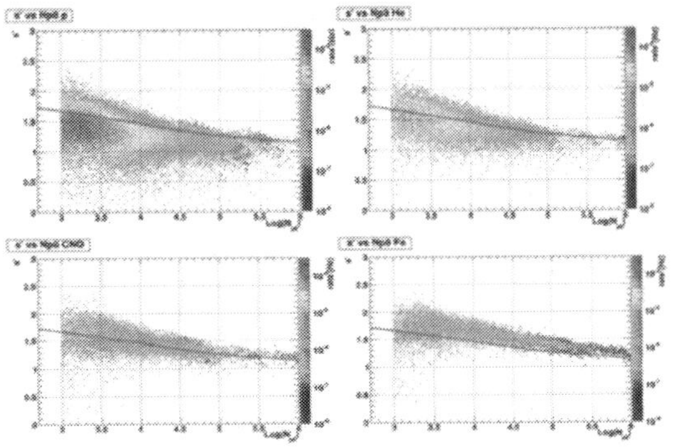

Figure 4. Relation between the LDF shape parameter s' and the truncated size N_{p8} for different nuclei. The p+He selection cut is shown by the lines.

(1) A selection of events in the $s' - N_{p8}$ space allowing to get a light-component sample of showers with a contamination of heavier nuclei less than about 15% (see Fig. 4).
(2) A Bayesian unfolding technique similar to that applied to measure the light-component spectrum up to 200 TeV [12, 13, 20].
(3) The ARGO-YBJ/WFCTA hybrid measurement [14] with a different selection procedure which increases the aperture of a factor 2.4, thus allowing the extension to larger energies (see [15] for a detailed description of the method and a discussion of the results).

Preliminary results of the three analyses (and the previous ARGO-YBJ measurement below 200 TeV) are summarized in Fig. 5. The systematic uncertainty on the flux is shown by the shaded area and the statistical one by the error bars. A systematic uncertainty on the energy scale at the level of 5–10% (depending on the analysis) has also been conservatively estimated (not shown in the plots). All three different analyses are consistent with low energy (direct) measurements and show a clear evidence for a bending at larger energies but below 1 PeV. With respect to a single power-law with a spectral index –2.62 the deviation is observed at a level of about 6 s.d. The results obtained with the two analysis of RPC charge readout data (label 1 and 2 in the previous list) are in fair agreement. They also agree with the

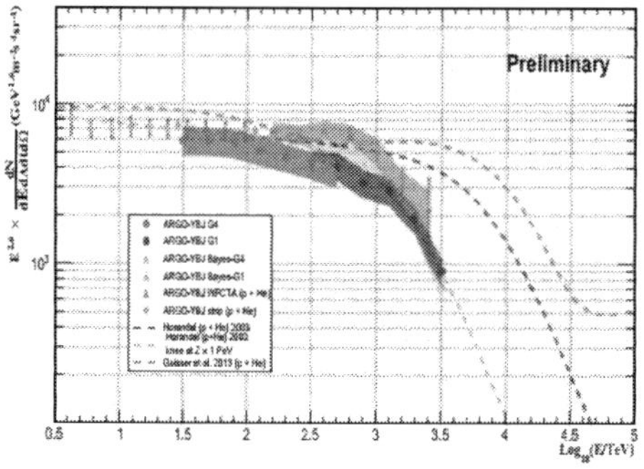

Figure 5. Light (i.e. p+He) component energy spectrum of primary CRs measured by ARGO-YBJ with four different analyses (see text).

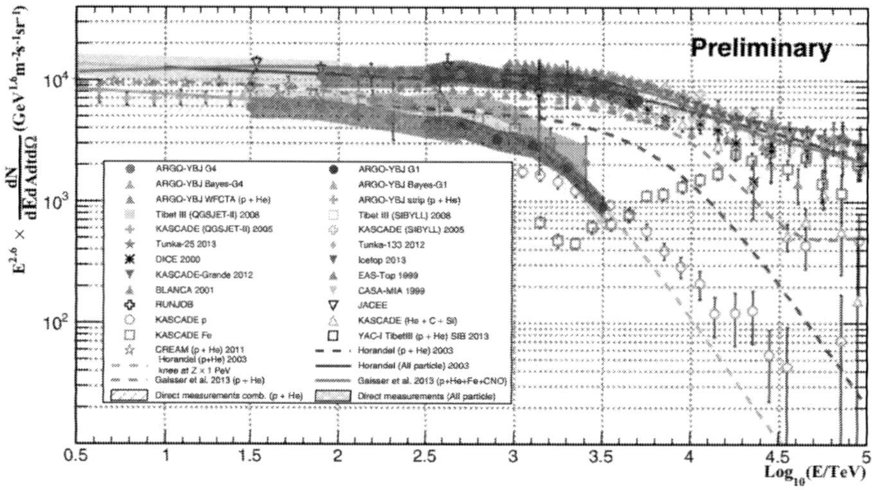

Figure 6. All particle and light (p+He) component energy spectra of primary CR measured by ARGO-YBJ and compared to different experimental results.

ARGO-YBJ/WFCTA hybrid measurement within systematic uncertainties and the possible difference in the energy scale. For comparison, the parametrizations of the light-component provided by [19] and [3] are shown by the blue and red dashed lines, respectively. A Hörandel-like spectrum with a modified knee at $Z\times 1$ PeV (a factor four lower in energy than in the original formulation) is also shown for comparison. Finally ARGO-YBJ results are compared to a compilation of several other measurements in Fig. 6.

4. Conclusions

The CR spectrum has been studied by the ARGO-YBJ experiment in a wide energy range (TeVs $\rightarrow$ PeVs). This study is particularly interesting because not only it allows a better understanding of the so called 'knee' of the energy spectrum and of its origin, but also provides a powerful cross-check among very different experimental techniques. The all-particle spectrum (measured in the energy range 100 TeV–3 PeV) is in good agreement with both theoretical parametrizations and previous measurements, making us confident about the selection and reconstruction procedures. The light-component (i.e. p + He) has been measured with high resolution up to about 5 PeV. The ARGO-YBJ preliminary result is in agreement with direct measurements and then show a clear indication of a bending below 1 PeV. Improvements of event selection with the full statistics and a complete analysis of systematic uncertainties is currently under way.

References

[1] H.S. Ahn et al., Astrophys. J. Lett. **714**, L89 (2010)
[2] Y.S. Yoon et al., Astrophys. J. **728**, 122 (2011)
[3] T.K. Gaisser, T. Stanev, S. Tilav, Front. Phys. **8**(6) (2013) 748
[4] D. Caprioli et al., Astrop. Phys. **34**, 447 (2011)
[5] W.D. Apel et al., Astrop. Phys. **31**, 86 (2009)
[6] M. Aglietta et al., Astrop. Phys. **21**, 223 (2004)
[7] M. Ambrosio et al. Phys. Rev. D **56** (1997) 1418
[8] M. Amenomori et al., Phys. lett. B **632**, 58 (2006)
[9] H. Tokuno et al., Astrop. Phys. **29**, 453 (2008)
[10] H. T. Freudenreich et al., Phys. Rev. D **41**, 2732 (1990)
[11] M.A.K. Glasmacher et al., Astrop. Phys. **12**, 1 (1999)
[12] I. De Mitri et al., Nucl. Instr. and Meth. in Phys. Res. A **742** (2014) 2–9
[13] B. Bartoli et al., Phys. Rev. D **85**, 092005 (2012)
[14] B. Bartoli et al., Chinese Phys. C **38**, 045001 (2014)
[15] Z. Cao et al., Proc. of the Vulcano workshop 2014, Frascati Physics Series, in publication
[16] B. Bartoli et al., Astrop. Phys. in press, doi: 10.1016/j.astropartphys.2015.01.007
[17] A. D'Amone et al., in Proc. 33rd ICRC (Rio de Janeiro, 2013) **ID: 0779**
[18] P. Bernardini et al., in Proc. 33rd ICRC (Rio de Janeiro, 2013) **ID: 0781**
[19] J.H. Hörandel, Astrop. Phys. **19**, 193 (2003)
[20] S. Mari et al., theese proceedings

The AMY experiment: Microwave emission from air shower plasmas

J. Alvarez-Muñiz[1], M. Blanco[2], M. Boháčová[3], B. Buonomo[4], G. Cataldi[5], M.R. Coluccia[5,6], P. Creti[5], I. De Mitri[5,6], C. Di Giulio[7,8,a], P. Facal San Luis[9], L. Foggetta[4], R. Gaïor[2], D. Garcia-Fernandez[1], M. Iarlori[6], S. Le Coz[11], A. Letessier-Selvon[2], K. Louedec[11], I.C. Maris[12], D. Martello[5,6], G. Mazzitelli[4], M. Monasor[13], L. Perrone[5,6], S. Petrera[10], P. Privitera[13], V. Rizi[10], G. Rodriguez Fernandez[7,8], F. Salamida[14], G. Salina[8], M. Settimo[2], P. Valente[4], J. R. Vazquez[15], V. Verzi[8], and C. Williams[13]

[1] Depto. de Fisica de Particulas, Universidad de Santiago de Compostela, Santiago de Compostela, Spain
[2] Laboratoire de Physique Nucléaire et de Hautes Energies (LPNHE), Universités Paris 6 et Paris 7, CNRS-IN2P3, Paris, France
[3] Institute of Physics, Academy of Sciences of the Czech Republic, Prague, Czech Republic
[4] Istituto Nazionale di Fisica Nucleare, Laboratori Nazionali di Frascati, via E. Fermi, 40, 00044 Frascati, Italy
[5] Sezione INFN, Lecce, Italy
[6] Dipartimento di Matematica e Fisica Ennio De Giorgi, Università del Salento, Lecce, Italy
[7] Dipartimento di Fisica, Università di Roma Tor Vergata , Roma, Italy
[8] Sezione INFN, Roma Tor Vergata, Italy
[9] University of Chicago, Enrico Fermi Institute Kavli Institute for Cosmological Physics, Chicago, USA
[10] Dipartimento di Fisica, Università dell'Aquila and sezione INFN, l'Aquila, Italy
[11] Laboratoire de Physique Subatomique et de Cosmologie (LPSC), Université J. Fourier Grenoble, CNRS-IN2P3, Grenoble, France
[12] Universidad de Granada and C.A.F.P.E., Granada, Spain
[13] University of Chicago, Enrico Fermi Institute Kavli Institute for Cosmological Physics, Chicago, USA
[14] Institut de Physique Nucléaire d'Orsay (IPNO), Université Paris 11, CNRS-IN2P3, France
[15] Universidad Complutense de Madrid, Madrid, Spain

Abstract. You The Air Microwave Yield (AMY) experiment investigate the molecular bremsstrahlung radiation emitted in the GHz frequency range from an electron beam induced air-shower. The measurements have been performed at the Beam Test Facility (BTF) of Frascati INFN National Laboratories with a 510 MeV electron beam in a wide frequency range between 1 and 20 GHz. We present the apparatus and the results of the tests performed.

1. Introduction

The cosmic rays at energies above 10^{18} eV could be detected using the Earth atmosphere as a calorimeter. This method called fluorescence technique provides a calorimetric measurement of the primary energy of the cosmic ray from the detection of the radiation produced by the

[a] Corresponding author: `claudio.digiulio@roma2.infn.it`

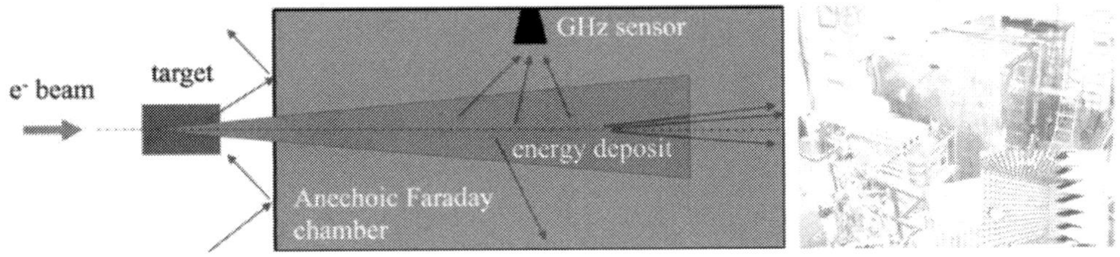

Figure 1. Left panel: sketch of the AMY experiment. Right panel: the AMY anechoic Faraday chamber mounted at the BTF, on the bottom right the inside view of the chamber.

de-excitation of the atmospheric nitrogen. The collection of this fluorescence light by the telescopes are used by the Pierre Auger Observatory [2] and Telescope Array [3], which are the largest detectors of cosmic rays that were ever built. In those detectors the fluorescence measurements are used to determine the energy scale [4] of the surface array detector only in small fraction of events detected by the surface array since the fluorescence light could be collected only during moonless nights with an overall duty cycle of about 10%.

An unconventional technique has been suggested after the first statement of the GHz radiation from air shower plasmas [1] in which the Molecular Bremsstrahlung Radiation (MBR) is predictable that to be emitted by low energy electrons throughout their interaction with the field of the neutral molecules. Similar the fluorescence emission, MBR should be isotropic and un-polarized but this radiation is released at frequencies in the GHz range respect to the UV range of the fluorescence. The major advantage of a telescope in the GHz frequency is that it would be comparable to the fluorescence one, but with a 100% of duty cycle.

This procedure was first developed at the Stanford Linear Accelerator Center (SLAC) using a 28 GeV electron beam. The measurements have been achieved in a 1 m^3 copper anechoic Faraday chamber in the frequency band 1.5–6.0 GHz. Afore enter into the chamber, the beam was collided with a 90% Al2O3–10% SiO2 target. The signal was found to decay exponentially with $\tau = 10$ ns and to be proportional with the square of the beam energy, demonstrating that the emission is fully coherent. The density flux of the radiation was estimated to be 4×10^{-16}W/m^2/Hz.

Based on those results the potentiality of a MBR technique telescope was investigated by prototypes of GHz telescopes as AMBER [1, 5], MIDAS [5, 6] and EASIER [5] at the Pierre Auger Observatory and CROME detector that has been setup within the KASCADE-Grande array [7]. Several shower candidates have been detected, but the measurements disfavors the hypothesis of coherent emission, which is in contrast to what has been reported in [1].

Other experiments using electron beam have been performed [8, 9] that suggests a linear scaling of the radiation intensity with the beam energy, which means that the emission is not radiated coherently; this could be due to the fact that MBR radiation emission depends strongly by the plasma density that is related to the source of electrons.

In the next sections we present the AMY experiment arrangement and the first outcomes of the analysis of the data.

2. Equipment and experimental method

In this Section particulars of the AMY anechoic Faraday chamber, a essential part of the experiment, are specified. The electron beam and the target setup provides the source for the MBR measurements, is also explained. In Fig. 1 a sketch of the all the components is shown.

The MBR emission is expected by the secondary electrons and its intensity should be proportional to the energy deposited as the fluorescence radiation. This radiation is detected with GHz sensors within the anechoic Faraday chamber that is mandatory to reduce the background and the echoed radiation produce by the BTF hall and from the beam entering into the chamber.

2.1 The accelerator beam and the interaction target

The BTF [11] is a part of the DAΦNE complex. The LINAC provide electron bunches with a charge up to 10^{10} electrons for pulse, in an energy range between 25 and 750 MeV, with a bunch size between 1–10 ns and a maximum repetition rate of 50 Hz. The bunch is made of micro-bunches with a FWHM of 14 ps, separated by 0.35 ns, which matches to the inverse of the LINAC frequency 2.856 GHz.

The interaction target contains six modules of 95% pure Al_2O_3 distantly managed by a pneumatic system. The maximum thickness is 45 cm that corresponds to about 7 radiation lengths (X_0). At the shower maximum, our energy deposit is almost a factor 6 larger than the energy deposit in [1].

The AMY collaboration has completed four measurement campaigns, one in 2011, two in 2012 and one in 2014. Following the first two tests we had improved the operation conditions and the experimental apparatus. The bunch length was 10 ns, 3 ns and the 1.5 ns only in the test of 2012.

2.2 The anechoic Faraday chamber and the data acquisition system

The anechoic Faraday chamber was manufactured at the mechanical shop of the physics department of the Roma Tor Vergata University. The sizes of the chamber (2 m width × 4 m long) are mainly constrained by the space accessible in the BTF hall. To enable its transportation to the BTF, the chamber has been built in 3 sections: two have a length of 1.5 m and one is 1 m long.

The left panel of Fig. 1 shows the chamber installed at the BTF. The outer surface is covered by copper and connected to ground. The internal side of the chamber (see in the bottom of the right panel of Fig. 1) has a cover of pyramidal RF absorbers (AEP-12 model). The shielding of the chamber from external radiation it is superior than 80 dB above 2 GHz and it falls quickly at lower frequencies (at 1 GHz it is $\sim$50 dB). The chamber can host the radio receivers in 5 different locations (matching to the aluminum boxes visible in Fig. 1). The radio receiver sustain allows the variation of the distance of the antennas from the beam axis and to exchange their polarization plane.

The receivers are two Rohde&Schwarz (R&S) HL050 log-periodic antennas and two RF Spin Double Ridged Waveguide Horn DRH20. Both types of antennas are linearly polarized and operate in a broadband range of input frequencies, from about 1 to 20 GHz. The antennas have been calibrated at the SATIMO StarLab calibration system [12] as shown in the left panel of Fig. 2 for the horn.

The signal from the antenna was amplified by about 26 dB using the Minicircuits wide band amplifier ZVA-213-S+ (800 MHz−21 GHz). The signal were lead through about 20 m low loss cables to the control room and acquired by a Lecroy SDA 830Zi-A oscilloscope with a 20 GHz real time bandwidth and a sampling rate of 40 GS/s. The channels were calibrated using a Rohde&Schwarz SMF100A signal generator (range 100 kHz−22 GHz) and a Rohde&Schwarz SFSV30 spectrum analyzer (range 9 kHz–30 GHz and 40 MHz bandwidth). The overall calibration with the amplifier and cables is shown in the right panel of Fig. 2. The loss at higher frequencies is due to the cables.

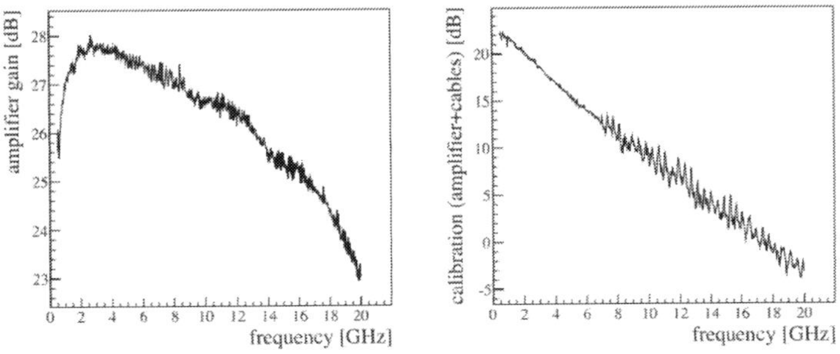

Figure 2. Left panel: gain of the horn as a function of the frequency. Right panel: calibration factor that includes the amplifier and the loss in the cables.

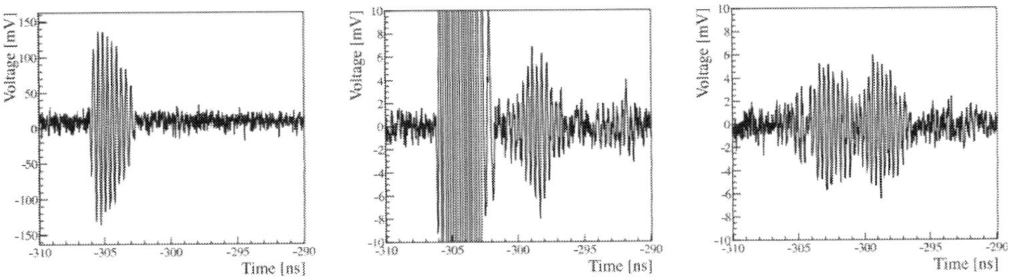

Figure 3. Left panel: gain of the horn as a function of the frequency. Right panel: calibration factor that includes the amplifier and the loss in the cables.

3. AMY data analysis

The significant argument of the measurement is the understanding of the enormous background result from the Cherenkov and transition radiation emitted by the relativistic electrons.

This radiation is linearly polarized in the plane including the beam axis and the direction of the radiation. While the antenna polarization plane is parallel to the beam axis (co-pol arrangement) the antenna signal is very elevated and dominated by the Cherenkov part as exposed in the left panel of Fig. 3. When the polarization plane is perpendicular to the beam axis (cross-pol arrangement) the signal is much lower as shown in the right panel of Fig. 3.

The signal in the cross-pol arrangement has a double peak structure and several tests performed do not support the reflections in the electronic chain and in the chamber as the source. In the central panel of Fig. 3 the co-pol signal of the left panel was acquired with the oscilloscope scale saturated and the second peak is shown. Contrary to the first peak, the second one is un-polarized which is the experimental signature of the MBR that distinguishes it from the Cherenkov radiation. Moreover it is delayed and this is consistent with MBR and Cherenkov signals respectively dominated by the emission at the center and the beginning of the chamber.

The cross-pol signal was found to be fully coherent. The power of the signal as a function of the number of electrons per bunch is well explained by a quadratic power low as presented in the left panel of Fig. 4. The power has been estimated by the root mean square of the oscilloscope trace in a time window around the signal and subtracting the background contribution estimated by the start of the trace, when the beam is not still existent in the chamber.

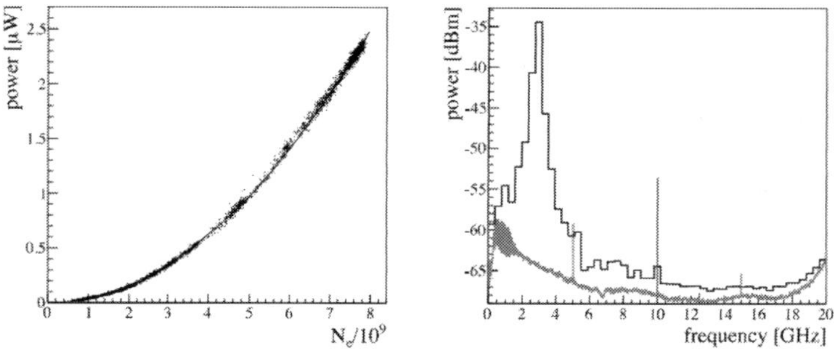

Figure 4. Left panel: power of the cross-pol signal as a function of the beam intensity. Right panel: frequency spectrum of the oscilloscope trace.

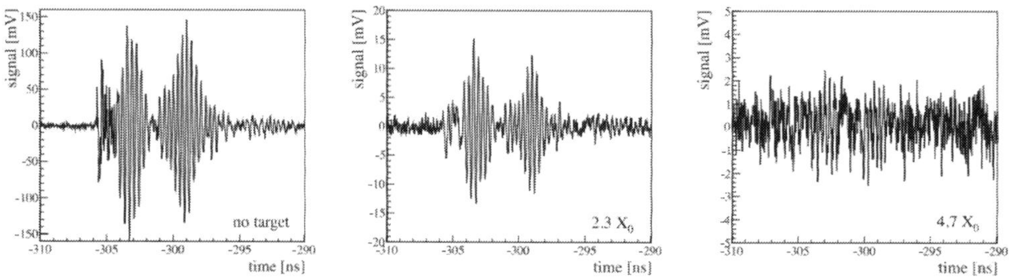

Figure 5. Left panel: cross-pol signal for different thicknesses of the interaction target. The number of particles $N_e = 2.3 \times 10^9$ and the bunch length duration is 1.5 ns for all signals.

The right panel of Fig. 4 shows the spectrum of the signal (black) by Fourier Transform of the oscilloscope traces and averaging over many events. The contribution from the background is shown in red and is dominated by the LINAC frequency ($f_L = 2.856$ GHz). This is estimated by the cross-pol signal with a target of 2.3 X_0. Repeating the analysis to the co-pol signal confirm the same feature.

We have developed a detailed simulation of the Cherenkov contribution according to the model developed in [13] and qualitatively reproduce the experimental spectrum. The peaks are due to the constructive interference of the emission of the micro-bunches and the power of the signal scales quadratically with the beam intensity. Consequently we comprehend that the coherence in the cross-pol arrangement is produced by the LINAC so we cannot exclude the Cherenkov as the origin.

Additional important item of the cross-pol signal is the dependence between its intensity and the thickness of the interaction target. Figure 5 shows the raw oscilloscope traces for $N_e = 2.3 \times 10^9$ when there is no target and for 2.3 and 4.7 X_0. The signal decreases from $\sim$100 mV to $\sim$1 mV which corresponds to a factor 10^4 in power. This fall of the power can not be explained by the change of the energy deposit, which is only a factor 5. A similar trend has been observed for the co-pol signal. This behavior is interpreted as the result of a loss of coherence, which is caused by a larger average distance between the electrons after their collision with the target.

4. Conclusions

The AMY experiment and the first results of the measurements done at BTF(Frascati) is discussed. The previous results obtained at SLAC [1] are not confirmed. The detected signal seems to be very prompt like the Cherenkov emission and the coherence is surely produced by the particular time structure of the bunch. Using the AMY data is possible to put an upper limit on the intensity of the MBR process. A preliminary limit can be found with the analysis of the data with the largest target thickness. We find a density flux $< 10^{-16}$ W/m^2/Hz using the signals with 4.7 X_0, the calibration of the detector and considering that the signal is mostly intense in a narrow bandwidth of 1 GHz around f_L.

References

[1] P. W. Gorham et al., Phys. Rev. D **78** (2008) 032007
[2] The Pierre Auger Collaboration, Nucl. Instrum. Meth. A **523** (2004) 50
[3] H. Sagawa, TA Collaboration, Proc. 33rd ICRC 2013, Rio de Janeiro, Brazil, icrc2013-0128
[4] V. Verzi, Pierre Auger Collaboration, Proc. 33rd ICRC 2013, Rio de Janeiro, Brazil, icrc2013-0928, arXiv:1307.5059 [astro-ph]
[5] I. Al Samarai, "Radio detection of Cosmic Rays in the GHz band at the Pierre Auger Observatory", ARENA 2014
[6] J. Alvarez-Muñiz et al., Phys. Rev. D **86** (2012) 051104(R)
[7] R. Smida et al., arXiv:1108.0588v2 [hep-ex]
[8] M. Monasor et al., arXiv:1108.6321
[9] E. Conti et al., arXiv:1408.5886 [astro-ph]
[10] I. Al Samarai, "An estimate of the spectral intensity expected from the molecular Bremsstrahlung radiation in extensive air showers", ARENA 2014 and I. Al Samarai et al., arXiv:1409.5051 [astro-ph]
[11] G. Mazzitelli et al., Nucl. Instr. and Methods A **515** (2003) 524
[12] L. J. Foged and A. Scannavini, Radioengineering **18**(4) (2009) 461
[13] J. Alvarez-Muñiz et al., Phys. Rev. D **81** (2010) 123009

Gamma-ray emission from young supernova remnants: Hadronic or leptonic?

Stefano Gabici[1,a] and Felix Aharonian[2,3]

[1] APC, Univ Paris Diderot, CNRS/IN2P3, CEA/Irfu, Obs de Paris, Sorbonne Paris Cité, France
[2] Dublin Institute for Advanced Studies, 31 Fitzwilliam Place, Dublin 2, Ireland
[3] Max-Planck-Institut für Kernphysik, Postfach 103980, 69029 Heidelberg, Germany

Abstract. The debate on the nature of the gamma-ray emission from young supernova remnants is still open. Ascribing such emission to hadronic rather than leptonic processes would provide an evidence for the acceleration of protons and nuclei, and this fact would fit with the very popular (but not proven) paradigm that supernova remnants are the sources of Galactic cosmic rays. Here, we discuss this issue with a particular focus on the best studied gamma-ray-bright supernova remnant: RX J1713.7-3946.

1. Introduction

The acceleration of protons at supernova remnant (SNR) shocks is accompanied by the emission of gamma-rays, resulting from the decay of neutral pions produced in interactions between the accelerated cosmic rays (CR) and the ambient interstellar matter (see e.g. [1] and references therein). It is widely believed that the efficiency of energy conversion into CRs should be at the $\approx 10\%$ level at SNR shocks, i.e. each SNR should convert $W_{\rm CR} \approx 10^{50}$ erg into CRs. The reason for that is phenomenological: it is the energy needed to sustain the observed intensity of CRs against their escape from the Galaxy [2]. An estimate of the very high energy (VHE) gamma-ray emission resulting from the interactions between the CRs accelerated in a SNR and the ambient gas of density $4 \times n_{gas}$ (the factor of 4 accounts for the compression of the gas at the SNR shock) is [1]:

$$E_\gamma^2 F_\gamma \approx \frac{W_{\rm CR}}{\ln(E_{\rm max}/E_{\rm min})\,\tau_{\pi^0}\,(4\pi d^2)} = 4 \times 10^{-11} \left(\frac{W_{\rm CR}}{10^{50}\,{\rm erg}}\right)\left(\frac{n_{gas}}{{\rm cm}^{-3}}\right)\left(\frac{d}{\rm kpc}\right)^{-2} {\rm erg/cm^2/s} \tag{1}$$

where $\tau_{\pi^0} \approx 1.6 \times 10^8 (n_{gas}/{\rm cm}^3)^{-1}$ yr is the energy loss time due to neutral pion production and d the distance to the SNR. In computing Eq. (1) we implicitly assumed that the CR spectrum is $\propto E^{-2}$ and extends from $E_{min} = 1$ GeV to $E_{max} = 4$ PeV, which is the energy of the CR knee. This is a large flux, well within the reach of current Cherenkov telescopes, unless the SNR is too far away or the ambient density is much lower than the average one in the Galaxy [3]. A spectrum much steeper than E^{-2} and/or a cutoff in the spectrum at an energy much lower than that of the knee would result is a dramatic suppression of the VHE gamma-

[a] e-mail: stefano.gabici@apc.univ-paris7.fr

[1] For convenience, all the estimates done in this section relies on the assumption of a E^{-2} spectrum for both protons and electrons. It is trivial to consider the more general case E^{-s} ($s \gtrsim 2$ seems to better agree with observations of SNRs and of CRs) and check that the main claims made here remain substantially correct.

ray flux, but these spectral feature would be in contradiction with the SNR paradigm for the origin of CRs. For this reason, the detection of SNRs in VHE gamma-rays was long awaited and considered a test for CR origin [3]. However, the detection of several SNRs in VHE gamma-rays cannot be considered a proof of CR acceleration because leptonic processes (most notably inverse Compton scattering) could as well explain the observed emission [4].

The contribution to the gamma-ray flux from inverse Compton scattering of electrons in the cosmic microwave background radiation can be computed by recalling that photons of energy of ≈ 1 TeV are produced by electrons of energy $E_e \approx 20$ TeV. At this energy the rate of production of TeV photons is $\tau_{\rm IC}^{-1} \approx (5 \times 10^4 {\rm yr})^{-1}$. If the spectrum of electrons accelerated at a SNR shock is K_{ep} times that of protons, then the contribution to the ≈ 1 TeV gamma-ray flux from inverse Compton scattering is $\approx K_{ep} W_{\rm CR}/2\tau_{\rm IC} \ln(E_{max}/E_{min})(4\pi d^2) \approx 2 \times 10^{-11}(W_{CR}/10^{50}\ {\rm erg})(d/{\rm kpc})^{-2}(K_{ep}/10^{-3})$ erg/cm^2/s. From which the ratio of leptonic to hadronic contribution to the VHE gamma-ray emission at 1 TeV can be deduced:

$$\phi_{\rm IC/pp}(1\ {\rm TeV}) \approx 0.5 \left(\frac{n_{gas}}{{\rm cm}^{-3}}\right)^{-1} \left(\frac{K_{ep}}{10^{-3}}\right) \qquad (2)$$

which indicates that in normal interstellar medium conditions the leptonic contribution dominates unless electrons are less than $\approx 0.1\%$ abundant than protons. In other words, inverse Compton scattering is much more efficient than proton-proton interactions in producing gamma rays and thus the leptonic emission from a SNR might well exceed the hadronic one even in the presence of a significant acceleration of protons. Hence, the difficulty of ascribing unambiguously the VHE gamma-ray emission from SNRs to hadronic processes.

The estimate given above is valid under the assumption that the shape of electron and proton spectra are identical. This is true only if the electron spectrum is not affected by energy loss processes[2]. In SNRs the main cooling channel for electrons is synchrotron radiation, which proceeds at a rate $\tau_{syn} \approx 6 \times 10^3 (B/10\ \mu{\rm G})^{-2}(E_e/20\ {\rm TeV})^{-1}$ yr. This implies that in young SNRs of age τ_{age} of the order of few thousand years the multi-TeV spectrum of electrons is affected by synchrotron losses if the magnetic field is $\gtrsim 10\ \mu$G. In this case, the ratio between leptonic and hadronic emission is significantly reduced with respect to the value computed in Eq. (2), because a break forms in the electron spectrum at an energy E_e^* that satisfies the condition $\tau_{age} \approx \tau_{syn}(E_e^*)$.

The spectrum of electrons accelerated at a SNR shock can thus be described as follows: at low energies, where both electrons and protons are unaffected by energy losses, the electron spectrum is identical in shape to the proton one, but is rescaled by the factor K_{ep} which accounts for the different acceleration efficiency of the two species. If the magnetic field strength is large enough, a break may appear in the electron spectrum at an energy $E_e^* \approx 1\ (B/100\ \mu{\rm G})^{-2}(\tau_{age}/{\rm kyr})^{-1}$ TeV. In gamma rays, the break in the inverse Compton spectrum would appear at a photon energy $E_\gamma^* \approx 4\ (B/100\ \mu{\rm G})^{-4}(\tau_{age}/{\rm kyr})^{-2}$ GeV. At energies larger than the break, the electron and gamma-ray spectra steepen by one power and half power of energy, respectively, and proceed until a maximum energy which can be computed by equating the energy loss time (determined by synchrotron radiation) to the acceleration time at the shock $D_B(E_e)/u_s^2$, where $D_B = (1/3) R_L c$ is the Bohm diffusion coefficient, R_L the particle Larmor radius, c the speed of light, and u_s the shock speed. This gives $E_e^{max} \approx 30\ (B/100\ \mu{\rm G})^{-1/2}(u_s/1000\ {\rm km/s})$ TeV, which corresponds to a gamma-ray photon energy of a few TeV in the inverse Compton spectrum. Thus, a cutoff in the electron spectrum at energies < 30 TeV would prevent the production of VHE gamma rays through

[2] Energy losses of protons have characteristic times well in excess than the SNR age and can be safely neglected.

inverse Compton scattering, unless soft radiation fields other than the cosmic microwave radiation (e.g. infrared, optical...) are present.

Having identified all the main physical quantities that shape the hadronic and leptonic contribution to the gamma-ray spectrum of SNRs, we can proceed discussing a specific object, RX J1713.7-3946, in an attempt to understand the origin of its gamma-ray emission.

2. The SNR RX J1713.7-3946

RX J1713.7-3946 is the best studied SNR in VHE gamma-rays and the hadronic or leptonic nature of its emission is still debated (for a non-exhaustive list of references see [5–10]). In the following we discuss this issue by listing the various arguments that have been brought forward in favor or a hadronic or leptonic origin of the emission.

2.1 Arguments in favor of an hadronic gamma-ray emission

2.1.1 Evidence for magnetic field amplification

According to shock acceleration theory, the acceleration of CRs at SNR shocks must be accompanied by the amplification of the magnetic field up to values of hundreds of milliGauss or even milliGauss. In order to have an effective amplification of the field, an efficient acceleration of CR hadrons is required [11, 12]. This finding received strong observational support with the detection of thin X-ray synchrotron filaments from several SNR shocks. In an amplified field, electrons accelerated at shocks are cooled much more rapidly than the rate at which they are advected downstream of the shock. Thus, they radiate synchrotron X-rays in the close proximity of the shock and their emission appears as a very thin filament. The thinness of filaments allows to constrain the magnetic field strengths in the hundreds of microGauss range from a number of young SNRs (see [13] for a review). Though the presence of X-ray filaments in RX J1713.7-3946 was revealed by Chandra observations [14], the most spectacular evidence for magnetic field amplification in this object came from the observation, also made by Chandra, of fast (year-scale) variability of the synchrotron emission of some small (sub-parsec scale) X-ray knots [14]. If the variability is interpreted as the result of the fast cooling of freshly accelerated electrons, a magnetic field of ≈ 1 mG is inferred [14]. Alternatively, the rapid variability of the X-ray emission can be explained by a fluctuating magnetic field. This has been indeed proposed and discussed in [15], where the authors, in order to explain the variability, adopted a 100 μG field, which is lower than that inferred in [14] but still significantly amplified.

It has to be stressed that the evidence for an amplified magnetic field is limited to a very small fraction of the SNR volume (filaments and knots) and thus it is difficult to reach strong conclusions on a global level. For example, it is unclear if CR protons are accelerated with large efficiencies in the whole shock surface or limited to a small fraction of it. Also, very little constrains are available for the strength of the magnetic field in the whole remnant. If the volume averaged value of the magnetic field is significantly larger ≈ 10 μG then a leptonic interpretation of the gamma-ray emission from RX J1713.7-3946 is problematic, because in this case electrons would radiate mainly through synchrotron radiation rather than inverse Compton scattering. On the other hand, if the magnetic field outside filaments and knots is at the level of ≈ 10 μG then the leptonic interpretation would be definitely favored.

To conclude, high resolution X-ray observations of SNRs, and in particular of RX J1713.7-3946, are generally interpreted as a very strong evidence for an efficient acceleration of CR hadrons from at least a part of SNR shocks. It is tempting to link such an efficient acceleration efficiency with a correspondingly high level of the hadronic gamma-rays from the remnants. However, the ambient gas density also plays a crucial role here (see Eq. (1)), and it can be easily shown that if the target density is as low as 0.1 cm^{-3} even an effective conversion of energy into CRs ($\approx 3 \times 10^{50}$ erg in form of CR protons) would not suffice to provide a significant hadronic contribution to the observed gamma-ray flux [16].

2.1.2 Spatial correlation of gamma-ray emission and gas density

The search for spatial correlations between gamma-ray brightness and gas column density is a powerful tool that can serve to discriminate amongst hadronic and leptonic emission. This is because a correlation with the ambient gas density is naturally expected in a hadronic scenario where CR protons produce gamma rays in interactions with the ambient gas. Such a correlation is not granted (but at the same time not excluded!) in leptonic models.

A spatial correlation of SNRs and dense gas is likely if the SNR progenitor is a supernova of type II, which is expected to explode inside a molecular cloud as a result of the very fast evolution of a very massive star [17]. Indeed, a type II progenitor for RX J1713.7-3946 seems to be suggested by observations [18], and a correlation with dense molecular gas has been claimed [19]. The significance of the correlation has been found to increase as a result of detailed studies that considered the total gas density (atomic plus molecular) in the region [20]. Also, the ambient gas within the SNR shell appears to be structured in small (sub-parsec) and dense (up to at least 10^4 cm^{-3}) clumps [21]. The presence of a significant amount of gas in the SNR shell might of course provide a target for CR proton-proton interactions and enhance the hadronic contribution to the gamma-ray emission [20].

2.2 Arguments in favor of a leptonic gamma-ray emission

2.2.1 Spectral shape of the gamma-ray emission

The spectral shape of the gamma-ray emission from RX J1713.7-3946 is probably the most popular argument presented in favor of a leptonic origin of the emission. As described in the Introduction, if the magnetic field is amplified, two features are expected in the leptonic gamma-ray spectrum of SNRs: *i)* a break at energy E_e^* due to the fact that particles above a given energy cool via synchrotron emission on a time scale shorter than the age of the remnant, and *ii)* a cutoff at an energy E_e^{max} which represents the maximum energy a particle can be accelerated to before losing energy via synchrotron radiation. If we consider an injection spectrum of electrons proportional to E_e^{-2}, then the electron (gamma-ray) spectrum is $\propto E_e^{-2}$ ($\propto E_\gamma^{-1.5}$) below the break and $\propto E_e^{-3}$ ($\propto E_\gamma^{-2}$) above it.

The picture described above changes if the magnetic field at the SNR is weak. This is because both the break and the cutoff energies increase for weaker magnetic fields, but in two different ways, i.e., $E_e^* \propto B^{-2}$ and $E_e^{max} \propto B^{-1/2}$, implying that a magnetic field strength exists below which the break in the spectrum disappears. Such a field is ≈ 10 μG (see Introduction), and for field strength smaller or equal to this value the electron (gamma-ray) spectrum becomes a simple power law in energy with index -2 (-1.5).

Indeed, a weak magnetic field is required for the leptonic interpretation to be viable. Synchrotron X-ray emission has been detected from RX J1713.7-3946 at a flux level of $E_X^2 F_X \approx 10^{-10}$ erg/cm^2/s, while the VHE gamma-ray emission $E_\gamma^2 F_\gamma$ lays roughly one order of magnitude below that. If both X-rays and gamma rays are produced by the same electrons through synchrotron and inverse Compton scattering in the cosmic microwave background, respectively, then their flux ratio should reflect the ratio between the energy density in magnetic field $\omega_B = B^2/8\pi \approx 2.5$ $(B/10\ \mu\mathrm{G})^2$ eV/cm^3 and in the cosmic microwave

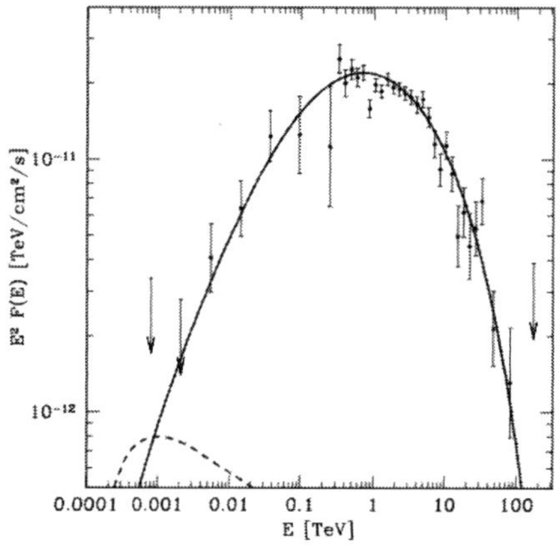

Figure 1. Hadronic model for RX J1713.7-3946, under the assumption that the remnant originated from a type II supernova explosion in a molecular cloud [10]. The SNR is assumed to expand in a rarefied ($\sim 2 \times 10^{-2}$ cm^{-3}) cavity inflated by the wind of the progenitor star. The wind swept away all the matter but the densest parts ($\gtrsim 10^3$ cm^{-3}), concentrated in small (~ 0.1 pc) clumps. Such dense clumps survive intact the passage of the SNR shock. CRs accelerated at the SNR shock diffusively penetrate into the clumps. Bohm diffusion in a $\approx 100\ \mu$G field is assumed to operate at the interface between the clumps and the diffuse medium in the SNR. The solid line represents the predicted emission from the clumps, while the dashed line that from the diffuse gas inside the SNR. Here, an explosion energy of 10^{51} ergs, a CR acceleration efficiency of 10% and a spectrum of accelerated CRs at the shock $\propto E^{-2.2}$ have been assumed. To match the data, the total mass in the clumps has to be $\approx 550\ M_\odot$. Data points are from FERMI [16] and HESS [34].

background $\omega_{CMB} \approx 0.25$ eV/cm^3:

$$\frac{F_X}{F_\gamma} \approx 10 \left(\frac{B}{10\ \mu\text{G}} \right)^2 \tag{3}$$

which implies that a field strength of $\approx 10\ \mu$G is required to fit data.

To summarize, a simple, one-zone model leptonic interpretation of the spectrum of RX J1713.7-3946 requires a weak magnetic field, and a rising spectral energy distribution of gamma-ray photons below the cutoff: $E_\gamma^2 F_\gamma \propto E_\gamma^{0.5}$. On the other hand, if the magnetic field is $\gg 10\ \mu$G, an hadronic interpretation seems favored, and for a typical $\propto E^{-2}$ injection spectrum of protons one would expect a flat spectral energy distribution of gamma-rays from neutral pion decay: $E_\gamma^2 F_\gamma \propto E_\gamma^0$. This is the reason why the detection by *Fermi* of a rising spectral energy distribution in the GeV domain from RX J1713.7-3946 (see [16] and Fig. 1) has been interpreted as a strong evidence for the fact that the gamma-ray emission is indeed leptonic. Following the same reasoning, the much steeper gamma-ray spectrum detected from the Tycho SNR has been interpreted as an evidence for hadronic emission [22, 23].

In fact, it has been soon realized that a one-zone leptonic model fails to obtain a good fit to the gamma-ray data for RX J1713.7-3946. The reason is that the expected spectrum has a peak much narrower than the observed one (see e.g. Fig. 3 in [4]). This problem can be solved either by assuming that two populations of relativistic electrons exist (e.g. accelerated at the forward and reverse shock, respectively?) [9, 24], or by invoking the presence of a strong infrared radiation field in the SNR that would provide a target for inverse Compton interactions distinct than the cosmic microwave background [25]. However, such a strong additional radiation field seems quite difficult to be justified (see e.g. [6]).

2.2.2 Absence of thermal X-ray emission

The absence of thermal Bremsstrahlung continuum emission in the X-ray spectrum of RX J1713.7-3946 was claimed to be an evidence against the hadronic origin of the gamma rays [26]. In order to have effective CR hadronic interactions the ambient gas density must not be too low, since the hadronic emission scales as the target gas density. On the other hand, if this is the case one should also expect a powerful X-ray thermal emission from the gas heated by the SNR shock (thermal emission scales as the gas density squared). In fact, it has been shown that the problem comes mainly from the non detection of X-ray thermal lines, rather than continuum emission. Lines should overcome the observed synchrotron emission of RX J1713.7-3946 unless the ambient density is very low, $\lesssim 0.05$ cm^{-3}. Since no thermal emission is observed (neither continuum nor lines) hadronic models are challenged [8].

The hadronic interpretation would still be viable if the gas is not heated to the high temperatures required for the production of X-rays. This could happen if the shock converts a large fraction (significantly above $\approx 10\%$) of the incoming kinetic energy into CRs, since in this case not enough energy would be left to heat the gas (e.g. [27–30]). However, recent simulations of the injection phase of particles at shocks suggest for the CR acceleration efficiency a value of $\approx 10\%$ [31], implying that significan gas heating must occur. Another interesting possibility to avoid gas heating involves the presence of dense clumps in the medium surrounding the SNR and will be discussed below.

2.3 A hadronic model for RX J1713.7-3946

It has been argued in [7] and [32] that the origin of the gamma-ray emission from RX J1713.7-3946 may well be hadronic if the SNR expands in a clumpy medium. The simulations performed in [32] suggests that this is a quite natural situation if the SNR progenitor is a type II supernova exploding in a molecular cloud, as RX J1713.7-3946 is believed to be [18]. The powerful wind from the progenitor star would sweep away the dense gas creating a cavity characterized by a gas density of the order of $\approx 10^{-2}$ cm^{-3}. However, the densest cores (of density above $\approx 10^3$ cm^{-3}) in the molecular cloud would survive into the cavity. Remarkably, such clumps would also survive the SNR shock passage. This happens because a SNR shock of velocity u_s that hits a dense clump generates a shock into the clump itself, with a velocity $u_c \approx u_s/a$, where $a \approx (10^3 \text{cm}^{-3}/10^{-2}\text{cm}^{-3})^{1/2} \approx 300$ is the square root of the ratio between the clump to inter clump density [33]. Thus, the SNR shock is virtually stalled at the clump border and the clump remains *unshocked*.

Remarkably, this fact solves one of the problems of hadronic models. The absence of thermal emission can be explained if the gas mass within the SNR is dominated by the contribution of dense clumps. Given that clumps remain unshocked and thus cold no appreciable X-ray thermal emission is expected.

The presence of clumps can also solve the problem related to the observed gamma-ray spectrum of RX J1713.7-3946, which is generally considered too hard to be explained by hadronic interactions. According to simulations, the interaction between the SNR and the clumps would lead to a strong amplification of the turbulent magnetic field at the interface between the clumps and the diffuse gas [32]. Such a turbulent magnetic field would prevent low energy CRs (accelerated at the SNR shock and confined within the SNR shell) to penetrate the clumps over a time scale shorter than the SNR age (≈ 1600 yr). On the other hand, high energy CRs would penetrate the clumps, their transport being characterized by a faster diffusion coefficient. The outcome of this is that the CR spectrum inside clumps is expected to be much harder than that of CRs in the SNR shell. If the total mass of the gas in the shell is dominated by the clumps, then the gamma-ray spectrum observed from the SNR would be hard, as the underlying CR spectrum inside clumps.

Following these arguments, a detailed model for RX J1713.7-3946 has been developed in [10], demonstrating that an excellent fit to the data can be obtained with a hadronic model (see Fig. 1).

3. Conclusions

The origin of the gamma-ray emission from RX J1713.7-3946 is not well established yet. A leptonic scenario requires a low magnetic field strength ($\approx 10\ \mu$G), and either two populations of relativistic electrons or two populations of target soft photons for inverse Compton interactions. On the other hand, a hadronic scenario implies a much stronger magnetic field and requires the presence of a structured ambient medium with dense clumps embedded in a tenuous diffuse gas. The detection or non detection of neutrinos will allow to discriminate amongst these two scenarios. For the hadronic scenario the detection of neutrino is well within the reach of a km^3-scale detector located in the northern hemisphere [35].

Also the detection of radiation from beyond the SNR shell would provide strong support to the hadronic scenario. Such a radiation is expected from the vicinity of the shell as the result of the interactions of escaping CR protons in the ambient gas [36, 37]. For RX J1713.7-3946 this would result in the presence of a faint diffuse gamma-ray emission surrounding the SNR, of brightness at the limits of detectability of current Cherenkov instruments, but within the reach of the future Cherenkov Telescope Array [38]. Remarkably, also X-ray telescopes like NuSTAR and Astro-H can contribute to the search of escaping CR protons by detecting the synchrotron emission produced by the secondary electrons produced by runaway CRs in interactions with the ambient gas [39].

SG acknowledges Paolo Piattelli and all the organizers of the RICAP conference for the invitation and the kind hospitality. He also acknowledges support under the LabEx UnivEarthS program at Sorbonne Paris Cité (ANR-10-LABX-0023/ANR-11-IDEX-0005-02) and the *Action Fédératrice CTA* at the Observatory of Paris.

References

[1] P. Cristofari, S. Gabici, S. Casanova, R. Terrier, E. Parizot, MNRAS **434**, 2748 (2014)
[2] T.K. Gaisser, *Cosmic rays and particle physics* (Cambridge University Press, 1990)
[3] L.O'C. Drury, F.A. Aharonian, H.J. Völk, A&A **287**, 959 (1994)
[4] F.A. Aharonian, APh **43**, 71 (2013)
[5] E.G. Berezhko, H.J. Völk, A&A **451**, 981 (2006)
[6] G. Morlino, E. Amato, P. Blasi, MNRAS **392**, 240 (2009)
[7] V.N. Zirakashvili, F.A. Aharonian, ApJ **708**, 965 (2010)
[8] D.E. Ellison, D.J. Patnaude, P. Slane, J. Raymond, ApJ **712**, 287 (2010)
[9] J.D. Finke, C.D. Dermer, ApJ **751**, 65 (2012)
[10] S. Gabici, F.A. Aharonian, MNRAS Lett. **445**, L70 (2014)
[11] A.R. Bell, MNRAS **353**, 550 (2004)
[12] A.R. Bell, K.M. Schure, B. Reville, G. Giacinti, MNRAS **431**, 415 (2013)
[13] J. Vink, A&A Rev. **20**, 49 (2012)
[14] Y. Uchiyama, F.A. Aharonian, T. Tanaka, T. Takahashi, Y. Maeda, Nature **449**, 576 (2007)
[15] A.M. Bykov, Y.A. Uvarov, D.C. Ellison, ApJ Lett. **689**, L133 (2008)
[16] A.A. Abdo, et al., ApJ **734**, 28 (2011)
[17] T. Montmerle, ApJ **231**, 95 (1979)

[18] P. Slane, et al., ApJ **525**, 357 (1999)
[19] Y. Moriguchi, et al., ApJ **631**, 947 (2005)
[20] Y. Fukui, et al., ApJ **746**, 82 (2012)
[21] N.I. Maxted, et al., MNRAS **422**, 2230 (2012)
[22] F. Giordano, et al., ApJ Lett. **744**, L2 (2012)
[23] G. Morlino, D. Caprioli, A&A **538**, A81 (2012)
[24] T. Tanaka, et al., ApJ **685**, 988 (2008)
[25] T.A. Porter, I.V. Moskalenko, A.W. Strong, ApJ Lett. **648**, L29 (2006)
[26] B. Katz, E. Waxman, JCAP **01**, 018 (2008)
[27] D.C. Ellison, AIP Conf. Proc. **528**, 383 (2000)
[28] P. Blasi, S. Gabici, G. Vannoni, MNRAS **361**, 907 (2005)
[29] L.O'C. Drury, F.A. Aharonian, D. Malyshev, S. Gabici, A&A **496**, 1 (2009)
[30] E.A. Helder, et al., Science **325**, 719 (2009)
[31] D. Caprioli, A. Spitkowsky, ApJ **783**, 91 (2014)
[32] T. Inoue, R. Yamazaki, S. Inutsuka, Y. Fukui, ApJ **744**, 71 (2012)
[33] R.I. Klein, C.F. McKee, P. Colella, ApJ **420**, 213 (1994)
[34] F.A. Aharonian, et al., A&A **464**, 235 (2007)
[35] M.L. Costantini, F. Vissani, Astropart. Phys. **23**, 477 (2005)
[36] F.A. Aharonian, A.M. Atoyan, A&A **309**, 917 (1996)
[37] S. Gabici, F.A. Aharonian, S. Casanova, MNRAS **396**, 1629 (2009)
[38] S. Casanova, et al., PASJ **62**, 1127 (2010)
[39] F.A. Aharonian et al., for the Astro-H Science Working Group, arxiv:1412.175

EPJ Web of Conferences **121**, 04002 (2016) DOI: 10.1051/epjconf/201612104002
RICAP-2014

MAGIC highlights

Rubén López-Coto[a] for the MAGIC Collaboration

Institut de Física d'Altes Energies, Spain

Abstract. The present generation of Imaging Air Cherenkov Telescopes (IACTs) has greatly improved our knowledge on the Very High Energy (VHE) side of our Universe. The MAGIC IACTs operate since 2004 with one telescope and since 2009 as a two telescope stereoscopic system. I will outline a few of our latest and most relevant results: the discovery of pulsed emission from the Crab pulsar at VHE, recently found to extend up to 400 GeV and along the "bridge" of the light curve, the measurement of the Crab nebula spectrum over three decades of energy, the discovery of VHE γ-ray emission from the PWN 3C 58, the very rapid emission of IC 310, in addition to dark matter studies. The results that will be described here and the planned deep observations in the next years will pave the path for the future generation of IACTs.

1. The MAGIC telescopes

The two MAGIC IACTs were built and are currently operated by a collaboration of institutions in Bulgaria, Croatia, Finland, Germany, Italy, Poland, Japan, Spain, and Switzerland [1].

MAGIC is located at the Roque de los Muchachos Observatory in the island La Palma (Spain). The first single telescope (MAGIC-I) started operations in 2004 and was at the time the largest IACT yet constructed (17 m diameter mirror), a fact which translated into a very low energy threshold for VHE γ-ray detection. MAGIC-I featured significant novelties in IACTs, such as the fastest sampling of Cherenkov signals (2 GSps) or active mirror control. Its ultralight carbon fiber frame and mirrors enable very fast repositioning of the telescope ($<$20 secs for half a turn), a crucial fact to study the prompt emission of GRBs.

The introduction of a second telescope, MAGIC-II, enabled the instrument to perform stereoscopic observations with significantly better sensitivity, and angular and spectral resolutions, starting in Fall 2009. As of today the MAGIC telescopes remain the IACT array with the largest mirrors in the world. MAGIC-II was built essentially as an improved copy of MAGIC-I. The main difference between the telescopes were their cameras and their readout electronics. MAGIC I camera was composed of 577 pixels of two different sizes. On the contrary, the MAGIC II camera was built with 1039 pixels of the same size (0.1°). The readout sampling speed was 2 GSps for both telescopes, but using very different electronics.

During the summers of 2011–2012 the instrument experienced a thorough upgrade [14]: the readout was replaced by a homogeneous system based on the DRS4 analog memory sampling chip, a clone of the MAGIC-II camera and L1 trigger were installed in MAGIC-I. The upgrade of the MAGIC-I trigger results in an enlarged trigger area for the system.

---

[a]e-mail: `rlopez@ifae.es`

[1] An updated list of collaboration members can be found at `http://magic.mpp.mpg.de`

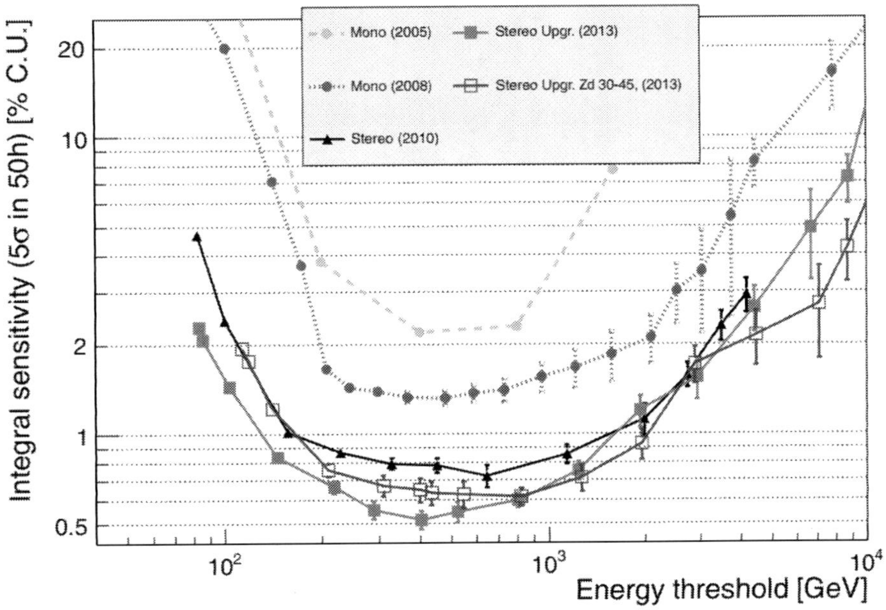

Figure 1. Integral sensitivity of the current MAGIC Stereo system and previous experimental setups, defined as the flux for which $N_{excess}/\sqrt{N_{bgd}} = 5$ after 50 h, and calculated using Crab data.

Besides, since both telescopes are now essentially identical, maintenance and operation are easier.

The trigger energy threshold (peak of the energy distribution of stereo recorded events) of the upgraded telescope has been estimated to be 50 GeV. In Fig. 1 we can see the sensitivity of the instrument as a function of energy. For energy above 300 GeV the sensitivity is 0.76% Crab units. There is a good agreement with the predictions from MC. Respect to single telescope observations, a factor ~2 improvement in sensitivity is achieved at a few hundred GeV and up to a factor ~3 at lower energies. The differential sensitivity remains acceptable (10% Crab units) below 100 GeV. Stereo observations result in a significant improvement both in angular and spectral resolutions. An angular resolution of 0.07° is reached at 300 GeV. The best energy resolution of 16% is reached at a few hundred GeV. Find more details about the instrument's performance in [15].

2. Extragalactic sources

Radiogalaxies (active galaxies displaying a radio jet) generate CR (electron or proton) bubbles in the intergalactic medium. The same process injects magnetic field into the intracluster medium. The total injected energy is huge ($10^{60} - 10^{61}$ erg): it represents a few % of the total energy of accretion into the central supermassive black hole. Relativistic electrons produce synchrotron which can be studied using radiotelescopes, but they also produce VHE γ-rays through Inverse Compton.

In fact radiogalaxies are interesting VHE sources because the emission is not so strongly beamed, i.e. the jet is not as aligned with the line of sight as in blazars, and because they are nearby objects, i.e. we can study them in more detail. IACTs have discovered four radiogalaxies at VHE: Cen-A, M 87, NGC 1275 and IC 310. MAGIC has discovered the last two sources, which actually belong to the same cluster of galaxies: Perseus.

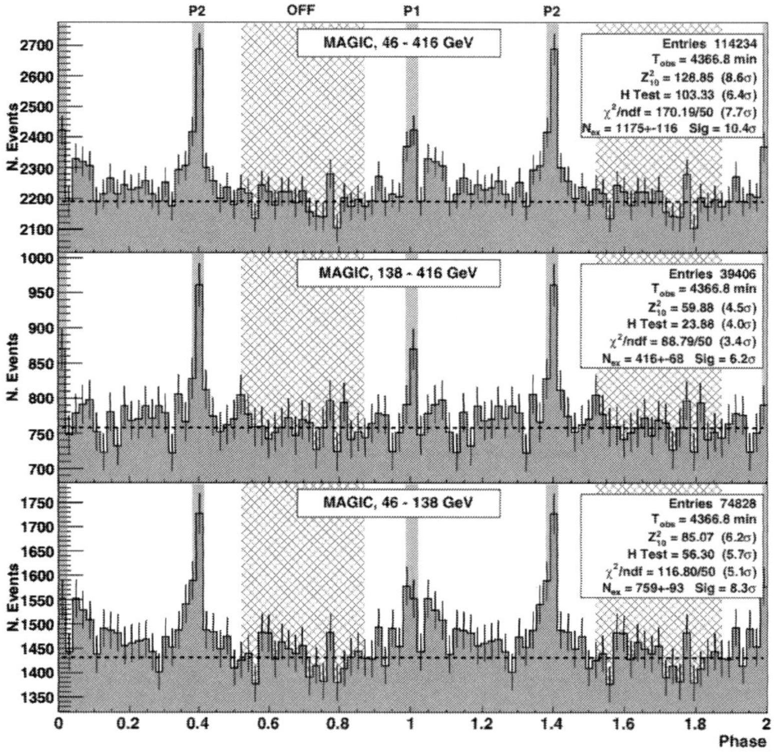

Figure 2. MAGIC folded light curves of the Crab pulsar for our total range in estimated energy and for two separate sub-bins. The shaded areas are the on-phase regions $P1_M$ and $P2_M$, the light shaded area is the off-region [0.52–0.87]. The dashed line is the constant background level calculated from that off-region.

MAGIC discovered the radiogalaxy IC 310 at VHE [4] during observations of Perseus. A flare was observed in 2011 which revealed that the source is variable from day to day [10]. The observation of a second flare in 2012 showed even faster variability with time scales of 1–10 minutes [12]. Even faster variability has been observed in blazars like Mrk 501 and PKS 2155-304, but emission in blazars is doppler-shifted by a larger factor than in a radiogalaxy like IC 310 for which the largest allowed Doppler factor is around 4. The intrinsic variability of the source may in fact be so fast that the emission region is smaller than the event horizon light-crossing time. Hardly any of the existent models can accommodate such a small emission region.

S3 0218+357 is a blazer discovered by Fermi that in 2012 revealed a series of flares with their counterparts after 11.46 $\pm$0.16 days delay, interpreted as due to the gravitational lensing effect [18]. On 2014 July 13 and 14 Fermi-LAT detected another flaring episode, but due to the full moon period, MAGIC could not observe it. However, observations scheduled at the expected time of arrival of the gravitationally lensed component led to the first significant detection of a gravitationally lensed blazar and the most distant source detected at VHE by an IACT [23].

3. Galactic sources

At VHE we study the particles with the highest energy which a pulsar is able to accelerate. MAGIC discovered emission at >25 GeV from the Crab pulsar [16]. VERITAS discovered

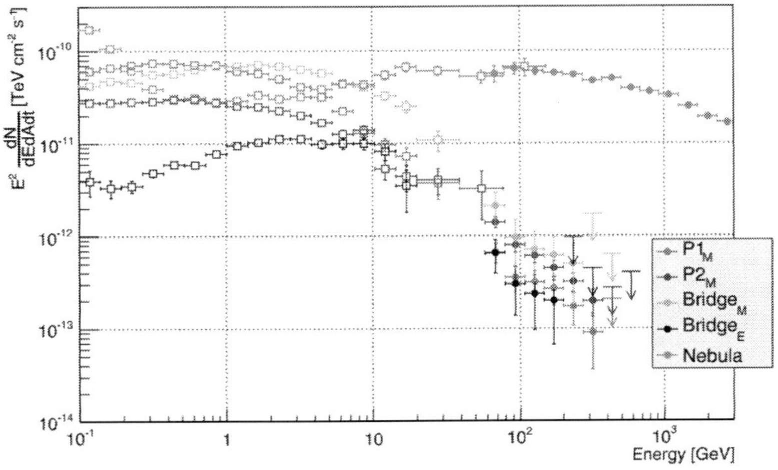

Figure 3. Spectral energy distributions of the Crab nebula, $P1_M$, $P2_M$, $Bridge_M$, and $Bridge_E$ measured with *Fermi*-LAT (below 50 GeV) and MAGIC (above 50 GeV). The flux values averaged over the rotation period are plotted.

that the pulsed spectrum extends to >100 GeV [17]. Some months later, MAGIC measured the spectrum of both peaks up to 400 GeV [6] (see Figure 2). MAGIC has recently discovered emission from the "bridge" (the region between P1 and P2) of the light curve at energies exceeding 100 GeV [8], measuring the separated spectra of both peaks and the bridge (see Fig. 3). The presence of the bridge and the fact that both peaks are very narrow is hard to explain within existing models. Aharonian et al [3] for instance propose that VHE γ-rays are not produced inside the magnetosphere but in the wind region. If true, VHE observations would allow to study the wind, which is totally dark at other wavelengths. This model is successful in producing bridge emission but predicts much broader peaks than those measured by MAGIC. Hirotani [19, 20] may be able to reproduce the shape of the light curve assuming that there is an additional toroidal component in the pulsar's magnetic field.

The largest population of sources in the HESS galactic plane survey are Pulsar Wind Nebulae (PWN). Energetically speaking pulsars are CR sources, that is, they spend most of their rotational power in accelerating particles. Photons, especially E<100 MeV, may well be considered as a "sideshow". MAGIC has only detected two PWN at VHE: Crab and 3C 58. These are however extreme PWNe. Crab is the brightest PWN, while 3C 58 is the weakest and least luminous. They are both the least efficient VHE PWN. 3C 58 has in fact a γ-ray luminosity which is as low as 10^{-5} of the pulsar spin-down power.

The latest MAGIC spectrum of the Crab Nebula is based on a 70 hour stereo observation spanning from 2009 to 2011 [13]. It extends from 50 GeV up to 30 TeV, with a statistical precision as low as 5% at E<100 GeV (see Figure 4). Combined with the Fermi-LAT data, these data yield the most precise measurement of the IC peak so far, at (52.5±2.6) GeV (statistical error only).

The spectrum has been fitted to two different models. A static, constant B-field model [22] predicts however too broad an IC peak. Most probably this implies that the assumption of the homogeneity of the magnetic field inside the nebula is incorrect. A time-dependent model [21] is successful in reproducing the spectral shape under the assumptions of a low magnetic field of less than hundred μG. However, this model fails to provide a good fit of the new spectral data if the observed morphology of the nebula is taken into account.

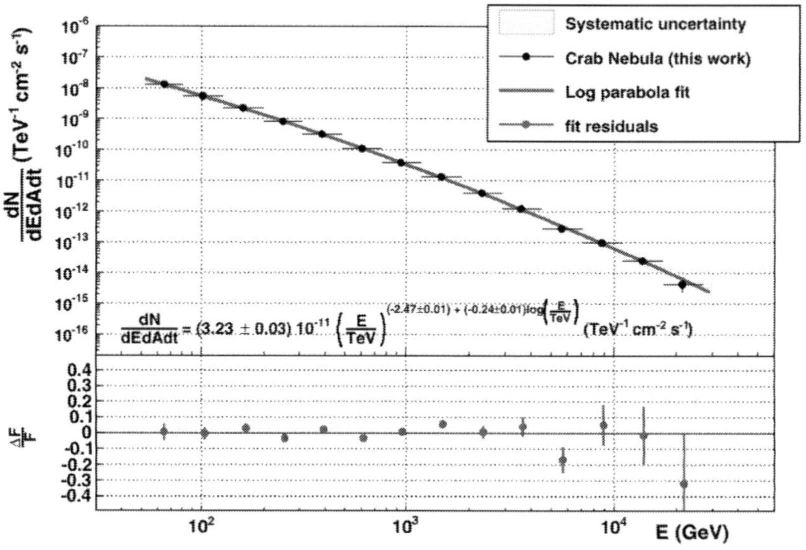

Figure 4. Differential energy spectrum of the Crab Nebula obtained with data recorded by the MAGIC stereoscopic system.

3C 58 is a PWN centered in PSR J0205+6449, one of the highest spin-down pulsars in the sky ($\dot{E} = 2.7 \times 10^{37}$ erg s^{-1}, or 2% of the Crab pulsar's $\dot{E}$). The distance and age of this PWN are controversial. The distance may range between 2 and 3.2 kpc. It may be very young and associated to the historical supernova SN1181 or as old as 7000 years. The X-ray thermal emission from the central objects seems to be too weak for a neutron star in this range of ages [24], so it has been speculated that it may not be a simple neutron star, but contains a more exotic sort of matter. Like Crab it shows a torus and a jet in X-rays. Fermi-LAT detected pulsed emission at E<4 GeV and steady emission up to ~100 GeV.

MAGIC discovered 3C 58 at VHE after a 81 h observation [9]. Its flux is 0.65% C.U., the weakest PWN detected at these energies. For existing models, only a short distance of 2 kpc or a high IR density can reproduce the data from radio to VHE (see Fig. 5). The IR density is probably unrealistically high, so the distance of 2 kpc is favored. The derived magnetic field by all the models fitting the γ-ray data is in any case smaller than 35 μG, very far from equipartition.

4. Dark matter

Discovering the nature of dark matter is one of the most exciting tasks of modern science. Among the targets suitable for dark matter searches, dwarf spheroidal galaxies are considered to be excellent candidates and, among them, Segue 1 stands out with mass-to-light ratio estimated to be of the order of 1000. We present the results of stereoscopic observations of the satellite galaxy Segue 1 with the MAGIC Telescopes, carried out between 2011 and 2013. With almost 160 hours of good-quality data, this is the deepest observational campaign on any dwarf galaxy performed so far in the very high energy range of the electromagnetic spectrum. Our results represent the most stringent constraints to the annihilation cross-section or decay lifetime obtained from observations of satellite galaxies, for masses above few hundredGeV. In particular, our strongest limit (95% confidence level), shown in Fig. 6 corresponds to a ~500 GeV dark matter particle annihilating into $\tau^+\tau^-$, and is of order $\langle\sigma_{\rm ann} v\rangle \simeq 1.2 \times 10^{-24}$ cm^3 s^{-1} – a factor ~40 above the $\langle\sigma_{\rm ann} v\rangle$ thermal value [7].

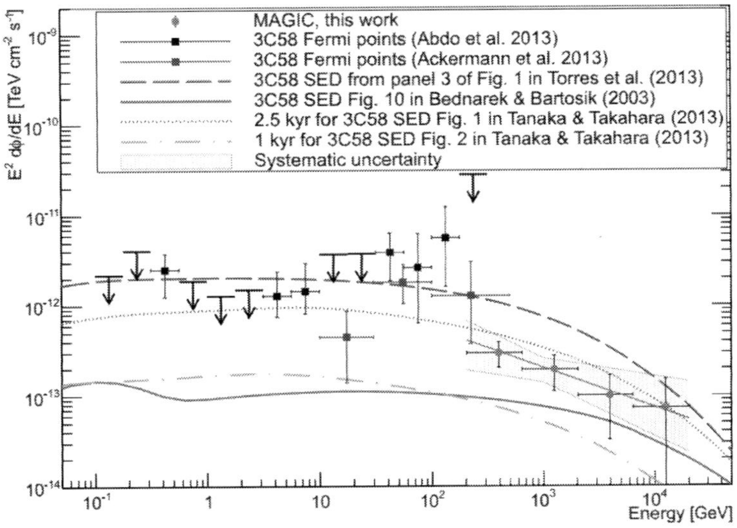

Figure 5. 3C 58 spectral energy distribution in the range between 0.1 GeV and 20 TeV. Red circles are the VHE points reported in this work. The best-fit function is drawn in red and the systematic uncertainty is represented by the yellow shaded area. Black squares and arrows are taken from the Fermi-LAT second pulsar catalog. Blue squares are taken from the Fermi-LAT high-energy catalog. The magenta, clear green dashed-dotted, dark green and blue dashed lines correspond to different models of the source. See [9] for details.

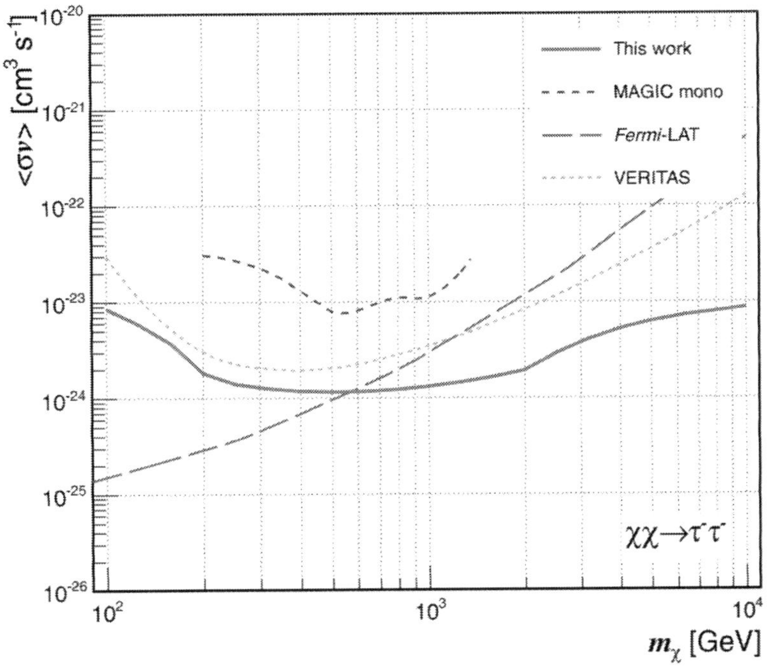

Figure 6. Upper limits on $\langle\sigma_{\rm ann} v\rangle$ for the final state channels $\tau^+\tau^-$ from the Segue 1 observations with MAGIC (solid lines), compared with the exclusion curves from *Fermi*-LAT (long-dashed lines), H.E.S.S. (dot-dashed line), VERITAS (dotted lines) and MAGIC-I (dashed lines).

5. Future prospects

These were just a few of the latest results of MAGIC. We plan to operate the telescopes during the next few years, for sure until the first CTA telescope start to operate. Since the current generation of IACTs have already studied most of the obvious candidates for VHE emission, future observations will probably go deeper into relatively few objects. In the last years we have setup a program of Key Observation Projects to identify and observe the most promising targets.

We would like to thank the Instituto de Astrofísica de Canarias for the excellent working conditions at the Observatorio del Roque de los Muchachos in La Palma. The support of the German BMBF and MPG, the Italian INFN, the Swiss National Fund SNF, and the ERDF funds under the Spanish MINECO is gratefully acknowledged. This work was also supported by the CPAN CSD2007-00042 and MultiDark CSD2009-00064 projects of the Spanish Consolider-Ingenio 2010 programme, by grant 127740 of the Academy of Finland, by the Croatian Science Foundation (HrZZ) Project 09/176 and the University of Rijeka Project 13.12.1.3.02, by the DFG Collaborative Research Centers SFB823/C4 and SFB876/C3, and by the Polish MNiSzW grant 745/N-HESS-MAGIC/2010/0.

References

[1] A. Abramowski et al., ApJ **746**, Issue 2, 151 (2012)
[2] V. A. Acciari et al., Science **325**, 444 (2009)
[3] F. Aharonian et al., Nature **482**, 507 (2012)
[4] J. Aleksić et al., ApJ **723**, L207 (2010)
[5] J. Aleksić et al., A&A **541**, 99 (2012)
[6] J. Aleksić et al., A&A **540**, A69 (2012)
[7] J. Aleksić et al., JCAP **02**, 008 (2014)
[8] J. Aleksić et al., A&A **565**, L12 (2014)
[9] J. Aleksić et al., A&A **567**, L8 (2014)
[10] J. Aleksić et al., A&A **563**, A91 (2014)
[11] J. Aleksić et al., A&A **567**, 15 (2014)
[12] J. Aleksić et al., submitted for publication
[13] J. Aleksić et al., submitted to JHEAp and arXiv:1406.6892
[14] J. Aleksić et al., submitted to Astroparticle Physics, arXiv:1409.6073
[15] J. Aleksić et al., submitted to Astroparticle Physics, arXiv:1409.5594
[16] E. Aliú et al., Science **322**, 1221 (2008)
[17] E. Aliú et al., Science **334**, 69 (2011)
[18] Cheung, T. et al., ApJ **782**, L14 (2014)
[19] K. Hirotani, ApJ **733**, L49 (2011)
[20] K. Hirotani, ApJ **766**, 98 (2013)
[21] J. Martin, D. F. Torres, N. Rea, MNRAS **427**, 415 (2012)
[22] M. Meyer, D. Horns, H. Zechlin, A&A **523**, A2 (2010)
[23] R. Mirzoyan et al., ATel 6349
[24] P. Slane et al., ApJ **571**, L45 (2002)

EPJ Web of Conferences **121**, 04003 (2016) DOI: 10.1051/epjconf/201612104003
RICAP-2014

TeV γ-ray astronomy with ground-based air-shower arrays

Miguel A. Mostafá[a]

Department of Physics, The Pennsylvania State University, University Park, 16802 PA, USA

Abstract. The TeV energy band is a very exciting window into the origin of high energy cosmic radiation, particle acceleration, and the annihilation of dark matter particles. Above a few hundred GeV, ground-based experiments of very large effective areas open a new domain to study extragalactic sources at intermediate redshifts, galaxy clusters, gamma ray bursts, AGN and their flaring states, extended sources and galactic diffuse emission, and to indirect searches for dark matter. In particular, ground arrays of particle detectors –that operate with high duty cycles and large fields of view– can extend to multi-TeV energies the measurements made with experiments on satellites, and complement the observations done with air Cherenkov telescopes on the ground. Key science goals of ground arrays include performing unbiased all-sky surveys, monitoring of transient events from known (and unknown) sources, and detecting extended regions of diffuse emission. In this paper, the status and most recent results from ARGO-YBJ, Tibet AS, HAWC, and LHAASO are presented.

1. Introduction

The measurement of the spatial and energy distribution of very high energy γ rays is an essential tool to identify and understand the cosmic accelerators of the highest energy particles. On the other hand, the time variability and energy spectra of the γ-ray emission can be used to study the environment of these accelerators and the acceleration mechanisms. Ground-based arrays of particle detectors play a key role in very high energy γ-ray astronomy due to their large active area ($>10^4$ m^2), high duty cycle (usually $>90\%$), and large field of view ($\sim$2 sr). Thus, ground arrays can extend the spectra of known γ-ray sources beyond the reach of space-based detectors, characterize spatially-extended sources that are difficult to be observed with telescopes, and trigger multi-wavelength observations of transient events without the need to steer.

The experimental techniques used for each ground array are described in Sect. 2. The most recent results and the current status of ARGO-YBJ, Tibet AS, HAWC, and LHAASO are presented in Sect. 3. The outlook of TeV γ-ray astronomy with ground-based arrays is summarized in Sect. 4.

[a]e-mail: `miguel@psu.edu`

2. Detection techniques

2.1 Resistive plate chambers

The Astrophysical Radiation with Ground-based Observatory at YangBaJing (ARGO-YBJ) experiment [1] is located at 4,300 m above sea level at the Yangbajing Cosmic Ray Observatory in Tibet, China (90.5° east, 30.1° north), and was designed for very-high energy γ-ray astronomy and cosmic-ray observations.

The ARGO-YBJ experiment is a full coverage Resistive Plate Chamber (RPC) array. The array consists of a single layer of 153 "*clusters*" (7.6 m ×5.7 m) of 12 RPCs each. To collect the induced charges when particles pass through the layer, RPCs are operated in streamer mode with a gas mixture of $C_2H_2F_4/Ar/i\text{-}C_4H_{10}$ = 75/15/10 and a voltage of 7.2 kV applied to the 2 mm gas gap. To provide a larger dynamic range for particle counting, each RPC is read out in 10 pads of 8 strips (62 cm ×7 cm) each. Signals from the strips are digitized by 18,360 time-to-digital converters (TDCs) to provide timing and position information of the particles. The signals from the strips of a given pad are sent to the same channel of the TDC. The arrival times of the particles are measured with a resolution of ~ 1.8 ns [1]. The central 130 clusters form the central carpet with an active area of ~92% of the 74 m × 78 m, and began recording data in July 2006. The surrounding 23 additional (partially instrumented) clusters (a.k.a. the "*guard ring*") were merged into the DAQ stream in November 2007. To extend the dynamical range up to PeV energies, the analog charge of each RPC is read out with two large pads (1.4 m × 1.2 m), each one covering half a chamber. These larger pads (also called "*big pads*") collect the total charge deposited by the particles hitting the detector [2]. The total area of the array is 110 m × 100 m. The installation of the RPCs started in 2001, and the central carpet was commissioned in June 2006. Data taking with the full detector started in November 2007, and ended in February 2013. ARGO-YBJ has collected over 5×10^{11} events in this period.

The ARGO-YBJ detector is operated by requiring the coincidence of at least 20 triggered pads ($N_{\rm pad}$) within 420 ns on the entire detector. The time (and location) of each triggered pad within a window of 2 μs are recorded. The trigger rate of ARGO-YBJ is 3.5 kHz with a dead time of 4% and an average duty-cycle higher than 86%. The shower core is estimated using a maximum likelihood method for fitting the lateral density distribution of the shower using a modified Nishimura-Kamata-Greisen (NKG) function. The core resolution (68% containment) is better than 10 m for events with $N_{\rm pad} > 100$. The arrival direction is reconstructed using a least-squares method assuming a conical shape of the shower front. The angular resolution for γ rays ranges from 1.7° to 0.2°, depending on the size of the shower quantified by the number of hit pads [3].

The effective area of the ARGO-YBJ array was estimated using Monte Carlo simulations [3], and found to be ~100 m^2 at 100 GeV and ~10,000 m^2 above 1 TeV for a zenith angle of 10°.

The number of triggered pads $N_{\rm pad}$ is the observable commonly used to infer the primary energy. But due to the large shower fluctuations and the dependance of the shower size with the position of the shower core with respect to the detector center, $N_{\rm pad}$ does not actually allow a precise determination of the primary energy on an event-by-event basis. Instead, the energy spectrum of a γ-ray source is determined from the comparison of the observed distribution of $N_{\rm pad}$ with simulations.

2.2 Scintillators

The Tibet Air-Shower (AS) experiment [4], also located at the Yangbajing Cosmic Ray Observatory in China, is an array of 761 scintillation counters placed on a lattice of 7.5 m spacing, and 28 scintillation counters on a lattice of 15 m spacing. The total area cover by the array is almost 37,000 m^2. Each counter has a plate of plastic scintillator, 0.5 m^2 in area and 3 cm thick, and a 2" PMT. Additionally, there are burst detectors (BDs) at the center of the array to measure the core region of the air showers. Each BD is composed of a plate of plastic scintillator (160 cm $\times$ 50 cm $\times$2 cm) and 4 photodiodes attached to each corner of the plate. They detect the high-energy particles in the core of the showers by converting them into electromagnetic cascade showers in a lead target. The total thickness of the lead plates is 3.5 cm (7 radiation lengths). The location of the burst can then be determined with a spatial resolution of 10 cm. A total of 100 BDs cover an area of 80 m^2 near the center of the air-shower array. The Tibet AS array has been taking data in this hybrid configuration since 1999.

The time and charge information from each triggered PMT are recorded to determine the direction and energy of the air shower. The energy of the showers is estimated from the shower size at ground, which is determined from the lateral charge distribution using an NKG-like function. The energy resolution is $\sim$20% around the knee. Arrival directions are determined using the timing information from the counters and assuming a conical shower front. The angular resolution, estimated with the Moon's shadow, is 0.2° for energies above 100 TeV. The trigger rate is $\sim$1.5 kHz, and the threshold energy is $\sim$3 TeV.

2.3 Water Cherenkov detectors

The High Altitude Water Cherenkov (HAWC) observatory [5] is located on a 200 m $\times$450 m plateau at 4,100 m above sea level inside the Pico de Orizaba national park in Mexico. The design of HAWC is based on the experience with the Milagro experiment [6]. Milagro was the first generation of γ-ray detectors using the water Cherenkov technology, and consisted of a big pond with a double layer of PMTs at the bottom. Instead, HAWC is an array of individual water Cherenkov detectors (WCDs) with a single deep layer of PMTs with a wider separation. This configuration provides a larger active area for the same photo-cathode area, and optical isolation for the WCDs.

The HAWC array will consist of 300 WCDs densely-spaced over an area of 22,000 m^2. The structural support of the WCDs is provided by corrugated steel tanks of 7.3 m diameter. There is a light-tight plastic bladder inside each tank, filled with almost 200,000 liters of ultra-pure water up to a height of 4.5 m. There are four upward-facing, baffled PMTs anchored to the bottom of each WCD. Three of them are 8" PMTs, previously used in Milagro, and located 6 ft from the center of the tank. The fourth, central tube is a high quantum-efficiency 10" PMT designed to increase the sensitivity of HAWC to air-showers with energies below 1 TeV.

The design of HAWC combines the water Cherenkov technique with a high altitude site. The higher altitude, the increased active area, and the optical isolation of the PMTs lead to a better angular and energy resolution, improved γ/hadron separation, and an order of magnitude increase in sensitivity with respect to Milagro. The full HAWC array will observed a 5σ signal from the Crab Nebula in a single 4-hr transit while maintaining this sensitivity over 2 sr. Charged cosmic rays are discriminated from γ rays by observing large energy depositions far (>40 m) from the shower core. The cosmic-ray rejection is estimated from air-shower simulations, and found to be better than 99% for energies above $\sim$3 TeV. As expected, the rejection efficiency improves with energy. The arrival direction of an air shower is determined from the timing information from the PMTs, and estimated (also from simulations) to be 0.1° for air showers with energies above 10 TeV.

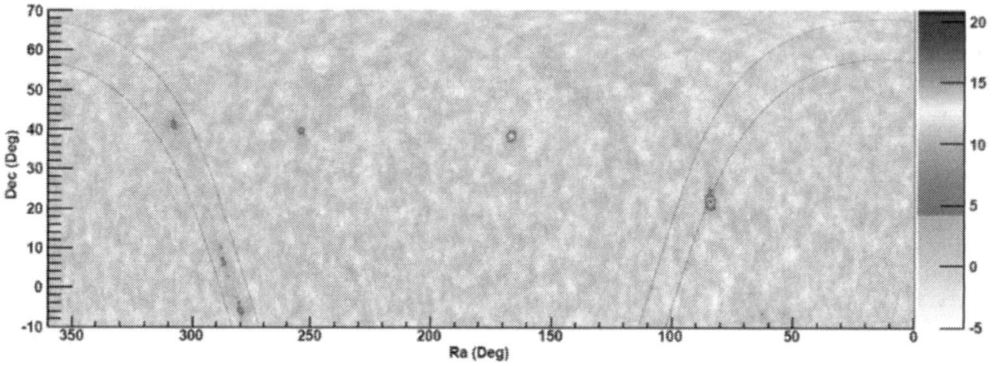

Figure 1. Figure 3 from Ref. [3].Survey of the Northern sky in γ-rays with the ARGO-YBJ experiment in the energy range from 10 GeV to 100 TeV. The six sources observed with a statistical significance greater than 5 s.d. are consistent in position with the Crab Nebula, the blazars Mrk 421 and Mrk 501, MGRO J2031+41 (and TeV J2032+4130), HESS J1841−055, and HESS J1908+063. The significances of the excesses, in terms of standard deviations, are shown by the color scale on the right side. The two dotted lines indicate the Galactic latitudes $b = \pm 5°$.

3. Status and results

3.1 ARGO-YBJ

The ARGO-YBJ detector carpet is connected to two DAQ systems that work independently. In the shower DAQ mode, the location and timing of each detected particle is recorded for each event. This mode allows the reconstruction of the arrival direction and of the lateral distribution for each shower. The threshold for this mode is $E_{\text{th}} \approx 300$ GeV. Instead, in the scaler DAQ mode the counting rate of each cluster is measured every 0.5 s. Thus, in this mode there is no information on either the charge distribution of the detected secondary particles on the ground, or the arrival direction of the shower. The threshold for the scaler mode is lower, $E_{\text{th}} \approx 1$ GeV.

The ARGO-YBJ collaboration performed a survey of the Northern sky ($-10° < \delta < 70°$) to search for steady VHE γ-ray emitters using more than five years of data. This corresponds to an integrated sensitivity ranging from 0.24 to $\sim$1 Crab units (depending on the declination). They have detected six sources with a statistical significance greater than 5 standard deviations (s.d.) [3]. These sources, shown in Fig. 1, are consistent in position with well known TeV γ-ray emitters: the Crab Nebula, the blazars Mrk 421 and Mrk 501, MGRO J2031+41 (and TeV J2032+4130), HESS J1841−055, and HESS J1908+063. They also reported evidence for possible TeV emission from five directions. (The locations of the excesses with significance greater than 4.0 s.d.) Two of these five excesses are not associated with any known counterpart and one of them is coincident in position with a high-mass X-ray binary.

The spectral energy distribution (SED) of the Crab Nebula (shown in Fig. 2) is fitted in the energy range from 0.1 TeV to 35 TeV. The spectral index of this fit is -2.62 ± 0.06 (stat). This SED derived from the ARGO-YBJ data is in agreement, within uncertainties, with the results obtained by other experiments, e.g. HEGRA, H.E.S.S., MAGIC, and Tibet AS [7–10]. The comparison is also shown in figure 2. (Only statistical uncertainties are shown.) The integral flux of this spectrum is 1.85×10^{-11} cm^{-2} s^{-1} (5.69×10^{-11} cm^{-2} s^{-1}) above 1 TeV (500 GeV). The systematic uncertainties on the flux for point sources are less than 30% [11].

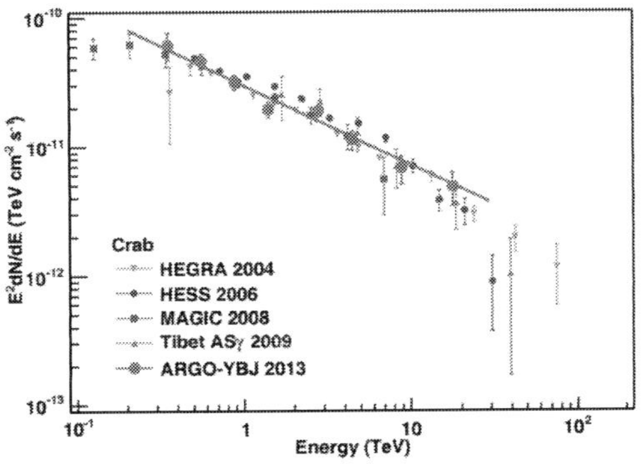

Figure 2. Figure 5 from Ref. [3]. Spectral energy distribution of the Crab Nebula measured with the ARGO-YBJ experiment, and comparison with the measurements from HEGRA, H.E.S.S., MAGIC, and Tibet AS [7–10]. The (red) line is the best fit to the ARGO-YBJ data using a power-law function. The spectral index is −2.6. (Only statistical uncertainties are shown.)

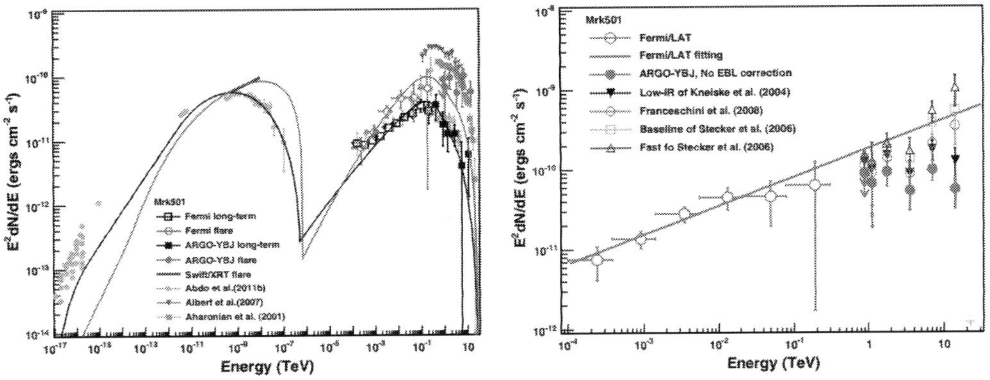

Figure 3. Left: spectral energy distribution of Mrk 501. (Fig. 6 from Ref. [13].) The solid black (red) line shows the best fit to the long-term (flaring) data using a one-zone SSC model. The best-fit parameters can be found in Ref. [13]. Right: spectra of Mrk 501 during the flaring period between October 2011 and April 2012. (Fig. 5 from Ref. [13].) The spectrum below 300 GeV is from the *Fermi*-LAT. Data from ARGO-YBJ has been corrected using four EBL models. Details can be found in Ref. [13].

The ARGO-YBJ collaboration also performed a 4-year long continuous monitoring of Mrk 501 between August 2008 and April 2012 [13]. Both the flux (0.3 Crab units above 1 TeV) and the spectral index (−2.6) measured with ARGO-YBJ between 450 GeV and 10 TeV are consistent with those obtained with other experiments. The fit to the long-term averaged SED of Mrk 501 over a very wide energy range (combining the radio-to-TeV campaign during the first 480 days of Fermi operation presented in Ref. [12] with the ARGO-YBJ data above 300 GeV) favors the γ-ray emission due to the inverse Compton scattering of the synchrotron radiation (synchrotron self-Compton, SSC) from the same population of relativistic electrons [13]. (The electron energy distribution is parameterized using a single power-law function with an exponential cutoff at its high energy end.) The best fit to the average broadband spectrum of Mrk 501 (from radio to TeV) in the framework of the one-zone SSC model is the black solid line in the left panel of Fig. 3.

A significant ($> 6\sigma$) excess in the event rate was detected by ARGO-YBJ in the direction of Mrk 501 between October 17 and November 22, 2011. This bright flaring episode represented a large ($\times 7$) increase in the γ-ray flux observed above 1 TeV with respect to the steady-state emission [13]. The hardness of the spectrum observed during the flare (shown in the right panel of Fig. 3) challenges the one-zone SSC model that is in good agreement with the long-term spectrum.

Previously, the ARGO-YBJ collaboration had also reported a long-term continuous monitoring of Mrk 421. The observation time reported in Ref. [14] (December, 2007 to February, 2010) covered both active and quiet phases. The γ-ray flux observed with the ARGO-YBJ experiment shows a good long-term correlation with the satellite-borne X-ray data. They also found that all the large X-ray flares have γ-ray counterparts in the ARGO-YBJ data. The temporal and spectral analysis of the Mrk 421 data also strongly favors the hypothesis that γ-rays and X-rays have a common origin as assumed in the SSC radiative mechanism.

The ARGO-YBJ collaboration had also reported (using data up to August, 2011 [11]) the observation of TeV γ-rays from a large part of the Cygnus region, which indicates a possible diffuse γ-ray emission. A statistical significant ($> 6\sigma$) excess was observed from the direction of MGRO J2031+41. The measured flux is consistent with the Milagro observation, but higher than the flux observed with MAGIC. No evidence of an emission above 3σ is found at the location of MGRO J2019+37 (inside the Cyg OB1 region) even considering extended sources. MGRO J2019+37, detected with the Milagro experiment at a median energy of 20 TeV [15], is the second most significant source in the Milagro data set after the Crab Nebula. The upper limits (90% c.l.) derived for MGRO J2019+37 do not confirm the spectrum determined with Milagro at energies below 5 TeV, and the exposure of ARGO-YBJ above 5 TeV is too low to reach a firm conclusion. Because the two observations were performed five years apart, variability in the VHE γ-ray flux of this source has been proposed as an explanation of this discrepancy. On the other hand, this would be in direct conflict with the identification of this source as a PWN. It is important to note that the upper limits from ARGO-YBJ are consistent with a new fit to the Milagro data assuming a power law with an exponential cutoff (~ 30 TeV) and a hard spectral index (-2.0) [16].

A more recent analysis of the ARGO-YBJ full data set (November, 2007 to January, 2013) [17] suggests that the location of the extended TeV γ-ray source ARGO J2031+4157 (or MGRO J2031+41) is consistent with the Cygnus Cocoon discovered by Fermi-LAT at GeV energies in the Cygnus superbubble [18]. The significance map around the Cygnus Cocoon as observed in the ARGO-YBJ full data set is shown in the left panel of Fig. 4. The most statistically significant ($> 6\sigma$) excess in this map corresponds to the position of ARGO J2031+4157, and has a similar angular size ($\sim 2^{\circ}$) than the Cygnus Cocoon as measured with Fermi-LAT. After subtracting the contribution of the overlapping TeV sources (details can be found in Ref. [17]), they find a differential energy spectrum with a spectral index -2.6 ± 0.3 in the energy range $0.2 - 10$ TeV. The spectral measurements with Fermi-LAT, ARGO-YBJ, and Milagro are shown in the right panel of Fig. 4. The flux determined with ARGO-YBJ appears consistent with the extrapolation of the spectrum of the Cygnus Cocoon as measured with Fermi-LAT in the $1 - 100$ GeV energy range. The similarity of the angular extensions and the consistency of the spectral measurements from 1 GeV to 10 TeV suggest that the TeV emission from ARGO J2031+4157 may be identified as the counterpart of the Cygnus Cocoon observed at GeV energies. The combined data from Fermi-LAT and ARGO-YBJ can be fitted with the power law function with a spectral index -2.16 ± 0.04 as shown by the dot-dashed line in the right panel of Fig. 4. (The Milagro data points refer to the source MGRO J2031+41 from Refs. [19–21].) The slope of the spectral measurements from 1 GeV to 10 TeV indicates that the particle acceleration inside a superbubble is similar to that in a Supernova Remnant. Assuming a hadronic model where the primary proton spectrum follows

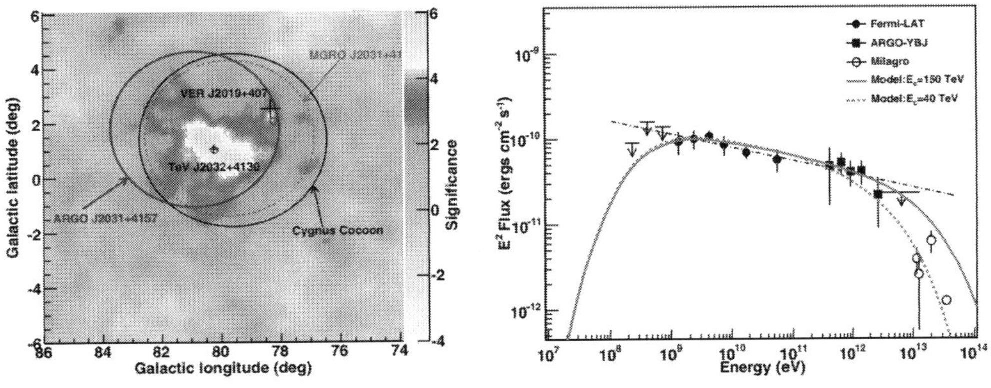

Figure 4. Left: significance map around PSR 2032+4127 in the Cygnus region as observed using ARGO-YBJ data. (Fig. 1 from Ref. [17].) The positions of other known TeV sources and of the Cygnus Cocoon are also marked. Right: spectral energy distribution of the Cygnus Cocoon region as measured with different experiments. (Fig. 3 from Ref. [17].) The measurements at 12, 20 and 35 TeV are reported by the Milagro collaboration for MGRO J2031+41 in Refs. [19–21]. (The lower data point at 12 TeV is the flux in Milagro after subtracting the contribution from TeV J2032+4130 [18].) The dot-dashed line corresponds to the best fit to the combined Fermi-LAT and ARGO-YBJ data assuming a simple power-law spectrum. The red lines represent hadronic models with two proton cutoff energies (at 150 TeV and 40 TeV).

a power law (with the same spectral index of γ-rays) and an exponential cutoff, the flux upper limits from Fermi-LAT and ARGO-YBJ suggest a change of slope below $\sim$1 GeV and a cutoff above $\sim$ 10 TeV.

Other important contributions from the ARGO-YBJ collaboration include the detection of diffuse γ rays from the Galactic plane (journal paper in preparation)[1], and the fluence upper limits to the emission from Gamma Ray Bursts (GRBs) in the interesting energy range between 1 and 100 GeV [23]. In a more recent analysis of the ARGO-YBJ data set up to January, 2013, no significant excess coincident with satellite detection was found for 206 GRBs inside the ARGO-YBJ field of view, the largest sample ever investigated with a ground-based detector.

3.2 Tibet AS

Using the Tibet-III array, that was completed in 1999, the Tibet AS collaboration observed γ-ray flares from Mrk 421 and found a correlation between TeV γ-ray and X-ray intensities [24]. An enlarged ($\sim 37,000$ m^2) Tibet-III array started operation in 2003. Using data collected with the Tibet-III array between 1999 and 2005, the measured differential energy spectrum of γ-rays from the Crab Nebula could be described with a spectral index of (-2.96 ± 0.14) in the energy range $2 - 40$ TeV [10]. This is consistent with data obtained with other experiments, and statistically consistent (within 2.6σ) with a previous Tibet result. No evidence was found for either time variability of flux intensity from the Crab Nebula, or pulsed γ-rays from the Crab pulsar at multi-TeV energies.

[1] The preliminary results for the TeV diffuse flux from the Cygnus region measured with ARGO-YBJ do not show the strong excess reported by the Milagro collaboration at a median energy of 15 TeV [22], and they hypothesize that the difference could be due to the Cygnus Cocoon which had not been discovered at the time of the Milagro measurement.

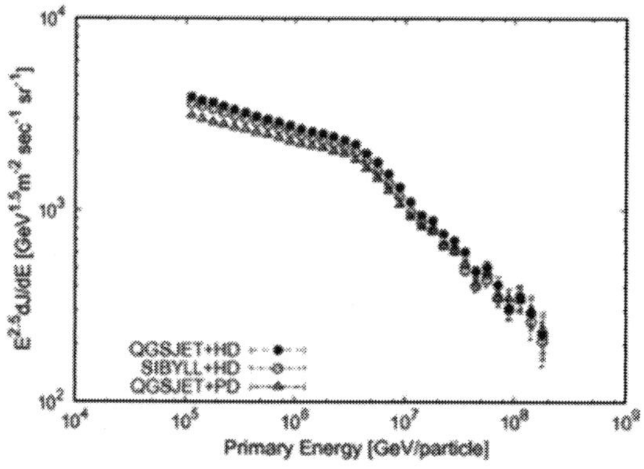

Figure 5. All particle spectrum measured with the Tibet-III array derived with three combinations of hadronic models and mass composition assumptions (QGSJET+HD, SIBYLL+HD and QGSJET+PD). Figure 11 from Ref. [4].

The Tibet collaboration has also measured the energy spectrum of primary cosmic rays in the region of the "knee" [25]. The chemical composition in this region is an important measurement to understand the cosmic-ray acceleration and propagation mechanisms. The primary cosmic-ray nuclei can be identified according to the energy and spatial distributions of the high-energy component of an air shower event detected with the burst detectors and emulsion chambers (at the center of the air shower array). The total energy of the individual γ-ray air showers is determined by the shower size, which is calculated by fitting the lateral particle density distribution to a given electromagnetic shower theory (a modified NKG structure function). Thus, the energy spectrum of each primary mass group can be determined simultaneously with the burst detectors, emulsion chambers and air shower array. The all-particle spectrum measured with the Tibet-III air-shower array between 10^{14} and 10^{17} eV is shown in Fig. 5. Two hadronic interaction models (QGSJET and SIBYLL) are used to estimate the detection efficiency to γ-rays, and two mass composition models are assumed (proton dominated, PD, and heavy dominated, HD) to estimate the contamination due to heavy primaries. The results shown in Fig. 5 were obtained using the combinations QGSJET+HD, SIBYLL+HD and QGSJET+PD. The systematic uncertainties below the knee are $\sim 20\%$ due to mass composition, and $\sim 10\%$ due to hadronic interaction models. The model dependence decreases with energy. The knee of the primary spectrum is observed at an energy of ~ 4 PeV. The measurements of proton and helium spectra with the Tibet experiment suggest the dominance of heavy nuclei around the knee. The sharp knee feature of the spectrum can be explained by either an additional component from a nearby source (or sources) dominated by heavy nuclei, or a nonlinear effect in the diffusive shock acceleration mechanism of cosmic rays that accelerates heavy primaries more efficiently than lighter elements.

3.3 HAWC

Once the array of 300 WCDs is completed in early 2015, HAWC will be the most sensitive γ-ray detector above 10 TeV, and will maintain unprecedented wide-field sensitivity to γ-rays above 100 GeV. The differential sensitivity of the HAWC array is shown in the left panel of Fig. 6 together with the sensitivity from other current instruments. The full HAWC array will be an order of magnitude more sensitive than Milagro. Thus, it observes a similar fraction of the sky ~100 times faster. The sensitivity of the HAWC array to differential E^{-2} spectra

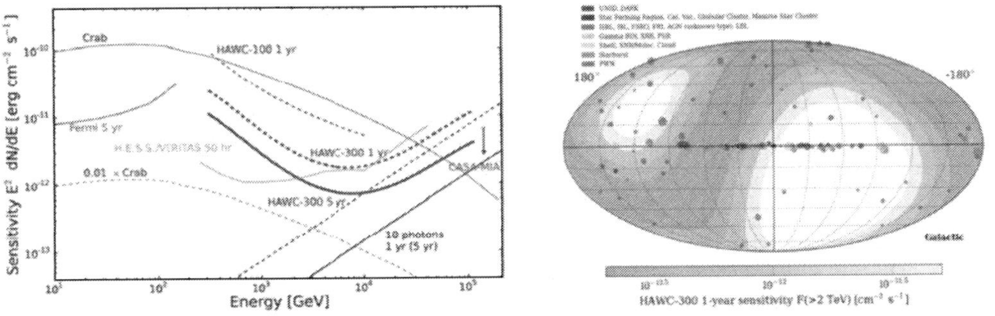

Figure 6. Left: differential sensitivity of HAWC compared to other instruments. (Fig. 4 from Ref. [26].) The sensitivity for the first 100 of the 300 HAWC WCDs is also shown for comparison. Right: sensitivity of the HAWC array to differential E^{-2} spectra along with known sources from the TeVCat TeV catalog [27] in Galactic coordinates. (Fig. 3 from Ref. [26].) The size of the symbols is representative of the flux of the source in the TeVCat catalog.

is shown in the right panel of Fig. 6 in Galactic coordinates along with sources from the TeVCat catalog [27]. The HAWC detector will be sensitive to pure power law spectra at a level of ~50 mCrab above 2 TeV over 5 sr (40%) of the sky in one year of exposure. If a source were to flare by a factor of N, the time required to see the source at fixed significance would be reduced by N^2.

With its wide field of view and sensitivity 10 times Milagro's, HAWC will survey the γ-ray sky measuring Galactic sources beyond 10 TeV, and daily monitoring half of the sky for flaring sources. The significance map of the full field of view of HAWC is shown in Fig. 7 using γ rays up to 60° in zenith angle recorded between August, 2013 and June, 2014. (This corresponds to 260 days of live-time.) The number of active WCDs in this period grew from 106 to 133, and this array is called HAWC-111. The median energy of this map is ~2 TeV. This corresponds to roughly one order of magnitude lower energy than Milagro's full sky map (for eight years of data). Maps are smoothed using a Gaussian function with a width proportional to the angular resolution (of the corresponding energy bin). The current angular resolution in the HAWC data is 2.3° for the lowest energy bin (mean energy ~ 300 GeV), and 0.35° for the highest energy bin (mean energy ~ 10 TeV). The energy scale is still preliminary, and this impacts the rate of events that satisfies the event selection. It is important to note that this does not appear to affect the efficiency of the gamma/hadron separation. It is clear that there is some source confusion in certain regions of the sky. The absolute pointing uncertainties and the instrument response are still under investigation.

Both the geometrical reconstruction and the gamma/hadron separation used in the current HAWC analyses have been optimized partly on the observed signal from the Crab Nebula. The significance map of the region around the Crab Nebula is shown in the left panel of Fig. 8. The Crab is observed at a significance level higher than 20σ in 260 days of HAWC-111. This is consistent with the sensitivity anticipated from simulations after correcting for the angular resolution observed in the HAWC data. This observation is also at the same level of significance of the Milagro detection of the Crab using all their data. It is important to mention that the Crab is observed significantly in all the energy bins. Mrk 421 is also observed in the HAWC-111 data at a significance level of 6.5σ (Fig. 7), and there is clear emission from multiple sources along the Galactic plane (right panel of Fig. 8). There is no significant excess observed from the Galactic Center which transits over the HAWC site at a zenith angle of ~48° at the edge of the HAWC sensitivity.

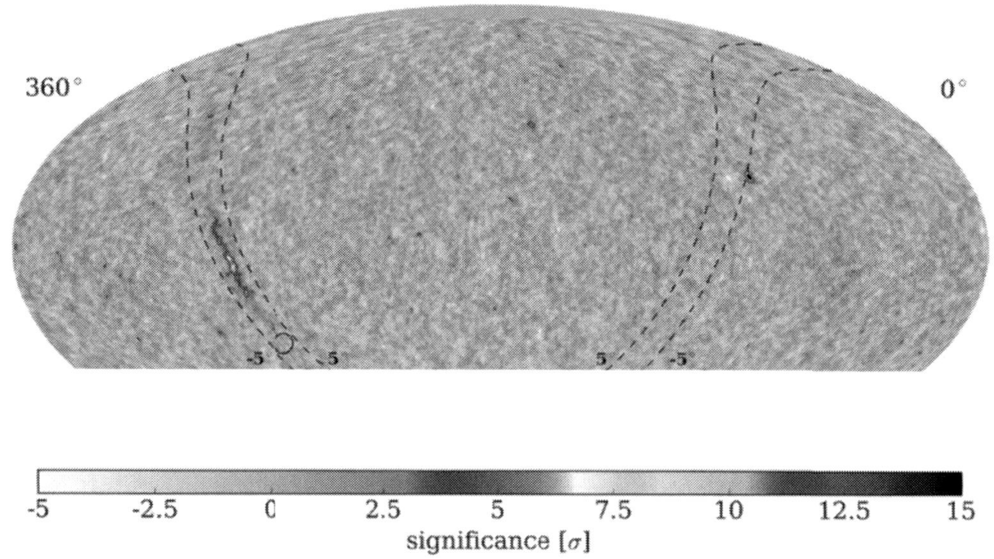

Figure 7. PRELIMINARY. Significance map of the HAWC field of view in equatorial coordinates using 260 days of live-time of HAWC-111. This map includes γ rays observed with zenith angles up to $60°$. The dashed lines indicate the Galactic latitudes $b = \pm 5°$, and the circle marks the location of the Galactic Center.

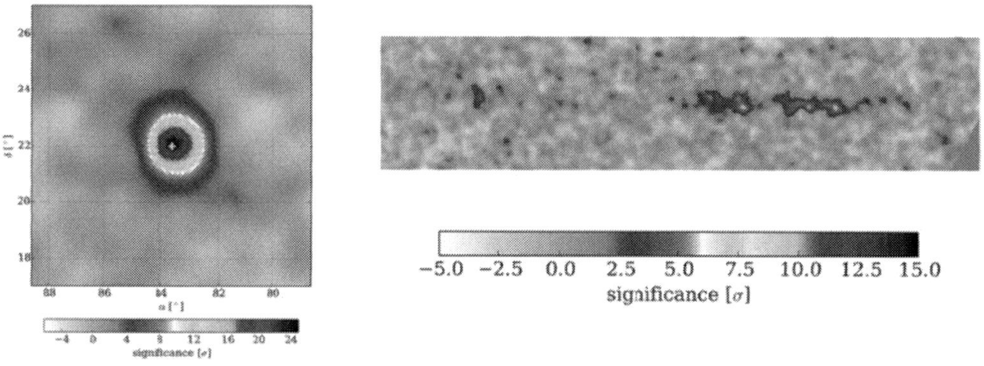

Figure 8. PRELIMINARY. Significance maps using 260 days of live-time of HAWC-111. Left: region around the Crab Nebula. Right: galactic plane $l = (0°, 90°)$, $b = (-10°, 10°)$.

HAWC data are recorded using two data acquisition (DAQ) systems. The main system is used to record the arrival time and the time over threshold (TOT) of the PMT pulses. This system provides the information necessary to reconstruct the shower core, arrival direction, and lateral distribution of the deposited charge at the ground level. A secondary system, called the scaler DAQ, operates in a simple counting mode, and is sensitive to transient events that produce a sudden increase (or decrease) in the counting rates of the PMTs. Both DAQ systems are sensitive to the high-energy emission from GRBs that extend to at least 30 GeV [28], and have energy thresholds that are low enough to observe events similar to GRB 090510 [29] and GRB 090902b [30] with the characteristics observed by Fermi LAT. While both systems can measure the spectra of GRBs, the sensitivity of the scaler and main DAQ systems to GRBs are complementary. The scaler DAQ reaches lower energies, while the main DAQ

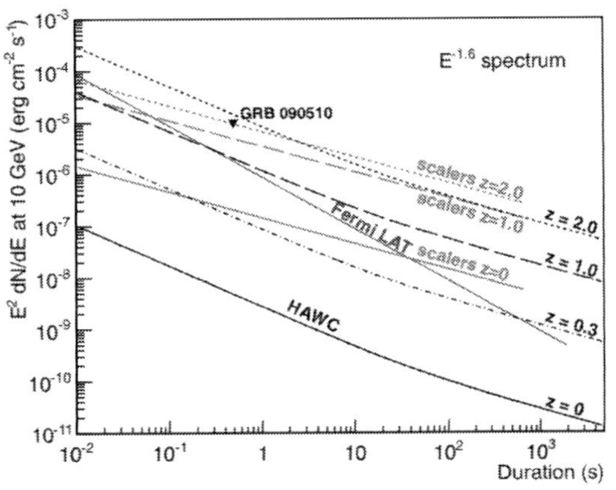

Figure 9. Sensitivity of the HAWC detector to GRBs using the scaler and main DAQ systems as a function of the burst duration. (Fig. 8 from Ref. [28].) The main DAQ uses a simple multiplicity trigger of 70 PMTs or more. The source position is set at a zenith angle of 20°, and the assumed spectrum is $E^{-1.6}$. The sensitivity lines represent the 5σ detection level for the scalers (red lines), and the 5σ discovery potential for the main DAQ (black lines). The blue line is the flux necessary for the observation of 1 γ-ray above 10 GeV with the Fermi-LAT, and the triangle represents GRB 090510 [29].

can reconstruct the energy and direction of individual γ-rays. The minimum flux required (for both DAQ systems) to make a 5σ detection of a transient source (at a zenith angle of 20°) as a function of the burst duration is shown in Fig. 9 assuming a source spectrum of $E^{-1.6}$. The effects of the extragalactic background light (EBL) (modeled as in Ref. [31]) are included for various values of redshift. (These are different for the scalers and main DAQ due to the energy response of each system.) The full HAWC array will be able to detect GRBs with characteristics similar to those already observed with the Fermi-LAT experiment. For example, if a burst such as GRB 090510 were to occur within a zenith angle of 26°, it would be detected as a 5σ (or greater) significant excess. Thus, HAWC will be particularly useful to study the high-energy emission of GRBs with hard non-Band power-law spectra.

The detection of γ-rays with energies above 100 GeV from GRBs is only possible for very bright, nearby bursts because these photons are attenuated in their interactions with the EBL. GRB 130427A is the most energetic burst detected so far with a redshift less than 0.5. Its prompt phase triggered the Fermi-GBM at $T_0 = $ 07:47:06.42 UTC. The GeV emission was delayed, and between $T_0 + 11.5$ s and $T_0 + 33$ s an additional power-law component was required in the spectral fit. The Fermi-LAT also detected an unprecedented long (about 20 hours, interrupted by occultations) high-energy (above 100 MeV) emission [32]. This burst also featured the most energetic photon ever detected from a GRB extending the observed energy spectrum to at least 95.3 GeV (128 GeV when corrected for redshift), well within the range accessible to HAWC. At T_0, only the scaler DAQ was taking data with 29 operational WCDs (called HAWC-30) and 115 PMTs. At that time, the local zenith angle of GRB 130427A was 57° and setting. The high zenith angle reduces the effective area by 2 orders of magnitude with respect to 20° for example. (Furthermore, the energy threshold is also significantly higher towards the horizon.) Then, the effective area to bursts overhead for HAWC-30 is also an order of magnitude worse than the full array, HAWC-300. Thus, no statistically significant excess of counts was found and upper limits have been placed [33]. The derived flux upper limits between 20 GeV and 1 TeV using the HAWC effective area for GRB 130427A in the 11.5 s to 33 s search window are shown in figure 10. The sensitivity of

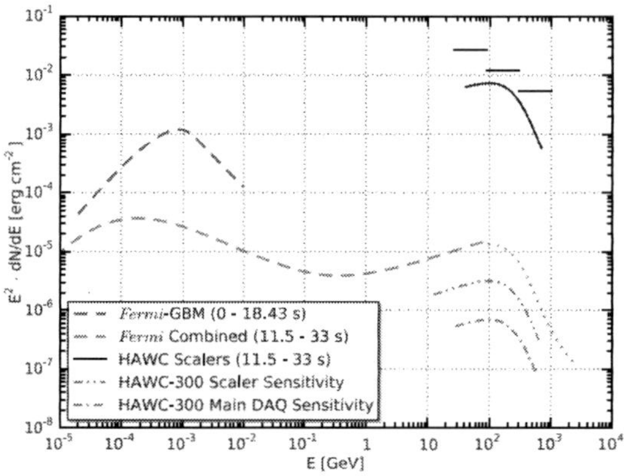

Figure 10. Flux upper limits for GRB 130427A in the $11.5 - 33$ s search window using the effective area of the HAWC array. (Fig. 4 from Ref. [33].) The blue (red) dashed line shows the spectral fit to the prompt GBM (combined GBM and LAT) data. The dotted red line shows the effect of EBL absorption. The black lines show the upper limits for the scaler DAQ assuming the fit to the combined Fermi data and EBL absorption for the integral limit, and assuming a spectrum of E^{-2} for the differential limits. The green dashed-dotted lines show the expected sensitivity to an overhead GRB for the two DAQ systems of the full HAWC array. (The main DAQ is more sensitive, while the scalers reach lower energies.)

the two DAQ systems of the full HAWC-300 array for a GRB within 20° from the zenith is also shown in the figure (green dashed-dotted lines). It is clear that a similar GRB close to zenith ($\theta < 26°$) would be easily detected by the full HAWC detector. Besides the unusually bright bursts like GRB 130427A, HAWC will be sensitive to other GRBs with a rate as high as one to two per year [34]. Observations with HAWC beyond the Fermi-LAT energy range will constrain the maximum energy produced by the GRB, and/or establish the existence of a cut-off, probing the γ-ray absorption either in the GRB (probing the bulk Lorentz factor), or along the line of sight due to the extragalactic background light.

While the main design goal of the HAWC observatory is to study γ-ray sources, the array is also sensitive to charged cosmic rays that are a large background to those γ-ray searches. But the large number of cosmic rays can also be used to make precise measurements of small deviations from isotropy in the distribution of arrival directions at TeV energies. This distribution was analyzed using the first $\sim$50 billion events recorded with HAWC-95 (95 working WCDs) and HAWC-111 between June, 2013 and February, 2014 [35]. This study revealed an anisotropy at the level of 10^{-4} on small angular scales ($\sim$10°). In a first step, a relative intensity map is obtained by comparing the data to the detector response to an isotropic cosmic-ray flux. Then, the angular power spectrum is determined from the relative intensity map. The map of pre-trial significances is shown in figure 11 after removing the sum of the monopole ($l = 0$), dipole ($l = 1$), quadrupole ($l = 2$), and octupole ($l = 3$) terms. (The lower order multipoles must be removed from the sky map to study the small angular scales.) A smoothing procedure is applied (with a radius of 10°) to improve the sensitivity to features on angular scales larger than the pixel size. After the elimination of the large-scale structure, the residual HAWC cosmic-ray sky map shows three regions of significant excess flux. The most significant (17σ) excess is found at $(\alpha, \delta) \approx (60°, -7°)$, and corresponds to the so-called Region A (Region 1) in the Milagro (ARGO-YBJ) sky map. Even though a detailed study of the energy dependence of the flux in the excess regions is not possible with HAWC at this point, the analysis presented in Ref. [35], using a simple energy proxy based on the number of PMTs in the event and the zenith angle of the cosmic ray, is sufficient to confirm that the energy spectrum of Region A is harder than the isotropic cosmic-ray spectrum. The relative

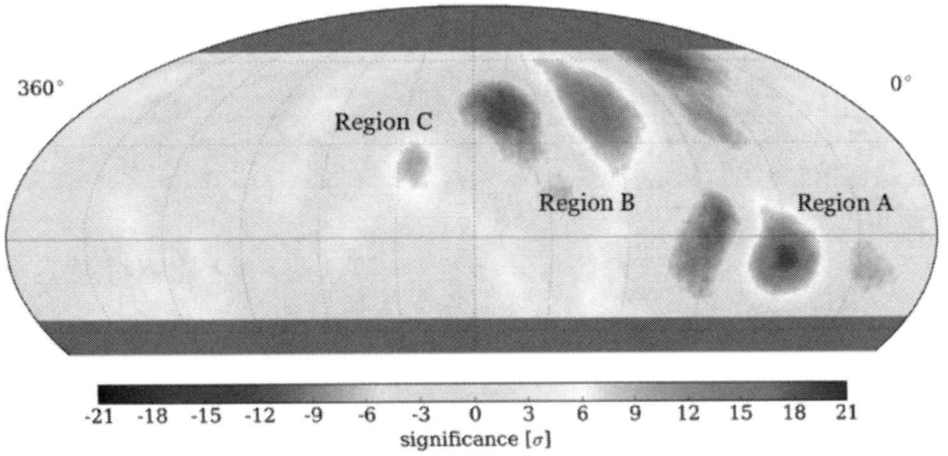

Figure 11. Map of pre-trial significances after subtracting the dipole, quadrupole, and octupole terms from the total cosmic-ray flux. (Fig. 5 from Ref. [35].) There is a 10° smoothing applied to this map.

intensity of the excess observed with HAWC agrees (within uncertainties) near the location of the centroid of Region A reported by Milagro in the two bins of highest median energy (14 and 19 TeV). The second most significant (11.2σ) excess around $\alpha = 120°$, identified as Region B (Region 2) in the Milagro (ARGO-YBJ) map, is the most extended essentially spanning almost the entire declination range visible to HAWC. The third excess (8.2σ) region, Region C in Fig. 11, is not significant in the Milagro data, but coincides with Region 4 in the ARGO-YBJ map. The median cosmic-ray energy at this declination is 2 TeV. With HAWC, this region is detected at almost twice the relative intensity observed by ARGO-YBJ, making it more significant in HAWC despite the considerably shorter observation period. A small excess in the HAWC map near Region 3 in the ARGO-YBJ map is currently not statistically significant (3.7σ). The study of the morphology and relative intensity of all excess regions as a function of energy will be continued with more data in the near future. The full HAWC-300 array will also have an improved energy resolution which will allow for a detailed study of these regions as a function of energy.

Data from the HAWC Observatory will also contribute to tests of Lorentz invariance with transient sources [36], searches for exotic signals such as massive relic particles (SUSY Q-Balls [37], WIMP dark matter [38]), and primordial black holes [39].

3.4 LHAASO

The Large High Altitude Air Shower Observatory (LHAASO) is a proposed project designed to understand the origin of high energy cosmic rays by studying gamma ray sources above 30 TeV [40]. The proposed experiment will cover an area of 1 km^2 using five types of detection techniques to measure the arrival direction of air showers, the total number of secondary particles, the muon content, the Cherenkov image, and the high energy gamma rays near the shower core. The proposed layout of LHAASO is shown in Fig. 12.

3.4.1 LHAASO-KM2A

The main array of LHAASO will cover an effective area of 1 km^2 (KM2A). It will consist of 6,300 scintillator detectors (of 1 m^2 each) arranged in a triangular grid with a 15 m spacing, and 1,220 muon detectors covering an active area of 40,000 m^2 with a 30 m spacing. Each

Figure 12. Proposed layout for LHAASO including a 1 km^2 main array with 6,300 scintillator and 1,220 muon detectors, a central array with 24 Cherenkov telescopes and 542 burst detectors, and a large water Cherenkov detector (90,000 m^2).

electromagnetic particle detector of KM2A will consist of 4×4 tiles (25 cm $\times$ 25 cm $\times$ 2 cm each) of plastic scintillators. Wavelength-shifting fibers guide the scintillating photons from each group of 16 tiles to a single photomultiplier tube per detector. Each muon detector of KM2A is a water Cherenkov detector in a concrete tank (3.6 m in radius and 1.2 m in height) covered by 2.5 m of dirt. Each tank will be equipped with one 8" PMT facing downward into a highly reflecting liner bag filled with pure water.

3.4.2 LHAASO-WCDA

The design of LHAASO also includes a Water Cherenkov Detector Array (WCDA) covering a total active area of 90,000 m^2. The WCDA consists of 4 water ponds, each one of them roughly the size of HAWC (150 m $\times$ 150 m), with a depth of ~ 4.5 m. Each pond will be subdivided into 900 cells (30×30 cells, 5 m $\times$ 5 m each), optically isolated with black plastic curtains that will prevent light contamination from neighboring cells. There will be an upward-looking 8" PMT at the bottom of each cell to collect the Cherenkov light produced by charged particles in the water.

3.4.3 LHAASO-WFCTA

Two more detection techniques are proposed for the central array of LHAASO. The additional components are the Wide Field of view Cherenkov Telescope Array (WFCTA) and the high threshold Shower Core Detector Array (SCDA). The 24 wide FOV telescopes of the WFCTA will be designed and arranged to measure the UV light from the air showers observed above the ground array from a distance of ~ 5 km. There will be telescopes covering elevations from 3° to 59° and others covering elevations from 3° to 31°. Finally, the 24 burst detectors of the SCDA covering an effective area of 5,000 m^2 will contribute to the identification of the cosmic-ray species.

4. Outlook

The current generation of ground arrays have established very high energy γ-ray astrophysics as an exciting new window into the electromagnetic spectrum. The large field of view and nearly 100% duty cycle of ARGO-YBJ, Tibet AS, and HAWC allow surveys of large fractions of the sky and observations in the prompt phase of flaring and transient events. The increased sensitivity of these instruments at energies beyond those covered by satellites have contributed to make the number of identified TeV sources over one hundred. It is not unreasonable to expect exciting new measurements and discoveries in the near future including TeV sources with no counterparts at other wavelengths, the study of the acceleration and propagation mechanisms in a wide range of accelerator environments, the identification of dark matter, constrains of the extragalactic background radiation, precise measurements of the diffuse background emission, etc.

Each RPC in ARGO-YBJ was instrumented with two large electrodes facing the other side of the gas volume and an analog charge readout over the entire central carpet in December 2009. The detector in this configuration can measure the particle density in the core region from tens to several thousands of particles per m^2. The analog readout of the RPCs allows to extend the energy range of the analysis of the existing ARGO-YBJ data to the 1–10 PeV region, and provides highly detailed images of the charged component at the core of those air showers.

A large underground water Cherenkov muon-detector array has been added to the Tibet AS experiment. This array has 80 detectors covering an area of $\sim$4,500 m^2, and started taking data in February 2014. Data from this experiment will contribute to the measurement of the differential energy spectra of primary cosmic-ray components in the region of the knee.

The full HAWC array with 300 detectors will start operations in early 2015. With an order of magnitude higher sensitivity with respect to the previous generation of water Cherenkov detectors, HAWC data is expected to significantly contribute to the measurement of extended and large-scale structures (e.g., galactic diffuse emission or the "Fermi bubbles"), the discovery of previously unidentified TeV sources, the detection of time variable sources of γ rays (such as AGN, GRBs, pulsars, and/or binary systems), and the detailed study of the cosmic-ray anisotropy.

Prototypes for the five detection techniques proposed for LHAASO are being tested mostly at the ARGO-YBJ site. Data from LHAASO will expand the searches for very high energy γ-ray sources beyond the sensitivity of current experiments, and will significantly contribute to our understanding of the origin of cosmic rays, and of the energy spectra and composition of cosmic rays over a wide range of energies.

References

[1] G. Aielli, et al. [ARGO-YBJ Collaboration], Nucl. Instrum. Meth. A **608**, 246 (2009)
[2] G. Aielli, et al. [ARGO-YBJ Collaboration], Nucl. Instrum. Meth. A **661**, S56 (2012)
[3] B. Bartoli et al. [ARGO-YBJ Collaboration], ApJ **779**, 27 (2013)
[4] M. Amenomori, et al. [Tibet AS Collaboration], Advances in Space Research **47**, 629 (2011)
[5] M. Mostafá for the HAWC Collaboration, Proceedings of the 33rd ICRC, Rio de Janeiro, Brazil (2013)
[6] R. Atkins, et al. [Milagro Collaboration], Phys. Rev. Lett. **95**, 251103 (2005)
[7] F. Aharonian, et al. [HEGRA Collaboration], ApJ **614**, 897 (2004)
[8] F. Aharonian, et al. [H.E.S.S. Collaboration], A&A **457**, 899 (2006)
[9] J. Albert, et al. [MAGIC Collaboration], ApJ **674**, 1037 (2008)
[10] M. Amenomori, et al. [Tibet AS Collaboration], ApJ **692**, 61 (2009)
[11] B. Bartoli, et al. [ARGO-YBJ Collaboration], ApJ **745**, L22 (2012)
[12] A. A. Abdo, et al., ApJ **727**, 129 (2011)
[13] B. Bartoli, et al. [ARGO-YBJ Collaboration], ApJ **758**, 2 (2012)
[14] B. Bartoli, et al. [ARGO-YBJ Collaboration], ApJ **734**, 110 (2011)
[15] A. A. Abdo, et al. [Milagro Collaboration], ApJ **664**, L91 (2007)
[16] A. A. Abdo, et al. [Milagro Collaboration], ApJ **753**, 159 (2012)
[17] B. Bartoli, et al. [ARGO-YBJ Collaboration], ApJ **790**, 152 (2014)
[18] M. Ackermann, et al. [Fermi-LAT Collaboration], Science **334**, 1103 (2011)
[19] A. A. Abdo, et al. [Milagro Collaboration], ApJ **658**, L33 (2007)

[20] A. A. Abdo, et al. [Milagro Collaboration], ApJ **664**, L91 (2007)
[21] A. A. Abdo, et al. [Milagro Collaboration], ApJ **700**, L127 (2009)
[22] A. A. Abdo, et al. [Milagro Collaboration], ApJ **688**, 1078 (2008)
[23] G. Aielli, et al. [ARGO-YBJ Collaboration], ApJ **699**, 1281 (2009)
[24] M. Amenomori, et al. [Tibet AS Collaboration], ApJ **598**, 242 (2003)
[25] M. Amenomori, et al. [Tibet AS Collaboration], Astrophys. Space Sci. Trans. **7**, 15 (2011)
[26] A. U. Abeysekara, et al. [HAWC Collaboration], Astroparticle Physics **50-52**, 26 (2013)
[27] S. Wakely and D. Horan, `http://tevcat.uchicago.edu/`.
[28] A. U. Abeysekara, et al. [HAWC Collaboration], Astroparticle Physics **35**, 641 (2012)
[29] M. Ackermann, et al. [Fermi-LAT Collaboration], ApJ **716**, 1178 (2010)
[30] A. A. Abdo, et al., ApJ **706**, L138 (2009)
[31] R. C. Gilmore, et al., MNRAS **399** (2009) 1694-1708
[32] M. Ackermann, et al. [Fermi-LAT Collaboration], Science **343**, 42 (2014)
[33] A. U. Abeysekara, et al. [HAWC Collaboration], accepted for publication in ApJ (2015)
[34] I. Taboada, and R. C. Gilmore, Nuclear Instruments and Methods in Physics Research A **742**, 276 (2014)
[35] A. U. Abeysekara, et al. [HAWC Collaboration], ApJ **796**, 108 (2014)
[36] L. Nellen for the HAWC Collaboration, Proceedings of the 33rd ICRC, Rio de Janeiro, Brazil (2013)
[37] P. Karn for the HAWC Collaboration, Proceedings of the 33rd ICRC, Rio de Janeiro, Brazil (2013)
[38] A. U. Abeysekara, et al. [HAWC Collaboration], Phys. Rev. D **90**, 122002 (2014)
[39] A. A. Abdo, et al. [Milagro and HAWC Collaborations], Astroparticle Physics **64**, 4 (2015)
[40] Z. Cao and M. Zha for the LHAASO Collaboration, Proceedings of the 33rd ICRC, Rio de Janeiro, Brazil (2013)

EPJ Web of Conferences **121**, 04004 (2016) DOI: 10.1051/epjconf/201612104004
RICAP-2014

The new gamma-ray observatory: CTA

John Carr for the CTA Collaboration
Aix Marseille Université, CNRS/IN2P3, CPPM UMR 7346, 13288 Marseille, France

Abstract. CTA is the next generation gamma-ray observatory and will have a factor 10 better sensitivity compared to existing facilities, as well as many other superior parameters. Aspects of array layout, performance and sites are presented. The broad range of forefront science which will be studied is described.

1. Introduction

The Cherenkov Telescope Array (CTA) is global effort, with > 1000 participating scientists and engineers from 28 countries, to deliver an observatory for photon astronomy at the highest energies (~20 GeV to ~300 TeV). CTA will dramatically improve on all aspects of performance with respect to the highly successful current generation of Cherenkov telescope arrays (see e.g. [1]). CTA will survey the sky ~200 times faster than any current instrument in the energy range above 100 GeV, have the best energy flux sensitivity and angular resolution of any instrument operating above the X-ray band and will outperform the Fermi space telescope by 4–5 orders of magnitude in sensitivity for sub-minute-timescale transient phenomena such as gamma-ray bursts. The project combines guaranteed scientific return, in the form of precision very-high-energy astrophysics, with considerable potential for major particle astrophysics discoveries. For example, CTA will conduct searches for dark matter annihilation, which complement direct detection and production (LHC) experiments, and test for Lorentz invariance violation at the Planck scale. The astrophysics of CTA includes new probes of extreme astrophysical environments, such as the neighbourhoods of black holes and neutron stars, relativistic jets, outflows and explosions, and a census of particle acceleration in the universe and the impact of cosmic ray feedback on all scales.

High energy gamma-rays offer the only sensitive probe of the ultra-relativistic protons and nuclei which may dominate the pressure in environments from supernova remnant shells to cluster-scale active galactic nucleus blown bubbles. The breadth and depth of the CTA science case has made it a key project in the field of particle astrophysics. CTA will be the first particle astrophysics instrument to be operated as an open observatory, serving a wide scientific community. The science of CTA is described in detail in a recent special issue of the journal Astroparticle Physics [2].

The technologies planned for CTA represent an evolution of those employed in the existing imaging atmospheric Cherenkov telescopes H.E.S.S [3], MAGIC [4] and VERITAS [5]. The key hardware elements are telescope structures, cameras, mirrors, and data-processing infrastructure. The concept for telescope structure is similar to that of MAGIC in the case of the LST, and the MST structure combines elements of the VERITAS and HESS designs. Camera technologies are considerably improved with respect to current instruments, with multiple viable technology options in some cases. Mirror technologies are likewise significantly improved.

The CTA Observatory will be located at two sites, one in the southern and one in the northern hemisphere, at each site combining Cherenkov telescopes of different sizes and designs to cost-effectively cover the very wide energy band. The sites will complement each other, providing full-sky coverage for Galactic and extragalactic sources. While the southern site, with privileged Milky Way views, will have energy coverage up to 300 TeV, allowing exploration of Galactic sources up to the highest energies, the northern site will have more limited high-energy capabilities.

At least three telescope types are required to cover the full CTA energy range in a cost-effective way: small, medium and large sized telescopes. An extension for the southern observatory is envisaged based on a fourth, medium-sized Schwarzschild Couder telescope type. The expected performance of CTA performance has been established using very detailed simulations, which have been verified by comparison to existing instruments. The required performance can be achieved at all CTA candidate sites investigated and for a range of implementation options, being rather insensitive to details of the telescope layout. Science performance has been demonstrated through a number of detailed case studies.

2. CTA arrays

To achieve full sky coverage, the CTA Observatory will be located on two sites, one in the southern and one in the northern hemisphere. Each telescope consists of a primary reflector and in some cases a secondary reflector, a camera and a drive system. Telescope control, array-level event-selection/triggering and data processing will be handled by centralised systems. The two sites will have different number and layout of telescopes. To cover a wide energy range in a cost-effective way, the arrays will contain several different types of telescope:

- The low energy instrumentation, efficiently measuring gamma-rays with energies below a few hundred GeV, will consist of 23 metre diameter, Large-Sized Telescopes (LSTs) with a field of view (FoV) diameter of at least 4.4°.
- The medium energy range, from around 100 GeV to more than 10 TeV, will be covered by 12 metre diameter Medium-Sized Telescopes (MSTs) with a FoV of at least 7°. The Schwarzschild-Couder Telescope (SCT) is a second type of medium-sized telescope, using a two-mirror design, which will be used to extend the area covered by medium-sized telescopes beyond the baseline configuration to more than 1 km^2 and improve angular resolution.
- The high energy instrumentation, mainly operating above 10 TeV, will consist of a large number of 4 metre diameter, Small-Sized Telescopes (SSTs) with a FoV of at least 8°.

The baseline layout of the various telescopes on the two sites is shown in Fig. 1, and schematics of the main dimensions of the different telescopes are shown in Fig. 2. Figure 1 includes the SCTs, extending the baseline MST array. Depending on the deployment timescales of the baseline MST telescopes, and of the additional SCTs, a "hybrid" configuration may be employed where MSTs and SCTs are interleaved.

3. Performance

The contribution of the different types of telescopes to the overall sensitivity for on-axis point sources is demonstrated for an example array layout in Fig. 3, for an observation time of 50 hours. The cross-over in sensitivity between the LST and MST components occurs at about 250 GeV and that between MSTs and SSTs at about 4 TeV. At the LST/MST as well as the MST/SST transition, the combined sensitivity is almost a factor of two better than

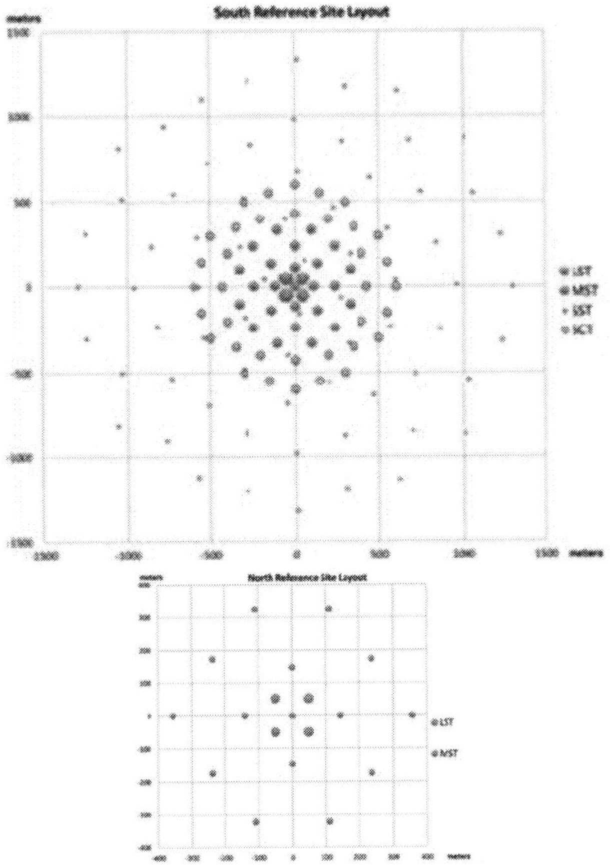

Figure 1. Example layouts of telescopes on the two sites in the southern hemisphere (top) and northern hemisphere (bottom) with different scales. The symbols indicate the locations of the telescopes as: red LST; blue MST; open purple SCT and green for SST. The southern site contains more telescopes and covers a larger area.

that of the individual components. However, close to 1 TeV, the onset of low-quality (close to threshold) SST data, with large effective area but poor gamma-hadron rejection can reduce the combined sensitivity in our simplest analysis method to close to the sensitivity of MST data alone. With the smaller SSTs in current designs, the MST/SST cross-over point is expected to occur at higher energies, but still below 10 TeV.

The expected CTA sensitivity is compared to that of other observatories in Fig. 5.

4. Candidate sites

Following a call for site proposals, a total of nine proposal documents were received with sites in several countries in the world. The map in Fig. 5 indicates the locations of all sites considered.

Among the candidate sites proposed, the one at Sutherland in South Africa was withdrawn by the proposers. The sites in India and China were considered low priority for study and have not been evaluated in detail.

The site in Spain is located 5 km from the Teide Observatory in Tenerife. The site in Mexico is 5 km from the San Pedro Martir Observatory in Baja California Mexico. The two

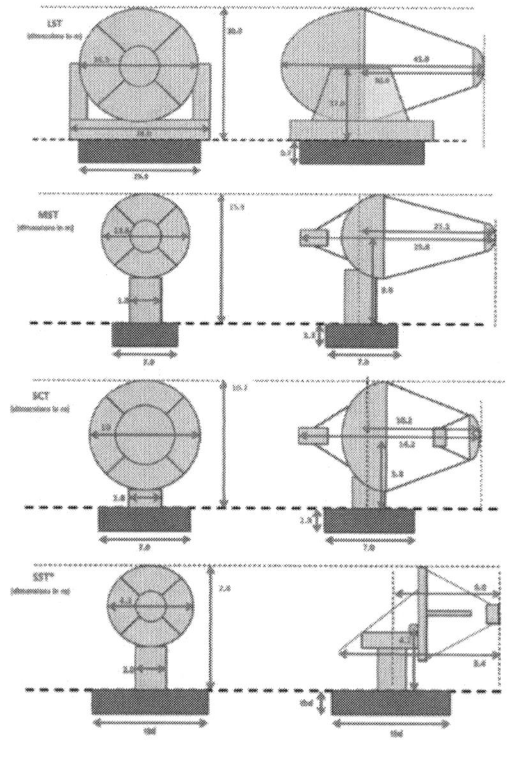

Figure 2. Indicative sketches of the four sizes of telescopes.

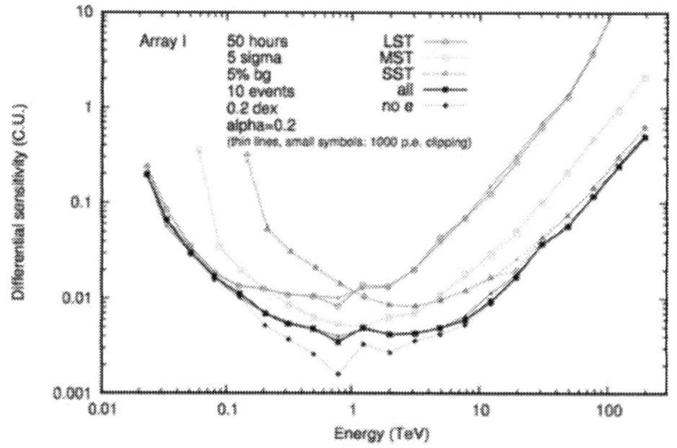

Figure 3. Point source sensitivity of for an example array layout (solid black line, filled squares) compared to that of its component sub-systems: 3 LSTs (red, open circles), 18 MSTs (green, open squares), and 56 SSTs (blue, open triangles). The importance of the background of cosmic ray electrons for the combined sensitivity is demonstrated by the dashed line with diamonds, where this background is ignored.

sites in the USA are located in Arizona Yavapai 160km west and Meteor Crater 70 km east of Flagstaff where the Lowell Observatory is located. In Argentina the Leoncito site is located within the CASLEO reserve in San Juan province and the San Antonio site is located near the city of San Antonio de los Cobres in Salta province. In Namibia, one site is adjacent to the location of the HESS Gamma-ray Telescope array, roughly 100km south of the capital

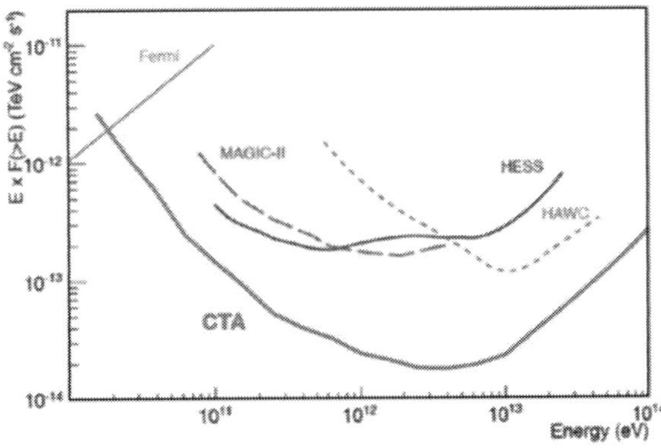

Figure 4. Integral sensitivity for CTA from MC simulations, together with the sensitivities in comparable conditions (50 h for IACTs, 1 year for Fermi-LAT and HAWC) for some gamma-ray observatories.

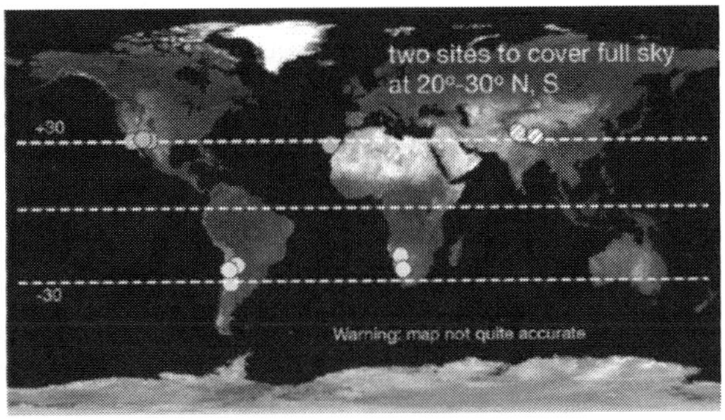

Figure 5. Approximate locations of all site candidates proposed.

city of Windhoek and another site is between Keetmanshoop and Lüderitz, about 20 km from the town of Aus. In Chile, the site is close to the Paranal Observatory about 120km south of Antofagasta.

Following detailed studies of the candidate sites a ranking was made by the collaboration. With advice from an independent committee, the CTA funding agencies choices the sites in Chile and Namibia as the preferred locations for the Southern site and negotiations have started to make the final site choice in this hemisphere by summer 2015. A similar process will take place for the Northern hemisphere during 2015.

5. Science

Ground-based gamma-ray astronomy is a young field with enormous scientific potential. The possibility of astrophysical measurements at teraelectronvolt (TeV) energies was demonstrated in 1989 with the detection of a clear signal from the Crab (Pulsar Wind) Nebula above 1 TeV with the Whipple Imaging Atmospheric Cherenkov Telescope (IACT). Since then, the instrumentation for, and techniques of, astronomy with IACTs have evolved to the extent that a flourishing new scientific discipline has been established, with the detection of 150 sources and a major impact in astrophysics and more widely in physics. The current major arrays of IACTs: HESS, MAGIC, and VERITAS, have demonstrated the huge physics potential at these energies as well as the maturity of the detection technique. Many astrophysical source classes have been established, some with many well-studied individual objects, but there are indications that the known sources represent the tip of the iceberg in terms of both individual objects and source classes. The Cherenkov Telescope Array will transform our understanding of the high-energy universe. An advanced member of the suite of forthcoming major astroparticle physics experiments and observatories; CTA will exploit synergies with projects seeking gravitational wave and astrophysical neutrino detection as well as with classical photon observatories. The synergy with the upcoming Square Kilometre Array (SKA) is perhaps the most exciting in terms of mutual scientific benefit. CTA will address a wide range of major questions in and beyond astrophysics, which can be grouped in to three broad themes:

Theme 1: Understanding the Origin and Role of Relativistic Cosmic Particles

- What are the mechanisms for cosmic particle acceleration?
- What are the sites of high-energy particle acceleration in the universe?
- What role do accelerated particles play in feedback on star formation and galaxy evolution?

Theme 2: Probing Extreme Environments

- What physical processes are at work close to neutron stars and black holes?
- What are the characteristics of relativistic jets, winds and explosions?
- How intense are radiation fields and magnetic fields in cosmic voids, and how do these evolve over cosmic time?

Theme 3: Exploring Frontiers in Physics

- What is the nature of Dark Matter? How is it distributed?
- Are there quantum gravitational effects on photon propagation?
- Do axion-like particles exist?

5.1 Understanding the origin and role of relativistic cosmic particles

Relativistic particles appear to play a major role in a very wide range of astrophysical systems, from supernova explosions to the galaxy cluster-scale outbursts of active galactic nuclei. Within the interstellar medium of our own galaxy these cosmic rays are close to pressure equilibrium with turbulent motions of gas and magnetic fields, yet the relationship between these three components, and the overall impact on the star-formation process and the evolution of galaxies, is very poorly understood. CTA will provide the first real precision measurements of cosmic ray protons and nuclei (rather than the energetically sub-dominant electrons that produce the non-thermal emission seen at radio and X-ray wavelengths) in astrophysical systems, providing insights in to the processes of acceleration, transport and the cosmic ray mode feedback mechanisms in these systems. Historically, non-thermal effects in astrophysical systems have largely been ignored or parameterized away due to a lack of high quality data. The insights from CTA will represent a major contribution to our deepening understand of the processes by which galaxies and clusters of galaxies evolve.

Cosmic accelerators

The primary goal of gamma-ray astrophysics thus far has been to establish in which cosmic sources particle acceleration takes place, and in particular, to establish the dominant contributors to the locally measured cosmic rays which are 99% protons and nuclei (usually hadrons). Huge progress has been made in the last decade in this area, with the combination of Fermi-LAT and IACT data proving extremely effective in identifying the brightest Galactic accelerators and providing strong evidence of hadron acceleration in a handful of sources. However, key questions remain unanswered: are supernova remnants (SNR) the only major contributor to the Galactic cosmic rays? Where in our galaxy are particles accelerated up to PeV energies? What are the sources of high-energy cosmic electrons? What are the sources of the ultra-high energy cosmic rays? CTA will address all of these questions and also the critical issue of the mechanisms for particle acceleration at work in cosmic sources, through two main approaches:

- **A census of particle acceleration in the universe**, with Galactic and Extragalactic surveys and deep observations of key nearby galaxies and clusters.
- **Precision measurements of archetypal sources**, where bright nearby sources will be targeted to provide resolved spectroscopy or very high statistics light curves to provide a deeper physical understanding of the processes at work in cosmic accelerators.

A general census is required to understand the populations of accelerators, and the evolution/life-cycle of these source classes. Deep observations of individual sources are required to acquire the very broad band spectra needed to unambiguously separate lepton and hadron acceleration and to test acceleration to the highest energies possible for Galactic accelerators. A key target of CTA will be the bright, young supernova remnant (SNR) RX J1713–3946 where the dominant gamma-ray emission mechanism is unclear from current measurements, but where CTA can resolve the ambiguity and resolve sub-structure within the SNR shell. Known TeV-emitting source classes where CTA data will have a transformational impact on our understanding include pulsar wind nebulae, gamma-ray binaries, colliding-wind binaries, massive stellar clusters, starburst galaxies and active galaxies. There is clearly huge potential for the discovery of new classes of accelerator, with emission on the scale of clusters of galaxies as one of the most exciting possibilities.

Propagation and influence of accelerated particles

Beyond the question of how and where particles are accelerated in the universe, is the question of what role these particles play in the evolution of their host objects and how they are transported up to larger and larger scales. On the scale of clusters of galaxies, cosmic rays with TeV-PeV energies are thought to be confined over a Hubble time, on smaller scales they typically escape from their acceleration sites and may impact upon their environments in a number of ways:

1. As a dynamical constituent of the medium
2. Through generation / amplification of magnetic fields
3. Through ionization and subsequent impact on the chemical evolution of (for example) dense cloud cores.

All these effects are relevant for the interstellar medium of our own galaxy, and are likely to be important in star-forming systems on all scales and also for the process of AGN feedback on the host galaxy cluster and growth of massive galaxies.

CTA will map extended emission around many gamma-ray sources and look for energy dependent morphology associated with diffusion (in the case of hadrons) or cooling (in the case of electrons). As the energy-dependence is expected to be opposite in the two cases, such mapping provides another means to separate emission from these two populations. It is CTAs unprecedented angular resolution, energy resolution and background rejection power that make this possible.

5.2 Probing extreme environments

Particle acceleration to very high energies is typically associated with extreme environments: close to compact objects such as neutron stars and black holes, or in relativistic outflows or explosions. VHE emission from accelerated particles can therefore act as a probe of these environments, providing access to time and distance scales which are inaccessible in other wavebands. VHE emission escapes from systems where UV and X-ray emission is absorbed, and provides information independent of assumptions on magnetic field strengths. In addition, VHE photons from distant objects can be used as a probe of the intervening space. Gamma-gamma pair production will allow us to measure the redshift evolution of the UV-IR background and hence the star-formation history of the universe, and probe magnetic fields in cosmic voids down to values many orders of magnitude below the reach of any other technique. CTA will establish if the VHE photons heat the gas in these under dense regions, suppressing the formation of dwarf satellite galaxies.

Black holes and jets

Active galactic nuclei are thought to harbour supermassive black holes (SMBHs), accreting material and producing collimated relativistic outflows by a still poorly-understood process. Similarly, accreting stellar mass black holes are known to produce jets and particle acceleration seems to be universally associated with BH-powered jets. Acceleration may occur extremely close to the SMBH (in the case of very low accretion rate systems) up to Mpc scales where the largest AGN jets finally terminate. Active galaxies are seen as one of the most likely sites of the acceleration of the ultra-high energy cosmic rays, with energies up to energies around 10^{20} eV, but so far there is an absence of any strong evidence for hadronic acceleration in AGN jets. Simultaneous broad-band data is needed to study variable jet emission in both galactic and extragalactic systems, with CTA data playing a key role: establishing the presence of very high energy particles, identifying the presence of hadrons and probing extremely short timescale variability that provides information on the smallest scales, probing the bulk ultra-relativistic motions of the inner jet. Time-resolved VHE spectral measurements are key to disentangling leptonic and hadronic emission scenarios, and to study jet power and dynamics, and probe magnetic fields in this extreme environment. The low luminosity AGN in the heart of our own galaxy, Sgr A*, is coincident with a TeV source and has associated non-thermal emission in the radio and X-ray. However, the sensitivity, resolution and pointing precision of current gamma-ray telescopes is insufficient to separate emission from very nearby sources and the diffuse emission around the Galactic Centre. CTA will map this region in unprecedented detail, probing the relationship between the central source and the diffuse emission, and on much larger scales, to the Fermi bubbles. There is evidence from current IACTs for TeV emission from a single system hosting a stellar mass black hole: Cygnus X-1. The sensitivity of CTA should allow this object to be studied in detail and provide new detections, opening the door to studies of high-energy non-thermal processes associated with stellar mass black holes and allow the first comparisons to be made with SMBH systems.

Neutron stars and relativistic outflows

CTA will probe the environment around neutron stars via pulsed gamma-ray emission from the magnetosphere of pulsars, and the ultra-relativistic outflows of these systems via mapping and spectral measurements of the associated synchrotron/IC nebula and (possibly and uniquely) the un-shocked pulsar wind. Young and energetic (in terms of available rotational energy) pulsars cluster tightly along the Galactic plane and hence the majority of objects will be covered. Two key objects: HESSJ1825–137 and the Vela pulsar (and associated Vela X nebulae), will be studied in the CTA Key Science Programmes. Binary systems including a pulsar provide a unique opportunity to study a relativistic outflow under changing physical conditions as the orbit progresses, via energy-dependent light-curve measurements. Merging neutron stars and other compact object mergers are the likely counterparts of short GRBs and are expected to give rise to detectable gravitational wave transients in the near future. CTA will rapidly respond to triggers from gravitational wave or GRB instruments, to probe the highest energy processes associated with such events.

Cosmic voids

Much of the universe consists of extremely under-dense regions known as cosmic voids. Very high energy photos interact within these voids, and allow us to probe the radiation fields and magnetic fields that they contain. The Extragalactic Background Light (EBL) is the integrated emission from stars and galaxies of all types through the evolution of the universe. As such it is an important tool for cosmology, but is extremely difficult to measure directly, due to very strong foregrounds from the solar system and the Milky Way. However, the EBL leaves an imprint on the measured spectra of gamma-ray sources, via the process of gamma-gamma pair production. The wide-band, high-quality spectra measured with CTA, for a large number of objects, will allow the EBL spectrum from the optical to the far infrared to be very precisely measured at redshift zero. Furthermore, with the expected large sample of blazars up to redshift ~ 1 detected with CTA, the evolution of the EBL with cosmic time can be probed for the first time. The pair-production of TeV photons in voids also offers the prospect of measuring the extremely weak magnetic fields thought to exist in these regions. Secondary gamma-rays are produced by the original $e^{\pm}$ pairs via inverse Compton scattering on the EBL. A cascade can then develop from further pair and IC interactions. Depending on the typical value of the intergalactic magnetic fields (IGMF) deflections of the secondary particles may either be small enough that secondary components may be observable as pair echoes, which arrive with a time delay relative to the primary emission, or as a pair halo, potentially resolvable extended emission around the primary source. The properties of the extended emission depend on the IGMF strength. Strong enough IGMF ($> 10^{-12}$ G) leads to full isotropisation of the cascade emission and formation of a physical pair halo, while weaker magnetic field leads to appearance of an extended emission with an IGMF-dependent size. If the IGMF strength is in the range, $B \sim 10^{-16}$ to 10^{-12} G, the spatially-extended emission may be detectable and resolvable by CTA by virtue of its high sensitivity and angular resolution; e.g., for a source at a distance of 100 Mpc, the extended emission would be on a $\sim 1°$ scale, and be comfortably contained within the CTA FoV. If, as has been recently suggested, TeV electrons produced in gamma-gamma interactions in the voids do not initiate cascades but rather heat the ultra-low dense plasma, CTA will allow this hypothesis to be proven and the heating rate to be very well constrained. As such heating could be the dominant means of heating in low density regions after redshift ~ 2, and could solve the problem of the missing dwarf satellite galaxies, this measurement would be a very valuable addition to cosmology.

5.3 Physics frontiers

The reach of CTA encompasses considerable discovery space in the area of fundamental physics. CTA will reach the expected thermal relic cross-section for self-annihilating dark matter for a wide range of dark matter particle masses, including those inaccessible to the LHC. The long travel paths of gamma-rays from extragalactic sources combined with their short wavelength make them a sensitive probe for energy dependent variation of the speed of light due to quantum-gravity induced fluctuations of the metric. CTA will be sensitive to such effects on their expected characteristic scale: the Planck scale. On their long journey, gamma-rays may couple to other light particles such as axions, under the influence of intergalactic magnetic fields. Such photon-axion oscillations effectively make the Universe more transparent to gamma-rays, and akin to neutrino oscillations introduce a spectral modulation. Each of these effects would represent a very major discovery, alone worth the effort of constructing and operating CTA. The major step in sensitivity and energy coverage that CTA represents brings such effects within reach, and will undoubtedly allow further issues in fundamental physics to be addressed.

Quantum gravity and axion search

The beams of photons with energies up to hundreds of TeV, produced by cosmic sources and detected with CTA, represent a powerful tool to explore departures from the Standard Model and search for a wide range of new physics. The two main areas in fundamental physics, beyond the dark matter search are the search for axion-like particles (ALPs) and evidence of Lorentz Invariance Violation (LIV) associated with the quantum nature of gravity. Blazars and gamma-ray bursts are thought to be the most promising (bright, distant) target classes for both these searches. It has been suggested that QG effects may induce time delays between photons with different energies travelling over large distances due to a non-trivial refractive index of the vacuum. High statistics measurements over a very wide energy range of GRBs and Blazars will allow CTA to make order of magnitude better constraints than possible with current IACTs, reaching deep in to the interesting parameter space for models.

Axions are expected to convert into photons (and vice versa) in the presence of magnetic fields. In the case of a very distant AGN, the ALP/photon coupling can result in a detectable enhancement of the TeV photon flux (in competition with the EBL absorption), depending on the ALP mass. CTA ALP limits will complement those from indirect astrophysical tests, from solar observations, and from X-ray telescopes.

Dark Matter

The nature of dark matter (DM) in the universe is one of the most compelling questions facing physics and astronomy at the present time. There are many programmes around the world that have capabilities or are devoted to searching for non-gravitational signals of DM. Among these, CTA has an almost unique capability during the next decade to explore the WIMP mass region in the many hundreds of GeV to multi-TeV regime, above the reach of the LHC and ton-scale direct detection experiments. To capitalize on this remarkable opportunity, it is imperative for CTA to devote substantial resources (both observing time and effort) towards a comprehensive and integrated DM program. If this is done, CTA will be a cornerstone of a global, multi-faceted, line of attack from different experiments probing with different methods to understand the nature of the dominant gravitational matter in the Universe.

The indirect DM search with CTA has several possible astrophysical targets each of them with its own inherent advantages and disadvantages. The Milky Way (MW) galaxy represents a natural place to look for DM signatures and its centre is expected to be the brightest source in the DM induced gamma-ray sky, although the exact magnitude is rather uncertain. The DM density profile in the MW should also lead to an annihilation signal all around us; however astrophysical galactic foregrounds coupled with the enormous spatial extent and the truly diffuse nature of this galactic DM emission make separation between signal and background problematic. On the other hand, nearby dwarf spheroidal galaxies should provide easier separation of signal and background but comparatively lower signals because of both the distance and lower DM content compared to the Milky Way, even though somewhat compensated by higher DM concentrations.

The concordance cosmological CDM model predicts that the formation of visible structures has been guided by gravitational accretion of baryons onto previously formed DM over-densities. The astrophysical structures of interest result from the hierarchical formation of DM halos from primordial DM over-densities. The subsequent evolution of these DM halos occurred in many different ways. In particular on galactic and sub-galactic scales, the process depends on halo parameters such as the halo mass and the mass density profile, the evolution history, and conditions set by the local galactic evolution environment.

Some of the resulting halos could have been sufficiently massive to accrete enough baryons to initiate star formation and give origin to galaxies, including the variety of satellite galaxies we actually observe in the Milky Way halo. In-falling dwarf galaxies (e.g., dwarf irregular galaxies) approaching more central parts of their host halo could have evolved to dwarf spheroidal galaxies (dSphs). These dwarf galaxies, being highly DM dominated and comparatively close by, form one of the primary targets for CTA observations of this proposal and have already been subject to extensive studies with currently operating IACTs. However, for various reasons including the halo size, location, encounters, the baryonic content of the environment, or the presence of a central Black Hole, less massive DM clumps could have evolved differently and not into a visible dSph. They would constitute DM-dominated over-densities purely observable in gamma-rays or cosmic-rays emerging from DM annihilations. While, in principle, secondary emission from inverse Compton or synchrotron scattering processes (dependent on ambient photon and magnetic fields, respectively) will lead to diffuse multi-wavelength (MWL) signatures, classical MWL survey techniques do not provide sufficient sensitivity for detecting their signatures. Such objects are commonly known as DM subhalos or DM clumps.

The centre of the Milky Way galaxy has in the past been considered as a target for DM searches but more recently, because of the rich field of VHE gamma-ray astrophysical sources in the region, the searches focus on the galactic halo excluding the central region of galactic latitude $b < 0.3°$. Even excluding the very central region, the total mass of DM in the galactic halo together with its proximity to Earth make it the most promising source for DM searches in CTA. The inconvenience of this target, however, is the fact that being a diffuse source, the integration over the inner halo, while yielding a large signal, gives a very large instrumental background from misidentified charged cosmic rays. Further, there are astrophysical backgrounds from various sources which must be understood, even with the very central region excluded from the analysis.

It is believed that the disadvantages of the MW target can be overcome with sufficient experimental effort to control systematic effects in background subtraction.

The Large Magellanic Cloud is a massive nearby satellite galaxy visible from the southern hemisphere, at Galactic latitude of $-32.9°$. It is located at a distance $d = 50$ kpc and it has the shape of a disk seen nearly face-on. The LMC hosts many interesting astrophysical sources: the largest star-forming region of the local group of galaxies; one of the densest stellar clusters ever known; the most massive stars ever observed; several tens of HII regions; more than a dozen super-bubbles; numerous giant shells; more than 50 catalogued SNRs and the remnant of the recent naked-eye supernova SN 1987A. For the search for DM, the mass of the LMC which is 1% of the MW makes it interesting even when its distance at $\times 6$ further away than the Galactic Centre gives a large signal reduction. Although smaller in subtended angle than the MW, with most of the emission within the CTA FoV, the LMC is also an extended source where the backgrounds are again a significant drawback.

The dwarf spheroidal galaxies (dSphs) of the Local Group could give a clear and unambiguous detection of DM. They are gravitationally bound objects and are believed to

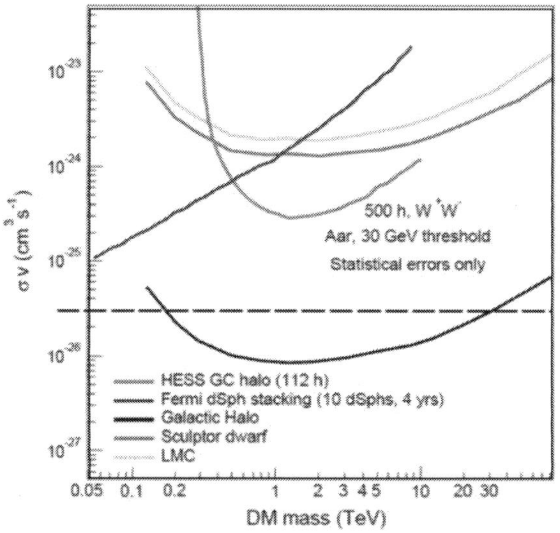

Figure 6. Comparison of predicted sensitivities in $\langle\sigma_{\chi\chi}v\rangle$ for the targets of: the Milky Way Halo; the Large Magellanic Cloud and the dwarf galaxy Sculptor. The CTA sensitivity curves use the same method and W^+W^- annihilation modes for each target and the NFW DM profile. The sensitivities for the three targets are all for 500 hours taking into account only statistics errors; for the MW and the LMC, the systematics of backgrounds must be very well controlled to achieve this statistically possible sensitivity. The comparisons with the HESS and Fermi-LAT results respectively with the bb annihilation modes as published.

contain up to $O(10^3)$ times more mass in DM than in visible matter, making them widely discussed as potential targets.

Being small and distant many of the dwarf galaxies will appear as near point sources in CTA and hence the nuisance of the instrumental background is much reduced. Further, they are also environments with a favourably low astrophysical gamma-ray background. Neither astrophysical gamma-ray sources (SNRs, pulsar wind nebulae, etc.) nor gas acting as target material for cosmic rays, have been observed in these systems.

Structure formation predicts gravitationally bound DM clumps down to much lower masses than observed for dSph galaxies. The low-mass cut-off of the clump distribution is related to the free-streaming scale of DM particles in the early universe and is expected between 10^{-12} M_{sun} and 10^{-3} M_{sun} for typical WIMP scenario. The number-count distribution of clumps in MW-like galaxies has been investigated with numerical N-body simulations, finding a power law: $dN/dM \propto M^{-\alpha}$ in clump mass M, with a high-mass cut-off at $M \sim 10^{10}$ M_{sun} and an index α being between 1.9 and 2.0. The total number of clumps orbiting within galactic halos can be as high as 10^{15}. For the MW this would imply a clumpiness factor of $\sim 18\%$ (referring to the relative contribution of the total mass in clumps to the total mass of the main halo). The spatial distribution of DM clumps is biased away from the smooth central DM distribution of the host halo, i.e., the majority of clumps populate outer halo regions ($r > 100$ kpc).

The subhalo population above 10^5 M_{sun} can be probed using gravitational milli-lensing of background objects while the statistical properties of lighter clumps in the solar neighbourhood can be measured from gravitational nano-lensing.

The Galactic Centre (GC) region is the prime target to look for DM due to its proximity and the high concentration of DM. The DM distribution in the halo centre inferred from N-body simulations without baryons follows a cuspy profile. However the place where the DM annihilation signal is expected to be the strongest is also the place where the baryons settle and perhaps dominate the overall gravitational potential well. Even if there is no consensus yet on how baryons affect the DM distribution, the GC is expected to be the brightest source of DM annihilations in the gamma-ray sky. However, the GC is a crowded region with several known high-energy gamma-ray emitters and more are expected to be detected thanks to the CTA observational capabilities. Given the standard astrophysical emission in this region, the quest for DM requires a deep and uniform exposure over several degrees around the central black hole Sgr A* to allow for both spectral and spatial morphological studies, a deep understanding of the instrumental and observational systematics, and precise measurements of the standard astrophysical emissions. The expected CTA angular and energy resolutions are key ingredients to disentangle a DM signal from standard astrophysical background contamination.

The sensitivity predictions for the Galactic Halo, the dwarf galaxy Sculptor and the Large Magellanic Cloud are compared in Fig. 6. Here it can be clearly seen that the sensitivity possible with the Galactic Halo observation is very much better than that which is possible with a single dwarf galaxy and the LMC. Also it can be seen that the sensitivity possible with Sculptor for 500h is better than that for the LMC. Although this plot does not show the effect of systematics, the relative rankings of the sensitivities is not changed by including them.

6. Summary

The design of CTA telescope and the site selection process is well advanced. It is expected that founding agreements will be signed during 2015 so construction can start soon afterwards.

We gratefully acknowledge financial support from the following agencies and organizations: Ministerio de Ciencia, Tecnología e Innovación Productiva (MinCyT), Comisión Nacional de Energía Atómica (CNEA), Consejo Nacional de Investigaciones Científicas y Técnicas (CONICET), Argentina; State Committee of Science of Armenia, Armenia; Conselho Nacional de Desenvolvimento Científico e Tecnológico (CNPq), Fundacão de Amparoà Pesquisa do Estado do Rio de Janeiro (FAPERJ), Fundacão de Amparo à Pesquisa do Estado de São Paulo (FAPESP), Brasil; Croatian Science Foundation, Croatia; Ministry of Education, Youth and Sports, MEYS LE13012, 7AMB12AR013, Czech Republic; Ministry of Higher Education and Research, CNRS-INSU and CNRS-IN2P3, CEA-Irfu, ANR, Regional Council Ile de France, Labex ENIGMASS, OSUG2020 and OCEVU, France; Max Planck Society, BMBF, DESY, Helmholtz Association, Germany; Department of Atomic Energy, Department of Science and Technology, India; Istituto Nazionale di Astrofisica (INAF), MIUR, Italy; ICRR, University of Tokyo, JSPS, Japan; Netherlands Research School for Astronomy (NOVA), Netherlands Organization for Scientific Research (NWO), Netherlands; The Bergen Research Foundation, Norway; Ministry of Science and Higher Education, the National Centre for Research and Development and the National Science Centre, Poland; MINECO support through the National R+D+I, CDTI funding plans and the CPAN and MultiDark Consolider-Ingenio 2010 programme, Spain; Swedish Research Council, Royal Swedish Academy of Sciences, Sweden; Swiss National Science Foundation (SNSF), Ernest Boninchi Foundation, Switzerland; Durham University, Leverhulme Trust, Liverpool University, University of Leicester, University of Oxford, Royal Society, Science and Technologies Facilities Council, UK; U.S. National Science Foundation, U.S. Department of Energy, Argonne National Laboratory, Barnard College, University of California, University of Chicago, Columbia University, Georgia Institute of Technology, Institute for Nuclear and Particle Astrophysics (INPAC-MRPI program), Iowa State University, Washington University McDonnell Center for the Space Sciences, USA. The research leading to these results has received funding from the European Union's Seventh Framework Programme (FP7/2007-2013) under grant agreement n° 262053.

References

[1] Hinton & Hofmann (2010), Teraelectronvolt Astronomy, Ann. Rev. Astron. Astrophys. **47**, 523
[2] Seeing the High-Energy Universe with the Cherenkov Telescope Array, Astroparticle Physics **43**, Hinton, Sarkar, Torres & Knapp (2013)
[3] HESS website: `http://www.mpi-hd.mpg.de/hfm/HESS`
[4] MAGIC website: `http://magic.mppmu.mpg.de/`
[5] VERITAS website: `http://veritas.sao.arizona.edu/`

EPJ Web of Conferences **121**, 04005 (2016) DOI: 10.1051/epjconf/201612104005
RICAP-2014

Future Extensive Air Shower arrays: From Gamma-Ray Astronomy to Cosmic Rays

Giuseppe Di Sciascio[a]

INFN, Sezione Roma Tor Vergata

Abstract. Despite large progresses in building new detectors and in the analysis techniques, the key questions concerning the origin, acceleration and propagation of Galactic Cosmic Rays are still open. A number of new EAS arrays is in progress. The most ambitious and sensitive project between them is LHAASO, a new generation multi-component experiment to be installed at very high altitude in China (Daocheng, Sichuan province, 4400 m a.s.l.). The experiment will face the open problems through a combined study of photon- and charged particle-induced extensive air showers in the wide energy range 10^{11} – 10^{18} eV. In this paper the status of the experiment will be summarized, the science program presented and the outlook discussed in comparison with leading new projects.

1. Introduction

Cosmic Rays (hereafter CR) are the messengers of the *"non-thermal Universe"*, generated in environments able to transfer energy and accelerate nuclei (and electrons) up to values not yet reached in our laboratories. Understanding their origin and propagation through the Interstellar Medium (ISM) is a fundamental problem which has a major impact on models of the structure and nature of the Universe.

The CR energy spectrum can be described by a power law with some features, the most important being the *"knee"* around 3×10^{15} eV and the *"ankle"* at about 10^{19} eV. In the so-called standard model CRs below 10^{17} eV are expected to be mainly Galactic, produced and accelerated in SuperNova Remnants (SNR), which can provide the necessary energy budget and naturally produce particles with a power law energy spectrum and spectral index close to the value inferred from the observations. The particles propagate diffusively in the ISM over very long time, under the effect of the Galactic Magnetic Fields (GMF) (for a review see, for example, [1] and references therein).

The study of CRs is based on two complementary approaches:

(1) Measurement of the energy spectrum, elemental composition and anisotropy of charged CRs.
(2) Search of their sources through the observation of neutral radiation (photons and neutrinos), which points back to the emitting sources not being affected by the magnetic fields.

The integrated study of charged CRs and of gamma rays and neutrinos, which should trace high energy hadronic interactions mainly nearby the still unidentified acceleration sites, is one of the most important (and exciting) fields in the so-called *"multi-messenger astronomy"*.

[a]e-mail: `giuseppe.disciascio@roma2.infn.it`

Despite large progresses in building new detectors and in the analysis techniques, the key questions concerning origin, acceleration and propagation of CRs are still open. In particular, the identification of sources in our Galaxy able to accelerate particles up to PeV energies, the so-called *"PeVatrons"*, is still missing. In fact, even there is no doubt that galactic CR are accelerated in SNRs, the capability of SNRs to accelerate CRs up to the knee of the spectrum and above is still under debate.

In this paper we briefly summarize the motivation for new generation experiments addressed to study Galactic CR physics and γ-ray astronomy. We will focus on Extensive Air Shower (EAS)-arrays, not discussing new imaging Cerenkov projects like Cerenkov Telescope Array, described, as an example, in [2].

2. Open problems in Galactic Cosmic Ray Physics

In the knee energy region an accurate measurement of the CR primary spectrum can be carried out only by ground-based EAS arrays. In fact, since the CR flux rapidly decreases with increasing energy and the size of detectors is constrained by the weight that can be placed on satellites/balloons, their collecting area is small and determine a maximum energy (of the order of 100 TeV/nucleon) related to a statistically significant detection. In addition, the limited volume of the detectors make difficult the containement of showers induced by high energy nuclei, thus limiting the energy resolution of instruments in direct measurements.

In the standard picture, mainly based on the results of the KASCADE esperiment, the knee is attributed to the steepening of the p and He spectra [3]. However, a number of experiments reported evidence that the bending of the light component (p+He) is well below the PeV and the knee of the all-particle spectrum is due to heavier nuclei [4–7]. Interestingly, increasing the altitude of the experiment locations the measured average mass at the knee increases as well. As an example, recent results obtained by the ARGO-YBJ experiment (located at 4300 m asl) with different analyses reported evidence that the knee of the light component energy spectrum starts at $\sim$700 TeV, well below the knee of the all-particle spectrum that is confirmed by ARGO-YBJ at about $3 \cdot 10^{15}$ eV [8–11].

At higher energies, the KASCADE-Grande experiment reported evidence for a hardening slightly above 10^{16} eV and a steepening at log10(E/eV) = 16.92 ± 0.10 in the CR all particle spectrum. A steepening at log10(E/eV) = 16.92 ± 0.04 in the spectrum of the electron poor event sample (heavy primaries) and a hardening at log10(E/eV) = 17.08 ± 0.08 in the electron rich (light primaries) one were observed. The slope of the heavy mass group spectrum above the knee-like feature is similar to the one of the light mass spectrum before the ankle-like feature [12].

A large number of theoretical papers discussed the highest energies achievable in SNRs and the possibility that protons can be accelerated up to PeVs [1]. The determination of the proton knee, as well as the measurement of the evolution of the heavy component across the knee, are the key components for understanding CR acceleration mechanisms and the propagation processes in the Galaxy, and to investigate the transition from Galactic to extra-galactic CRs. As mentioned, experimental results in the knee range are still conflicting. New high resolution measurements of the energy spectrum and elemental composition in the range from 10^{12} to 10^{18} eV are required to investigate in detail the role of different EAS observables and their dependence upon hadronic interaction models.

The measurement of the anisotropy in the arrival direction distribution of CRs is a complementary way to understand the origin and propagation of the radiation. It is also a tool to probe the structure of the magnetic fields through which CRs travel.

As CRs are mostly charged nuclei, their trajectories are deflected by the action of GMF they propagate through before reaching the Earth atmosphere, so that their detection carries directional information only up to distances as large as their gyro-radius. If CRs below 10^{15} eV are considered and the local GMF ($\sim 3\,\mu$G) is accounted for, gyro-radii are so short that isotropy is expected. At most, a weak di-polar distribution may exist, reflecting the contribution of the closest CR sources.

However, a number of experiments observed an energy-dependent *"large scale"* anisotropy in the sidereal time frame with an amplitude of about 10^{-4}–10^{-3}, revealing the existence of two distinct broad regions: an excess distributed around 40° to 90° in Right Ascension (commonly referred to as "tail-in" excess) and a deficit (the "loss cone") around 150° to 240° in Right Ascension. In recent years different experiments reported also the observation of a "medium" scale anisotropy in both hemispheres (for a review see [13]). Therefore, the observed pattern cannot be reproduced with a simple dipole, which implies the need at least two harmonics to fit data projected in sidereal time. In addition, results obtained by EASTOP and IceCube/IceTop showed that the anisotropy dramatically changes topology above 100 TeV and that the dipole component seems to disappear at PeV energies and beyond.

From the theoretical viewpoint, the anisotropy of CRs is important as a direct trace of potential sources. Nonetheless, models failed to explain the whole set of CR anisotropy observations. So far, no theory of CRs in the Galaxy exists which is able to explain the origin of these different anisotropies leaving the standard model of CRs and that of the local GMF unchanged at the same time. The anisotropy problem is the most serious challenge to the standard model of the origin of galactic CRs from diffusive shock acceleration [14].

A measurement of the evolution of anisotropy with the energy across the knee and the determination of the chemical nature of excesses and deficits, are crucial to disentangle between different models of CR propagation in the Galaxy. Moreover, the study of the anisotropy can clarify the origin of the knee. Indeed, if the knee is due to an increasing inefficiency in CR containment in the Galaxy a change in anisotropy is expected. If, on the contrary, the knee is related to the limit of the acceleration mechanism, we do not expect a change of anisotropy across the knee.

3. Motivation for a Multi-TeV (>10 TeV) γ-ray detector

Actually TeV γ-rays have been observed from a number of SNRs, demonstrating that in SNRs some kind of acceleration occurs. However the question whether TeV gamma rays are produced by the decay of π^0 from protons or nuclei interactions (*"hadronic"* mechanism), or by a population of relativistic electrons via Inverse Compton scattering or bremsstrahlung (*"leptonic"* mechanism), still need a conclusive answer.

Recently AGILE and Fermi observed GeV photons from two young SNRs (W44 and IC443) showing the typical spectrum feature around 1 GeV (the so called *"π^0 bump"*, due to the decay of π^0) related to hadronic interactions [15, 16]. This important measurement however does not demonstrate the capability of SNRs to accelerate CRs up to the knee of the spectrum and above. In fact, the identification of sources able to accelerate particles up to PeV energies, *"PeVatrons"*, is still missing.

In *pp* interactions, 20% of the proton energy goes into each pion flavor on average. Therefore, each pionic gamma carries approximately 10% of the energy of the proton primary. Thus the search for PeVatrons implies the observation of hundreds TeV photons.

The study of the multi-TeV (>10 TeV) emission is of particular importance in discriminating between different mechanisms. A typical *leptonic* (electron-produced) γ-ray spectrum would cut off earlier than typical *hadronic* (CR-produced) spectra because of Klein-Nishina effects in the inverse-Compton cross section. Therefore, a power law energy spectrum extending up to hundreds TeV with no breaks can be reliably attributed to the hadronic mechanism.

To extend the energy range at PeVs a very large effective area is required. The most sensitive experimental technique for the observation of multi-TeV γ-rays is the detection of EAS via large ground-based arrays. The muon content of photon-induced showers is very low, therefore these events can be discriminated from the large background of CRs via a simultaneous detection of muons that originate in the muon-rich CR showers. Their large field of view (FoV) ($\Omega_{FoV} \sim$ O(sr)) and high duty cycle ($> 90\%$) make these observational technique particularly suited to perform unbiased all-sky surveys (not simply of limited regions as the Galactic plane) and to monitor the sky for the brightest transient emission from Active Galactic Nuclei (AGN) and Gamma Ray Bursts (GRB), and search for unknown transient phenomena.

Among 150 TeV sources, only less than 10 have been observed above 30 TeV, and the data are affected by large uncertainties. Even the most luminous TeV source, the Crab Nebula, is well studied only up to about 20 TeV. In fact, for higher energies, the sensitivity of current instruments is not enough to determine clearly the spectral shape.

It is known that the ensembles of the brightest sources in the GeV and TeV ranges do not coincide exactly, with some bright sources detected by Fermi relatively faint at TeV energy and vice-versa. It is therefore possible that the brightest sources in the 100 TeV ranges are unexpected. In this sense the large sky coverage of an EAS detector is well suited to discover the emission from unknown sources. It has also been proposed, that the highest energy particle produced in astrophysical accelerators can escape rapidly the accelerator, and therefore that the highest energy emission is not from point-like or quasi point-like sources, but it could have a broader extension, mapping the interstellar gas distribution in the vicinity of the accelerators, and perhaps also the structure of the regular magnetic field in the region, if the diffusion is strongly anisotropic. In these circumstances the identification of the PeVatrons is more difficult, but the wide FoV of the EAS-arrays could be decisive to trace the γ-ray emission.

The space distribution of diffuse γ-rays can trace the location of the CR sources and the distribution of interstellar gas. In fact, this emission can be produced by protons and nuclei via the decay of π^0 produced in hadronic interactions with interstellar gas. In addition, the observation of a knee in the energy spectrum of diffuse galactic gamma-rays would provide a complementary way to investigate the elemental composition in the knee range. The recent claim of a proton knee starting well below the PeV by the ARGO-YBJ experiment implies that a corrsponding knee in the diffuse energy spectrum is expected below hundred TeV and, in principle, could be detected by next generation γ-ray detectors. Observing a location dependence of the knee energy would provide important clues on the nature of the knee, as do similar measurement for individual sources of CRs.

4. Future EAS-arrays

A number of new experiments or projects are under way in the coming years to study gamma-ray astronomy and CR physics. As mentioned, in this Section we focus on non-imaging Cherenkov and particle detectors.

DOI: 10.1051/epjconf/201612104005

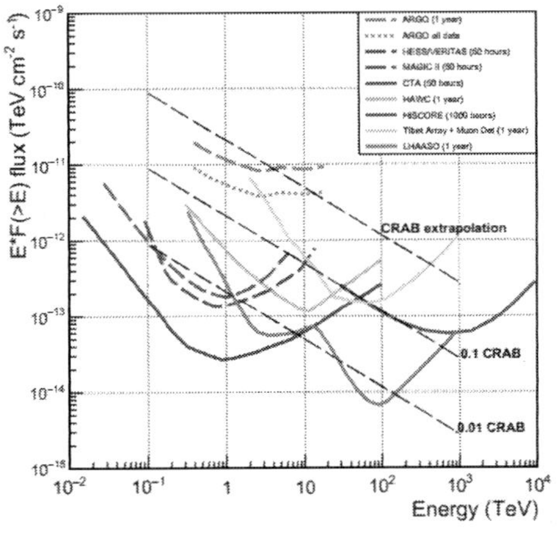

Figure 1. Sensitivity of ground-based γ-ray detectors to a Crab-like point source.

4.1 Tibet ASγ

The Tibet air shower array is located at Yangbajing (Tibet, P.R. China, 4300 m asl) and consists of (i) a high density array of over 500 scintillation counters with 7.5 m spacing, together with wider spaced outriggers, spread on a 370,000 m^2 instrumented area; (ii) a close packed array of 124 burst detectors distributed over a 500 m^2 area (YAC, Yangbajing Air shower Core detector array); (iii) 5 underground water Cerenkov muon detectors [17]. The burst detectors are designed to detect the high-energy particles in the air shower core by converting them into electromagnetic cascade showers (burst) using a lead absorber 7 r.l. in thickness above a scintillation counter of the same area and 2 cm in thickness. In the first phase, named "YAC-I", 16 YAC detectors each having the size 40 cm$\times$50 cm are distributed in a grid with an effective area of 10 m^2. YAC-I is used to check hadronic interaction models. The second phase of the experiment, called "YAC-II", consists of 124 YAC detectors. The inner 100 detectors of 80 cm$\times$50 cm each are deployed in a 10$\times$10 matrix 1.9 m apart and the outer 24 detectors of 100 cm$\times$50 cm each are distributed around them to reject events with the core outside the YAC-II array [17]. YAC-II is used to study the primary cosmic-ray composition in the range 5×10^{13}–10^{16} eV. The array has been operating in various configurations since 1990, and provides an angular resolution of $\sim$0.9° at 10 TeV. This experiment is mainly devoted to CR studies but the installation of muon detectors will allow the selection of high-energy ($>$ 10 TeV) photon-induced showers based upon the shower muon content. Data-taking with 5 of the planned 12 muon detectors and with YAC-II is started in 2014.

The expected sensitivity of Tibet ASγ to point–like gamma ray sources [18] is shown in Fig. 1 that plots, as a function of the threshold photon energy E_{min}, the minimum integral flux $\Phi_\gamma(E_{\mathrm{min}})$ that is detectable with a 5 σ statistical significance after one year (50 h) of operation for EAS-arrays (Cerenkov telescopes). The curve refers to the final layout with12 muon detectors.

4.2 HiSCORE

HiSCORE (Hundred*i Square-km Cosmic ORigin Explorer) will consist of an array of wide-angle ($\Omega\sim$ 0.6-0.85 sr) light-sensitive detector stations, distributed over an area of the order of 100 km^2. A HiSCORE detector station consists of four photomultiplier tubes, each equipped with a light-collecting Winston cone of 30° half-opening angle pointing to the zenith [19]. The primary goal of this non-imaging Cherenkov detector is gamma-ray astronomy in the 10 TeV to several PeV range. A prototype array of 9 wide-angle optical stations, spread on a 300×300 m^2 area, has been deployed in the Tunka Valley near Lake Baikal since October 2013, and technical tests are underway. A 1 km^2 engineering array is planned for deployment in 2015, aiming at proof-of-principle measurements and first physics results. The sensitivity of the HiSCORE standard configuration to point-sources is shown in Fig. 1. The curve has been calculated for 1000 h of exposure time, corresponding to 5 years of standard mode operation and roughly 1.4 years of tilted mode operation (see [19] for details).

4.3 HAWC

The HAWC (High Altitude Water Cherenkov Observatory) experiment, currently under construction at 4,100 m asl on the Sierra Negra volcano in Mexico, is the evolution of Milagro. The final array will consist of 300 water Cherenkov tanks of 7.3 m diameter and 4.5 m deep, covering an instrumented area of about 22,000 m^2 (the actual tank coverage is 12,550 m^2 with a coverage factor less than 60%). Deployment and commissioning are ongoing, with science operations already underway, and completion of the full array is expected in 2015. Detector performance and first physics results are described in [20]. The expected sensitivity to a Crab-like point-source is shown in Fig. 1 [21].

5. The LHAASO project

LHAASO (Large High Altitude Air Shower Observatory) is a new generation instrument strategically built to act simultaneously as a wide aperture (~sr), continuosly-operated gamma ray telescope in the energy range between 10^{11} and 10^{15} eV and as a high resolution CR detector in the broad energy range from 10^{13} to 10^{18} eV.

To achieve its scientific goals, the first phase of LHAASO will consist of the following major components [22]:

- 1 km^2 array (LHAASO-KM2A), including 5635 scintillator detectors, with 15 m spacing, for electromagnetic particle detection.
- An overlapping 1 km^2 array of 1221, 36 m^2 underground water Cherenkov tanks, with 30 m spacing, for muon detection (total sensitive area 40,000 m^2).
- A close-packed, surface water Cherenkov detector facility with a total area of 90,000 m^2 (LHAASO-WCDA), four times that of HAWC.
- 24 wide field-of-view air Cherenkov (and fluorescence) telescopes (LHAASO-WFCTA).
- 452 close-packed burst detectors, located near the centre of the array, for detection of high energy secondary particles in the shower core region (LHAASO-SCDA).

LHAASO will be located at high altitude (4400 m asl, 600 g/cm^2, 29° 21′ 31″ N, 100° 08′15″ E) in the Daochen site, Sichuan province, P.R. China. The start of data taking is expected 2 – 3 years after the start of installation planned at the beginning of 2016. The completion of installation in 6 years.

The sensitivity of LHAASO to point–like gamma ray sources is shown in Fig. 1 where is compared to other experiments [23]. The sensitivity curve has been calculated for a Crab Nebula like energy spectrum (power law with exponent -2.63) extending to PeVs without any cutoff. LHAASO is capable of observing sources with a brightness below ~1 % of the Crab flux in the energy ranges ~1–10 TeV and ~30–150 TeV. The LHAASO sensitivity curve shows a structure with two minima, reflecting the fact that the observation and identification

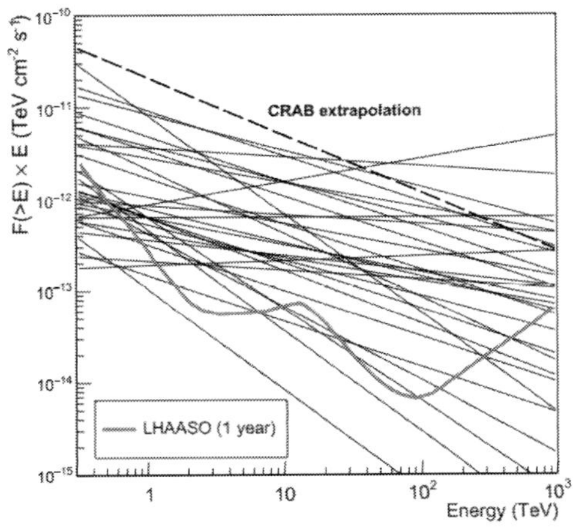

Figure 2. Integral spectra of galactic γ-ray sources visible by LHAASO extrapolated to 1 PeV, compared to the LHAASO sensitivity (red curve) [25].

of photon showers in different energy ranges is controlled by different components of the detector: water Cherenkov detector (WCDA) in the range ~0.3–10 TeV and KM2A array above 10 TeV.

For comparison of sensitivities, it is important to note that the LHAASO sensitivity shown is the point-source survey sensitivity for ~ 2 sr of the sky. While the sensitivities for EAS-arrays (ARGO-YBJ, HAWC, LHAASO, HiSCORE and Tibet ASγ) are also valid for surveys, the sensitivities for Cerenkov telescopes (CTA, HESS, MAGIC and VERITAS) are given for pointed observations of 50 h "on source" in a small FoV of the order of $\pi/100$ sr. The choice of different conventions is inevitable because of the different operation modes of the two detection techniques. Cherenkov telescopes work only during clear moonless nights, with a total observation time of about 1000–1500 hours per year (depending on the location), and have a FoV of a few degrees of radius. This implies that they can observe only one (or very few) sources at the same time, and only in the season of the year when the source culminates during night time. Fifty hours is a typical time that a Cherenkov telescopes will dedicate to a selected source in one year.

In contrast, the sky region observed by an EAS detector is completely determined by its geographical location. The detector observes nearly continuously a large fraction of the celestial sphere (spanning 360 degrees in right ascension and about 90 degrees in declination). Sources located in this portion of the sky are in the FoV of the detector, either always, or for several hours per day, depending on their celestial declination. This situation is ideal to perform sky surveys, discover transients or explosive events (such as GRBs), and monitor variable or flaring sources such as AGNs.

One of the most interesting aspects of LHAASO is its sensitivity to gamma rays above 30 TeV. Inspecting Fig. 1 one can see that in this energy region LHAASO, thanks to the KM2A array, will be the most sensitive gamma–ray telescope. To give a more quantitative idea of the LHAASO performance, we can compare the detector sensitivities with the fluxes of the known TeV sources visible by LHAASO. Figure 2 reports the spectra of galactic sources [24] extrapolated up to 1 PeV, using the same spectral slope measured in the TeV region, compared to the LHAASO one-year sensitivity [25]. The figure clearly shows that almost all fluxes are above the LHAASO sensitivity. Probably a consistent fraction of real sources spectra will not reach 1 PeV, but LHAASO will provide a unique information by studying in detail the spectral steepening or cutoff of galactic γ-ray sources.

As it is well known, an EAS detector is better suited than Cherenkov telescopes to survey large portion of the sky. If we consider a survey of the Galactic Plane in the latitude band $-6° < b < +6°$, for a longitude interval of 200° (corresponding to the fraction of the Galaxy plane visible from the LHAASO site with zenith angle $\theta \lesssim 40°$), in one year LHAASO can observe each source of this region for an average time of ∼1800 hours (∼5 hours per day). A Cerenkov telescope, with its limited FoV, must scan the whole region with different pointings. The number of pointings determines the maximum observation time that can be dedicated to any source. Assuming a FoV of radius 5° with a decrease of sensitivity of about 50% at a distance of 3° from the center and a step for different pointings of 4°, the galactic survey requires approximately 150 pointings. Assuming a total observation time of ∼1300 hours/year, a full year dedicated to the survey allows an exposure of ∼9 hours for each galactic source. Rescaling, as an example, the CTA sensitivity for an observation time of 9 hours, one finds that galactic survey performed by LHAASO in 1 year reaches a sensitivity better than CTA for E $\gtrsim$ 2 TeV. The difference grows with E, and is better a factor of ∼30 (170) for $E \gtrsim 30(80)$ TeV.

The difference in sky survey capabilities between EAS-arrays and Cerenkov telescopes is more impressive in case of an *all sky survey*. Assuming a region of 7 sr to be scanned, the number of pointings is approximately ∼1400, allowing only 0.9 hours/year of exposure per point in the sky. Rescaling, as an example, the CTA sensitivity for an observation time of 0.9 hours, one finds that an all sky survey performed by LHAASO in 1 year is better than CTA by a factor of approximately 270 (1600) for $E \gtrsim 30$ (80) TeV.

In conclusion, the two techniques (EAS detectors and Cherenkov telescopes) have both great scientific potential, and should be seen as complementary approaches in the exploration of different aspects, and different energy ranges, of the γ-ray emission. On the contrary, CR studies are a peculiar characteristics of EAS-arrays

6. Conclusions and outlook

The LHAASO experiment is expected to be the most sensitive detector to study the multi-TeV γ-ray sky opening the PeV range to the direct observations of the high energy CR sources. The installation of an unprecedented muon detection area (40,000 m^2), together with arrays of wide field of view Cerenkov telescopes and burst detectors, will allow high resolution detection of nuclei-induced showers in the range 10^{12}–10^{18} eV.

As mentioned in the previous Section, all new wide field of view γ-ray experiments under way in the coming years will be located in the Northern hemisphere. To maximize the scientific return for Galactic sources, a future instrument should be located at sufficiently Southern latitude to continuously monitor the Galactic Center and the Inner Galaxy. To lower the energy threshold down to hundred GeV range such an instrument would require very high altitude location (as an example, the 4800 m asl of the Alto Chorrillos region, hosting the Long Latin American Millimeter Array). The detection of very small low-energy showers with high efficiency and good angular resolution requires a detector exploiting a full coverage approach (with a coverage factor >90%) with high segmentation of the readout. In fact, being precluded the γ/hadron discrimination due to the absence of muons in the sub-TeV showers, the angular resolution is the main parameter to improve the sensitivity. The possible detection of PeVatrons in the Northern hemisphere by LHAASO (or HAWC) would require the increase of the collecting area with a large array. In the near future such an instrument will be paired with the coming km^3 Mediterranean neutrino detector.

References

[1] Blasi P., Astron. Astrophys. Rev. **21**, 70 (2013)
[2] Carr J., these proceedings
[3] Apel W.D. et al., Astropart. Phys. **31**, 86 (2009)
[4] Aglietta M. et al., Astropart. Phys. **21**, 223 (2004)
[5] Amenomori M. et al., Phys. lett. B **632**, 58 (2006)
[6] Glasmacher M.A.K. et al., Astropart. Phys. **12**, 1 (1999)
[7] Ambrosio M. et al., Physics Rev. D **56**, 1418 (1997)
[8] Di Sciascio G., Int. J. Mod. Phys. D **23**, 1430019 (2014)
[9] Di Sciascio G. et al., *Proc. "Frontier Objects in Astrophysics and Particle Physics" (Vulcano)*, Frascati Physics Series **58** (2014), arXiv:1408.6739
[10] De Mitri I. et al., these proceedings
[11] Montini P. et al., these proceedings
[12] Chiavassa A. et al., J. of Phys.: Conf. Ser. **531** 012001 (2014)
[13] Di Sciascio G. and Iuppa R., *"On the observation of the Cosmic Ray Anisotropy below 10^{15} eV"*, in "Homage to the Discovery of Cosmic Rays, the Meson-Muons and Solar Cosmic Rays", Chapter 9, p. 221-257 (Nova Science Publishers, Inc., New York, 2013). Preprint: arXiv:1407.2144
[14] Hillas A.M. 2005, J. Phys. G **31**, R95 (2005)
[15] Giuliani A. et al., Astrophys. J. Lett. **742**, L30 (2011)
[16] Ackermann M. et al., Science **339**, 807 (2013)
[17] Liu J.S. et al., arXiv:1501.06327 (2015)
[18] Sako T.K. et al., Astrop. Phys. **32**, 177 (2009)
[19] Tluczykont M. et al., Astrop. Phys. **56**, 42 (2014)
[20] Mostafa M., these proceedings
[21] Abeysekara A.U. et al., Astrop. Phys. **50**, 26 (2013)
[22] Zha M. et al., Int. J. Mod. Phys.: Conf. Ser. **10**, 147 (2012)
[23] Cui S. et al., Astrop. Phys. **54**, 86 (2014)
[24] `http://tevcat.uchicago.edu/`
[25] Vernetto S., private comunication

The ASTRI mini-array within the future Cherenkov Telescope Array

Stefano Vercellone[1,a] for The ASTRI Collaboration[2] , The CTA Consortium[3]

[1] INAF/IASF Palermo, via U. La Malfa 153, 90146 Palermo, Italy
[2] `http://www.brera.inaf.it/astri/`
[3] `https://portal.cta-observatory.org`

Abstract. The Cherenkov Telescope Array (CTA) is a large collaborative effort aimed at the design and operation of an observatory dedicated to very high-energy gamma-ray astrophysics in the energy range from a few tens of GeV to above 100 TeV, which will yield about an order of magnitude improvement in sensitivity with respect to the current major arrays (H.E.S.S., MAGIC, and VERITAS). Within this framework, the Italian National Institute for Astrophysics is leading the ASTRI project, whose main goals are the design and installation on Mt. Etna (Sicily) of an end-to-end dual-mirror prototype of the CTA small size telescope (SST) and the installation at the CTA Southern site of a dual-mirror SST mini-array composed of nine units with a relative distance of about 300 m. The innovative dual-mirror Schwarzschild-Couder optical solution adopted for the ASTRI Project allows us to substantially reduce the telescope plate-scale and, therefore, to adopt silicon photo-multipliers as light detectors. The ASTRI mini-array is a wider international effort. The mini-array, sensitive in the energy range 1–100 TeV and beyond with an angular resolution of a few arcmin and an energy resolution of about 10–15%, is well suited to study relatively bright sources (a few $\times 10^{-12}$ erg cm^{-2} s^{-1} at 10 TeV) at very high energy. Prominent sources such as extreme blazars, nearby well-known BL Lac objects, Galactic pulsar wind nebulae, supernovae remnants, micro-quasars, and the Galactic Center can be observed in a previously unexplored energy range. The ASTRI mini-array will extend the current IACTs sensitivity well above a few tens of TeV and, at the same time, will allow us to compare our results on a few selected targets with those of current (HAWC) and future high-altitude extensive air-shower detectors.

1. Introduction

The very high-energy (VHE) portion of the electromagnetic spectrum (above ≈ 100 GeV) is currently being investigated by means of ground-based imaging atmospheric Cherenkov telescopes (IACTs, see [1] for a recent review). In order to dramatically boost the current IACT performance and to widen the VHE science, a new Cherenkov telescope array (CTA) has been proposed, as described in [2] and more recently in [3]. The wide energy range covered by the CTA (from a few tens of GeV to above 100 TeV) requires different kinds of telescopes. Four large size telescopes (LSTs, D$\sim$23 m) will be placed at the center of the array, to lower the energy threshold down to a few tens of GeV. A few tens of medium size telescopes (MSTs, D$\sim$12 m, SCTs, D$\sim$9.5 m) will cover approximately 1 km^2, to improve by

[a] e-mail: `stefano.vercellone@iasf-palermo.inaf.it`

Figure 1. *Left panel*: the ASTRI SST-2M prototype at the Serra La Nave observing station. *Right panel*: Artistic concept (not to scale) of the ASTRI mini-array.

a factor of ten the sensitivity in the energy range 0.1–10 TeV. Finally, 70 small size telescopes (SSTs, primary mirror D∼4 m, Aeff∼5–10 m^2) covering about 10 km^2 will enhance Galactic plane source studies in the energy range beyond 100 TeV. A detailed review of the CTA project is given in [4].

2. The ASTRI Project and its end-to-end prototype

Within the CTA framework, the Italian National Institute for Astrophysics (INAF) is leading the "Astrofisica con Specchi a Tecnologia Replicante Italiana" (ASTRI) Flagship Project [5] of the Ministry of Education, University and Research. Primarily, INAF has designed and developed an end-to-end prototype of the CTA small-size telescope in a dual-mirror configuration (SST-2M). This prototype is currently being tested under field conditions at the INAF "M.C. Fracastoro" observing station in Serra La Nave (Mount Etna, Sicily). The ASTRI SST-2M prototype was recently inaugurated during the CTA Consortium Meeting in September 2014. Figure 1 (left panel) shows the prototype in front of the domes. Currently all telescope sub-systems are in place and properly working. The camera is being assembled, an it will be installed at the telescope focal plane at the beginning of 2016.

A new dual-mirror, Schwarzchild-Couder (SC) aplanatic design has been proposed [6]. In the SC telescope, the focal plane is located in-between two aspherical mirrors, close to the secondary mirror. No Cherenkov telescope has adopted this optical system before. The dual-mirror optical system will reduce the dimension, the weight, and the cost of the camera at the focal plane of the telescope, and will obtain a more compact and stiffer mechanical structure, and an optimal imaging resolution across a wide field of view. Moreover, due to the reduced plate-scale, silicon-based photo-multipliers (SiPMs) can be adopted as light detectors.

The ASTRI SST-2M prototype adopts a segmented 4.3 m primary mirror (M1) composed of 18 facets, a monolithic 1.8 m secondary mirror (M2, with a radius of curvature of 2.2 m), a focal length F=2.15 m, a field of view FoV∼9.6°, for a ratio F/D_1=0.5. The mirror manufacturing process is the "glass cold shaping" technique, specifically developed by INAF for Cherenkov mirrors [7, 8]. The curved focal plane (∼1 m of radius of curvature) hosts 1984 logical pixels (6.2 mm × 6.2 mm, 0.17°). The current photo-sensors are Hamamatsu S11828-3344M silicon-based photo-multipliers, but other sensors are also being tested [9]. The ASTRI camera [10] is extremely compact (∼ 50 cm × 50 cm × 50 cm) and light (∼ 50 kg). Contrary to other CTA telescopes adopting a signal-sampler front-end electronics (FEE), the

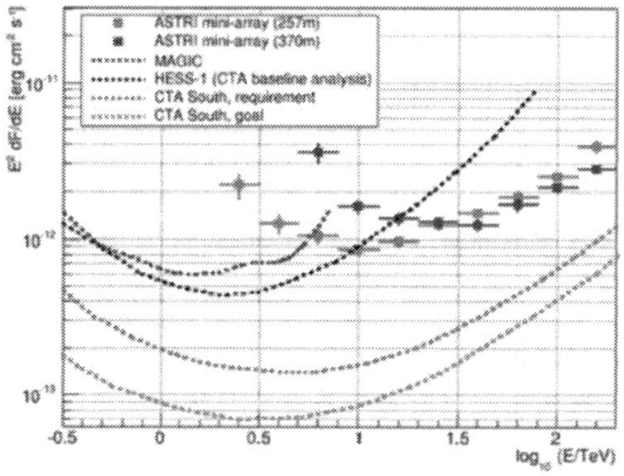

Figure 2. ASTRI mini-array differential sensitivity for 9 telescopes (blue and red points, for two different telescope relative distances) compared to the MAGIC, H.E.S.S.-I, and CTA (requirements and goal) ones (blue, black, red, and magenta dashed lines, respectively). See [13] and references therein.

ASTRI camera adopts as FEE the CITIROC, a customized version of the EASIROC [11] ASIC signal-shaper manufactured by Omega[1].

We are completing the commissioning phase during which we tested all the mechanical and optical subsystems, as well as the control software. The commissioning of the camera will take place in Spring 2016. A science verification phase, during which we will observe Mrk 421, Mrk 501, and afterwards the Crab Nebula, will start at the end of Spring 2016. The results of these observations will allow us to cross-check the prototype performance with our Monte Carlo simulations. We estimate that, in the sensitivity range ($E \geq 2$ TeV), we can detect a flux level of 1 Crab at 5σ in a few hours [12].

3. The ASTRI mini-array

A remarkable improvement in terms of performance could come from the operation, in late 2016, of a mini-array, composed of nine SST-2M telescopes (see Fig. 1, right panel, for an artistic concept) to be placed at the final CTA southern site. The ASTRI mini-array is a collaborative effort led by INAF in synergy with the University of São Paulo and FAPESP (Brazil) and the North-West University (South-Africa). Preliminary Monte Carlo simulations [13] yield an improvement in sensitivity that, for nine telescopes, could surpass the H.E.S.S. sensitivity above 10 TeV, extending it up to about 100 TeV, as shown in Fig. 2. The ASTRI mini-array will be able to study in detail relatively bright (a few $\times 10^{-12}$ erg cm^{-2}s^{-1} at 10 TeV) sources with an angular resolution of a few arcmin and an energy resolution of about 10–15%.

Thanks to the array approach, it will be possible to verify the wide FoV performance to detect very high energy showers with the core located at a distance up to 500 m and to compare the mini-array performance with the Monte Carlo expectations by means of deep observations of a few selected targets. Moreover, it will be possible to perform the first CTA science, with its first solid detections during the first year of operation, as described in [14]. Prominent sources such as extreme blazars (KUV 00311−1938), nearby well-known BL Lac objects (Mrk 501) and radio-galaxies (M 87), galactic pulsar wind nebulae (Crab Nebula, Vela-X), supernovae remnants (Vela-junior, RX J1713.7−3946), as well as

[1] `http://omega.in2p3.fr/`; manufactured under INAF intellectual property.

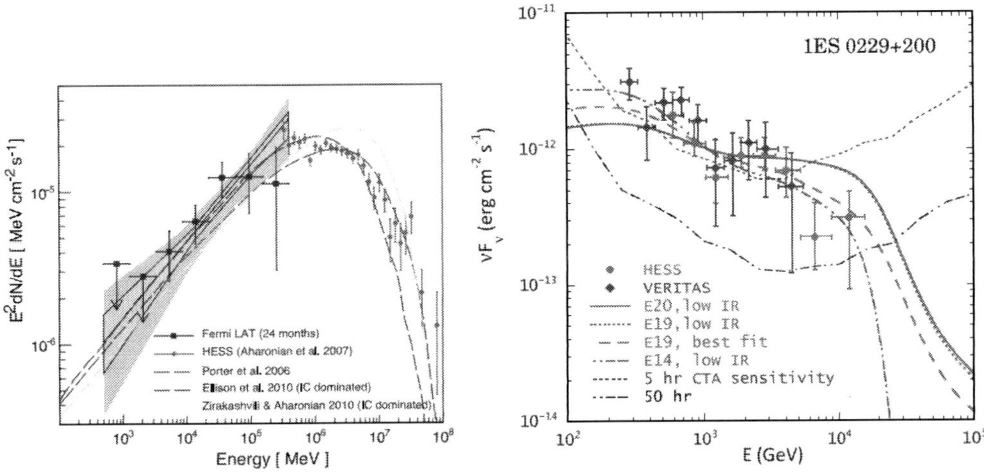

Figure 3. Left panel: supernova remnant RX J1713.7−3946 (see [15] for details). We expect for the ASTRI mini-array sensitivity at least comparable to (or slightly better than) the H.E.S.S. one above a few TeV. Right panel: spectral energy distribution of the extreme blazar 1ES 0229+200 (see [16] for details).

the Galactic Center can be observed in a previously unexplored energy range, in order to investigate the electron acceleration and cooling, relativistic and non relativistic shocks, the search for cosmic-ray (CR) PeVatrons, the study of the CR propagation, and the impact of the extragalactic background light on the spectra of the sources.

On the Galactic plane, one of the best targets is RX J1713.7−3946. Its detection by *Fermi* [15] and the combined study with H.E.S.S. (Fig. 3, left panel) shows that the high-energy and very high-energy (VHE) emission could be interpreted in the framework of a leptonic scenario. The improved and uniform sensitivity (within a few degrees off-axis) and the comparable angular resolution of the ASTRI mini-array at $E \geq 10$ TeV with respect to the current IACTs could allow us to investigate the VHE emission in the different regions of this source, studying their spectra, and to extend the current spectral energy distribution (SED) well above a few tens of TeV, searching for possible spectral cut-offs. Figure 3 (right panel) shows the SED of the extreme blazar 1ES 0229+200 with superimposed different theoretical SED fits (both hadronic and leptonic) assuming different EBL models (see [16] for a detailed discussion) and the 5-σ differential sensitivity for 5 and 50 hr observations with CTA (configuration E, as reported in [2]). Because of the uncertainty in EBL models, it is not easy to distinguish between the hadronic and leptonic scenarios at $\sim$1–10 TeV energies. At higher energies, however, UHECR-induced cascade emission becomes harder than the gamma-ray one. A detection of $\geq$25 TeV gamma-rays from 1ES 0229+200 would only be consistent with an hadronic scenario.

4. Synergies and conclusions

The ASTRI mini-array will operate when the other IACTs will still be active. Compared to them, the ASTRI mini-array will extend the sensitivity up to 100 TeV and beyond, a never-explored energy range by IACTs. Moreover, it will benefit from a much larger field of view which will allow us to study in detail extended sources at energies about a decade higher than what is currently being explored and to monitor simultaneously a few close-by sources during the same pointing. Long exposures will be preferred, restricting the number of possible targets, and extending the observations also during moon light periods, thanks to the use of a SiPMs-based camera.

The lower imaging energy threshold of current and future extended air-shower (EAS) detectors (~100 GeV) and the wider energy range of the ASTRI mini-array (beyond 100 TeV) will allow us a direct comparison of scientific data (spectra, light-curves, integral fluxes) of those sources which could be monitored simultaneously, although on different integration time-scales. Moreover, the high-energy boundary of both EAS and the ASTRI mini-array will allow us to study the VHE (E $\geq$10 TeV) emission from extended source such as SNRs and PWN, and to investigate the presence of spectral cut-offs. In summary, the ASTRI mini-array could be considered as the first CTA *seed*, allowing the entire CTA Consortium to start seminal studies on both Galactic and extra-galactic sources.

This work was partially supported by the ASTRI "Flagship Project" financed by the Italian Ministry of Education, University, and Research (MIUR) and led by the Italian National Institute of Astrophysics (INAF). We acknowledge partial support by the MIUR Bando PRIN 2009 and TeChe.it 2014 Special Grants. We also acknowledge support from the Brazilian Funding Agency FAPESP (Grant 2013/10559-5) and from the South African Department of Science and Technology through Funding Agreement 0227/2014 for the South African Gamma-Ray Astronomy Programme. We gratefully acknowledge support from the agencies and organizations listed in this page: `http://www.cta-observatory.org/?q=node/22`.

References

[1] J.A. Hinton, W. Hofmann, Annual Review of Astronomy & Astrophysics **47**, 523 (2009)

[2] M. Actis, G. Agnetta, F. Aharonian, A. Akhperjanian, J. Aleksić, E. Aliu, D. Allan, I. Allekotte, F. Antico, L.A. Antonelli et al., Experimental Astronomy **32**, 193 (2011)

[3] B.S. Acharya, M. Actis, T. Aghajani, G. Agnetta, J. Aguilar, F. Aharonian, M. Ajello, A. Akhperjanian, M. Alcubierre, J. Aleksić et al., Astroparticle Physics **43**, 3 (2013)

[4] J. Carr, This Volume (in press)

[5] G. Pareschi, the ASTRI Collaboration, the GATE Collaboration, the CHEC Collaboration, the CTA Consortium, Proc. 33rd ICRC (2013)

[6] V. Vassiliev, S. Fegan, P. Brousseau, Astroparticle Physics **28**, 10 (2007)

[7] R. Canestrari, G. Bonnoli, E. Cascone, P. Conconi, G. Crimi, M. Fiorini, E. Giro, N. La Palombara, G. Pareschi, L. Perri et al., *The ASTRI SST-2M prototype for the CTA: manufacturing of the structure and the mirrors*, in *SPIE Conference Series* (2014), Vol. 9145

[8] R. Canestrari, G. Bonnoli, G. Crimi, M. Fiorini, E. Giro, N. La Palombara, G. Pareschi, L. Perri, G. Rodeghiero, G. Sironi et al., *The glass cold-shaping technology for the mirrors of the CTA*, in *SPIE Conference Series* (2014), Vol. 9151

[9] G. Bonanno, D. Marano, M. Belluso, S. Billotta, A. Grillo, S. Garozzo, G. Romeo, M. Timpanaro, Sensors Journal, IEEE **14**, 3567 (2014)

[10] O. Catalano, M.C. Maccarone, C. Gargano, G. La Rosa, A. Segreto, G. Sottile, V. De Caprio, F. Russo, M. Capalbi, P. Sangiorgi et al., *The camera of the ASTRI SST-2M prototype for the Cherenkov Telescope Array*, in *SPIE Conference Series* (2014), Vol. 9147

[11] S. Callier, C.D. Taille, G. Martin-Chassard, L. Raux, Physics Procedia **37**, 1569 (2012), proceedings of the 2nd International TIPP Conference

[12] C. Bigongiari, the ASTRI Collaboration, Proc. 33rd ICRC (2013), `arXiv:1307.5006`

[13] F. Di Pierro, C. Bigongiari, A. Stamerra, P. Vallania, the ASTRI Collaboration, the CTA Consortium, Proc. TAUP 2015 (in press)
[14] S. Vercellone, the ASTRI Collaboration, Proc. 33rd ICRC (2013), arXiv:1307.5671
[15] A.A. Abdo, M. Ackermann, M. Ajello, A. Allafort, L. Baldini, J. Ballet, G. Barbiellini, M.G. Baring, D. Bastieri, R. Bellazzini et al., Astrophysical Journal **734**, 28 (2011)
[16] K. Murase, C.D. Dermer, H. Takami, G. Migliori, Astrophysical Journal **749**, 63 (2012)

EPJ Web of Conferences **121**, 04007 (2016) DOI: 10.1051/epjconf/201612104007
RICAP-2014

Discovery of TeV gamma-ray emission from the pulsar wind nebula 3C 58 by MAGIC

Rubén López-Coto[a] for the MAGIC Collaboration

Institut de Física d'Altes Energies, Spain

Abstract. The pulsar wind nebula (PWN) 3C 58 is one of the historical very-high-energy (VHE; E> 100 GeV) gamma-ray source candidates. It has been compared to the Crab Nebula due to their morphological similarities. This object was detected by Fermi-LAT with a spectrum extending beyond 100 GeV. We analyzed 81 hours of 3C 58 data taken with the MAGIC telescopes and we detected VHE gamma-ray emission for the first time at TeV energies with a significance of 5.7 sigma and an integral flux of 0.65% C.U. above 1 TeV. According to our results 3C 58 is the least luminous PWN ever detected at VHE and the one with the lowest flux at VHE to date. We compare our results with the expectations of time-dependent models in which electrons up-scatter photon fields. The best representation favors a distance to the PWN of 2 kpc and Far Infrared (FIR) comparable to CMB photon fields. Hadronic contribution from the hosting supernova remnant (SNR) requires unrealistic energy budget given the density of the medium, disfavoring cosmic ray acceleration in the SNR as origin of the VHE gamma-ray emission.

1. General description

The supernova remnant 3C 58 (SNR G130.7+3.1) has a flat radio spectrum and is brightest near the center, therefore it was classified as a pulsar wind nebula [PWN; 1]. It is centered on PSR J0205+6449, a pulsar discovered in 2002 with the *Chandra* X-ray observatory [2]. It is widely assumed that 3C 58 is located at a distance of 3.2 kpc [3], but recent H I measurements suggest a distance of 2 kpc [4]. The age of the system is estimated to be ~ 2.5 kyr [5] from the PWN evolution and energetics, however this is a matter of debate. The pulsar has one of the highest spin-down powers known ($\dot{E} = 2.7\times10^{37}$erg s^{-1}). The PWN has a size of $9'\times6'$ in radio, infrared (IR), and X-rays [6–9]. Its luminosity is $L_{0.5-10\,\mathrm{keV}} = 2.4\times10^{34}$ erg s^{-1} in the X-ray band, which is more than 3 orders of magnitude lower than that of the Crab nebula [10]. 3C 58 has been compared with the Crab because the jet-torus structure is similar [8]. Because of these morphological similarities with the Crab nebula and its high spin-down power (5% of Crab), 3C 58 has historically been considered one of the PWNe most likely to emit γ rays.

The PWN 3C 58 was previously observed in the VHE γ-ray range by several IACTs. The Whipple telescope reported an integral flux upper limit of 1.31×10^{-11} cm^{-2}s^{-1} $\sim$ 19% C.U. at an energy threshold of 500 GeV [11], and VERITAS established upper limits at the level of 2.3% C.U. above an energy of 300 GeV [12]. MAGIC-I observed the source in 2005 and established integral upper limits above 110 GeV at the level of 7.7×10^{-12} cm^{-2}s^{-1} ($\sim$4%

[a]e-mail: rlopez@ifae.es

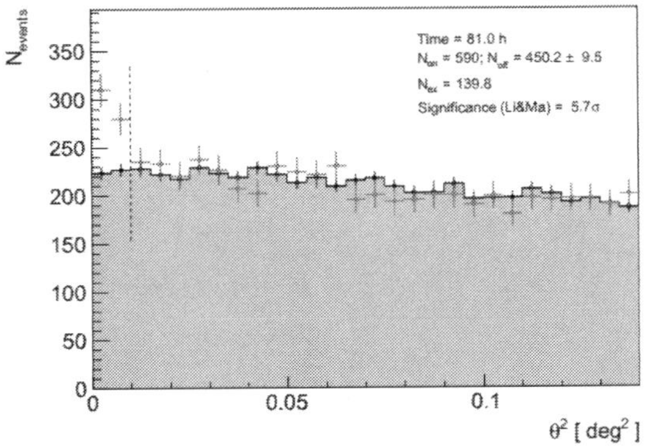

Figure 1. Distribution of squared angular distance, θ^2, between the reconstructed arrival directions of gamma-ray candidate events and the position of PSR J0205+6449 (*red points*).

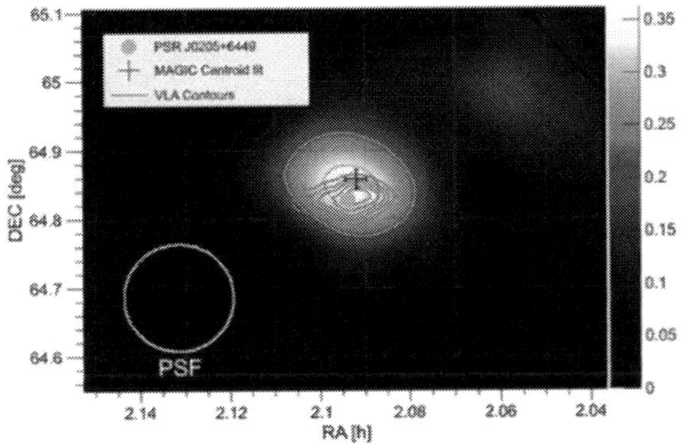

Figure 2. Relative flux (excess/background) map for MAGIC observations. The cyan circle indicates the position of PSR J0205+6449 and the black cross shows the fitted centroid of the MAGIC image with its statistical uncertainty. In green we plot the contour levels for the TS starting at 4 and increasing in steps of 1. The magenta contours represent the VLA flux at 1.4 GHz [16], starting at 0.25 Jy and increasing in steps of 0.25 Jy.

C.U.) [13, 14]. The improved sensitivity of the MAGIC telescopes with respect to previous observations and the *Fermi*-LAT results motivated us to perform deep VHE observations of the source.

2. MAGIC observations and results

MAGIC observed 3C 58 in the period between 4 August 2013 to 5 January 2014 for 99 hours, and after quality cuts, 81 hours of the data were used for the analysis. The data were analyzed using the MARS analysis framework [15].

The applied cuts yield an energy threshold of 420 GeV. The significance of the signal, calculated with the LiMa formula, is 5.7σ, which establishes 3C 58 as a γ-ray source. The θ^2 distribution is shown in Fig. 1. As the five OFF positions were taken for each of the wobble positions, the OFF histograms were re-weighted depending on the time taken on each wobble position.

We show in Fig. 2 the relative flux (excess/background) skymap, produced using the same cuts as for the θ^2 calculation. The TS significance, which is the LiMa significance applied on

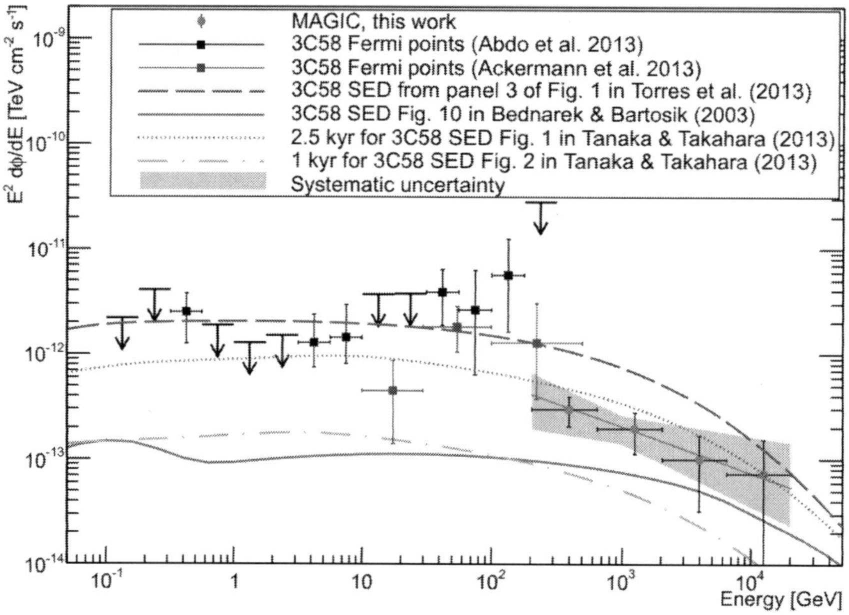

Figure 3. 3C 58 spectral energy distribution in the range between 0.1 GeV and 20 TeV. Red circles are the VHE points reported in this work. The best-fit function is drawn in red and the systematic uncertainty is represented by the yellow shaded area. Black squares and black arrows are taken from the *Fermi*-LAT second pulsar-catalog results [17]. Blue squares are taken from the *Fermi* high-energy LAT catalog [18]. The magenta line is the SED prediction for 3C 58 taken from Fig. 10 of [21]. The clear green dashed-dotted line is the SED predicted by [22], assuming an age of 1 kyr, and the dark green dotted line is the prediction from the same paper, assuming an age of 2.5 kyr. The blue dashed line represents the SED predicted by [23] assuming that the Galactic FIR background is high enough to reach a flux detectable by the MAGIC sensitivity in 50h.

a smoothed and modeled background estimate, is higher than 6 at the position of the pulsar PSR J0205+6449. The excess of the VHE skymap was fit with a Gaussian function. The best-fit position is RA(J2000)=2 h 05 m $31(09)_{\rm stat}(11)_{\rm sys}$ s; DEC (J2000)= $64°51'(1)_{\rm stat}(1)_{\rm sys}$. This position is statistically deviant by 2σ from the position of the pulsar, but is compatible with it if systematic errors are taken into account. In the bottom left of the image we show the point spread function (PSF) of the smeared map at the corresponding energies, which is the result of the sum in quadrature of the instrumental angular resolution and the applied smearing (4.7′ radius, at the analysis energy threshold). The extension of the signal is compatible with the instrument PSF. The VLA contours are coincident with the detected γ-ray excess.

Figure 3 shows the energy spectrum for the MAGIC data, together with published predictions for the gamma-ray emission from several authors, and two spectra obtained with three years of *Fermi*-LAT data, which were retrieved from the *Fermi*-LAT second pulsar-catalog [2PC, 17] and the *Fermi* high-energy LAT catalog [1FHL, 18]. The 1FHL catalog used events from the *Pass 7 Clean class*, which provides a substantial reduction of residual cosmic-ray background above 10 GeV, at the expense of a slightly smaller collection area, compared with the *Pass 7 Source class* that was adopted for 2PC [19]. The two γ-ray spectra from 3C58 reported in the 2PC and 1FHL catalogs agree within statistical uncertainties. The differential energy spectrum of the source is well fit by a single power-law function $d\phi/dE = f_0(E/1\ \mathrm{TeV})^{-\Gamma}$ with $f_0 = (2.0 \pm 0.4_{\rm stat} \pm 0.6_{\rm sys})10^{-13}\mathrm{cm}^{-2}\mathrm{s}^{-1}\mathrm{TeV}^{-1}$,

$\Gamma = 2.4 \pm 0.2_{\rm stat} \pm 0.2_{\rm sys}$ and χ^2=0.04/2. The systematic errors were estimated from the MAGIC performance paper [20] including the upgraded telescope performances. The integral flux above 1 TeV is $F_{E>1\,{\rm TeV}} = 1.4 \times 10^{-13}{\rm cm}^{-2}{\rm s}^{-1}$. Taking into account a distance of 2 kpc, the luminosity of the source above 1 TeV is $L_\gamma, E > 1\, TeV = (3.0 \pm 1.1)\times 10^{32} d_2^2$ erg s^{-1}, where d_2 is the distance normalized to 2 kpc.

3. Discussion

Several models have been proposed that predict the VHE γ-ray emission of PWN 3C 58. [24] presented a one zone model of the spectral evolution of PWNe and applied it to 3C 58 using a distance of 3.2 kpc. The VHE emission from this model consists of IC scattering of CMB photons and optical-to-IR photons, and also of pion decay. The flux of γ rays above 400 GeV predicted by this model is about an order of magnitude lower than the observation.

[21] proposed a time-dependent model in which positrons gain energy in the process of resonant scattering by heavy nuclei. The VHE emission is produced by IC scattering of leptons off CMB, IR, and synchrotron photons and by the decay of pions due to the interaction of nuclei with the matter of the nebula. The age of 3C 58 is assumed to be 5 kyr, using a distance of 3.2 kpc and an expansion velocity of 1000 km s^{-1}. According to this model, the predicted integral flux above 400 GeV is $\sim 10^{-13}$ cm^{-2}s^{-1}, while the integral flux above 420 GeV measured here is 5×10^{-13}cm^{-2}s^{-1}. Calculations by [25], using the same model with an initial expansion velocity of 2000 km s^{-1} and considering IC scattering only from the CMB, are consistent with the observed spectrum. However, the magnetic field derived in this case is $B \sim 14\mu$G and it underestimates the radio emission of the nebula, although a more complex spectral shape might account for the radio nebula emission.

[26] developed a time-dependent model of the spectral evolution of PWN including synchrotron emission, synchrotron self-Compton, and IC. They evolved the electron energy distribution using an advective differential equation. To calculate the observability of 3C 58 at TeV energies they assumed a distance of 2 kpc and two different ages: 2.5 kyr and 1 kyr [22]. For the 2.5 kyr age, they obtained a magnetic field $B \sim 17\ \mu$G, while for an age of 1 kyr, the magnetic field obtained is B=40 μG. The emission predicted by this model is closer to the *Fermi* result for an age of 2.5 kyr.

[27] presented a different time-dependent leptonic diffusion-loss equation model without approximations, including synchrotron emission, synchrotron self-Compton, IC, and bremsstrahlung. They assumed a distance of 3.2 kpc and an age of 2.5 kyr to calculate the observability of 3C 58 at high energies [23]. The predicted emission, without considering any additional photon source other than the CMB, is more than an order of magnitude lower than the flux reported here. It predicts VHE emission detectable by MAGIC in 50 hours for an FIR-dominated photon background with an energy density of 5 eV/cm^3. This would be more than one order of magnitude higher than the local IR density in the Galactic background radiation model used in GALPROP [$\sim$0.2 eV cm^{-3}; 28]. The magnetic field derived from this model is 35 μG. To reproduce the observations, a large FIR background or a revised distance to the PWN of 2 kpc are required. In the first case, a nearby star or the SNR itself might provide the necessary FIR targets, although no detection of an enhancement has been found in the direction of the PWN. As we mentioned in Sect. 1, a distance of 2 kpc has recently been proposed by [4] based on the recent H I measurements of the Canadian Galactic Plane Survey. At this distance, a lower photon density is required to fit the VHE data.

4. Conclusions

We have for the first time detected VHE γ rays up to TeV energies from the PWN 3C 58. Following the assumptions in [29], it is highly unlikely that the measured flux comes from hadronic emission of the SNR. The measured luminosity and flux make 3C 58 into an exceptional object. It is the weakest VHE PWN detected to date, a fact that attests to the sensitivity of MAGIC. On the other hand, it is also the least luminous VHE PWN, far less luminous than the original expectations. Its ration $L_{VHE}/\dot{E} \simeq 10^{-5}$ is the lowest measured, similar to Crab, which makes into a very inefficient γ-ray emitter. Only a closer distance of 2 kpc or a high local FIR photon density can qualitatively reproduce the multiwavelength data of this object in the published models. Since the high FIR density is unexpected, the closer distance with FIR photon density comparable with the averaged value in the Galaxy is favored. The models that fit the γ-ray data derived magnetic fields which are very far from equipartition.

References

[1] K.W. Weiler, N. Panagia, A&A **70**, 419 (1978)
[2] S.S. Murray, P.O. Slane, F.D. Seward, S.M. Ransom, B.M. Gaensler, ApJ **568**, 226 (2002), astro-ph/0108489
[3] D.A. Roberts, W.M. Goss, P.M.W. Kalberla, U. Herbstmeier, U.J. Schwarz, A&A **274**, 427 (1993)
[4] R. Kothes, A&A **560**, A18 (2013)
[5] R.A. Chevalier, ApJ **619**, 839 (2005), arXiv:astro-ph/0409013
[6] M.F. Bietenholz, N.E. Kassim, K.W. Weiler, ApJ **560**, 772 (2001), arXiv:astro-ph/0106334
[7] F. Bocchino, R.S. Warwick, P. Marty, D. Lumb, W. Becker, C. Pigot, A&A **369**, 1078 (2001), astro-ph/0102087
[8] P. Slane, D.J. Helfand, E. van der Swaluw, S.S. Murray, ApJ **616**, 403 (2004), arXiv:astro-ph/0405380
[9] P. Slane, D.J. Helfand, S.P. Reynolds, B.M. Gaensler, A. Lemiere, Z. Wang, ApJ **676**, L33 (2008), 0802.0206
[10] K. Torii, P.O. Slane, K. Kinugasa, K. Hashimotodani, H. Tsunemi PASJ **52**, 875 (2000), astro-ph/0006034
[11] T.A. Hall, S.P. Wakely, VERITAS Collaboration, *Observations of Galactic Pulsars and Shell-Type SNRs with the Whipple 10 m Imaging Atmospheric Cherenkov Telescope*, in *International Cosmic Ray Conference* (2001), Vol. 6 of *International Cosmic Ray Conference*, p. 2485
[12] E. Aliu, *Search for VHE γ-ray emission in the vicinity of selected pulsars of the Northern Sky with VERITAS*, in *American Institute of Physics Conference Series*, edited by F.A. Aharonian, W. Hofmann, F. Rieger (2008), Vol. 1085 of *American Institute of Physics Conference Series*, pp. 324–327, 0812.1415
[13] H. Anderhub, L.A. Antonelli, P. Antoranz, M. Backes, C. Baixeras, S. Balestra, J.A. Barrio, D. Bastieri, J. Becerra González, J.K. Becker et al., ApJ **710**, 828 (2010)
[14] E. Aliu (2007), vHE gamma-ray observations of Northern sky pulsar wind nebulae with the MAGIC telescope, PhD. thesis
[15] R. Zanin, E. Carmona, J. Sitarek, P. Colin, K. Frantzen, M. Gaug, S. Lombardi, for the MAGIC collaboration, Proceedings of the ICRC 2013, id 773 (2013), 0907.0943
[16] J.J. Condon, W.D. Cotton, E.W. Greisen, Q.F. Yin, R.A. Perley, G.B. Taylor, J.J. Broderick, AJ **115**, 1693 (1998)

[17] A.A. Abdo, M. Ajello, A. Allafort, L. Baldini, J. Ballet, G. Barbiellini, M.G. Baring, D. Bastieri, A. Belfiore, R. Bellazzini et al., ApJ **208**, 17 (2013), 1305.4385
[18] The Fermi-LAT Collaboration, ArXiv e-prints (2013), 1306.6772
[19] M. Ackermann, M. Ajello, A. Albert, A. Allafort, W.B. Atwood, M. Axelsson, L. Baldini, J. Ballet, G. Barbiellini, D. Bastieri et al., ApJ **203**, 4 (2012), 1206.1896
[20] J. Aleksić, E.A. Alvarez, L.A. Antonelli, P. Antoranz, M. Asensio, M. Backes, J.A. Barrio, D. Bastieri, J. Becerra González, W. Bednarek et al., Astroparticle Physics **35**, 435 (2012), 1108.1477
[21] W. Bednarek, M. Bartosik, A&A **405**, 689 (2003), arXiv:astro-ph/0304049
[22] S.J. Tanaka, F. Takahara, MNRAS **429**, 2945 (2013), 1211.7266
[23] D.F. Torres, A.N. Cillis, J. Martín Rodriguez, ApJ **763**, L4 (2013), 1212.1995
[24] N. Bucciantini, J. Arons, E. Amato, MNRAS **410**, 381 (2011)
[25] W. Bednarek, M. Bartosik, Journal of Physics G Nuclear Physics **31**, 1465 (2005), arXiv:astro-ph/0509872
[26] S.J. Tanaka, F. Takahara, ApJ **715**, 1248 (2010), 1004.3098
[27] J. Martín, D.F. Torres, N. Rea, MNRAS **427**, 415 (2012), 1209.0300
[28] T.A. Porter, I.V. Moskalenko, A.W. Strong, ApJ **648**, L29 (2006), astro-ph/0607344
[29] L.O. Drury, F.A. Aharonian, H.J. Voelk, A&A **287**, 959 (1994), astro-ph/9305037

EPJ Web of Conferences **121**, 04008 (2016) DOI: 10.1051/epjconf/201612104008
RICAP-2014

Search for ultra high energy primary photons at the Pierre Auger Observatory

Roberta Colalillo[1,2,a] for the Pierre Auger Collaboration[3,b]

[1] Universitää Federico II di Napoli, Italy
[2] INFN, Sezione di Napoli, Italy
[3] Observatorio Pierre Auger, Av. San Martín Norte 304, 5613 Malargüe, Argentina

Abstract. The Pierre Auger Observatory, located in Argentina, provides an unprecedented integrated aperture in the search for primary photons with energy above 10^{17} eV over a large portion of the southern sky. Such photons can be detected in principle via the air showers they initiate at such energies, using the complement of Auger Observatory detectors. We discuss the results obtained in diffuse and directional searches for primary photons in the EeV energy range.

1. Introduction

Photons are possible primary cosmic messengers in the flux of particles with energy larger than 10^{18} eV. The Pierre Auger Observatory and other experiments have observed a suppression in the cosmic ray energy spectrum at $E_{GZK} \sim 6 \times 10^{19}$ eV. This feature could be related to the GZK effect [1, 2], but also to the photo-disintegration of heavy nuclei or to the maximum energy reached by the accelerating sources. The GZK effect predicts that cosmic rays with energies above the threshold energy E_{GZK} interact with cosmic microwave background photons, γ_{CMB}, and produce pions via the Δ resonance: $\gamma_{CMB} + p \rightarrow \Delta^+ \rightarrow p + \pi^0$; $\gamma_{CMB} + p \rightarrow \Delta^+ \rightarrow n + \pi^+$. GZK photons come from the π^0 decay, hence the observation of these Ultra High Energy (UHE) photons would be a proof of the GZK cut-off and a way to constrain source and propagation models. Moreover, photons are not deflected in magnetic fields, so the study of UHE photons is important in searching for astronomical sources, and the presence or lack of UHE photons could have an impact on the measurements of the energy spectrum, cross sections, mass composition of cosmic rays, and on fundamental physical processes.

The Pierre Auger Observatory, located near Malargüe, Argentina, consists of a surface detector array (SD) [3] of 1660 water Cherenkov stations spread over an area of 3000 km^2 and overlooked by 27 air fluorescence telescopes (FD) [4]. It is optimized to study Ultra High Energy Cosmic Rays (UHECR). The longitudinal profile of the showers is observed by the FD, while the distribution of shower particles at the ground is evaluated with the SD. Showers induced by photons develop deeper in the atmosphere compared to hadron showers, so the depth of shower maximum, X_{max}, is larger for photon-induced events. In addition, they are also characterized by a smaller number of secondary muons and their signal in the SD

[a]e-mail: colalillo@na.infn.it
[b]Full author list: http://www.auger.org/archive/authors_2014_09.html

has peculiar characteristics. Using these properties of photon-induced showers, several upper limits on the photon fraction and flux were published by the Pierre Auger Collaboration.

2. Upper limit on the cosmic ray photon flux using the surface detector

Upper limits on the photon flux above 10, 20, and 40 EeV were derived using data recorded by the surface detector [5]. The approach is based on observables sensitive to the longitudinal shower development and to the muon content, the signal risetime and the curvature of the shower front. The radius of curvature for photon-initiated showers is smaller than that for hadronic showers, but the risetime is larger for deeper showers. Data recorded from 2004 to 2006 were analyzed.

The criteria to select well-reconstructed events are: the station with the largest signal is surrounded by six active stations; ≥ 5 stations are used in the fitting of the lateral distribution function out of which ≥ 4 stations have a non-saturated signal of ≥ 10 VEM (vertical equivalent muons); the reduced χ^2 from the fit to estimate the radius of curvature is < 10. The photon efficiency decreases to small values below ~ 10 EeV. At higher energy, near-vertical photons can also fail the station multiplicity cut due to their deep development. Therefore, the analysis is restricted to primary energies ≥ 10 EeV and primary zenith angles of $30 - 60°$.

The first step in the search for photon candidates is the construction of the deviation Δ_x of the observable x ($x = t_{1/2}$ or R referring to risetime or radius of curvature, respectively) from the mean value $\bar{x}_\gamma$ predicted for photons from simulated photon-initiated showers in units of the spread $\sigma_{x,\gamma}$ of the observable x:

$$\Delta_x = \frac{x - \bar{x}_\gamma(S(1000), \theta)}{\sigma_{x,\gamma}(S(1000), \theta)},$$

where $S(1000)$ is the time-integrated energy deposit at 1000 m from the shower core and is used as an energy estimator.

Then $\Delta_{t_{1/2}}$ and Δ_R are combined to perform a principal component analysis [6]. A 5% fraction of the real events, as well as the results from photon simulations are used to derive the principal component, actually the axis with the largest variance. The remaining 95% of the data are then projected onto the principal axis together with the data from simulated photons as shown in Fig. 1(a). The simulations have been performed with the AIRES package using the QGSJET hadronic interaction model, and a power law spectrum of index −2.0 has been assumed. In the projected distribution, a cut has to be established to select photon candidate events. The mean of the distribution for photons is chosen as the cut, so the efficiency of this cut is $f = 0.5$ by construction. Any real event falling above this cut is considered a photon candidate.

Finally, an upper limit on the number of photons N_γ^{CL} at the 95% confidence level (CL) is calculated from the number of photon candidate events N_γ above a minimum energy, E_{min}. The upper limits on the flux and fraction of photons are based on N_γ^{CL}. The flux limit is given by:

$$\Phi_{CL}(E > E_{min}) = \frac{N_\gamma^{CL}(E_\gamma > E_{min}) \times \frac{1}{f} \times \frac{1}{\epsilon}}{0.95A},$$

where $f = 0.5$ is the photon selection cut efficiency, ϵ is the integrated efficiency of accepting photons and A is the exposure of the detector. The factor 0.95 originates from the fact that only 95% of the data are used to determine the number of photon candidate events. The upper

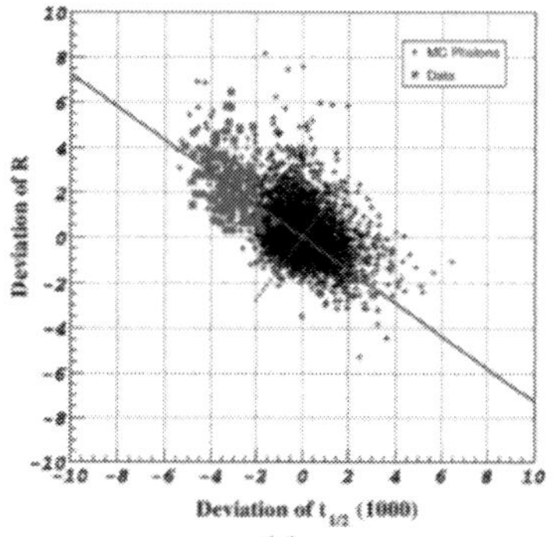

(a)

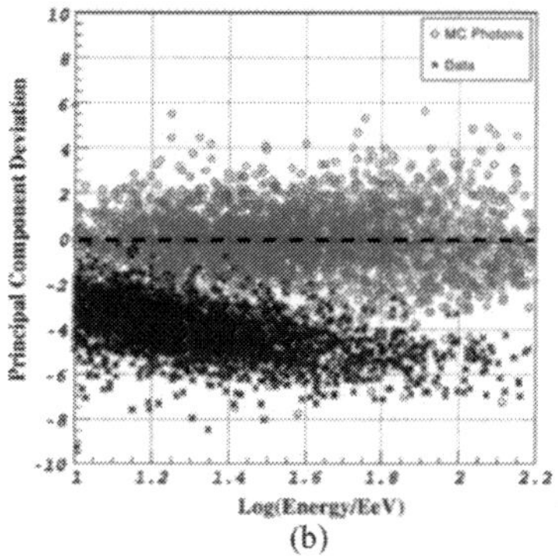

(b)

Figure 1. a) Deviation from a photon prediction for 5% of the data (closed squares) and for simulated photon showers (crosses). The solid grey line is the principal component while the dashed line is the axis perpendicular to the principal component. The minimum energy is 10 EeV. b) Deviation of data (black crosses) and photons (open red circles) from the principal component as a function of the primary energy (photon energy scale). The dashed line indicates the mean of the distribution for photons, chosen as the selection criterion. Data lying above this line are photon candidates. No event meets this requirement.

limit on the photon fraction is given by:

$$F_{CL}(E > E_{min}) = \frac{N_{\gamma}^{CL}(E_{\gamma} > E_{min}) \times \frac{1}{f} \times \frac{1}{\epsilon}}{N_{\gamma}(E_{\gamma} > E_{min}) + N_{non-\gamma}(E_{non-\gamma} > E_{min})},$$

where $N_{non-\gamma}$ represents the number of non-photon candidate events in the data set for the fraction limit.

No event passes the photon candidate cut, as shown in Fig. 1(b). The upper limits on the photon flux above 10, 20, and 40 EeV are then, respectively, 3.8×10^{-3}, 2.5×10^{-3}, and 2.2×10^{-3} km^{-2}sr^{-1}yr^{-1} (at the 95% CL). The limits on the photon fraction are 2.0%, 5.1%, and 31% (at the 95% CL) above 10, 20, and 40 EeV, respectively. The new limits are compared to previous results and to theoretical predictions in Fig. 3(b) for the photon flux and in Fig. 2 for the photon fraction. A discovery of a substantial photon flux could be interpreted as a signature of "top-down" models, which predict that UHECRs come from the decay of exotic particles directly at the energies observed. But these experimental limits put strong constraints on such models.

3. Search for ultra high energy photons using hybrid events

Hybrid events refer to air showers observed simultaneously with the surface array and fluorescence telescopes of the Auger Observatory, on dark, clear nights. From the calorimetric observation of the longitudinal development of ultra high energy air showers with the Auger Observatory, upper limits of 3.8%, 2.4%, 3.5%, and 11.7% (at the 95% CL) are obtained on the fraction of cosmic-ray photons above 2, 3, 5, and 10 EeV, respectively [7]. These are the first experimental limits on ultra high energy photons below 10 EeV, so they complete the results obtained with data recorded by the surface detector (Sect. 2). The depths of shower maxima, X_{max}, observed by fluorescence telescopes in hybrid mode was used to derive these upper limits. To improve the discrimination between photon-induced showers and showers initiated by protons, that are the hadronic showers most similar to photon-induced showers, and so to also improve the upper limits, another analysis was performed combining the X_{max} with a variable derived from the signal in the stations of the surface detector [8]. This variable,

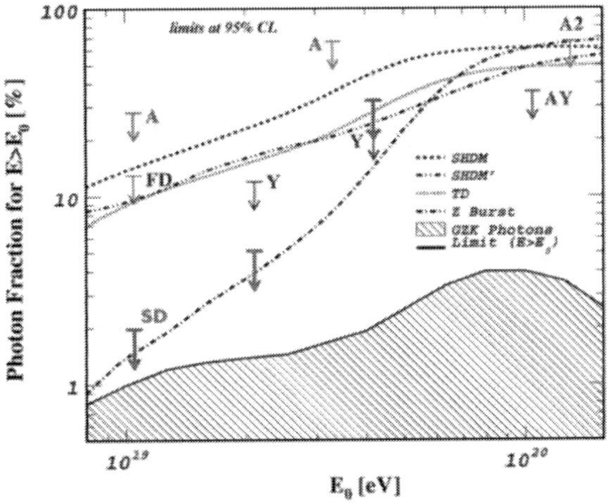

Figure 2. Upper limits on the fraction of photons derived in the work described in Sect. 2 (red arrows) along with previous experimental limits (A, A2: AGASA [11, 12], AY: AGASA-Yakutsk [13], Y: Yakutsk [14]; FD: Auger hybrid limit [16]). There is also the comparison with "top-down" models (TD, Z-Burst, SHDM from [17] and SHDM' from [18]).

called S_b, is defined as [9]:

$$S_b = \sum_i S_i \left(\frac{R_i}{R_{ref}} \right)^b ,$$

where the sum runs over the triggered stations, S_i is the signal recorded by the i-th station placed at a distance R_i from the hybrid shower core, and R_{ref} is a reference distance equal to 1000 m for this analysis. The exponent b is chosen equal to 4 to maximize the separation power between photons and hadrons. S_b is very sensitive to the difference in the muon component, and in fact combines the different amplitudes of the signals in the SD stations and the sharper distributions of the signal as a function of distance from the axis expected for photon-induced showers. In conclusion, these showers have a smaller S_b than hadronic showers.

A Fisher analysis [10] is performed in bins of 0.5 in the logarithm of energy using photon and proton showers simulated with CORSIKA following a power law spectrum between 10^{17} and 10^{20} eV. QGSJET-II and FLUKA are used as hadronic interaction models. Events with zenith angle smaller than 60° and with a good geometry reconstruction are selected for the analysis. The conditions to ensure a reliable profile reconstruction are: a reduced χ^2 of the Gaisser-Hillas fit to the longitudinal profile smaller than 2.5, the χ^2 value obtained from a linear fit to the longitudinal profile exceeding the χ^2 value from the Gaisser-Hillas fit by at least a factor of 1.1, the X_{max} observed within the field of view of the telescopes, any Cherenkov light contamination smaller than 50% and the uncertainty of the reconstructed energy less than 20%. Moreover, only time periods with the sky not obscured by clouds and with a reliable measurement of the vertical optical depth of aerosols are used. To avoid an underestimation of S_b, that would mimic the behavior of a photon event, at least 4 active stations within 2 km from the hybrid reconstructed axis are required. The Fisher response for simulations with energy between 10^{18} and $10^{18.5}$ eV (the first energy bin, the analysis is performed in three energy bins until $10^{19.5}$ eV) is shown in Fig. 3(a). The cut was chosen at the mean of the Fisher distribution for photons so the photon detection efficiency is 50% by

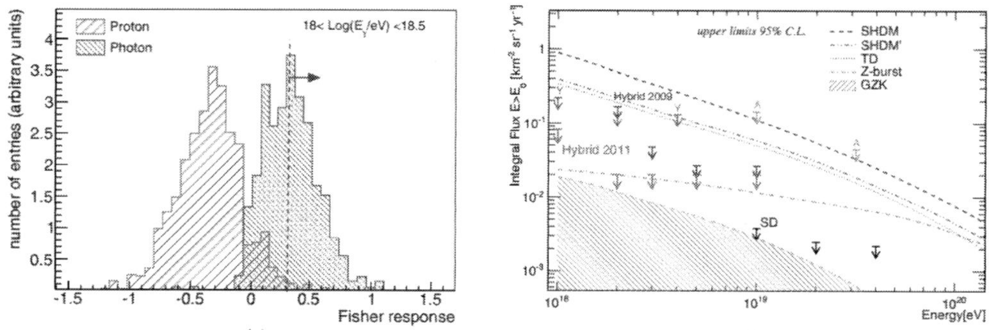

(a) (b)

Figure 3. a) Fisher distribution for simulated protons (red) and photons (blue) with energy between 10^{18} and $10^{18.5}$ eV. Photon-candidate events have a Fisher value larger than X_{cut} (dashed line, median of photon Fisher distribution). b) Upper limits on the photon flux above 1, 2, 3, 5, and 10 EeV derived in the work described in Sect. 3 (red arrows) compared to previous limits from Auger (SD (work described in Sect. 2) and Hybrid 2009 [7]), from AGASA (A) [11] and Yakutsk (Y) [15]. The shaded region and the lines give predictions for the GZK photon flux [17] and for "top-down" models (TD, Z-Burst, SHDM from [17] and SHDM' from [18]).

construction. In the lowest energy interval, the expected hadron contamination is about 1% and it becomes smaller for increasing energies.

The data used for this analysis are hybrid showers collected between January 2005 and September 2010. Applying the cuts described above to data, 6, 0, 0, 0, and 0 photon candidates are found for energies above 1, 2, 3, 5, and 10 EeV, respectively, but the observed number of photon candidates is consistent with the expectation for nuclear primaries under the assumption of a mixed composition.
The 95% CL upper limits on the photon flux $\Phi_{\gamma}^{0.95}$ integrated above an energy threshold E_0 is given by:

$$\Phi_{\gamma}^{0.95} = \frac{N_{\gamma}^{0.95}(E_{\gamma} > E_0)}{\epsilon_{\gamma,min}},$$

where E_{γ} is the reconstructed energy using the photon energy scale, $N_{\gamma}^{0.95}$ is the number of photon candidates above E_0 at the 95% CL, and $\epsilon_{\gamma,min}$ is the minimum value (to be conservative) of the exposure of the hybrid detector. Upper limits on the integral photon flux of $8.2 \times 10^{-2}\,\mathrm{km}^{-2}\mathrm{sr}^{-1}\mathrm{yr}^{-1}$ above 1 EeV and $2.0 \times 10^{-2}\,\mathrm{km}^{-2}\mathrm{sr}^{-1}\mathrm{yr}^{-1}$ above 2, 3, 5, and 10 EeV are derived and are shown in Fig. 3(b). The results disfavor "top-down" models also in the lowest energy region. Using the Auger spectrum, upper limits on the fraction of photons of about 0.4%, 0.5%, 1.0%, 2.6%, and 8.9% are obtained for energies above 1, 2, 3, 5, and 10 EeV, respectively.

4. Search for point sources of EeV photons

The identification of photons coming from discrete sources is a direct way to identify the origins of cosmic rays; so an analysis was performed to search for point sources of EeV photons in the exposed sky using the hybrid events collected between January 2005 and September 2011 at the Pierre Auger Observatory [19]. The considered declination band goes from $-85°$ to $+20°$ and the energy range from $10^{17.3}$ eV to $10^{18.5}$ eV. With this last choice, high event statistics is ensured and shower development processes that may introduce a bias

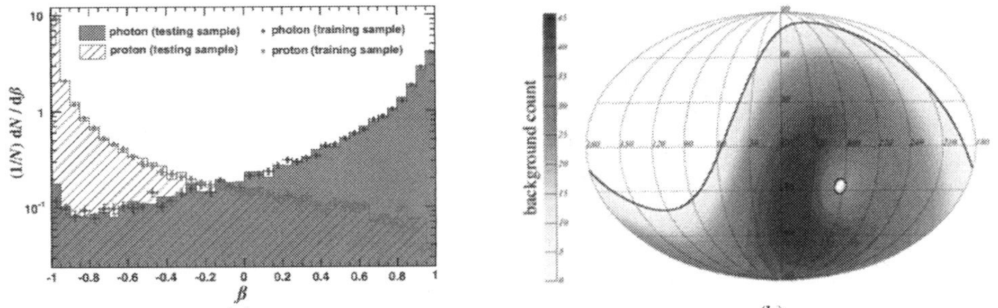

(a) (b)

Figure 4. a) MVA output value, β, for photon and proton primaries using BDTs. The first two steps of this technique are training and testing, so the MC sample is split half into a training (filled circles) and half into a testing sample (solid line). b) Celestial map in Galactic coordinates of the expected background contribution. The solid black lines indicate the declination range covered in this analysis.

at highest energies (pre-shower and LPM [20]) are avoided. The goal of this analysis is to derive directional upper limits on the photon flux over all exposed celestial directions using a multivariate analysis, a Boosted Decision Tree (BDT) analysis [21, 22], in which FD and SD variables are combined to select photon candidate events.

The three selected FD variables are: the depth of shower maximum X_{max}; the reduced χ^2 from the fit to the longitudinal profile with a Greisen function (optimized to describe the longitudinal profile of pure elecromagnetic showers); and the ratio E_{Gr}/E_{GH}, where E_{Gr} is the energy obtained with the Greisen fit and E_{GH} is that obtained by the fit with the Gaisser-Hillas function, the typical function used to describe the shower longitudinal profile.

The two SD observables used in the analysis are: S_b, already described in Sect. 3, but in the energy region of interest for this analysis, the exponent that maximizes the photon-hadron discrimination being $b = 3$; and the shape parameter, that is the ratio of the early-arriving to the late-arriving integrated signal as a function of time measured in the SD stations with the strongest signal:

$$\text{ShapeP}(r,\theta) = \frac{S_{early}(r,\theta)}{S_{late}(r,\theta)}.$$

The early and late signals are separated by a scaled time defined as:

$$t_i^{scaled}(r,\theta) = t_i \cdot \frac{r_0}{r} \cdot \frac{1}{c_1 + c_2 \cdot cos(\theta)},$$

where t_i is the real time of bin i and r_0 is a reference distance equal to 1000 m. Here $c_1 = -0.6$ and $c_2 = 1.9$ are scaling parameters to average traces over different inclination angles.

The separation is the quantity that evaluates the performance of BDTs. It is zero when signal and background have identical shapes and it is one when there is no overlap between the two shapes. The training and testing steps of BDTs are performed with photon and proton showers simulated using CORSIKA version 6.900. They are generated according to a power-law spectrum of index -2.7, using QGSJET01 as the high-energy and GHEISHA as the low-energy interaction models. The MVA output variable, β, is shown in Fig. 4(a). In this analysis, events with zenith angle smaller than 60° are used. The necessary cuts to have a reliable profile reconstruction and a good reconstruction of S_b are the same as the ones described in Sect. 3. The only difference is that it is not required that X_{max} have been observed within the field of view because the aim is to maximize the acceptance for photon showers. Measured data are used to estimate the expected isotropic background using the scrambling

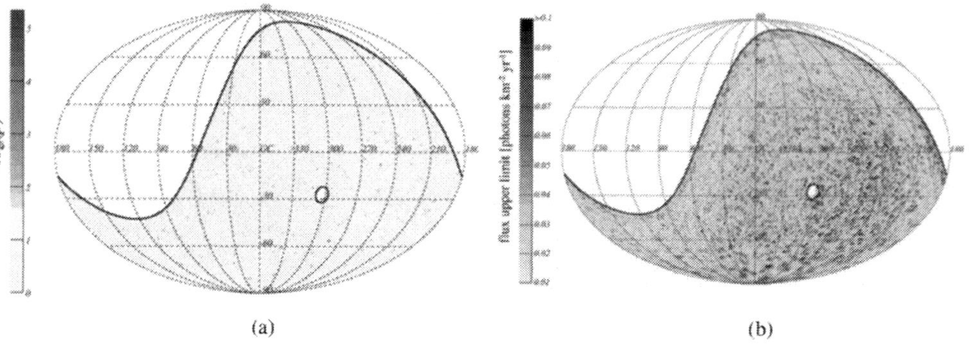

Figure 5. a) Sky map of -log(p) values in Galactic coordinates. b) Sky map of the photon flux in Galactic coordinates.

technique [23]. The sky map of directional background contribution for the covered search period is shown in Fig. 4(b).

To select photon candidate events, it is necessary to establish a cut on the β variable. It is optimized by taking into account the expected number of background events $n_b(\alpha, \delta)$, which is a function of the celestial coordinates α (right ascension) and δ (declination). The upper limit of photons, n_s, from a point source in a given direction is calculated under the assumption that n_{data}, the observed number of events, is equal to the expected number of events that pass the β cut, $n_b^{\beta}(\alpha, \delta) = n_b(\alpha, \delta) \cdot \epsilon_{data}^{\beta}(\delta)$ where ϵ_{data}^{β} is the data efficiency. This number is typically less than 4 events for the β_{cut} values finally chosen. The procedure of Zech [24] is used to define the upper limit on the photon number at a given confidence level in the presence of a Poisson-distributed background:

$$P(\leq n_{data}^{\beta} | n_b^{\beta} + n_s^{Zech}) = \alpha_{CL} \cdot P(\leq n_{data}^{\beta} | n_b^{\beta})$$

where $\alpha_{CL} = 1 - CL = 0.05$ and n_b^{β} is the expected background contribution. At this point, the probability p of obtaining a test statistic has to be calculated: $p = Poiss(\geq n_{data}^{\beta} | n_b^{\beta})$ is the Poisson probability to observe n_{data}^{β} or more events given a background expectation after β_{cut} of n_b^{β}. The celestial map distribution of the p-value is shown in Fig. 5(a). The directional upper limit on the photon flux from a point source is defined as:

$$f^{UL} = \frac{n_s^{Zech}}{n_{inc} \cdot \epsilon_{\beta}}$$

where n_s^{Zech} is the limit on the number of photons from a given direction, ϵ_{β} the directional acceptance from the same target at the 95% CL and n_{inc} is a correction term. Figure 5(b) shows the celestial map of upper limits on the photon flux. The p-values are referred to a local probability that the data is in agreement with a uniform distribution. The minimum p-value observed is $p_{min} = 4.5 \times 10^{-6}$. The probability to observe this value anywhere in the sky is $p_{chance} = 36\%$. The mean value of the photon flux is 0.035 photons km^{-2} yr^{-1}. This value is not a signature of a directional photon flux, in fact for the median exposure, a flux of 0.14 photons km^{-2} yr^{-1} or greater would yield an excess of at least 5σ.

5. Conclusions

The analyses performed until now by the Pierre Auger Collaboration to observe ultra high energy photons have not found candidates, but imposed severe constraints on "top-down" models. The GZK photons could be observed in the near future; the idea is to enhance the sensitivity with new analyses, an upgraded detector and accumulating more statistics.

Also the "blind analysis", that is the analysis performed in all parts of the exposed sky, to search for a directional photon flux has not detected any photon point sources. If some real photon fluxes are responsible for some of the lower p-values seen, additional exposure should increase the significance of those excesses. Moreover, targeted searches, choosing one or more directions of significance, are ongoing.

References

[1] K. Greisen, Phys. Rev. Lett. **16**, 748–750 (1966)
[2] G.T. Zatsepin and V.A. Kuz'min, Journal of Experimental and Theoretical Physics Lett. **4**, 78–80 (1966)
[3] The Pierre Auger Collaboration, Nucl. Instr. Meth. Phys. Res. A **613**, 29–39 (2010)
[4] The Pierre Auger Collaboration, Nucl. Instr. Meth. Phys. Res. A **620**, 227–251 (2010)
[5] The Pierre Auger Collaboration, Astropart. Phys. **29**, 243–256 (2008)
[6] E. Oja, Int. J. Neural Syst. **1**, 61–68 (1989)
[7] The Pierre Auger Collaboration, Astropart. Phys. **31**, 399–406 (2009)
[8] M. Settimo for The Pierre Auger Collaboration, 32nd ICRC, Beijing 2011, HE1.3, 0393
[9] G. Ros et al., Astopart. Phys. **35**, 140–151 (2011)
[10] R. A. Fisher, Annals Eugenics **7**, 179–188 (1936)
[11] K. Shinozaki et al., Astrophys. J. **571**, L117–L120 (2002)
[12] M. Risse et al., Phys. Rev. Lett. **95**, 171102:1-4 (2005)
[13] G.I. Rubtsov et al., Phys. Rev. D **73**, 063009:1-8 (2006)
[14] A. Glushkov et al., Journal of Experimental and Theoretical Physics Lett. **85**, 163–167 (2007)
[15] A. Glushkov et al., Phys. Rev. D **82**, 041101:1-5 (2002)
[16] The Pierre Auger Collaboration, Astropart. Phys. **27**, 155–168 (2007)
[17] G. Gelmini, O. Kalashev, D. Semikoz, J. Exp. Theor. Phys. **106**, 1061–1082 (2008)
[18] J. Ellis et al., Phys. Rev. D **74**: 115003:1-11 (2006)
[19] The Pierre Auger Collaboration, Astrophys. J. **789**, 160–171 (2014)
[20] M. Risse, P. Homola, Mod. Phys. Lett. A **22**, 749–766 (2007)
[21] L. Breiman, J. Friedman, R. A. Olshen, and C. J. Stone, *Classification and Regression Trees* (Wadsworth and Brooks, Monterey (CA), 1984)
[22] R. E. Schapire, Mach. Learn. **5**, 197–226 (1990)
[23] G. L. Cassiday et al., Nucl. Phys. Proc. Suppl. **14**A, 291–298 (1990)
[24] G. Zech, Nucl. Instrum. Meth. A **277**, 608–610 (1989)

EPJ Web of Conferences **121**, 05001 (2016) DOI: 10.1051/epjconf/201612105001
RICAP-2014

Neutrinos from Gamma Ray Bursts in the IceCube and ARA Era

Dafne Guetta[1,2,a]

[1] ORT Braude College, Carmiel, Israel
[2] INAF-OAR Monteporzio Catone, Italy

Abstract. In this review I discuss the ultra-high energy neutrinos (UHEN) originated from Cosmic-Rays propogation (GZK neutrinos) and from Gamma Ray Bursts (GRBs), and discuss their detectability in kilometers scale detectors like ARA and IceCube.

While GZK neutrinos are expected from cosmic ray interactions on the CMB, the GRB neutrinos depend on the physics inside the sources. GRBs are predicted to emit UHEN in the prompt and in the later "after-glow" phase.

I discuss the constraints on the hadronic component of GRBs derived from the search of four years of IceCube data for a prompt neutrino fux from gamma-ray bursts (GRBs) and more in general I present the results of the search for high-energy neutrinos interacting within the IceCube detector between 2010 and 2013.

1. Introduction

The detection of MeV neutrinos emitted from the sun and from the supernovae allowed the understanding of the physics of these astrophysical objects [1, 2]. MeV neutrino "telescopes" are capable of detecting neutrinos from sources close to our galaxy up to a distances < 100 kp. The main goal of the construction of high energy, >1 TeV, neutrino telescopes [3] is the extension of the distance accessible to neutrino astronomy to cosmological scales [4].

The existence of extra-Galactic high-energy neutrino sources is implied by cosmic-ray observations. The cosmic-ray spectrum extends to energies $\sim 10^{20}$ eV, and is likely dominated beyond $\sim 10^{19}$ eV by extra-Galactic sources. The origin of Cosmic Rays (CR) has been a tantalizing mystery ever since their discovery by Hess [5] nearly a century ago. While "lower energy" CRs of up to 10^{16} eV are believed to originate from supernova explosions in our galaxy [6], the source of the more energetic CR whose energies can exceed 10^{19} eV remains unknown [7]. Although ultrahigh energy cosmic rays (UHECR) are produced throughout the Universe, those observed at Earth must have been produced locally (within 50 Mpc) since they lose energy while propagating through the Cosmic Microwave Background (CMB). This process (discussed by Greisen [8] and Zatsepin and Kuzmin [9] – "GZK") not only causes energy loss for the primary, but also creates secondary particles of extremely high energy [10, 11], including neutrinos above 10^{17} eV. The secondary particles, particularly Ultra High Energy Neutrinos (UHEN), can be used to explore the origins of UHECR [12]. With no electric charge, neutrinos experience no scattering or energy loss, and so provide a probe of the source distribution even to high redshift. Since UHECRs and the

[a] e-mail: guetta@mporzio.astro.it

expected "GZK cutoff" have been observed, the expectation of a GZK neutrino flux is on very strong footing [13]. UHE neutrinos are also likely to be created at the acceleration sites of the UHECRs, pointing directly to the source from earth, and providing information on the role of hadrons in the acceleration.

Only Active Galactic Nuclei [14] and Gamma Ray Bursts, GRBs [15–17], are believed to be capable of accelerating CRs to such enormous energies [18]. Gamma Ray Bursts (GRBs) are powerful explosions, and are among the highest-redshift point sources observed. The most common phenomenological interpretation of these cosmological sources is through the so called fireball model [19–21]. In this model, part of the energy is carried out (e.g., from a collapsed star) by hadrons at highly-relativistic energies, some of which is dissipated internally and eventually reconverted into internal energy, which is then radiated as γ-rays by synchrotron and inverse-Compton emission by shock-accelerated electrons. As the fireball sweeps up ambient material, it energizes the surrounding medium through, e.g., forward shocks, which are believed to be responsible for the longer-wavelength afterglow emission [20].

If the GRB jet comprises PeV protons, it should produce energetic neutrinos through photon-hadron interactions. The photons for this process can be supplied by the GRB gamma rays during its prompt phase, or during the afterglow phase [22, 23]. These lead to the production of charged pions, which subsequently decay to produce neutrinos. Within this picture, GRBs should produce neutrinos with energies of $\sim$100 TeV from the same region in which the GRB photons are produced [24]. These neutrinos, if present, could be readily detected. Hence, the detectability of TeV to PeV neutrinos depends on the presence of ($>$) PeV protons and on the efficiency at which their energy is converted into neutrinos, as compared to how much of the energy is in electrons, which is manifested primarily in the prompt GRB photon emission.

Neutrino astronomy has steadily progressed over the last half century, with successive generations of detectors achieving sensitivity to neutrinos with increasingly higher energies. With each increase in neutrino energy, the required detector increases in size to compensate for the dramatic decrease of the flux with energy. The high-energy neutrinos from GRBs and the GZK neutrinos discussed above should be detected by large neutrino telescopes, such as IceCube, the Askarian Radio array (ARA) and in the future KM3NeT[1]. All these detectors look for the Cherenkov radiation initiated by the neutrino interactions in the ice using either optical detection (IceCube, KM3Net) or Radio Frequency (RF) detection (ARA).

2. Results from IceCube

IceCube, is a Cherenkov detector [25] with photomultipliers (PMTs) at depths between 1450 and 2450 meters in the Antarctic ice designed specifically to detect neutrinos at TeV-PeV energies. Since May 2011 [26], IceCube has been working with a full capacity of 86 strings (IC86). IceCube analyses include a model-independent search for GRB neutrinos [27], and for other diffuse and point sources. A search for high-energy neutrinos interacting within the IceCube detector between 2010 and 2013 provided the first evidence for a high-energy neutrino flux of extraterrestrial origin [28]. In the full 988-day sample, IceCube detected 37 events (Fig. 1) in the TeV-PeV range. A purely atmospheric explanation for these events is strongly disfavored by their properties [29]. The high galactic latitudes of many of the highest energy events (Fig. 2) suggest at least some extragalactic component. Moreover the intensity associated with the neutrino excess is much higher than that expected to

---

[1] `http://km3net.org/home.php`

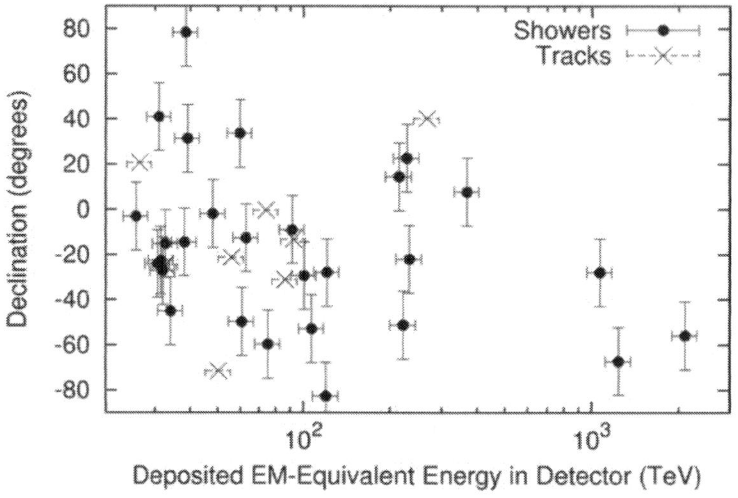

Figure 1. Arrival angles and deposited energies of the 37 events detected by Icecube.

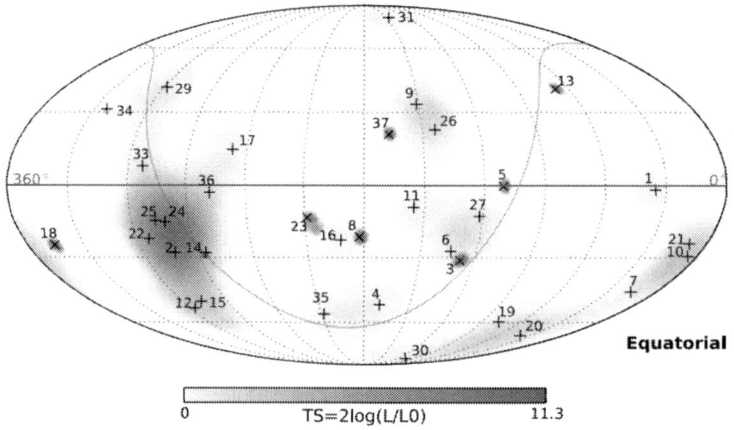

Figure 2. Arrival directions of the events in galactic coordinates. Shower-like events (median angular resolution $\sim 15^0$) are marked with + and those containing muon tracks ($< 1^0$) with x. The grey line denotes the equatorial plane. Colors show the test statistic (TS) for the point source clustering test at each location. No significant clustering was observed.

originate from interaction of cosmic-ray protons with interstellar gas in the Galaxy [30]. The flux, spectrum and angular distribution of the excess neutrino signal detected by IceCube between 50 TeV and 2 PeV suggest that the sources of these neutrinos are unlikely to be (unknown) Galactic sources. If the sources were galactic the events were expected to be strongly concentrated along the galactic disk [31]. Results are consistent with an astrophysical flux in the 100 TeV – PeV range at the level of 10^{-8}GeV cm^{-2}s^{-1}sr^{-1} per flavor. The data are consistent with expectations for equal fluxes of all three neutrino flavors and with isotropic arrival directions, suggesting either numerous or spatially extended sources. No evidence of neutrino emission from point-like or extended sources was found in four years of IceCube data. Searches for emissions from point-like and extended sources anywhere in the sky, from a pre-defined candidate source list and from stacked source catalogs all returned results consistent with the background-only hypothesis. 90% C.L. upper limits on the muon

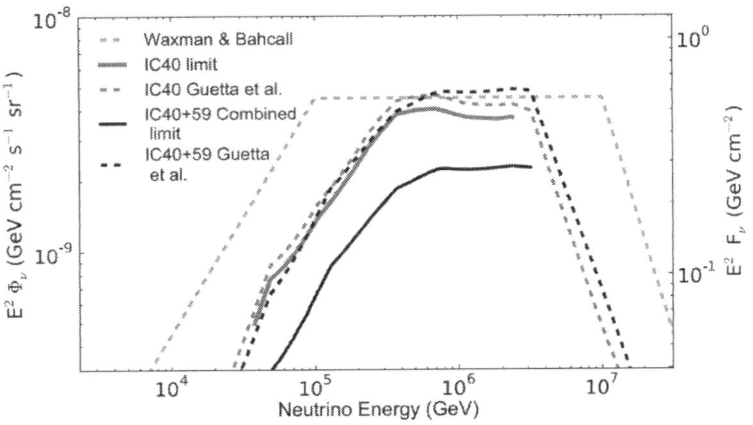

Figure 3. Comparison of results to predictions based on observed gamma-ray spectra. The summed flux predictions normalized to gamma-ray spectra [24, 36] is shown in dashed lines; pay attention that the cosmic ray normalized Waxman-Bahcall flux [22] shown here [34] for reference is not correct and should be a factor 10 smaller.

neutrino fluxes for models from a variety of sources have been calculated and compared to predictions [32]. The coincidence of the excess, $E_\nu^2\phi_\nu = 3.6 \pm 1.2\,10^{-8}\,\mathrm{GeV/cm^2 sr\ s}$, with the Waxman-Bahcall (WB) bound, $E_\nu^2\phi_{WB} = 3.6 \pm 1.2\,10^{-8}\,\mathrm{GeV/cm^2 sr\ s}$, is probably a clue to the origin of IceCubes neutrinos. The most natural explanation of this coincidence is that both the neutrino excess and the ultra-high energy, $> 10^{19}$ eV, cosmic-ray (UHECR) flux are produced by the same population of cosmologically distributed sources [31]. The coincidence with the WB flux also support the hypothesis that the detected neutrinos have an extra-Galactic origin. The $\nu_\mu{:}\nu_e{:}\nu_\tau = 1:1:1$ flavor ratio is consistent with that expected for neutrinos originating from pion decay in cosmologically distant sources, for which oscillations modify the original $1:2:0$ ratio to a $1:1:1$ ratio [33]. In order to identify the sources of UHE neutrinos a multiwavelength analysis is needed.

3. Neutrinos from Gamma-Ray-Bursts

The high-energy neutrinos from GRBs should be detected by large neutrino telescopes, such as IceCube and in the future KM3NeT[2]. Since GRB neutrino events need to be correlated both in time and in direction with the gamma-rays, they are sought after in small angular and short time windows. Therefore the search for neutrino events should be coordinated with the gamma-ray telescopes that provide the GRB trigger. In this context, IceCube has recently developed a powerful model-independent analysis tool for neutrinos detection, which is coincident in direction and in time to within 1,000 seconds with GRB flares reported by the gamma ray satellites. IceCube reported no detection of any GRB-associated neutrino in a data set taken from April 2008 to May 2010 [34]; None of the high energy neutrinos reported for the next two years [35] is GRB-associated either, and as far as we know no neutrino event has been associated with any GRB to date. This non-detection is in conflict with earlier models [22, 24, 36–38], all of which predicted the detection of approximately ten GRB neutrinos by IceCube during this period. Those earlier estimates were largely calibrated based on the fireball hypothesis, and were motivated by the assumption that UHECRs are produced

[2] `http://km3net.org/home.php`

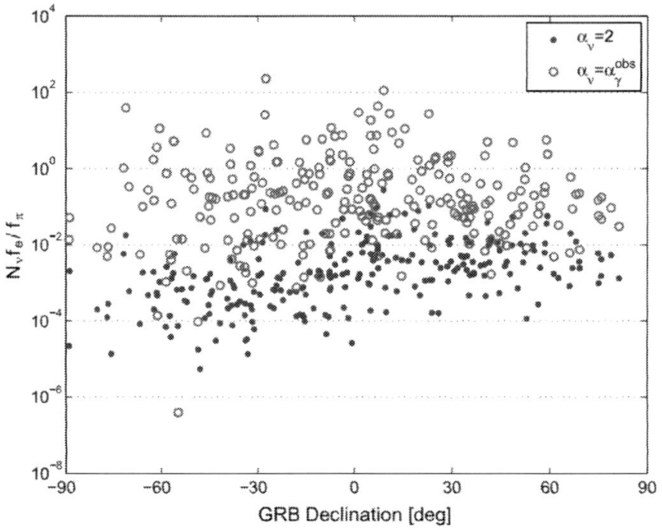

Figure 4. Expected number of ν_μ from 250 GBM detected GRBs, which have a measured spectral slope, as a function of declination, and factored by the unknown electron to pion energy ratio $f_e f_\pi$. The neutrino spectral slope α_ν is assumed to be that of the photons α_γ. The results for $\alpha_\nu = 2$ are reported for comparison [43].

primarily by GRBs. As shown in Fig. 3 the IceCube results thus appear to rule out GRBs as the main sources of UHECRs [34, 37]. This implies either that GRBs do not have the (>)PeV protons, hypothesized in the fireball model, or that the efficiency of neutrino production from these protons is much lower than what have been estimated [39–41]. Recently the Icecube collaboration set the most stringent limits yet on GRB neutrino production using four years of IceCube data [42]. They constrain parts of the parameter space relevant to the production of UHECRs in the latest models. Because of the very low on-time and on-source background rate, the analysis grows more sensitive almost linearly with time.

Recently Yacobi et al. (2014) [43] have used the data from the GRB Monitor (GBM) on board Fermi to calibrate the photon (representing electrons) energy content of the GRB jet. Subsequently, they have compared this with the upper limit on proton (turned pion) energy content, given the non-detection of GRB neutrinos and derived a constraint on the fraction of energy that goes into hadrons to the one that goes into electrons, the ratio f_π/f_e. The results for each individual GRB are plotted in Fig. 4, which shows many GRBs yielding high values of $N_\nu(f_e f_\pi)$, and thus tightening the upper limit on the f_π/f_e by more than two orders of magnitude. Considering all the GBM sample Yacobi et al. 2014 found that the lack of detected neutrinos from Fermi/GBM GRBs since 2008 points to $f_\pi f_e \lesssim 0.24$ with a 95% CL. The obtained value of $f_\pi f_e \lesssim 0.2$ is still consistent with the values of $f_e \approx 1$ [22], and of $f_\pi \approx 0.2$ [24].

Another constraint on the hadronic fractio of the GRB jet can be obtained using the data of the Large Area Telescope (LAT) on board Fermi. The first catalog of LAT includes 35 GRBs with gamma ray emission above 100 GeV [44]. Several models have been proposed to explain this high energy emission [20, 45, 46] including hadronic models [47, 48]. The same photon-hadron process that produces the charged pions and subsequently the 100 TeV neutrinos, would also generate neutral pions that decay to photons of similar energy. These high-energy photons have been hypothesized to cascade through pair production processes

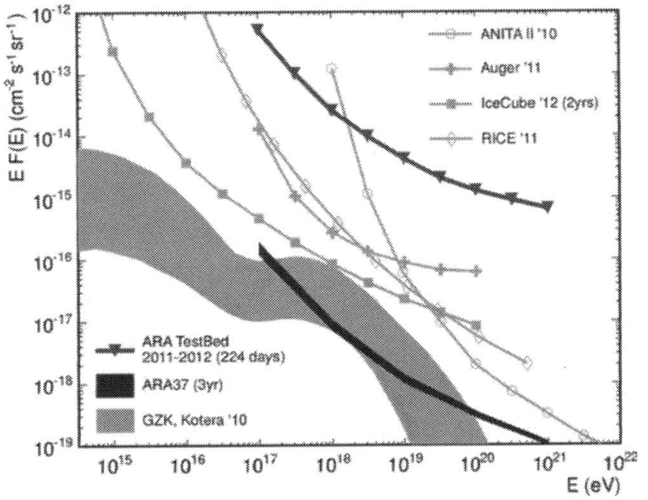

Figure 5. The limits placed compared with the projected ARA37 trigger-level sensitivity and results from other experiments.

down to the GeV regime, where they can escape the jet and be observed by LAT. Within this scenario, Yacobi et al. (2014) use the observed GeV burst fluence to put another upper limit on the energy content of the protons in the jet.

The most conservative estimation for the hadronic contribution to the GeV photon fluence measured by LAT is $f_{\mathrm{Had}} \leq 1$, i.e. all LAT fluence is hadronic (via pair-photon cascades). Using the typical ratio $F_{\mathrm{GBM}}/F_{\mathrm{LAT}} \approx 10$ [44] Yacobi et al. (2014) constrain the typical GRB hadronic fraction to be $f_\pi/f_e \lesssim 0.3$.

Since the flux of neutrinos depends on the energy fraction of protons in the FB for a given burst energy, the flux or limit of the neutrino emission could constrain the hadronic component in the GRB jet and the hadronic emission models. Therefore a multi-wavelength analysis is mandatory in particular the synergy between the Swift, LAT, IceCube and ARA is fundamental to give constraints to the FB model and the hadronic emission models.

4. GZK neutrinos and ARA

Although ultrahigh energy cosmic rays (UHECR) likely originate throughout the Universe, those observed at Earth must be produced locally since such UHECR lose energy while propagating through the CMB ($p\gamma- > n\pi^+- > n\mu + \nu_\mu$, etc.) producing GZK neutrinos. The detection of UHE GZK neutrinos is an experimental challenge at the frontier of neutrino astronomy, which has progressed over the last half century.

The flux of GZK neutrinos incident on earth ranges from $\sim 0.01/\mathrm{km}^2/\mathrm{year}$ [49] to $\sim 6/\mathrm{km}^2/\mathrm{year}$ [50], depending on the model considered. This flux can be detected in dense, RF-transparent media such as ice via the Askaryan effect [51]. There is approximately one UHE neutrino interaction occurring in each km^3 in the ice per year. The long RF attenuation length in the Antarctic ice allows a more efficient area coverage that makes it possible to construct detectors of order tens to hundreds of km^2, and several small-scale pioneering efforts to develop this approach exist [52–54]. Two years ago, a two-phased experiment named ARA (Askaryan Radio Array) was initiated, designed to ultimately accumulate hundreds of GZK neutrinos [55]. The goal of Phase 1 of ARA (2010–2016) is to make the first definitive observation of cosmogenic neutrinos; Phase 2 of ARA (37 stations) would then accumulate the statistics necessary to carry out an expanded astrophysics and particle-physics science program. The primary goal of ARA is to discover GZK neutrinos and to establish the spectrum.

The detection of GZK neutrinos from ARA will allow to understand the origin of the UHECR cutoff confirmed by the recent data of Auger and the nature of the UHECR composition.

I thank the Scientific Organization Committee for having invited me to give this plenary talk. This research is supported by a grant from the U.S. Israel Binational Science Foundation.

References

[1] For review see J. N. Bahcall, Proc. Nobel Symp. **129**: Neutrino Phys. (hep-ph/0412068)
[2] For review see G. Raffelt, Proc. Nobel Symp. **129**: Neutrino Phys. (hep-ph/0501049)
[3] For review see T. K. Gaisser, F. Halzen & T. Stanev, Phys. Rep. **258**, 173 (1995)
[4] E. Waxman, Nuclear Physics A **827**, 15 (2009)
[5] V. F. Hess, Uber Beobachtungen der durchdringenden Strahlung bei sieben Freiballonfahrt, Physikalische Zeitschrift **13**, 10841091 (1912)
[6] P. Blasi, arXiv:0801.4534 (2008)
[7] J. W. Cronin, Nucl. Phys. Proc. Suppl. **138**, 465 (2005)
[8] K. Greisen, Phys. Rev. Lett. **16**, 748 (1966)
[9] G. T. Zatsepin, V. A. Kuzmin, JETP Lett. **4**, 78 (1966)
[10] V. S. Beresinsky, G. T. Zatsepin, Phys. Lett. B **28**, 423 (1969)
[11] F. W. Stecker, Astrophys. Space Sci. **20**, 47 (1973)
[12] D. Seckel, T. Stanev, Phys. Rev. Lett. **95**, 141101 (2005)
[13] K. Mannheim, R. J. Protheroe, J. P. Rachen, Phys. Rev. D **63**, 023003 (2001)
[14] J. P. Rachen, P. L. Biermann, Astron. Astrophys. **272**, 161 (1993)
[15] M. Milgrom, V. Usov, Astrophys. J. **449**, L37 (1995)
[16] M. Vietri, Astrophys. J. **453**, 883 (1995)
[17] E. Waxman, Physical Review Letters **75**, 386 (1995)
[18] S. Wick, C. Dermer, A. Atoyan, Astropart. Phys. **21**, 2, 125 (2004)
[19] T. Piran, Phys. Rep. **333**, 529 (2000)
[20] P. Mészáros, A&AS **138**, 533 (1999)
[21] P. Mészáros, Rep. Prog. Phys. **69**, 2259 (2006)
[22] E. Waxman & J. N. Bahcall Phys. Rev. Lett. **78**, 2292 (1997)
[23] C. D. Dermer, ApJ **574**, 65 (2002)
[24] D. Guetta, D. Hooper, J. Alvarez-Miniz J.,F. Halzen, & E. Reuveni Astroparticle Physics **20**, 429 (2004)
[25] F. Halzen & S. Klein Rev. Sci. Instrum. **81**, 081101 (2010)
[26] Aartsen et al. ApJ **779**, 132 (2013a)
[27] IceCube, R. Abbasi et al., Nature **484**, 351 (2012)
[28] Aartsen et al., Phys. Rev. Lett. **113**, 101101 (2014)
[29] Aartsen, M. G. et al., Icecube collaboration Phys. Rev. Lett. **113**, 101101 (2014)
[30] Katz B. et al., astro-ph/1311.0287 (2013)
[31] Waxman E. astro-ph/1312.0558 (2013)
[32] Aartsen, M. G. et al., Icecube collaboration ApJ **796**, 109A (2014)
[33] J. G. Learned & S. Pakvasa, Astropar. Phys. **3**, 267 (1995); H. Athar, C. S. Kim & J. Lee, Modern Physics Letters A **21**, 1049 (2006)

[34] Abbasi R. et al., Nature **484**, 351 (2012)
[35] Aartsen et al., Science **342**, 6161 (2013b)
[36] Rachen J. P. & Mészáros, P. arXiv:astro-ph/9811266 [in C.A. Meegan, R.D. Preece, & T.M. Koshut (ed.), American Institute of Physics Conference Series **428**, 776]
[37] M. Ahlers, M. C. Gonzalez-Garcia, & F. Halzen, Astroparticle Physics **35**, 87 (2011)
[38] He et al., ApJ **752**, 29 (2012)
[39] P. Baerwald, S. Hümmer, & Winter W., Phys. Rev. D **83**, 067303 (2011)
[40] S. Hümmer, P. Baerwald, & W. Winter, Phys. Rev. Lett. **108**, 231101 (2012)
[41] B. Zhang & P. Kumar, arXiv:1210.0647
[42] Aartsen et al., Icecube collaboration arXiv:1412.6510
[43] L. Yacobi, D. Guetta & E. Behar, AJ **793**, 48, 6 (2014)
[44] M. Ackermann, et al., ApJ **209**, 11 (2013)
[45] D. Guetta & J. Granot, ApJ **585**, 885 (2003)
[46] D. Guetta, E. Pian, & E. Waxman, A&A **525**, 53 (2011)
[47] N. Gupta & B. Zhang 2007, MNRAS **380**, 78
[48] M. Böttcher & C. D. Dermer 2000, Proceedings of the International School of Cosmic Ray Astrophysics 20th Anniversary, 11th Course. Held 15–23 July 1998 in Erice, Italy.
[49] T. Stanev Nucl. Instrum. Meth. A **588** 215–220 (2008)
[50] V. Barger, P. Huber, P. Marfatia, Phys. Lett. B **642**, 333 (2006)
[51] G. A. Askaryan, Soviet Physics JETP **14**, 441 (1962). G.A. Askaryan, Soviet Physics JETP **21**, 658 (1965)
[52] I. Kravchenko et al., Phys.Rev. D **85**, 062004 (2012)
[53] IceCube, H. Landsman et al., L. Ruckman,G.S.Varner, Nucl. Instrum. Meth. Suppl. A **604**: 70–75 (2009)
[54] ANITA, P. W. Gorham et al., Phys. Rev. D **85**, 049901 (2012)
[55] ARA, P.Allison et al. Astropart. Phy. **35**(7), 457 (2012)
[56] Li Z. 2013, ApJ **770**, 40
[57] J. K. Becker, M. Stamatikos, F. Halzen & W. Rhode Astroparticle Physics **25**, 118–128 (2006) arXiv:astro-ph/0511785

EPJ Web of Conferences **121**, 05002 (2016) DOI: 10.1051/epjconf/201612105002
RICAP-2014

ANTARES and Baikal: Recent results from underwater neutrino telescopes

Fabian Schüssler[a] on behalf of the ANTARES Collaboration
IRFU, CEA-Saclay, 91191 Gif-sur-Yvette, France

Abstract. Two Northern hemisphere neutrino telescopes are currently searching for astrophysical neutrinos in the TeV/PeV range: ANTARES and Baikal. Both observatories utilize various signatures like a high energy excess over the atmospheric neutrino flux, searches for localized neutrino sources of various extensions and multi-messenger analyses based on time and/or space coincidences with other cosmic probes. We here review the status of both experiments and discuss a selection of recent results.

1. Introduction

The quest for the origin of cosmic rays (CRs) is, a century after their discovery, still ongoing. Despite tremendous theoretical and experimental efforts over several decades we still haven't located the astrophysical sources able to accelerate hadronic particles to energies surpassing all man-made accelerators. Various techniques are being employed for this search: from studies of clustering in the CR arrival directions to the analysis of secondary particles like gamma rays and neutrinos. Over 150 TeV gamma-ray sources have been detected over the last decade. Unfortunately the emission of most of them can be modeled equally well by leptonic (accelerating mainly electrons) as well as hadronic (accelerating predominantly hadronic CRs) scenarios. This ambiguity does not exist for high-energy neutrinos as they can be created efficiently only in hadronic interactions via the decay of charged pions. The detection of a high-energy astrophysical neutrino source would therefore be the tell-tale sign of accelerated hadrons and would be a major breakthrough in the century old cosmic ray puzzle. In addition neutrinos interact only weakly with matter, which makes them insensitive to radiation fields and provides access to cosmological distance scales. Unfortunately the low cross-section is at the same time making their detection challenging. In addition, high-energy neutrinos are produced copiously in the Earths atmosphere via CR induced extensive air showers. These "atmospheric neutrinos" are an important background for the search of astrophysical neutrino sources. In the following we'll discuss various attempts to reduce this background and the status of the searches for astrophysical neutrinos and their sources.

2. Underwater neutrino telescopes

Several very sensitive instruments searching for astrophysical neutrino sources are currently in operation: IceCube at the South Pole [1], ANTARES in the Mediterranean Sea and Baikal-NT200+ in Lake Baikal.

[a] e-mail: `fabian.schussler@cea.fr`

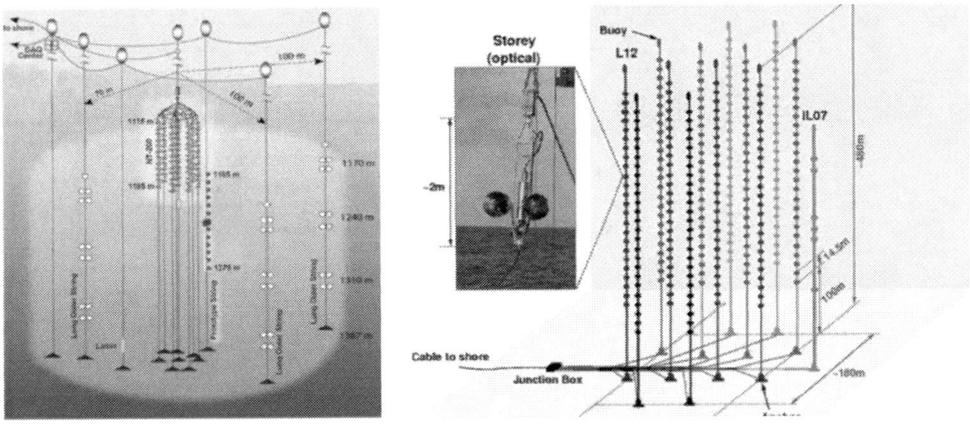

Figure 1. Left panel: layout of the Baikal-NT200+ detector. Right panel: layout of the ANTARES detector. The inset shows a photograph of one of the optical storeys holding 3 optical modules.

The fundamental building blocks of all (underwater) neutrino telescopes are similar: highly sensitive optical detectors called *optical modules* (OMs) are attached to a backbone cable providing the electrical supply and the readout of the data. These *strings* are anchored at the sea bottom and hold in an upright positions by buoys. The neutrino detection relies on the emission of Cherenkov light by high energy muons originating from charged current neutrino interactions or from electromagnetic cascades induced by interactions of electron and tau neutrinos inside or near the instrumented volume. All detected signals are transmitted via an optical cable to a shore station, where a farm of CPUs filters the data for coincident signals or *hits* in several adjacent OMs. The muon direction is then determined by maximising a likelihood which compares the timing of the hits with the expectation from the Cherenkov signal of a muon track or an electromagnetic cascade. The angular uncertainties for the track reconstructions are at the order of 0.3−0.5 deg.

2.1 Baikal

Pioneering the neutrino detection in water, the Baikal collaboration is operating a neutrino telescope in the worlds deepest body of freshwater, Lake Baikal in Siberia. The deployed detector underwent several phases and consecutive enhancements. One of the main building blocks is the Baikal-200 detector which is in operation since 1998. Is is comprised of 192 OMs on 8 strings of 72 m height which cover a circular footprint with a radius of 20 m. This setup has been extended to the current Baikal-200+ setup by adding 3 additional lines (each 200 m in height and holding 36 OMs) at a distance of 100 m around the original telescope. The layout is shown in the left panel of Fig. 1. This latest stage is in operation since 2005 and has mainly been used as testbed for the next phase, Baikal-GVD [2].

2.2 ANTARES

Whereas physics data taking started already during the deployment phase, the ANTARES detector [3] became fully equipped and operational in 2008. The detector is composed of 12 detection lines anchored at a depth of 2475 m off the French coast near Toulon. The detector lines are about 450 m long and hold a total of 885 OMs, consisting in 17″ glass spheres housing each a 10″ photomultiplier tube. The OMs look downward at 45° in order

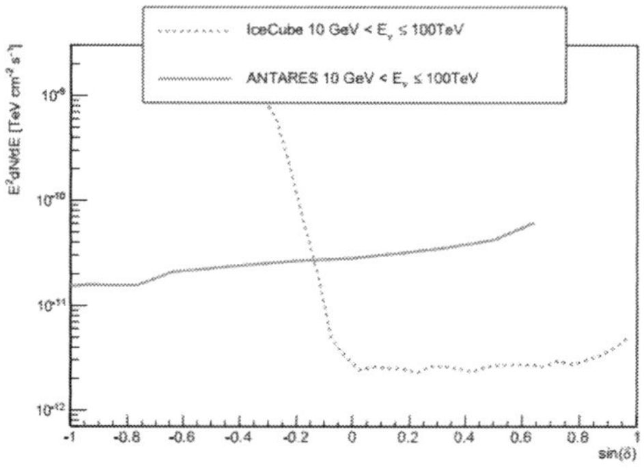

Figure 2. Sensitivity of ANTARES and IceCube to neutrino induced muon tracks following an E^{-2} energy spectrum within the range 10 GeV – 100 TeV.

to optimize the detection of upgoing, i.e. neutrino induced, events. The geometry and size of the detector make it sensitive to extraterrestrial neutrinos in the TeV-PeV energy range. A schematic layout of the detector is shown in the right panel of Fig. 1.

2.3 Comparison with IceCube

In terms of its instrumented volume the IceCube detector at the South Pole is significantly larger than the neutrino telescopes on the Northern hemisphere, but for a direct comparison other effects have to be taken into account: the underwater detectors benefit for example from an improved angular resolution due to the reduced light scattering in water and their denser line spacing increases the sensitivity towards lower energies. In addition, their location provides a high sensitivity towards the bulk of the Galactic Plane, i.e. towards the region containing most high-energy gamma-ray sources. In Fig. 2 the sensitivities of IceCube and ANTARES are compared for events induced by charged current muon neutrino interactions. The simulated neutrinos follow an E^{-2} energy spectrum in the interval between 10 GeV – 100 TeV as function of the declination. As can be seen, ANTARES is providing a better sensitivity to these track-like events for a significant part of the sky.

3. Diffuse flux searches

The main backgrounds for the search for astrophysical neutrinos are of atmospheric origin: downgoing atmospheric muons, which have been mis-reconstructed as upgoing, and atmospheric neutrinos originating in cosmic ray induced air showers at the opposite side of the Earth. Depending on the requirements of the analysis both backgrounds can at least partially be discriminated using various parameters like the quality of the event reconstruction or an estimator of the energy of the muon or the electromagnetic cascade. Especially the latter, being strongly correlated with the energy of the original neutrino, allows good discrimination of atmospheric origin from neutrinos produced in astrophysical sources. Atmospheric neutrinos have a much softer energy spectrum ($\propto E^{-3.7}$) compared to the generic E^{-2} spectrum expected from Fermi acceleration in astrophysical sources.

3.1 Atmospheric neutrinos

A precise measurement of the energy spectrum of atmospheric neutrinos is therefore an important pre-requisite for searches trying to detect an excess over this flux at high-energies.

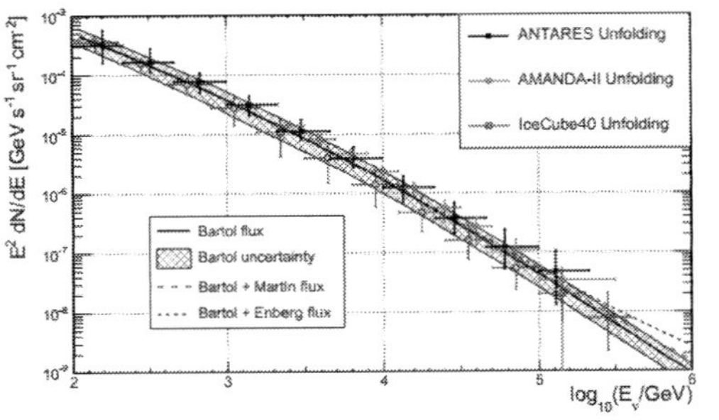

Figure 3. The energy spectrum of atmospheric neutrinos measured by ANTARES compared with earlier measurements and theoretical predictions [5].

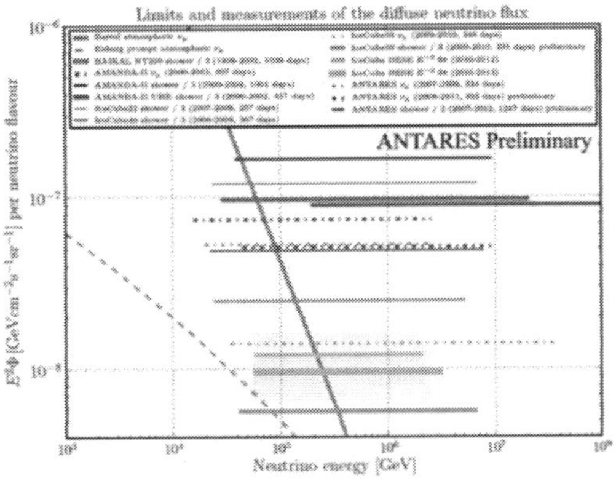

Figure 4. Summary of searches for diffuse, high-energy neutrino fluxes.

A recent measurement by the ANTARES collaboration is relying on several algorithms to estimate the energy of the events (e.g. [4]) and is employing unfolding algorithms to correct for biases induced by the limited resolution of these estimators. The resulting energy spectrum (cf. Fig. 3 and Ref. [5]) shows good agreement with previous in-ice measurements and a 25% higher normalization than the conventional model based on the flux and mass composition of cosmic rays [6].

3.2 Shower-like events

Recently the IceCube collaboration reported the observations of a high-energy excess over the atmospheric backgrounds at the level of 5.6 σ and with a flux at the level of $E^2\Phi = 0.95 \cdot 10^{-8}$ GeVcm^{-2}s^{-1}sr^{-1} [7]. The majority of the detected events are induced by electromagnetic cascades. Similarly 5 years (1247 effective days) of data recorded with the ANTARES telescope have been recently used to search for similar events. A dedicated reconstruction algorithm has been developed and extensive Monte Carlo simulations where used to optimize the event selection criteria. The number of background events has been estimated to $4.9^{+2.9}_{-3}$. After un-blinding the data set, 8 events have been found and an upper limit of $E^2\Phi_{90\%} = 4.9 \cdot 10^{-8}$ GeVcm^{-2}s^{-1}sr^{-1} could be derived. Details of this result and further searches for diffuse emission from interesting regions like the Galactic Plane and the Fermi Bubbles are discussed in [8]. An overview over the recent status of searches for a diffuse flux of high-energy neutrinos is shown in Fig. 4.

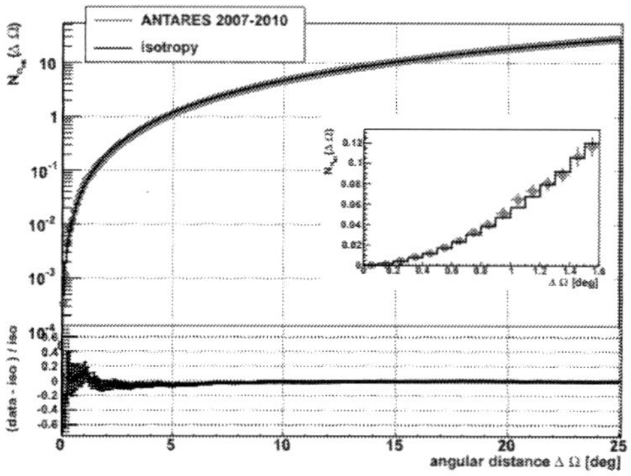

Figure 5. Autocorrelation analysis [12]: ANTARES data (red markers) compared to the reference distribution expected for an isotropic dataset (black histogram).

4. Searches for clusters

In addition to exploiting the different energy spectra, one can use the reconstructed arrival directions of the recorded events to search for an excess over the uniform atmospheric backgrounds. Exploiting a combination of both features, an enhanced two-point correlation method has been used recently to search for clusters in the arrival directions of neutrinos observed by ANTARES. The introduced novel method exploits the energy of the individual events and searches for the cumulative signal from clusters over the full accessible sky. No a-priori assumption on the underlying source morphologies are required. The method has been applied to 4 years of ANTARES data (813 days of effective lifetime; 3058 events) without the detection of a significant excess [12] (cf. Fig. 5).

Complementing the analysis, the developed two-point correlation method has also been used to search for multi-messenger cross-correlations. The ANTARES data set has been correlated with high energy gamma-ray sources given in the 2nd Fermi-LAT source catalog [9], massive black holes [10] and nearby galaxies [11]. None of these searches showed a significant excess.

5. Searches for point-like sources

The main goal of neutrino telescopes is the localization of astrophysical sources of high-energy neutrinos. Typically the data quality selection criteria are optimized on simulated data sets injecting point-like sources in addition to the atmospheric backgrounds and assuming an E^{-2} energy spectrum. Within the ANTARES collaboration these Monte Carlo simulations reproduce the actual, time-dependent data taking conditions and allow to describe the real data very accurately. Extensive comparison between data and Monte Carlo simulations are performed using blinded events where the relevant information for the final analysis (e.g. the arrival directions of the events) is hidden. Only once the data selection procedure is finalized, the data is unblinded and a skymap of the neutrino arrival directions is build. An example of such a map showing 5 years of data (1038 effective days) from the Baikal-200 detector is shown in Fig. 6 [13, 14].

The search for localized excesses is then using maximum likelihood algorithms taking into account the estimated energy of the events to further suppress the influence of atmospheric backgrounds. Monte Carlo pseudo-experiments are used to derive the probability of the analyzed cluster to originate from background only. Based on six years of data from

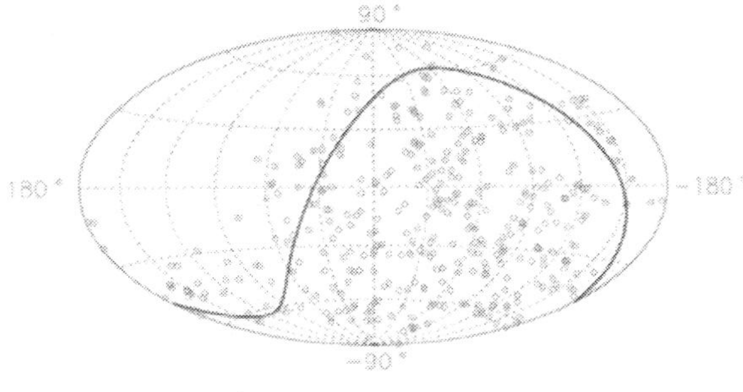

Figure 6. Arrival directions in galactic coordinates of neutrino candidates observed in 5 years of data from the Baikal-NT200 detector [13, 14].

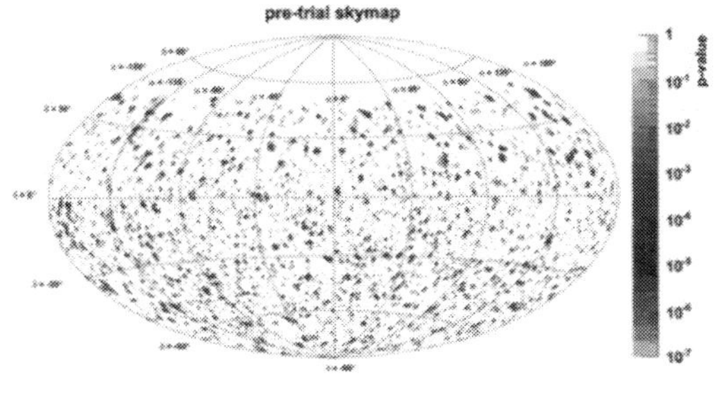

Figure 7. Sky map in equatorial coordinates showing the pre-trial p-values obtained searching for point-like clusters in 6 years of data from the ANTARES detector [15].

the ANTARES telescope which corresponds to 1338 live days and 5516 selected neutrino candidates, an example for such a distribution is shown in Fig. 7. It should be noted that the shown p-values are not corrected for the trial factors related to the scanning of large areas of the sky. Using pseudo-experiments to do so, the most significant cluster of events has been found around $(\alpha, \delta) = (-46^\circ.8, -64^\circ.9)$ with a post-trial p-value of 2.7% (significance of 2.2σ using the two-sided convention) [15].

This region has shown an increased number of neutrino candidates already in searches using an earlier subset of the data [16]. Although not significant on its own, it does currently constitute the *hottest* region in the neutrino sky and has therefore triggered multi-messenger follow-up analyses and observations. Analyzing 4 years of GeV gamma-ray data from Fermi-LAT and 2 hours of dedicated observations of TeV gamma rays obtained with the H.E.S.S. observatory, no source could be found within the region identified by ANTARES [17].

The current status of searches for a point-like high-energy neutrino excess is summarize in Fig. 8 where the obtained upper limits on the neutrino flux are shown. In addition to the all-sky limits, the ANTARES collaboration also searched for extended sources in the region around the Galactic Center [15]. None has been found and the derived limits are able to exclude the possibility that the accumulation of seven events reported by IceCube near the Galactic Center is produced by a single point-like or slightly extended source (cf. [18]).

6. Correlations in space and time

The use additional, complementary information helps to increase the sensitivity of the searches for neutrinos of astrophysical origin. On of the most promising attempts is to narrow

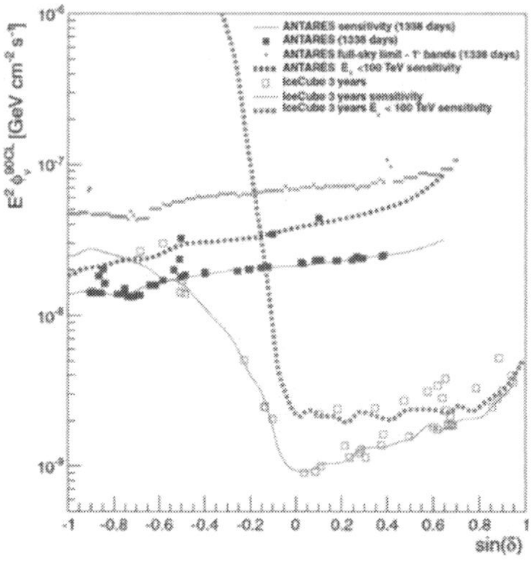

Figure 8. Limits on the high-energy flux obtained in searches for point-like sources [15].

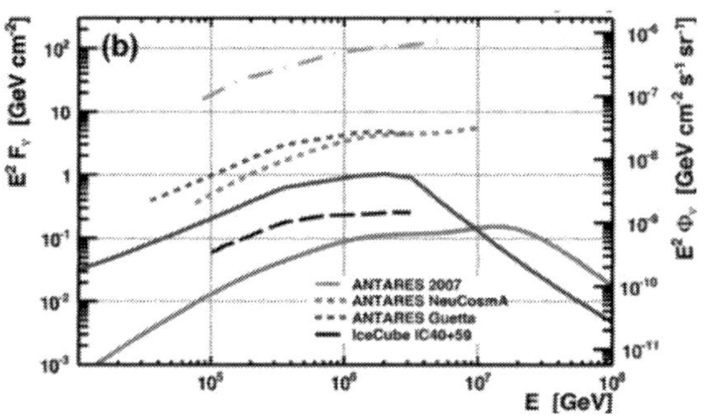

Figure 9. Expected total muon neutrino flux from 296 individual gamma-ray-bursts (solid lines) and upper limits derived from observations by the ANTARES telescope (dashed lines) [22].

down the time window of the searches. This can be done very effectively by analyzing coincidences with transient phenomena detected with other messengers.

6.1 Gamma ray bursts

Gamma ray bursts (GRBs) are one of the prime candidates of transient sources of high-energy cosmic rays (e.g. [19]) and are therefore candidates for searches of contemporaneous emission of high-energy neutrinos. In these searches neutrino events are typically selected in a time window around the detection time of the GRBs and spatial coincident with its location, information provided by X and gamma-ray satellites. All neutrinos telescopes are actively performing these searches, but so far no significant detection has been made. For example data from the Baikal-NT200 detector has been scanned for coincidences with 303 GBRs detected by the BATSE instrument between 1998 and 2000 [20]. Recently the ANTARES collaboration performed an analysis based on 296 GRBs detected by the Swift and Fermi satellites [22]. This analysis used for the first time a novel full Monte Carlo simulation [21] to derive the expected neutrino flux from each individual GRB. The results are summarized in Fig. 9.

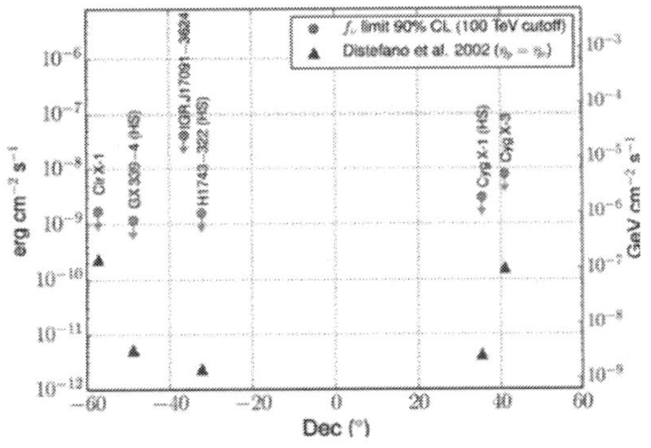

Figure 10. Upper limits on the neutrino flux from microquasars derived from ANTARES observations (red arrows) compared to model predictions (blue triangles) [23].

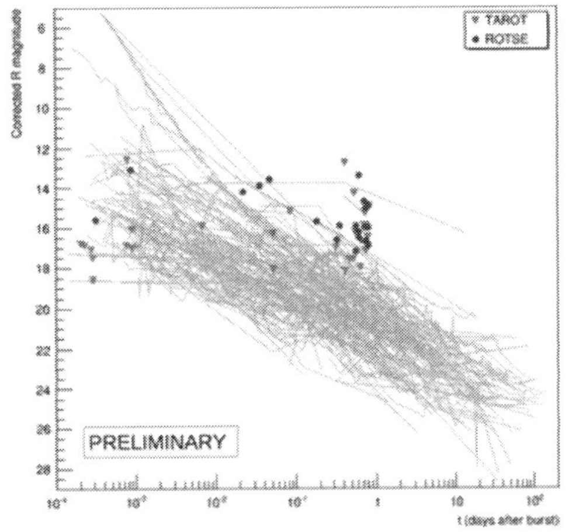

Figure 11. Limits on the optical magnitude of optical data obtained after a neutrino trigger sent by ANTARES (colored triangles) in comparison with the optical afterglows of detected GRBs (gray lines).

6.2 Microquasars

Microquasars are Galactic binary objects that show variable X and gamma-ray emission related to the formation of jet-like outflows. Due to their similarity to the jet-formation around massive black holes at the center of active galactic nuclei emission of high-energy radiation and potentially of neutrinos can be expected. 4 years of data from the ANTARES neutrino telescope have been scanned in a dedicated analysis searching for coincidences between the supposed periods of jet-formation of 6 microquasars and high-energy neutrinos from the same region. No statistically significant excess has been observed, and thus upper limits on the neutrino fluences have been derived. A comparison of these limits with model predictions is shown in Fig. 10 and constraints have been put on the ratio of the proton to electron luminosity in the jets [23].

6.3 Optical and x-ray follow-up

Another very promising method in the search for transient neutrino sources is the emission of alerts from neutrino telescopes triggering follow-up observations of a variety of observatories at different wavelengths. The TAToO project [24] within the ANTARES Collaboration is sending alerts to a range of optical telescopes including the Tarot and Rotse networks and to the Swift satellite. An extension of the system to the H.E.S.S. high-energy gamma-ray observatory is in preparation. The obtained optical images are analysed in search for transient sources at various timescales (e.g. GRBs, supernovae). An example of the obtained limits on the optical magnitudes of GRB afterglows is shown in Fig. 11. Details on the ANTARES multi-messenger program are given in [25].

7. Summary and conclusion

Thanks to their good angular resolution, their location offering access to the bulk of the Galactic Plane and an extensive physics analysis program including many multi-messenger and transient searches, the current neutrino telescopes on the Northern hemisphere Baikal-NT200 and ANTARES continue to produced a wide range of significant results. Future observatories (e.g. GVD and KM3NeT whose deployment started recently) will further enhance the scientific reach and provide important information about the high-energy universe.

References

[1] C. Finley on behalf of the IceCube Collaboration, these proceedings
[2] Z. Dzhilkibaev on behalf of the GVD Collaboration, these proceedings
[3] M. Ageron et al. (ANTARES Collaboration), NIM A **656** (2011) 11–38
[4] F. Schüssler on behalf of the ANTARES Collaboration, 33rd ICRC, Rio de Janeiro (2013), ID 421
[5] S. Adrián-Martá-nez et al. (ANTARES Collaboration), Eur. Phys. J. C **73** (2013) 2606
[6] G. D. Barr et al., Phys. Rev. D **70** (2004) 023006
[7] M. G. Aartsen et al. (IceCube Collaboration), PRL **113** (2014) 101101
[8] L.A. Fusco on behalf of the ANTARES Collaboration, these proceedings
[9] P. L. Nolan et al. (Fermi-LAT Collaboration), ApJ Supplement Series **199** (2012) 31
[10] L.I. Caramete and P.L. Biermann, The mass function of nearby black hole candidates, A&A **521** (2010) A55
[11] D.J. White, E.J. Daw and V.S. Dhillon, Class. Quantum Grav. **28** (2011) 085016
[12] S. Adrian-Martinez et al. (ANTARES Collaboration), JCAP **05** (2014) 001
[13] R. Wischnewski on behalf of the Baikal Collaboration, 29th ICRC, Pune (2005)
[14] O.V. Suvorova and T.A. Ovsiannikova, arXiv:1406.2478
[15] S. Adrian-Martinez et al. (ANTARES Collaboration), ApJ Letters **786** (2014) L5
[16] S. Adrian-Martinez et al. (ANTARES Collaboration), ApJ **760** (2012) 53
[17] F. Schüssler et al. on behalf of the H.E.S.S. Collaboration, 33rd ICRC, Rio de Janeiro (2013), arXiv:1307.6074
[18] M.C. Gonzalez-Garcia, F. Halzen and V. Niro, APP **57–58** (2014) 39
[19] E. Waxmann, PRL **75** (1995) 386
[20] A.V. Avrorin et al. (Baikal Collaboration), Astro. Letters **37** (2011) 692
[21] S. Hümmer et al., ApJ **721**, (2010) 630
[22] S. Adrian-Martinez et al. (ANTARES Collaboration), A&A **559** (2013) A9
[23] S. Adrian-Martinez et al. (ANTARES Collaboration), JHEAP **3–4** (2014) 9
[24] M. Ageron et al. (ANTARES Collaboration), APP **35** (2012) 35
[25] G. de Bonis on behalf of the ANTARES Collaboration, these proceedings

EPJ Web of Conferences **121**, 05003 (2016) DOI: 10.1051/epjconf/201612105003
RICAP-2014

Status and perspectives of the BAIKAL-GVD project

A.D. Avrorin[1], A.V. Avrorin[1], V.M. Aynutdinov[1], R. Bannash[7], I.A. Belolaptikov[2], D.Yu. Bogorodsky[3], V.B. Brudanin[2], N.M. Budnev[3], I.A. Danilchenko[1], G.V. Domogatsky[1], A.A. Doroshenko[1], A.N. Dyachok[3], Zh.-A.M. Dzhilkibaev[1,a], S.V. Fialkovsky[5], A.R. Gafarov[3], O.N. Gaponenko[1], K.V. Golubkov[1], T.I. Gress[3], Z. Honz[2], K.G. Kebkal[7], O.G. Kebkal[7], K.V. Konischev[2], E.N. Konstantinov[3], A.V. Korobchenko[3], A.P. Koshechkin[1], F.K. Koshel[1], A.V. Kozhin[4], V.F. Kulepov[5], D.A. Kuleshov[1], V.I. Ljashuk[1], M.B. Milenin[5], R.A. Mirgazov[3], E.R. Osipova[4], A.I. Panfilov[1], L.V. Pan'kov[3], A.A. Perevalov[3], E.N. Pliskovsky[2], M.I. Rozanov[6], V.Yu. Rubtzov[3], E.V. Rjabov[3], B.A. Shaybonov[2], A.A. Sheifler[1], A.V. Skurihin[4], A.A. Smagina[2], O.V. Suvorova[1], V.A. Tabolenko[3], B.A. Tarashansky[3], S.A. Yakovlev[7], A.V. Zagorodnikov[3], V.A. Zhukov[1], and V.L. Zurbanov[3]

[1] Institute for Nuclear Research, 60th October Anniversary pr. 7A, Moscow 117312, Russia
[2] Joint Institute for Nuclear Research, Dubna 141980, Russia
[3] Irkutsk State University, Irkutsk 664003, Russia
[4] Skobeltsyn Institute of Nuclear Physics MSU, Moscow 119991, Russia
[5] Nizhni Novgorod State Technical University, Nizhni Novgorod 603950, Russia
[6] St. Petersburg State Marine University, St. Petersburg 190008, Russia
[7] EvoLogics GmbH, Berlin, Germany

Abstract. The neutrino telescope Baikal-GVD in Lake Baikal will be a research infrastructure aimed mainly at studying astrophysical neutrino fluxes. The telescope will consist of *clusters* of strings – functionally independent sub-arrays. The deployment of the first demonstration cluster has been started in April 2013. In 2014 the deployment of the second stage of the demonstration cluster has been performed. We describe the configuration and design of the first GVD cluster and review the current status of cluster deployment in Lake Baikal.

1. Introduction

The next generation neutrino telescope Baikal-GVD in Lake Baikal will be a research infrastructure aimed primarily at studying astrophysical neutrino fluxes and particularly at mapping the high-energy neutrino sky in the Southern Hemisphere including the region of the galactic center. Baikal-GVD is a 3-dimensional lattice of optical modules (OMs) – photomultiplier tubes enclosed in transparent pressure spheres. The OMs are arranged on vertical load-carrying cables to form strings. The telescope will consist of clusters of strings – functionally independent sub-arrays, which are connected to shore by individual electro-optical cables. Each cluster comprises seven peripheral strings uniformly arranged at a 60 m distance around a central one. The OMs are spaced by 15 m along each string and facing downward. The OMs along a string are combined in sections – the functional detection unit of the telescope. The distances between the central strings of neighboring clusters are 300 m.

[a] Corresponding author: `djilkib@yandex.ru`

Two telescope configurations with 2304 OMs (GVD) and 10368 OMs (GVD*4) have been optimized. For GVD*4 the muon effective area rises from 0.3 km^2 at 1 TeV to about of 1.8 km^2 asymptotically and the angular resolution (median angle between generated and reconstructed muon directions) is about 0.25 degree. The shower effective volumes of GVD*4 above 10 TeV range from 0.4 to 2.4 km^3. The accuracy of shower energy reconstruction is about 20–35% depending on shower energy, and the angular resolution is about 3.5–6.5 degrees (median value).

The site chosen for the experiment is in the southern basin of Lake Baikal. Here, the combination of hydrological, hydro-physical, and landscape factors is optimal for the deployment and the operation of the neutrino telescope. The water depth is about 1360 m at distances greater than three kilometers from the shore. The water transparency is characterized by an absorption length of about 20–25 m and a scattering length of 30–50 m [1]. The water luminescence is moderate at the detector site.

Prototypes of the GVD elements and systems have been tested in Lake Baikal in 2008–2010 as a part of the NT200+ detector [2–5]. The prototyping/early construction phase of the Baikal-GVD project which aims at construction and operation of the first demonstration cluster has been started in 2011 by deployment and operation of autonomous engineering arrays [6–9]. The next important step in the realization of the GVD project was made in 2013 by the deployment of an enlarged engineering array – the first stage of the demonstration cluster, which comprises 72 OMs arranged on three 345 m long strings [10] and successfully operated from April 2013 to February 2014.

2. Demonstration cluster

The demonstration cluster will comprise a total of 192 optical modules arranged on eight 345 m long strings, as well as an acoustic positioning system [8] and an instrumentation string with equipment for the array calibration and the monitoring of environmental parameters. Seven side strings will be located at 60 m distances from a central one. Each string comprises 24 OMs spaced by 15 m at depths between 900 m and 1250 m below the surface. Each OM consists of a pressure-resistant glass sphere with 43.2 cm diameter which holds the OM electronics and the PMT which is surrounded by a high permittivity alloy cage for shielding it against the Earth magnetic field. A large photomultiplier tube Hamamatsu R7081-100 with a 10-inch hemispherical photocathode and quantum efficiency up to 35% has been selected as light sensor. Besides the PMT, an OM comprises a high voltage power supply unit (HV), a fast two-channel preamplifier, and a controller. For temporal and amplitude calibration of the measuring channel, two LEDs are installed in the optical module. The OM controller is intended for HV control and monitoring, for PMT noise measurements, and for time and amplitude calibration.

The optical modules are grouped into sections – the detection units of the array [11]. Each section includes 12 OMs and the central module (CeM). PMTs signals from all OMs are transmitted through 90 m long coaxial cables to the CeM of the section, where they are digitized by custom-made ADC boards with 200 MHz sampling rate. The waveform information from all measuring channels of the section is transferred to the Master board located in the CeM. The Master board provides the readout of the ADC data, the connection via local Ethernet to the cluster DAQ-center, the control of the section operation and the section trigger logic [11]. A request of the section trigger is transferred from the Master board to the cluster DAQ-center, where a global trigger for all sections is generated. The global trigger initiates data transmission from all sections to shore through a 6 km long electro-optical cable.

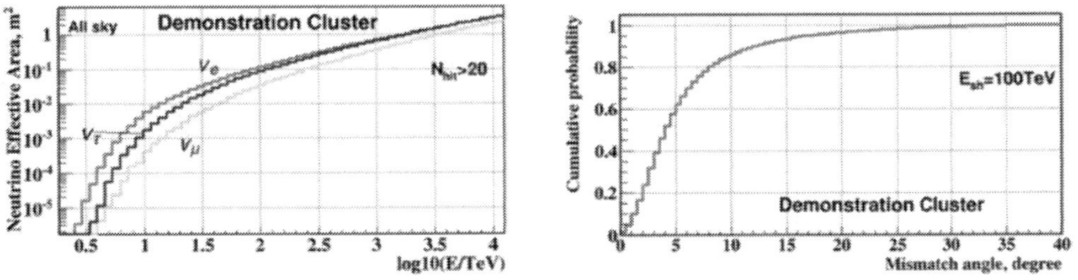

Figure 1. Left: neutrino effective areas for each flavor averaged over all arrival angles. The effective area includes effects from attenuation of neutrinos in Earth. Right: the fraction of reconstructed showers with mismatch angle less than a given value.

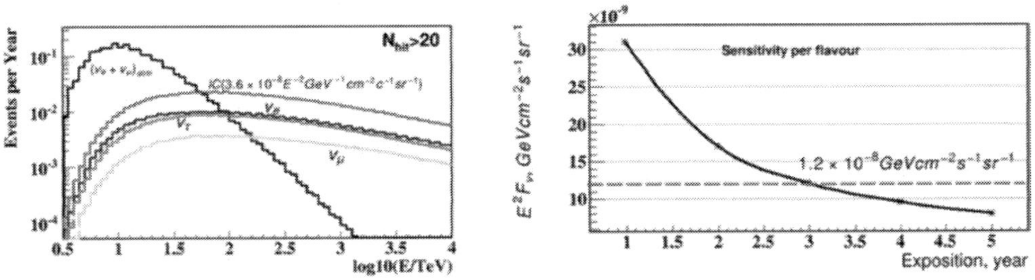

Figure 2. Left: expected distributions of events per year from astrophysical fluxes obtained by IceCube. Also shown is the expected distribution of background events from atmospheric neutrinos. Right: cluster sensitivity for one flavor neutrino flux with an E^{-2} spectrum as function of the observation years. The long-dashed line indicates the one flavor neutrino flux value obtained by IceCube.

The demonstration cluster has the potential to detect astrophysical neutrinos with a flux value measured recently by IceCube [12, 13]. The search for high-energy neutrinos is based on the selection of cascade events generated by neutrino interactions in the sensitive volume of array. After applying an iterative procedure of vertex reconstruction followed by the rejection of hits contradicting the cascade hypothesis on each iteration stage [14], events with a final multiplicity of hit OMs $N_{hit} < 20$ are selected as high-energy neutrino events. With this event selection, the neutrino effective areas for each flavor assuming an equal flux of neutrinos and antineutrinos and averaged over all arrival angles have been derived (see, Fig. 1 left). These areas are about ten times smaller than the corresponding areas of IceCube [12]. The accuracy of shower direction reconstruction is about 4 degree (median value), which is substantially better than the 10-15 degree accuracy for IceCube [12]. The fraction of shower events ($E_{sh} = 100\,\text{TeV}$) with mismatch angles between generated and reconstructed muon directions less than a given value is shown in Fig. 1 (right). Energy distributions of the expected number of shower events per year for IceCube astrophysical fluxes for different flavors and all-flavor, as well as the distribution of expected background events from atmospheric neutrinos are shown in Fig. 2 (left).

The expected background events from atmospheric neutrinos are strongly suppressed for energies higher than 100 TeV. We expect about one event per year with $E_{sh} < 100\,\text{TeV}$ from an equal-flavor astrophysical flux with a per-flavor normalization $E^2F = 1.2 \cdot 10^{-8}$ GeV $\text{cm}^{-2}\text{s}^{-1}\,\text{sr}^{-1}$ [12] in the demonstration cluster, compared to about 10 events in IceCube. The sensitivity of one cluster for one flavor neutrino flux with an E^{-2} spectrum as function of the observation years is shown in Fig. 2 (right). Three year exposition allows sensitivity at a level of flux value measured by IceCube.

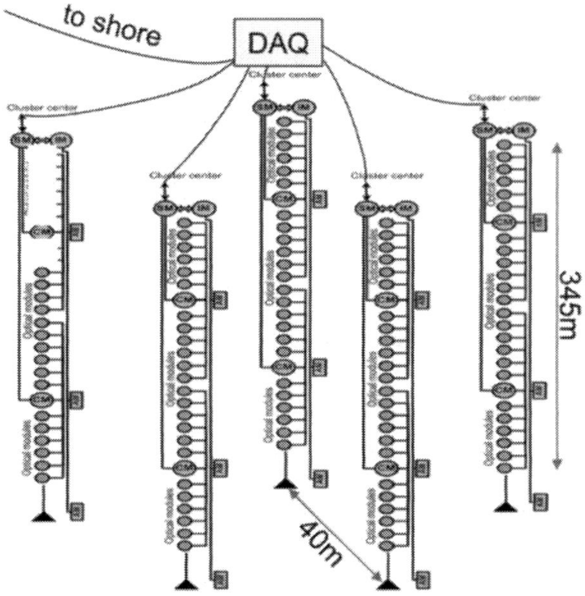

Figure 3. Schematic drawing of the second stage of the demonstration cluster.

3. Current stage of the demonstration cluster

The next important step in the realization of the GVD project was made in 2014 by deployment of the second stage of the demonstration cluster, which comprises 112 OMs arranged on five strings, as well as an instrumentation string with array calibration and environment monitoring equipment. The schematic view of this array is shown in Fig. 3. The vertical spacing of OMs is 15 m and the horizontal distance between strings is about 40 m. In addition to the OMs, each string comprises the communication module (CoM), and the two central modules (CeM) of the two sections of the string. Also each string comprises four transceivers of the acoustic positioning system (AM). The modified cluster DAQ-center is located at the central string of the demonstration cluster and is connected to shore by the electro-optical cable.

The instrumentation string is located at 60-100 m away from the measuring strings with the OMs. It comprises the calibration laser source, eight optical modules, as well as 10 acoustic sensors of the positioning system. The calibration laser source is located at 1215 m depth and is used for time synchronization between the OMs on different strings. The high intensity of the laser source (up to $6 \cdot 10^{13}$ photons/pulse) allows the illumination of OMs at distances more than 200 m away from the source. The acoustic sensors are arranged along the instrumentation string starting from 50 m depth to the bottom of the string and perform monitoring of the string displacements at different depths caused by deep or/and surface water currents. Eight optical modules housing two R8055 and six XP1807 PMTs are arranged at depths from 600 m to 900 m on the instrumentation string and aim at monitoring of the light background at these depths.

The second stage of the demonstration cluster is successfully operating since April 2014 in several testing and data taking modes. In data taking mode a total of $2.96 \cdot 10^8$ events has been recorded which corresponds to 166 days of live time from April 10 till December 30. Long-term control and monitoring of the array measuring system behaviour, as well as background conditions during the array operation in 2014 have been performed.

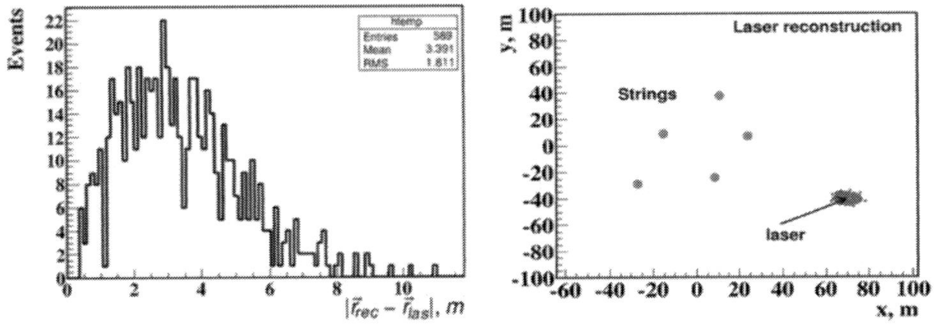

Figure 4. Left: deviation of reconstructed laser coordinates from those obtained by the acoustic positioning system. Right: top view of reconstructed laser positions. The star indicates the laser position obtained by the acoustic system.

The main contribution to the recorded counting rates comes from radiation produced by chemiluminescence in the deep water. During April – December 2014 the PMTs counting rates are about of 15–40 kHz, which are comparable with values obtained in 2012–2013.

One of the main goals of the array operation in testing modes is the estimation of the ability of in-situ calibration procedures. Calibration of the array recording system includes charge and timing calibrations of the measuring channels and time synchronization of OMs on different sections. All these calibration procedures are based on the usage of the OM's internal calibration LEDs and the external laser light source.

The performance and quality of the calibration procedures have been verified by the reconstruction of position and intensity of the calibration laser. An external calibration laser provides five series of 480 nm light pulses at five fixed intensity levels ranging from approximately 10^{12} to $6 \cdot 10^{13}$ photons/pulse, which roughly corresponds to shower energies from 10 PeV to 600 PeV. This allows testing the array capability for high-energy cascade reconstruction. The position of the laser source was reconstructed using the arrival times of photons detected by the PMTs, taking into account the timing calibration of the PMTs by LEDs. Results of the reconstruction were compared with the laser coordinates obtained by the acoustic positioning system. Differences between reconstructed coordinates and those obtained by the acoustic system are shown in Fig. 4 (left). The reconstruction accuracy (median value) is about 3 m. Shown in Fig. 4 (right) are strings and laser positions obtained by the acoustic system, as well as reconstructed laser positions. The obtained results demonstrate the expected quality of the array calibration procedures.

4. Conclusion

The ambition of the Baikal collaboration is the construction of a km3-scale neutrino telescope – the Gigaton Volume Detector in Lake Baikal. During the R&D phase of the GVD project in 2008–2010, its basic elements – new optical modules, FADC readout units, underwater communications and trigger systems – have been developed, produced and tested in situ by long-term operating prototype strings. The prototyping phase of the GVD project has been started since April 2011 and aims at completion of the first demonstration cluster of GVD in spring 2015. The first stage of the GVD-cluster comprising three strings was deployed and successfully operated in 2013. In 2014 the second stage of the demonstration cluster with 112 OMs at 5 strings has been deployed and is presently taking data.

This work was supported by the Russian Found for Basic Research (grants 13-02-12221, 14-02-00175, 14-02-00972).

References

[1] A. Avrorin et al., Nucl. Instr. and Meth. A **693**, 186 (2012)
[2] V. Aynutdinov et al., Nucl. Instr. and Meth. A **602**, 227 (2009)
[3] A. Avrorin et al., Nucl. Instr. and Meth. A **626**, S13 (2011)
[4] A. Avrorin et al., Nucl. Instr. and Meth. A **639**, 30 (2011)
[5] A.V. Avrorin et al., Instruments and Experimental Techniques **54**, 649 (2011)
[6] A. Avrorin et al., Nucl. Instr. and Meth. A **692**, 46 (2012)
[7] A. Avrorin et al., Nucl. Instr. and Meth. A **725**, 23 (2013)
[8] A.V. Avrorin et al., Instruments and Experimental Techniques **56**, 449 (2013)
[9] A. Avrorin et al., J. of Phys. Conf. Series. **409**, 012141 (2013)
[10] A. Avrorin et al., Nucl. Instr. and Meth. A **742**, 82 (2014)
[11] A.D. Avrorin et al., Instruments and Experimental Techniques. **57**, 262 (2014)
[12] M. G. Aartsen et al. [IceCube Collaboration], Science. **342**, 1242856 (2013)
[13] M. G. Aartsen et al. [IceCube Collaboration], Phys. Rev. Lett. **113**, 101101 (2014)
[14] A. Avrorin et al., Astronomy Lett. **35**, 651 (2009)

Status of KM3NeT

G. Riccobene[a] for the KM3NeT Collaboration
Laboratori nazionali del Sud, INFN, via S. Sofia 62, 95123 Catania, Italy

Abstract. The recent observation of cosmic neutrinos by IceCube has pushed the quest towards the identification of cosmic sources of high-energy particles. The KM3NeT Collaboration is now ready to launch the massive construction of detection units to be installed in deep sea to build a km-cubic size neutrino telescope. The main elements of the detector, the status of the project and the expected perfomances are briefly reported.

1. Introduction

The detection of high energy cosmic neutrinos is considered as the ultimate smoking gun to identify the sources of cosmic rays in our Galaxy and beyond. High energy γ-rays, succesfully detected by air Cherenkov telescopes can be, indeed, produced in sources where either hadronic or purely leptonic acceleration mechanisms occur. Moreover, the observed extragalactic γ-rays ($E_\gamma > 1\,\text{TeV}$) arrive to the Earth with a reduced flux and a distorted energy spectrum, due to their interaction with the intergalactic environment. The high energy (HE) cosmic rays at enegies $\leq 10^{19}$ eV are deflected by the intergalactic magnetic fields and cannot be used to point back to the sources, at larger energies the GZK effect suppresses the CR flux from extragalactic sources [1]. Neutrinos, neutral, light, weakly interacting particles, are neither deflected nor absorbed in their journey to the Earth, thus they are considered as optimal astrophysical probes for high energy astronomy. The discovery of a neutrino flux originated outside the Solar System by the IceCube Collaboration [2] , has paved the way to high-energy neutrino astronomy. Whether these events have a Galactic or extragalactic origin is still an open question, since the sample collected so far (37 events) does not allow firm statistical analysis. IceCube will continue data taking for the next years, while the KM3NeT project plans to deploy, in the Northern Hemisphere of the Earth another km^3-scale neutrino telescope (the ARCA project: Astroparticle Research using Cosmics in the Abyss) aiming to larger detection area and better angular resolution, exploiting the optical characteristics of deep seawaters. The detection of cosmic high energy neutrinos produced in galactic and extragalactic sources is expected to provide missing information on the production mechanisms of high energy particles in astrophysical sources, being a clear signature of hadronic acceleartion of high energy particles. Neutrino sources could be also identified, using statistical analysis of the direction of arrival of the events. In this case a multi-messenger approach will be a powerful tool to study the correlation between high energy neutrinos, high energy cosmic rays and photons.

The most mature detection technique to identify neutrinos in the energy range between 100 GeV and 100 PeV is the underwater (or underice) Cherenkov technique, that is the use of a large array of optical sensors deployed in a transparent natural medium in known,

[a] e-mail: `riccobene@lns.infn.it`

or measurable, positions in order to reconstruct the Cherenkov wavefront radiated by high energy particles outgoing a neutrino interaction in water, or ice, inside or close to the detector volume. As foreseen by several models and now clearly shown by IceCube, the detector size must be close, or better exceed, the cubic-kilometer scale to identify a statistically significant number of events in the foreseen life of the detector (not less than 10 years). The golden channel for neutrino astronomy is the detection of HE muons produced in charged current interactions of ν_μ in water. Thanks to the muon long range (from few hundreds meters to several km, at such energies) the muon track direction can be reconstructed with sub-degree accuracy, allowing for source pointing, being the muon and the neutrino almost co-linear at such energies. Other neutrino flavours (both ν_e and oscillated ν_τ) and neutral currents are identified as short, but intense, glows of light, produced by showers outgoing the neutrino interaction. To suppress the background due to downgoing atmospheric muons, falsely reconstructed as up-going (good) tracks, neutrino telescopes have to be shielded using a thick layer of natural medium. In the case of KM3NeT, three sites in the Mediterranean Sea have been candidate for detector installation: Capo Passero (3500 m depth), Toulon (2500 m depth) and Pylos (3000÷4500 m depth). Located in the Northern Hemisphere of the Earth, KM3NeT will optimally look for neutrinos from point-like Galactic sources. In fact at the latitude of the Mediterranean Sea it is possible to detect neutrinos with a low energy threshold (about 100 GeV) in a large field of view (about 3.6π sr at Mediterranean Sea latitude) with an almost complete view of the Galactic plane. The collaboration has a defined the technical project and has started the construction of a research infrastructure hosting the high energy neutrino telescope and auxiliary nodes for Earth and Sea sciences. More recently the collaboration has proposed to extend the physics case to the measurement of the neutrino mass hierarchy using atmospheric neutrinos. A feasibility study, called ORCA, has started. From this study the best sensitivity to measure the neutrino mass hierarchy is obtained with a dense Mega-ton scale detector that can measure atmospheric neutrinos in the GeV region. For ORCA the same KM3NeT technology will be used with a much denser granularity of optical modules. An experimental proposal for a multisite experiment is underway: a Gigaton detector for high energy astronomy is foreseen in the Capo Passero italian site (KM3NeT-It) while a Mton-scale, denser, detector for neutrino mass hierarchy research is foreseen in the Toulon French site (KM3NeT-Fr). In this paper the detector technology, the status of the project and the performance of the Gigaton detectors for the neutrino astronomy will be briefly described. A description of the ORCA feasibility study can be found in [3].

2. The KM3NeT technology

The KM3NeT detector is a three dimensional array of Digital Optical Modules (DOMs) designed to detect Cherenkov photons radiated by charged particles in water. Each DOM, the basic unit of the detector, is made of 31 PMTs (3″ diameter) coupled with their active bases, providing High Voltage and equipped with the PMT read-out system. The "time over threshold" information of PMT signal, sampled at 1 GHz, is provided to shore for analysis. The threshold, adjustable, is typically set to the equivalent of 0.3 photo-electrons to reject most of the PMT dark noise. PMTs are hosted in a 3D-printed support structure and surrounded by a reflector ring that increases the collection efficiency per PMT by about 27% [4]. The DOM is contained in a pressure resistant 17″ glass sphere divided in two halves. The optical contact between PMTs and the glass is assured by optical gel. The lower hemisphere of each DOM has 19 of the PMTs, which are thus downward-looking, whereas the other 12 PMTs look upwards. The DOM contains a Central Logic Board (CLB), an FPGA based electronics board that provides control from shore and data transmission to shore over

Figure 1. Drawing of the DOM.

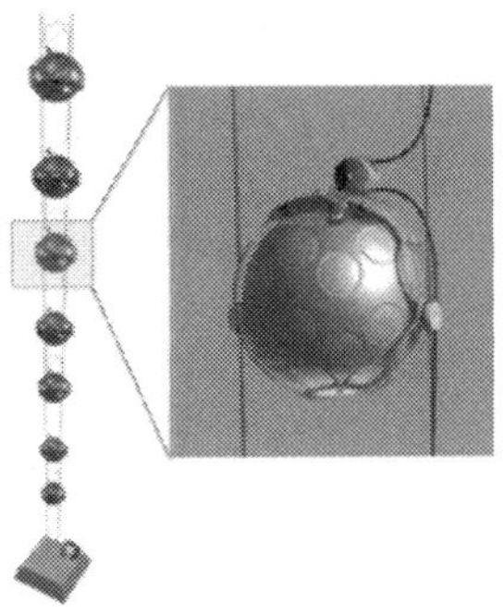

Figure 2. Drawing of DU and details of the DOM mounted into the string.

a fiber optic network via an SFP transceiver, coupled with 2 octopus boards, extenders that allow connection with the PMT bases. The DOM also contains auxiliary calibration sensors: a LED nano-beacon for time calibration and water properties study, that illuminates the neighbour DOMs; a compass and tiltmeter to monitor the absolute orientation of the DOM in water; a temperature and a humidity sensor; an acoustic piezo sensor glued to the inner surface aimed at the detection of acoustic signals for detector positioing. A drawing of the DOM is shown in Fig. 1. The DOM design has several advantages with respect to the large area PMT (10″ diameter) used in NEMO [6] and ANTARES [7]. The photo-cathode area in a single DOM is about three times that one of an ANTARES optical module and it has an almost uniform angular coverage in the solid angle. Because the photocathode is segmented, the identification of more than one photon arriving at the DOM can be done with extremely high efficiency and purity allowing a good optical background rejection and providing a directional information on the arrival photons. These feautures have been proven analysing the data collected in the first DOM prototype that has been integrated in the instrumented line of the ANTARES detector [8]. Moreover each DOM is a fully independent module, that does not need additional vessels for hosting communication and power electronics. From the optical communication point of view each DOM is characterised by the specific color, provided by its SFP transceiver (laser and receiver), defined in the Dense Wavelenghth Division Multiplexing standard grid. Using the DWDM techinque, data from about 80 DOMs can be transmistted over the same fiber. The DOMs are displaced on a vertical structure called Detection Units (DUs, see Fig. 2).The DUs are flexible slender strings anchored to the sea floor, kept vertical by buoys. Each DU hosts 18 DOMs starting 100 m above the sea floor and with 36 m of vertical distance between adjacent DOMs. The definition of the distances between DUs and between DOMs is the result of a detailed optimization performed by means of Monte Carlo simulations, focused on the observation of neutrinos from Galactic point-like sources. The DUs are mechanically sustained by two parallel Dyneema ropes. A vertical electro-optical backbone is used to connect the DOMs to the base of the DU. The backbone is a flexible, oil-filled plastic hose, in equi-pressure with the sea water, and it contains 18 optical fibers for data transport and two copper wires for the provision of electrical power to the DOMs. For each DOM, a break-out box provides connection to one fibre and two wires. Mechanical penetrators are used to route fibers and conductors from the backbone to the inner part of the DOM. For the deployment, the detection unit is wrapped on a spherical metallic frame with diameter of about 2 m (the Launcher of Optical Modules, LOM) which is deployed on the seabed, using a winch, and then unfurls in a rotating upwards movement.

After the unfurling, the DU is vertically displaced, and the LOM rises to the sea surface, where it is collected for reuse. The anchor hosts an additional vessel, the DU base, equipped with a CLB, an hydrophone for positioning, the power distribution and fiber distribution systems. The anchor has an electro-optical ROV mateable interface to the subsea network receiving 375 VDC from shore and connecting 2 fibers. One fiber is used bi-directionally to transmit the master clock (from shore to the sea) and slow controls to the DU base and to the DOMs, and to receive the detector control feedbacks from the DU base. The second fiber is used to transmit the DOMs' data to shore. The communication between shore and sea is based on the White Rabbit [5] precision time protocol, a new synchronous ethernet-based system that permits sub-nano second synchronisation of devices interconnected by an optical fiber link. All DUs are connected to shore by a network of electro-optical cables, that is adapted to the installation site as a function of the distance to shore (about 35 km for Toulon and 90 km for Capo Passero) and of the number of DUs to be served. In order to ensure scalability and future expansions, DUs are arranged in clusters: an array 115 DUs will constitute a detector building block. In the aim of neutrino astronomy research the average distance between neighboring DU is set in the range 90÷120 m and the single building block is made of 115 DUs. The building block concept allows also for a distributed detector deployment and management. The choice for a distributed research infrastructure made of several building blocks with common detector technology, management, data handling and operation control was based on both technical and funding arguments. Simulation results show that the building block concept does not affect the expected sensitivity with respect to a single larger detector. Moreover, the distributed infrastructure is the preferred choice of the Sea and Earth Science community that profits from KM3NeT infrastructures to install a distributed network of sensors in the Mediterranean Sea.

Neutrino events are reconstructed detecting Cherenkov light at each optical sensor and combining the arrival time and amplitude information with the detector position information recovered by the detector positioning system. Each photon hit time is time-stamped by the CLB, that recovers the absolute time information using the GPS signal distributed from shore. The CLB time is synchronised and synthonised (that is has a constant and measured phase) to the GPS time, thanks to the White Rabbit gear running in the CLB and in the shore station DAQ fabric. The number of photons detected by each PMT is recovered through the time-over-threshold information that is proportional to the PMT signal amplitude. Time-over-threshold data from all PMTs are sent to shore, where neutrino event candidates are selected by online filters running on a computer farm. The detector control system permits also setting of HV and threshold of the PMTs from shore. The position of each PMTs is measured acoustically and via a compass and tilt board: the AHRS (Altitude, Heading, Reference System) board. The acoustic positioning system is formed by three sub-systems: an array of receivers placed on the DOMs (piezos) and on the DU base (hydrophones); a long baseline of acoustic emitters (beacons) and receivers (hydrophones), placed in the installation field in geo-referenced positions; a farm of computers on shore running the position reconstruction algorithms [9]. The long baseline is formed by a synchronous and phased network of acoustic beacons which emit a known pulse at a known time in the detector clock reference and digital hydrophones placed close to digital hydrophones, whose read-out is time-stamped by the CLB. Hydrophones and beacons are placed on special bases, called calibration bases, whose position is geo-referenced during deployment operation from the ship. Once the long baseline is calibrated, that is the position of all emitters is reconsructed with the desired accuracy (< 10 cm), the positions of the DOM can be effectively reconstructed measuring the time of arrival of the beacon acoustic signals to each piezo in the DOM and via multi-lateration. The expected resolution of the position

Figure 3. The PPM-DOM installed in Toulon in 2013, and connected to the ANTARES instrumentation line.

Figure 4. The PPM-DU was deployed in Capo Passero in 2014.

reconstruction is close to 10 cm. The DOM orientation, pitch and roll is also measured in real time, the expected accuracy is 3 degrees in orientation (yaw) and few tenth of degree in pitch and roll. Combining hit time information with measured DOM positions it is possible to reconstruct the trajectory of the muons producing Cherenkov light with a very good angular resolution (few tenths of a degree). The amount of detected light can provide information on the energy of the particle (see next sections).

3. Status of the KM3NeT project

The design of the KM3NeT detector has followed a severe validation procedure and prototyping activity. In parallel with the design and test of the full architecture, two prototypes (defined as Pre Production Models, PPM) have been deployed in Toulon and Capo Passero. The first one, the PPM-DOM (see Fig. 3), deployed in April 2013, consisted of a full DOM attached to an ANTARES line, the data transmission and power systems where harmonised to the ANTARES infrastrucure, nevertheless, the PPM-DOM included all the sub-systems of a final DOM: 31 PMT (3 inches diameter), electronics and auxiliary calibration systems, including an external hydrophone. The PPM-DOM, provided a technological proof of concept and a large dataset for analysis. The optical noise of the Toulon site (induced both by bioluminescence and ^{40}K) was monitored and atmospheric muon tracks were clearly identified over the optical noise, using time-coincidence among PMTs. Thanks to the DOM segmentation, and different PMT orientation, a clear "top-down" space-time sequence of hits is observed for high multiplicity events, associated -after comparison with simulations- to atmospheric muon tracks [8, 10].

A PPM-DU was deployed in May 2014 in Capo Passero. The PPM-DU (see Fig. 4) consisted of 3 DOMs assembled on a 160 m long string (using standard spacing) and with a DU base. The DOMs were built using both ETEL D783FLA PMTs (DOM 1 and 2) and Hamamatsu R12199-02 (DOM3, the uppermost). The PPM-DU was connected to the Capo Passero deep sea network and it is still active and taking data. On each DOM, atmospheric muon tracks are clearly separated from the, very low, optical noise, while the analysis of coincidence among DOM PMTs is in progress [10]. Following the success of the demostrators the collaboration has defined the detector a construction plan. The KM3NeT infrastructure will be constructed in three phases. The phase 1, fully funded, will consist of 24 DUs deployed in Capo Passero and 7 strings in Toulon and it will have a volume of about

0.1 km^3. The next phase 2.0 foresees the construction of ARCA: 2 building blocks of 115 DUs, with a volume of about 1÷1.6 km^3 (depending on the final DU spacing), at this stage KM3NeT will have a size comparable to IceCube and it will provide a benchmark for the IceCube cosmic neutrino signal. Meanwhile the ORCA project will be implemented. The neutrino astronomy era will be opened when the full KM3NeT detector will be completed – phase 3- with a size of about 3÷4.8 km^3.

In the Italian site, together with the standard 24 DUs, additional 8 DUs à la NEMO (flexible tower concept) will be deployed. A prototype tower has been already deployed in 2013 in Capo Passero and took data for about one year. From the data collected in this period the first measurement of vertical atmospheric muon flux at a depth of 3500m was made [12] and published in [11]. The long term measurements of the environmental parameters confirms that the bioluminescence optical noise is very low at the KM3NeT-It site [13]. The Capo Passero infrastructure is now ready to host the KM3NeT phase 1: the shore laboratory is refurbished with a large data acquisition room equipped with computing and power systems, the electro-optical cables of the submarine network have been purchased, the first (over a total of three) junction boxe has been deployed together with the first tower. The shore laboratory is connected with a dedicated optical fiber and backup copper links to the Laboratori Nazionali del Sud and, from there, to all the computing and storage resources of the collaboration. KM3NeT phase 1 in Capo Passero will be the most sensitive neutrino telescope in the Northern Hemisphere and it will be equivalent to about 10% of the IceCube detector. In the French site (KM3NeT-Fr) the first DU will be soon deployed, and six more strings will be constructed, according to the outcome of the ORCA feasibility study, with an interspacing between DUs of the order of 20 m and between DOMs of about 6 m, thus providing a demonstrator for the neutrino mass hierarchy experiment. The first full DU (equipped with 18 DOMs) is already assembled (see Fig. 5) and will be installed in Toulon in Spring 2015. The Toulon site has been refurbished with a new electro optical cable – already deployed- and with a new primary junction box (node), equipped with control systems, power and fiber distribution systems and electro-optical ROV mateable connectors for the connection of the DUs. After the deployment of the node, the first DU will be installed and connected. The construction of the 18 DOMs of the first DU allowed the set up of assembly, test and calibration procedures, that are now documented and distributed to the full collaboration to allow massive construction of the DUs. All the PMTs will be tested and characterised in a dedicated set-up, which permits testing of about 60 PMTs in parallel in a few hours, all the electronics boards, mechanical and optical parts and auxiliary sensors will be tested (and calibrated), prior the DOM assembly. After assembly, the full DOM is tested and released for integration. The phase 1 production plan foresees assembly and test of five DOMs per week in four different integration sites. To build a DU, DOMs are connected to the electro-optical backbone and to the DU-base, the DU is thus tested and calibrated. Eventually the DU is assembled on the anchor and folded in the LOM ready for the deployment (see Fig. 6). All the procedure is supervised via QA/QC system.

4. Physics performance

The IceCube discovery of a cosmic neutrino flux corresponds to a flux of about $3.6 \times 10^{-8}\,\mathrm{GeV\,cm^{-2}\,s^{-1}\,sr^{-1}}$ with a cutoff at about 3 PeV, was measured in the energy range between 10 TeV to about 1 PeV and, a more recent analysis [14] extended the analysis at lower energies, but with a cut on the detector fiducial volume. Despite the muon channel has been condidered as the "golden" channel for nutrino detection, the IceCube analysis showed that most of these events are cascade events (showers induced by ν neutral current interacion

Figure 5. The first KM3NeT Detection Unit assembled, under test at CPPM.

Figure 6. The first DU: Anchor and LOM under assembly.

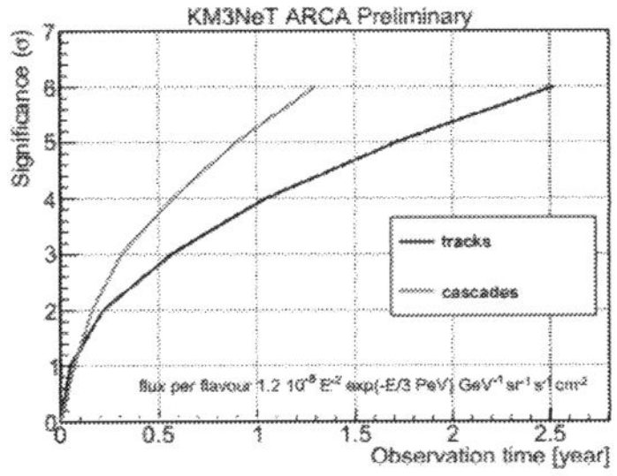

Figure 7. Significance as a function of the observation time for the detection of a neutrino diffuse flux corresponding to the signal reported by IceCube, in the up-going muon (black) and cascade (red) channels.

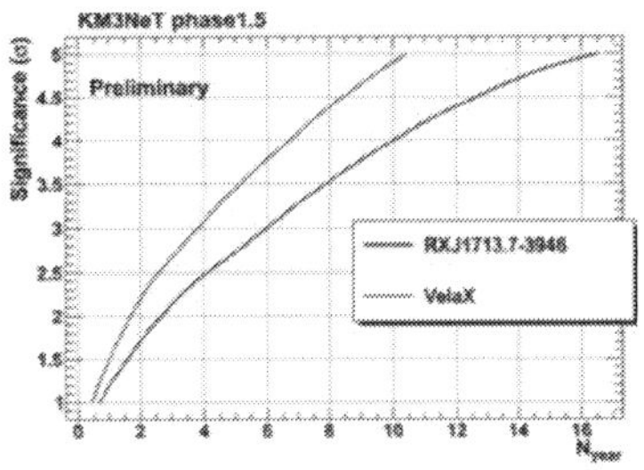

Figure 8. Significance as a function of the observation time for the detection of a neutrino diffuse from the RXJ1713 (blue) and the VelaX (red) sources.

and charged current interaction producing either e or τ). These events are detected with a poor angular resolution ($10^\circ \div 15^\circ$) strongly reducing the capabilities of source pointing. This discovery has led the KM3NeT collaboration to plan the construction of ARCA. Thanks to a better angular resolution with respect to IceCube, provided by design and seawater optical properties, larger field of view and, effective area, ARCA is expected to provide more information on the origin of the IceCube signal. In Fig. 7 the expected performance of ARCA are shown in terms of significance as a function of the observation time for the flux observed by IceCube. The significance, both for the cascade and track channels, includes all the neutrino flavors and it has been performed selecting tracks reconstructed in the full angular range (4π). The significance for the muon channel has been estimated for tracks reconstructed as up-going plus a small region of 10 degrees above the horizon. A significance of 5σ is obtained in about one year in the cascade channel and after about 2 years in the muon channel, for the flux reported by IceCube. A refined cascade reconstruction algorithm [15] shows also that a resolution of 3 degrees can be achieved is at 20 TeV, reaching a value of about 1 degree at higher energies.

For what concerns point source reconstruction, the collaboration focused its studies on the detection of Super Nova Remnants RXJ1713, Vela Junior and, the Pulsar Wind Nebula VelaX: the most intense high energy gamma-ray sources -thus possible neutrino sources- in the Galactic plane [16]. The expected KM3NeT-phase1.5 detection capability is reported for the RXJ1713 and VelaX in Fig. 8. These results were estimated in the hypotheses that the sources have a neutrino spectrum derived from the high energy γ-ray spectrum following the hypotheses in [17] and in [18, 19] and that the spatial extension of the neutrino emission region is the same of the measured γ-rays. ARCA is expected to detect the point-source signal with a significance level of 3σ in about 5 years of observation. The neutrino astronomy era will be ultimately open with the full KM3NeT detector that is expected to allow signal detection with a significance at 5σ in a couple of years of observation. KM3NeT phase 3 will complement and overlap the field of view of IceCube exceeding, by more than one order of magnitude, the IceCube sensitivity in the Southern Hemisphere.

5. Summary

The construction of the KM3NeT detector has started and it will be arranged in three phases. The refurbiushment of the shore and deep sea infrastructures of Capo Passero and Toulon in view of phase 1 is almost completed. In both infrastructures two prototypes of KM3NeT detectors have been sucessfully installed and operated. The collaboration has also completed the design and set up all the tools and actions for the mass production of DUs in view of the completion of KM3NeT phase 1. Twenty-four DU strings and 8 towers will be deployed in Capo Passero (Sicily) and 7 strings in Toulon. The first DU is now assembled and under test, to be ready for deployment in Spring in Toulon. The other six strings in Toulon will be arranged in a denser configuration to proof the KM3NeT ORCA concept aiming at neutrino mass hierarchy studies. The way to KM3NeT ARCA is paved and the detector is expected to confirm the IceCube signal with improved source pointing performance. Eventually, with KM3NeT phase 3 the Collabroation will operate a distributed infrastructure for high energy neutrino astromomy and neutrino physics reasearch.

References

[1] F. Halzen and U. Katz, Advances in High Energy Physics **Article ID 680584** (2013)
[2] M.G. Aartsen et al. (IceCube Coll.), Science **342**, 1242856 (2013)
[3] U. Katz (KM3NeT collaboration), Feb. 2014, available on `http://arxiv.org/abs/1402.1022` (2014)
[4] S. Adrián-Martínez et al. (KM3NeT collaboration), JINST **8**, T03006 (2013)
[5] P.P.M. Jansweijer, H.Z. Peek and E. de Wolf, Nucl. Inst. Meth. A **725**, 488 (2013)
[6] S. Aiello et al. (NEMO Collaboration), Astrop. Phys. **33**, 263 (2010)
[7] M. Ageron et al. 2011 (ANATRES collaboration), Nucl. Instrum. Meth. A **656**, 11 (2013)
[8] S. Adrián-Martínez et al. (KM3NeT Collaboration), Eur. Phys. J. C **74**, 3056 (2014)
[9] F. Ameli et al. 2011 (KM3NeT collaboration), Nucl. Instrum. Meth. A **626**, 211 (2010),
[10] S. Biagi (KM3NeT Collaboration), these proceedings
[11] S. Aiello et al. (NEMO collaboration), Astropart. Phys. **66**, 1 (2015)
[12] C. Distefano (NEMO Collaboration), these proceedings
[13] M.G. Pellegriti (NEMO Collaboration), these proceedings
[14] M.G. Aartsen et al. (IceCube collaboration), Phys. ReV. D **91**, 022001 (2015)
[15] L. Fusco (KM3NeT Collaboration), these proceedings
[16] A. Trovato (KM3NeT Collaboration), these proceedings
[17] S.R. Kelner, F.A. Aharonian and V.V. Bugayov, Phys. Rev. D **74**, 063007 (2006)
[18] F.A. Aharonian et al., Astronomy and Astrophysics **548**, A38 (2012)
[19] F.L. Villante and F. Vissani, Phys. Rev. D **78**, 103007 (2008) and references therein
[20] M.G. Aartsen et al. (IceCube collaboration), Astroph. J. **779**, 132 (2013)

EPJ Web of Conferences **121**, 05005 (2016) DOI: 10.1051/epjconf/201612105005
RICAP-2014

Galactic sources of high energy neutrinos: Expectation from gamma-ray data

N. Sahakyan[a]

ICRANet-Yerevan, Marshall Baghramian Avenue 24a, Yerevan 0019, Republic of Armenia

Abstract. The recent results from ground based γ-ray detectors (HESS, MAGIC, VERITAS) provide a population of TeV galactic γ-ray sources which are potential sources of High Energy (HE) neutrinos. Since the γ-rays and ν-s are produced from decays of neutral and charged pions, the flux of TeV γ-rays can be used to estimate the upper limit of ν flux and vice versa; the detectability of ν flux implies a minimum flux of the accompanying γ-rays (assuming the internal and the external absorption of γ-rays is negligible). Using this minimum flux, it is possible to find the sources which can be detected with cubic-kilometer telescopes. I will discuss the possibility to detect HE neutrinos from powerful galactic accelerators, such as Supernova Remnants (SNRs) and Pulsar Wind Nebulae (PWNe) and show that likely only RX J1713.7-3946, RX J0852.0-4622 and Vela X can be detected by current generation of instruments (IceCube and Km3Net). It will be shown also, that galactic binary systems could be promising sources of HE ν-s. In particular, ν-s and γ-rays from Cygnus X-3 will be discussed during recent gamma-ray activity, showing that in the future such kind of activities could produce detectable flux of HE ν-s.

1. Introduction

The neutrinos are unique messengers which carry astrophysical/cosmological information [1]. Since they can escape from much denser environments (high photon field and/or matter) than γ-rays, it provides a possibility to examine the sources which are impossible by γ-ray detectors. The detection of HE ν-s will help to understand the origin of cosmic rays since they are produced in hadronic interactions, thus bring direct information on the hadronic component accelerated in the sources. Many galactic and extragalactcic astrophysical object classes are able to accelerate particles above TeV energies which are confirmed by the recent observations with ground based γ-ray detectors (HEGRA, HESS, MAGIC, VERITAS and etc). While the acceleration of leptons (electrons) is limited by relatively short cooling times, the acceleration of hadrons goes beyond TeV/PeV energies. The interaction of these hadrons (with photon fields or matter) produces HE neutrinos ($>$ TeV). Recently, the IceCube detector has detected PeV neutrinos [2], which are probably produced in extragalactic objects (e.g. active galactic nuclei, the source of gamma-ray burst and etc). However, the galactic sources (especially supernova remnants) have been historically considered more favorable sources for HE neutrino detectors. This is justified considering that the galactic sources are responsible for acceleration of cosmic rays at energies below the so-called knee. Perhaps, the chance to detect the first galactic sources will increase with the arrival of KM3Net detector considering the northern hemisphere is more suitable to study galactic sources [3].

[a] e-mail: narek@icra.it

Current knowledge on the population and intensity of galactic nonthermal sources (provided by γ-ray detectors) allows to make quantitative estimation on the expectations for HE ν detectors. Indeed, γ-rays and neutrinos are produced almost in the same rate in pp interaction, thus γ-ray data can be used as a guide for neutrino searches. In [3], it was shown that the most relevant energy range for γ-ray observations (which will provide one event in neutrino detector) is around 20 TeV, where the intensity of γ-ray is at the narrow range of $I_\gamma \sim (2-6) \times 10^{-15}$ TeV^{-1} cm^{-2} s^{-1} [3]. However, for now only limited data are available at 20 TeV, which will be highly explored with the future planned CTA instrument, therefore we try to obtain a limit at somewhat lower energies which are currently better investigated (e.g. $\sim$ 1 TeV).

Let's suppose accelerated protons interact with low energy protons (pp interaction) and produce γ-rays. If the source is transparent for produced γ-rays, one can derive a limit on the γ-ray flux which accompanying HE ν flux can be detected by current instruments. The sensitivity of nowadays HE ν detectors (e.g. IceCube) at 1 TeV are:

$$F_{sens}(E_\nu = 1\,\mathrm{TeV}) = \begin{cases} 1.7 \times 10^{-11}\ \ \mathrm{TeV^{-1}cm^{-2}s^{-1}} & \text{– after 1 year} \\ 4.9 \times 10^{-12}\ \ \mathrm{TeV^{-1}cm^{-2}s^{-1}} & \text{– after 5 years} \end{cases} \tag{1}$$

On the other hand, the ratio between γ-ray and ν fluxes produced in pp interaction for broad proton distribution ($dN_p/dE_p \sim E_p^{-\alpha}$, $\alpha = (1.5 \div 3.5)$) corresponds to: $\frac{F_\gamma}{F_\nu} \approx (0.6\text{–}2.5)$ (see Fig. 2 in [4]). Taking into account this ratio and assuming that the HE ν flux corresponds to the threshold value presented in Eq. 1, it can be calculated the corresponding γ-ray luminosity for a given distance (d). This luminosity corresponds to:

$$L_\gamma(E_\gamma = 1\,\mathrm{TeV}) = \begin{cases} (1.95 \div 8.13) \times 10^{33} \left(\frac{d}{1\,kpc}\right)^2\ \ \mathrm{erg\ s^{-1}} & \text{– after 1 year} \\ (5.6 \div 23.3) \times 10^{32} \left(\frac{d}{1\,kpc}\right)^2\ \ \mathrm{erg\ s^{-1}} & \text{– after 5 years} \end{cases} \tag{2}$$

A γ-ray source can be also detected by current ν detectors, if at the distance $d = 1$ kpc it has a γ-ray luminosity exceeding the limits presented in the Eq. (2). Accordingly, measuring the γ-ray flux at 1 TeV (and a corresponding luminosity) by comparing with the values presented in Eq. (2) we can find relevant sources for observations with HE ν detectors. The limit presented in Eq. (2) was obtained assuming the source is transparent for produced γ-rays. The absorption of the γ-rays introduces uncertainties, between γ-ray and ν fluxes and the limits presented in Eq. (2) are not applied anymore. In particular, such conditions are satisfied for galactic binary systems which we discuss separately.

2. The sample

In principle above mentioned limits can be applied for both galactic and extragalactic sources. Poor understanding of physics of extragalactic sources (e.g. AGN and GRBs) and intrinsic and/or intergalactic absorption hardens the prediction for extragalactic objects. Instead, the situation is more clear with the galactic sources: comparatively better information on physical processes responsible for γ-ray and ν emissions. The recent observations in GeV/TeV band clearly demonstrated that several classes of galactic sources, such as supernova remnants, pulsar wind nebulae and microquasars (binary systems), are effective particle accelerators above TeV energies. However, in some cases both leptonic and hadronic scenarios can successfully reproduce the observed HE γ-ray data. Without going into details, here we assume that the observed γ-rays have a hadronic origin and we discuss the perspectives of detecting HE neutrinos from these sources.

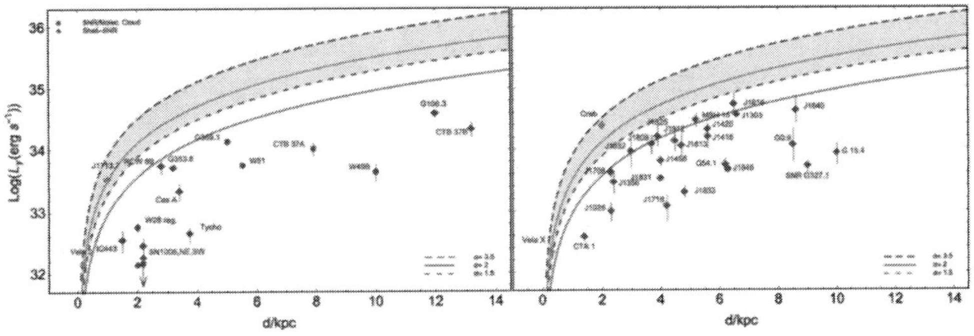

Figure 1. Gamma-ray luminosity of SNRs (on the left) and PWNe (on the right) compared with the minimal γ-ray luminosity required to produce detectable HE ν flux. The filled area corresponds to different power-law indicis of initial protons ($\alpha = (1.5 \div 3.5)$).

All the results from observations in HE energy band (including all instruments) are listed in *TeVCat*. Currently it contains around 150 sources which include both galactic and extragalactic objects. We consider only sources belonging to SNRs and PWNe classes, which are powerful nonthermal emitters up to very high energies.

Supernova Remnants: SNRs were considered as the main sources of galactic cosmic rays after the arguments by Baade and Zwicky in 1934 [5]. A modest efficiency of $\sim$ 10% in converting the kinetic energy of supernova shocks into particle acceleration and the rate of SNRs in our Galaxy can explain the observed flux of cosmic rays. There was no observation proof of proton acceleration in SNRs shock until recently, since both leptonic and hadronic modelings give a satisfactory description of the γ-ray data. However, recently AGILE and Fermi-LAT measured pion bumps in SNRs IC 443 and W 44 [6–8] which show indirect evidence of the acceleration of hadronic particles in SNRs. If so, it is natural to consider SNRs as potential targets for HE ν observations.

Currently, there are 10 SNRs interacting with molecular clouds and 13 shell type SNRs which are listed in *TeVCat*. Unfortunately, for some sources there are missing measurements of flux at 1 TeV (e.g. photon index) and only 20 SNRs are included in the sample. We assume, that the γ-rays have pure hadronic origin and we compare the luminosity at 1 TeV with limits obtained in Eq. (2). Figure 1 (on the left) illustrates the luminosity of shell type SNRs (circle) and SNRs interacting with molecular clouds (square) compared with the γ-ray luminosity. The filled area corresponds to different initial proton distributions changing from very soft $\alpha = 3.5$-long dashed to very hard $\alpha = 1.5$-dashed (solid line represents the power-law index $\alpha = 2$). As one can see, most of sources have γ-ray flux nearly two order of magnitude below than required threshold. It appears only for two powerful SNRs, the measured γ-ray luminosity is in the order of presented limit. These are SNRs RX J1713.7-3946 and RX J0852.0-4622 which have been extensively discussed as potential targets for HE ν detectors. Different estimations show that in principle several events in a year exposure can be detected if the produced γ-rays have hadronic origin (however, up to now the leptonic origin can not be disfavored). The location of these two SNRs is not covered by IceCube, but they are perfect targets for future planned KM3Net instrument.

The red solid line in Fig. 1 corresponds to necessary γ-ray luminosity in order to detect HE ν-s after 5-years of observations. It is noticeable, that even in this case the situation is not improved: only above mentioned SNRs will be detected. The γ-ray observations should find new SNRs otherwise the expectations from currently known SNRs are not promising.

Pulsar wind nebulae: PWNe are another class of galactic sources which accelerate particles up to VHEs. If one assumes the fraction of accelerated particles are protons, then from the inelastic interactions between accelerated protons and the ambient gas HE/VHE γ-rays and ν-s are produced. Of course, the presence of bayronic component in pulsar winds (coexisting) with the leptonic one is still an open question and moreover the pulsar must efficiently convert rotational energy into proton acceleration. However, considering the difficulties with the leptonic modeling, the hadronic scenario remains attractive for γ-ray emission from PWNe. This makes PWNe another class of galactic sources producing HE ν-s.

The number of PWNe detected in TeV band and presented in *TeVCat* corresponds to 35 but for 7 sources no data are available. For remaining 28 sources, we assume that the produced γ-rays have pure hadronic origin and compare with the limits presented in Eq. (2) (see Fig. 1). Similar to the previous case, the luminosity of almost all sources is significantly below than the required limit and only two sources, Crab nebulae and Vela X, have flux satisfying mentioned limit. One of these sources, crab nebulae – a very well studied object in all wavelengths, emits γ-rays up to TeV band which are most likely produced from inverse Compton scattering of soft photon fields inside the nebula. In case if γ-rays are produced from protons, from this source also HE ν can be detected. However, so far no evidence for proton contribution seen from the observations. On the other hand, Vela X PWN, an active pulsar (PSR B0833-45) associated with nebula, can be easily detectable with the current instrument (γ-ray flux corresponds to $L_\gamma \sim 1.7 \times 10^{32}$ erg s^{-1} exceeding the obtained limits). The origin of unusual TeV γ-rays (photon index $\Gamma \approx 1.32$) [9] is under debates but considering the difficulties for modeling within leptonic scenarios, the hadronic origin can not be disfavored. All this arguments show that Vela region (Vela X PWN and Vela Junior SNR) is a prime candidate for galactic ν sources to be detected with KM3Net.

It should be noted, that after 5-years of exposure, some of the sources can be detected with current HE ν instruments (red solid line in Fig. 1). Compared with the previous case, this is improved, but still few events in 5-years of observations do not make these source potentially interesting for ν astronomy.

Binary systems: For a long time, binary systems were considered as potential γ-ray emitters, however unavoidable absorption of HE γ-rays (by optical photons in the systems) makes hard their detection. The improved sensitivity of current instruments results the detection of GeV/TeV γ-rays from binary systems (microquasars) confirmed that their emission extends up to TeV energies (or even higher energies). If the γ-rays are produced from protons, the flux of HE ν-s can significantly exceed the one of γ-rays (by a factor of e^τ where τ is the optical depth) considering that they escape from the region without absorption unlike the γ-rays. Due to the unknown parameter τ no acceptable limit can be drown in this case. However, below it is demonstrated that using the observed γ-ray flux from Cygnus X-3 (well known galactic microquasar) and under reasonable assumptions for absorption it can be detected by neutrino detectors.

It has been shown, that γ-ray flaring activity from microquasar Cygnus X-3 can be well explained within hadronic scenario [10]: protons accelerated via jet interact with the wind particles from companion Wolf–Rayet star. Normalizing the proton spectrum using AGILE data (unabsorbed) and taking into account the absorption above GeV energies (in order not to overcome MAGIC upper limits) the expected HE ν flux is only by a factor of about 3 lower than the 1-year IceCube sensitivity at $\sim$10 TeV. Considering a prolonged "soft X-ray state" (which is characterized by γ-ray emission), the ν flux from this source might be close to being detectable by cubic-kilometer ν telescopes such as IceCube. Moreover, the estimations for HE ν-s from another galactic binary system, LS 5039, show that the flux of HE ν-s can be as large as 1.6×10^{-11} cm^{-2} s^{-1} (for energy greater than 1 TeV) [11], above the sensitivity threshold of experiments in the Mediterranean Sea (KM3Net). This shows that probably binary systems are a class of galactic objects, from which the first galactic HE ν-s will be detected.

3. Conclusion

The cubic kilometer scale HE ν detectors with the existing HE γ-ray detectors provide an opportunity to study nonthermal processes in both galactic and extragalactic objects in great detail. While the recent detection of PeV ν by IceCube demonstrate the beginning of extragalactic ν astronomy, the first galactic sources of HE ν still have to be detected. There are several classes of galactic sources, which *i)* can accelerate the particles (hadrons) up to hundreds of TeV energies and *ii)* are established to be TeV emitters. Clearly, the first galactic source which will be detected, belongs to one of these well established classes of TeV emitters.

The flux of TeV γ-ray allows to estimate the expected flux of HE ν, if the γ-rays have a hadronic origin and produced γ-rays do not suffer from absorption in the sources. Indeed, the minimal detectable flux of HE ν implies the minimal flux of accompanying γ-rays. Such estimation shows, that from already detected SNRs and PWNe, only two SNRs, RX J1713.7-3946 and RX J0852.0-4622, one PWN- Vela X, have significantly high HE ν flux, which can be detected with current ν detectors. Even longer exposure time (5-years) will not significantly increase the number of potentially detectable sources. However, if the γ-rays from above mentioned sources have a hadronic origin they will provide several events in kilometer-cube scale size detector in a year exposure.

On the other hand, the unexpected high flux of HE ν-s can be detected from galactic binary systems (recently established strong TeV emitters). The effective absorption of TeV γ-rays, means that the flux of HE ν can exceed the γ-ray one by several orders of magnitude (depending on the optical depth τ). The estimation of HE ν flux from several binary systems (e.g., Cygnus X-3, LS 5039) shows that it is either in the same order or exceeds the sensitivity of the current instruments. Moreover, considering that the TeV γ-rays can be totally absorbed in the system (high density of soft photons), the ν observations might discover new binary systems which are not "seen" by γ-ray instruments. Therefore, these are the best targets for HE ν detectors.

References

[1] Aharonian, F., Nuclear Physics B Proceedings Supplements **221**, 5 (2011)
[2] Aartsen, M. G. et al., Phys. Rev. Lett. **111**, 021103 (2013)
[3] Vissani, F. Aharonian, F. and Sahakyan, N., Astroparticle Phys. **34**(10), p. 778–783, 1 (2013)
[4] Kelner, S., Aharonian, F. and Bugayov, V., Phys. Rev. D **74**, 3, id. 034018 (2006)
[5] Baade, W. and Zwicky, F. Proceedings of the National Academy of Science **20**, 259 (1934)
[6] Giuliani, A., Cardillo, M. et al., ApJL **742**, 2, L30, 5 pp. (2011)
[7] Tavani, M., Giuliani, A. et al., ApJ **710**, 2, L151 (2010)
[8] Ackermann, M., Ajello, M. et al., Science **339**, 6121, 807–811 (2013)
[9] Abramowski, A., Acero, F. et al., A&A **548**, A38, 11 (2012)
[10] Sahakyan, N., Piano, G. and Tavani, M., ApJ **780**,1, 29, 7 (2014)
[11] Aharonian, F. et al., JPCS **39**, 1, 408 (2006)

EPJ Web of Conferences **121**, 05006 (2016) DOI: 10.1051/epjconf/201612105006
RICAP-2014

Analysis of the cumulative neutrino flux from *Fermi* LAT blazar populations using 3 years of IceCube data

Thorsten Glüsenkamp[a] for the IceCube Collaboration

DESY Zeuthen, Platanenallee 6, 15738 Zeuthen, Germany

Abstract. The recent discovery of a diffuse neutrino flux up to PeV energies raises the question of which populations of astrophysical sources contribute to this diffuse signal. One extragalactic candidate source population to produce high-energy neutrinos are Blazars. We present results from a likelihood analysis searching for cumulative neutrino emission from Blazar populations selected with the 2nd *Fermi* LAT AGN catalogue (2LAC) using an IceCube data set that has been optimized for the detection of individual sources. In contrast to previous searches with IceCube, the investigated populations contain up to hundreds of sources, the biggest one being the entire Blazar sample measured by the Fermi-LAT. No significant neutrino signal was found from any of these populations. Some implications of this non-observation for the origin of the observed PeV diffuse signal will be discussed.

1. Introduction

In 2013, the IceCube collaboration reported on the detection of an astrophysical diffuse neutrino flux in the 40 TeV–2 PeV energy range [1]. Since then it has been measured down to about 10 TeV in energy and the spectrum has been shown to be consistent with a power law of index -2.5 [2]. The origin of the observed neutrino signal is so far unidentified. A prominent candidate is the population of Blazars, active galactic nuclei (AGN) with their relativistic jets pointing along the line-of-sight. Neutrino emission from Blazars has been discussed extensively in hadronic jet models originating from p-p or p-γ interactions (e.g. [3]) which also produce gamma rays from neutral pion decays. The extragalactic non-thermal radiation background at GeV energies and above (EGB) is dominated by the emission from Blazars [4]. An experimental approach is presented here to answer the question if the Blazars that dominate the high energy EGB emission also produce a large fraction of the astrophysical neutrino flux. The idea is to look for directional clustering in a large sample of muon tracks collected by IceCube, around the directions of gamma-ray sources associated with Blazars in the *Fermi*-2LAC catalog [5].

This analysis is performed as a maximum likelihood stacking analysis looking for a cumulative signal from multiple sources. In contrast to previous stacking searches which looked at 10–30 sources [6, 7], the populations in this work usually comprise more than 100 sources, the biggest population being 862 sources when all Blazars from the 2LAC catalog are combined. Any search for a cumulative signal from multiple sources needs to make

---

[a]e-mail: thorsten.gluesenkamp@desy.de

assumptions about the relative contributions ("weights") of the individual sources within the population. Here, this weighting scheme is set up in a way to be minimally biased with respect to γ–ν correlation assumptions. Section 2 covers the motivation for this type of search and the technical setup. In Sect. 3 the sample of muon tracks used in this analysis is described. In Sect. 4 the results of this search are summarized. Conclusions are presented in Sect. 5.

2. Motivation and method

In previous stacking searches for a cumulative neutrino signal from Blazars in IceCube data [6, 7], the measured gamma-ray flux was taken as an estimate for the expected neutrino flux. However, the γ and ν spectra may not be strongly correlated: Other contributions to the gamma flux almost surely exist, e.g. inverse Compton photons [8] from high-energy electrons or synchrotron photons from muons and protons [9]. Moreover, radiation fields at the sources and in the interstellar medium absorb and reprocess the gamma rays originally produced together with the neutrinos. In order to overcome this bias as much as possible, two modifications with respect to previous IceCube stacking searches have been implemented:

- **There is no sub-selection of sources within the catalog based on their gamma-ray flux** The populations tested for a neutrino signal are only defined via their spectral classification, the largest one being just all Blazars of the 2LAC catalog. Due to the competing production processes and different environmental conditions described above, a source with a relatively low gamma-ray flux could still contribute significantly to the neutrino signal.
- **Two complementary weighting schemes are used, including a uniform relative weight for all sources**

1. $w_{source} \propto F_\gamma$
   Note that F_γ is the energy flux. Gamma-rays produced in neutral pion decays at multi-TeV energies (i.e. by the same process that also produces a neutrino signal) are absorbed quickly by IR photons present in the source environment and reprocessed to lower energies. In this process they would approximately maintain their bolometric luminosity.
2. $w_{source} = 1$
   Equal weighting is used as the most unbiased weighting scheme, completely ignoring potential correlations. Although it is in a strict sense unrealistic, because no population consist of sources which all emit equally strong, the resulting limits are quite independent of the unknown correlation between neutrino and gamma-ray flux. They even hold for the case that no correlation exists.

The analysis is performed with an unbinned maximum likelihood Ansatz. The log-likelihood function is constructed as

$$\ln(L) = \sum_i^N \ln\left(\frac{n_s}{N} \cdot S_i + \left(1 - \frac{n_s}{N}\right) \cdot B_i\right) \tag{1}$$

where N is the total number of neutrino events, S_i and B_i are the signal probability distribution function (PDF) and background PDF evaluations for event i. The background PDF follows the data distribution in declination, is uniform in right ascension, and includes an energy term. The signal PDF is defined as

$$S_i = \frac{\sum_j^{N_{\text{sources}}} w_{\text{j,tot}} \cdot S_j(x_i; \sigma_i) \cdot \varepsilon_j(E_i)}{\sum_j^{N_{\text{sources}}} w_{\text{j,tot}}}, \quad w_{tot} = w_{\text{source}} \cdot w_{\text{dec}}. \tag{2}$$

Table 1. Definitions of Blazar populations.

Type	No. of sources	Motivation
All 2LAC Blazars	862	No bias
FSRQ[1]	310	BLR radiation [11]
LSP[2]	308	FSRQ and LSP-BLLAC might be intrinsically similar [12]
ISP | HSP[2]	301	HSP objects seem to evolve differently [13]
LSP & BLLAC[2]	62	Motivated by work in [9]

[1] FSRQ/BL-LAC based on optical line equivalent width.
[2] Low/Intermediate/High Synchroton Peaked Object (LSP/ISP/HSP), based on position of synchroton peak.

where j denotes the individual sources within the population. The term $S_j(x_i;\sigma_i)$ denotes the point spread function (PSF) of source j evaluated at celestial coordinate x_i, depending on the event-specific angular error estimate σ_i, while the term $\varepsilon_j(E_i)$ denotes the energy PDF of source j evaluated for the reconstructed energy E_i. The PSF is locally modeled as an analytic 2-d radially symmetric gaussian while the energy PDF ε is estimated from Monte Carlo simulations. Each individual source within the signal PDF is weighted with a declination-dependent detector response weight $w_{\text{dec.}}$ and the relative source weight w_{source} described above. Equation (1) is minimized with respect to n_s and the likelihood ratio with respect to $n_s = 0$ is calculated for each population and each weighting scheme. It is tested if the observed likelihood ratio is compatible with statistical fluctuations of the background events by repeating the process for many randomized sky maps. A randomized sky map is generated by randomizing the right ascension of the observed events (a more detailed description of this procedure is found in [7]). In case no significant deviations from background fluctuations are observed, upper limits on the neutrino flux of the tested population are calculated via the CL_s method [10]. This is performed twice, once for each weighting scheme.

3. Data and source catalogs

3.1 Muon Track Data

A muon-track selection for the years 2009–2011 (IC59-IC86/1) is used, which is similar to the one described in [7][1]. The median angular resolution of muon-neutrino tracks with this selection is around 1° at 1 TeV and 0.5° at 100 TeV. One additional cut is applied which keeps only events that have an estimated angular resolution better than 5 degrees. In the southern hemisphere (downgoing events in IceCube) this sample is dominated by muons from air showers, while in the upgoing region the majority of events come from atmospheric muon-neutrinos.

3.2 Populations tested

The populations to be tested are based on the 2nd Fermi AGN catalog (2LAC) [5] and are classified purely by spectral properties. Only sources that do not suffer from potential confusion are considered (CLEAN flag set to True). The populations are listed in Table 1. It should be noted that some of the populations are overlapping and the p-values correlated (the largest overlap exists between the LSP and FSRQ populations, which is about 60%).

[1] The IC79 dataset is the reprocessed one also used in the Blazar stacking search in [7].

Table 2. Results of the Blazar population tests for both weighting schemes.

	p-values	
	$w_{source} \propto F_\gamma$	$w_{source} = 1$
All 2LAC Blazars	36%	6%
FSRQs	34%	34%
LSPs	36%	28%
ISP/HSPs	>50%	11%
LSP-BLLACs	13%	7%

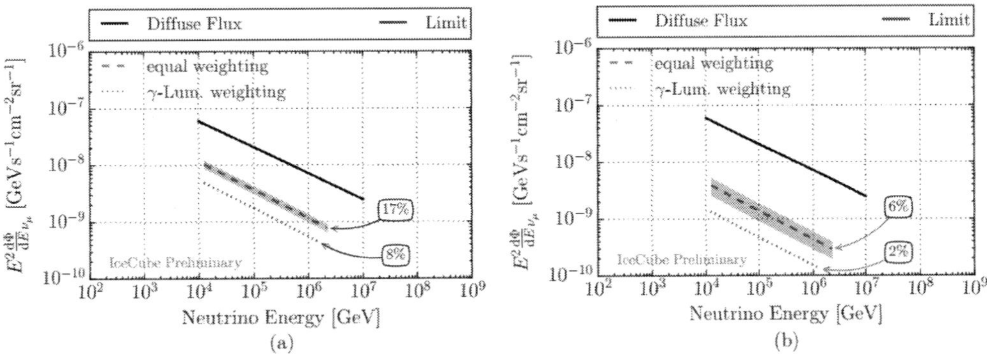

Figure 1. Neutrino flux upper limits for an $E^{-2.5}$ spectrum (blue) compared to diffuse bestfit (black solid) from [2]. a) shows this comparison for the "All 2LAC Blazar" sample, b) for the FSRQs. Percentages with arrows denote the fraction with respect to the diffuse flux.

4. Results

For each population two tests were performed, one for each of the two different source weighting schemes. Table 2 summarizes the results[2]. All test outcomes are compatible with statistical fluctuations of the background. The smallest p-value is 6% for the "All 2LAC Blazar" sample when tested with the "equal weighting" scheme. Figure 1 shows the resulting neutrino flux upper limits for an $E^{-2.5}$ spectrum for the "All 2LAC-Blazars" population and for the FSRQ population. Notice that the flux upper limit using the equal-weighting scheme is represented as a band, instead of a line. It serves as an indicator of how robust this flux limit is towards changes of the actual source count distribution over the sky, which is unknown. Random realizations are created assuming a source count distribution for the neutrino emission that has the same shape as the source count distribution in gamma rays (parametrization from [14]). For each source in the population a relative source strength is drawn from the source count distribution and assigned to the source. The band encompasses the central 90% of the flux upper limits found from different such realizations. In comparison to the limits the best fit diffuse flux above 10 TeV is shown from [2] with a spectral index of -2.46. Assuming an isotropic flux and 1/1/1 composition of flavors, the maximal contribution from the Blazar populations with respect to the diffuse flux can be read off. This contribution to the diffuse neutrino flux is at most around 20%, even if there is a very weak correlation with the observed gamma flux. For FSRQ's, these numbers are lower as the population size is smaller.

[2] No trial-correction for testing multiple populations is included.

5. Conclusion

In summary, the Blazars in the 2LAC catalog or any of the tested sub-samples are not responsible for the majority of the observed diffuse astrophysical neutrinos. This is in particular interesting since these Blazars dominate the high-energy extragalactic gamma-ray sky. The result is also valid if only a weak neutrino flux correlation with the observed gamma ray emission is assumed and expected to get worse at most by a factor of two for harder spectral indices (E^{-2}+cutoff scenario). Furthermore, one can go beyond the 2LAC sources, since extrapolation of the gamma-ray luminosity function shows that the 2LAC Blazars (FSRQ's) already resolve more than 60% [15] (70% [16]) of the EGB contribution from all Blazars (FRSQ's) in the observable universe. Thus, under the condition that the correlation between the γ and ν energy flux holds (i.e. hadronic processes dominate the gamma-ray emission), one could further extend the "γ-Lum. weighting" limits to apply to the total corresponding population by dividing by 0.6 (0.7).

References

[1] M.G. Aartsen et al. (IceCube) Science **342**, 1242856 (2013), 1311.5238
[2] M.G. Aartsen, M. Ackermann, J. Adams, J.A. Aguilar, M. Ahlers, M. Ahrens, D. Altmann, T. Anderson, C. Arguelles, T.C. Arlen et al., Phys. Rev. D **91** (2015)
[3] K. Mannheim. Astroparticle Physics **3**, 295–302 (1995)
[4] Y. Inoue (2014), 1412.3886
[5] M. Ackermann, M. Ajello, A. Allafort, E. Antolini. W.B. Atwood, M. Axelsson, L. Baldini, J. Ballet, G. Barwick, et al. The Astrophysical Journal **743**, 171 (2011)
[6] A. Achterberg, M. Ackermann, J. Adams, J. Ahrens, D. Atlee, J. Bahcall, X.Bai, B. Baret, M. Bartelt, S. Barwick et al., Astroparticle Physics **26**, 282–300 (2006)
[7] M.G. Aartsen, M. Ackermann, J. Adams, J.A. Aguilar, M. Ahlers, M. Ahrens, D. Altmann, T. Anderson, M. Arguelles, T.C. Arlen et al., The Astrophysical Journal **796**, 109 (2014)
[8] S. D. Bloom, A. P. Marscher, The Astrophysical Journal **461**, 657 (1996)
[9] A. Mücke, R. Protheroe, R. Engel, J.P. Rachen, and T. Stanev, Astroparticle Physics **18**, 593–613 (2003)
[10] K.A. Olive, Chinese Physics C **38**, 090001 (2014)
[11] A. Atoyan, C. Dermer, Physical Review Letters **87** (2001)
[12] P. Giommi, P. Padovani, G. Polenta, S. Turriziani, V. D'Elia, S. Piranomonte. Monthly Notices of the Royal Astronomical Society **420**, 2899–2911 (2012)
[13] M. Ajello, R. W. Romani, D. Gasparrini, M. S. Shaw, J. Bolmer, G. Cotter, J. Finke, J. Greiner, S. E. Healey, O. King, The Astrophysical Journal **780**, 73 (2013)
[14] A. A. Abdo, M. Ackermann, M. Ajello, E. Antolini, L. Baldini, J. Ballet, G. Barbiellini, D. Bastieri B.M. Bauhman, K. Bechtol et al., The Astrophysical Journal **720**, 435–453 (2010)
[15] Yoshiyuki Inoue and Tomonori Totani. The Astrophysical Journal **728**(1), 73 (2011)
[16] M. Ajello, M. S. Shaw, R. W. Romani, C. D. Dermer, L. Costamante, O. G. King, W. Max-Moerbeck, A. Readhead, A. Reimer, J. L. Richards et al., The Astrophysical Journal **751**, 108 (2012)

EPJ Web of Conferences **121**, 05007 (2016) DOI: 10.1051/epjconf/201612105007
RICAP-2014

ANTARES constraints on a Galactic component of the IceCube cosmic neutrino flux

Maurizio Spurio[1,2,a]

[1] Dipartimento di Fisica e Astronomia dell'Università di Bologna, Viale Berti Pichat 6/2, 40127 Bologna, Italy
[2] Istituto Nazionale di Fisica Nucleare, Sezione di Bologna, Viale Berti Pichat 6/2, 40127 Bologna, Italy

Abstract. The IceCube evidence for cosmic neutrinos has inspired a large number of hypothesis on their origin, mainly due to the poor precision on the measurement of the direction of showering events. A North/South asymmetry in the present data set suggests the presence of a possible Galactic component. This could be originated either by single point-like sources or from an extended Galactic region. Expected fluxes derived from these hypotheses are presented. Some values have been constrained from the present available upper limits from the ANTARES neutrino telescope.

1. The IceCube cosmic neutrinos and a possible Galactic component

The recent IceCube (IC) evidence for extraterrestrial high-energy neutrinos [1, 2] opened new windows in the field of astroparticle physics [3]. With the present statistics, the High Energy Starting Events (HESE) flux observed by IC is compatible with flavor ratios $\nu_e : \nu_\mu : \nu_\tau = 1 : 1 : 1$, as expected from charged meson decays in cosmic ray (CR) accelerators and neutrino oscillation on their way to the Earth. The non-observation of events beyond 2 PeV suggests a neutrino flux with a power law $\Phi(E) \propto E^{-\Gamma}$ with hard spectral index, e.g. $\Gamma \simeq 2.0$, and an exponential cutoff, or an unbroken power law with a softer spectrum, e.g. $\Gamma \simeq 2.3$.

The majority of HESE are downgoing; as the IC detector is at the South Pole, this corresponds to a larger flux from the Southern sky, where most of the Galactic plane is present. Table 1 (columns from 2 to 5) reports for HESE with deposited energy $E_{dep} >$ 60 TeV [2]: the number of events; the estimated background; the number of cosmic neutrinos (i.e. the *signal*); and the number of expected cosmic νs assuming the best-fit hypothesis. Values are given separately for the North/South sky regions.

Recently, IC presented a new search for neutrinos interacting in the instrumented volume and with energy between 1 TeV and 1 PeV, using 641 days of livetime [4]. Table 1 (columns from 6 to 10) reports, for the events in this new sample having $E_{dep} >$ 25 TeV, the same quantities defined above for the HESE. No hypothesis test on HESE as reported in [2] yielded at present statistically significant evidence of clustering or correlations, in particular from the Galactic Center or the Galactic Plane. The same for the neutrino sample studied in [4].

In the data presented in Table 1, an excess of downgoing (Down) events with respect to expectation seems however be present. The Northern sky, inducing upgoing (Up) events

[a]e-mail: maurizio.spurio@bo.infn.it

Table 1. Column 2 to 5: number of HESE with $E_{dep} > 60$ TeV (Data); number of background events (Bck); number n_{IC} of signal events (Data-Bck); expected number N_{IC} of signal events from the best-fit. Column 7 to 10: the same quantities for sample with $E_{dep} > 25$ TeV in [4]. Quantities are given for upgoing (from the Northern hemisphere) and downgoing (from the Southern hemisphere) events, and for the whole sky (All).

HESE [2] $E_{dep} > 60$ TeV	Data	Bck	n_{IC}	N_{IC} E^{-2}	New ν sample [4] $E_{dep} > 25$ TeV	Data	Bck	n_{IC}	N_{IC} $E^{-2.46}$
Up (North)	5	1.4	3.6	6.7	Up ($\sin\delta > 0.06$)	11	5.3	5.7	12.1
Down (South)	15	1.3	13.7	11.5	Down ($\sin\delta < -0.06$)	29	4.8	24.2	15.0
All	20	2.7	17.3	18.2	All	43	11.7	31.3	29.1

in IceCube, contains only a small fraction of the Galaxy. For this reason, let us assume that the 3.6 estimated HESE from cosmic neutrinos (n_{IC}) arising from the Northern sky are all of extragalactic origin. Assuming a $E^{-2.0}$ spectrum, a symmetric contribution from the North/South extragalactic sources and taking into account the different $A^{\nu_i}_{IC}$ for events coming from the North and South hemispheres, then 6.2 events are expected from the South (they reduce to 5.8 events for a $E^{-2.5}$ flux). This excess of $\sim$7.5 events in the Southern sky corresponds to $\sim 40\%$ of the total signal.

An even stronger excess from the South is derived from data in [4], using events with $E_{dep} > 25$ TeV. Here, to the 5.7 signal events coming from the North with $\sin\delta > 0.06$ (assuming the same considerations of above) should correspond $\sim$ 7.2 events from the South with $\sin\delta < -0.06$. As 24.2 events are observed, more than 50% of the number of signal events in the whole sky seems to be produced by a non-isotropic cosmic component, likely of Galactic origin. A possible contribution from transient extragalactic objects located in the Southern sky can be considered as well.

The above conclusions are derived for $\Gamma = 2.0$; however similar results are obtained using softer spectral indexes (i.e. $\Gamma > 2.0$) for the cosmic neutrino flux. This is relevant because different models involving Galactic, extragalactic or exotic origin of the IC signal exist in the literature. The neutrino flux predicted by each model has a preferred value, usually ranging in the interval $\Gamma = 2.0 \div 2.7$.

In the following, the effects of the hypothesis that a sizeable fraction of the cosmic neutrinos observed by IC is originated in our Galaxy is considered. Following the methods reported in [5], the neutrino flux from point-like or extended sources compatible with the above evaluated Galactic fraction of the IC signal is derived for different values of Γ.

A signal originating from the Southern sky region can be observed by the ANTARES neutrino telescope [6], located in the Mediterranean Sea. Existing ANTARES upper limits derived under the hypothesis of a $\Gamma = 2.0$ neutrino spectrum are used to infer upper limits for $\Gamma > 2.0$. The ANTARES expected sensitivities for extended sources are used to discuss the conditions under which an IceCube hot spot can be observed.

2. The ANTARES and IceCube effective areas

The neutrino effective area at a given energy, $A_{eff}(E)$, is defined as the ratio between the neutrino event rate in a detector (units: s^{-1}) and the neutrino flux (units: cm^{-2} s^{-1}) at that energy. The effective area depends on the flavor and cross-section of neutrinos, on their absorption probability during the passage through the Earth, and on detector-dependent efficiencies. Detector efficiencies are correlated to each particular analysis, referring to the criteria used to trigger and to reconstruct the events, and to the cuts applied to reduce the background. A fraction of the irreducible background due to atmospheric neutrinos

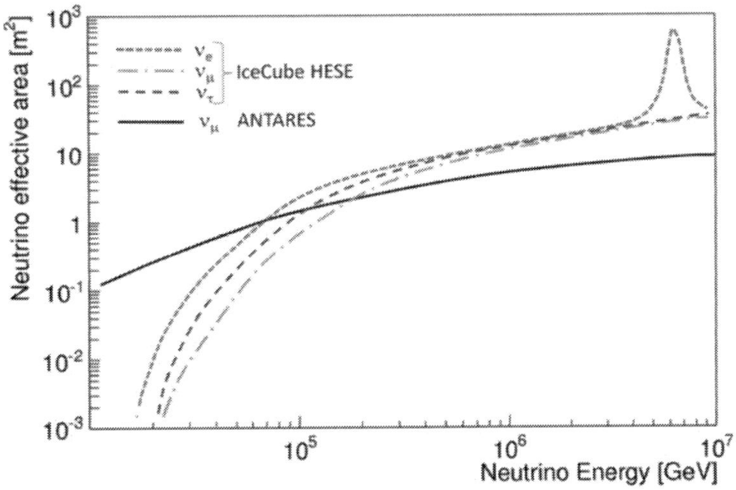

Figure 1. Full black line: ANTARES ν_μ effective area [7]. The neutrino track is determined with a median angle of $\lesssim 0.4°$. Colored dashed lines: IceCube ν_e, ν_μ and ν_τ effective areas [1] from the analysis yielding the HESE. The backgrounds due to atmospheric muons and neutrinos is largely suppressed above few tens of TeV.

contaminates in any case the signal, with a percentage depending on the strength of cuts used to define $A_{eff}(E)$.

Figure 1 shows the ANTARES effective area ($A^{\nu_\mu}_{ANT}$, full black line) for the ν_μ flavor as derived in the analysis [7] for the search for cosmic neutrino point sources in the declination band containing the Galactic Center. The red ($A^{\nu_e}_{IC}$), green ($A^{\nu_\mu}_{IC}$) and blue ($A^{\nu_\tau}_{IC}$) lines refer to the IceCube analysis on a 4π sr yielding the HESE.

Despite the fact that the ANTARES instrumented volume is much smaller than 1 km^3, Fig. 1 shows that the ANTARES ν_μ effective area is larger than $A^{\nu_e}_{IC}$, $A^{\nu_\mu}_{IC}$ and $A^{\nu_\tau}_{IC}$ below $\sim$ 60 TeV. At the highest energies where neutrinos were detected, 2 PeV, $A^{\nu_\mu}_{IC}$ is a factor of two larger than $A^{\nu_\mu}_{ANT}$ while the total IC effective area for HESE ($A^{\nu_e}_{IC} + A^{\nu_\mu}_{IC} + A^{\nu_\tau}_{IC}$) is only 7.3 times larger than $A^{\nu_\mu}_{ANT}$.

These values of the A_{eff} of the two experiments are largely dominated by the different criteria hidden in the analysis. Strong cuts are used in the IC analysis to select a high-purity sample of diffuse high-energy cosmic neutrinos with interaction vertex inside the instrumented volume. The estimated angular resolution is $\sim 1°$ for ν_μ and $\sim 10° - 15°$ for neutrino interactions producing showers (mainly from charged current interactions of ν_e, ν_τ). The criteria used in the ANTARES analysis allow a larger contamination of lower-energy atmospheric neutrinos, but enable to reconstruct ν_μ events with superior angular resolution, $\sim 0.4°$. The consequence is that ANTARES has equivalent (or superior, depending on the signal spectral index) capability to extract a signal if the cosmic source is located in the Southern sky, and it is point-like or confined in a region seen within a small solid angle $\Delta\Omega$ by the detector.

3. Normalization factors for different cosmic spectral indexes

The standard diffusive shock acceleration model yields a $\Gamma = 2.0$ spectral index for primary CRs, and consequently for secondary γ-rays and neutrinos. However, most γ-ray sources observed in the GeV and TeV range show spectral indexes larger than 2.0. The reason why

γ-ray spectra from supernovae remnants are observed with spectral indexes $\Gamma \simeq 2.2 - 2.3$ remains unclear. A softer spectral index ($\Gamma \simeq 2.4 - 2.5$) is consistent with the theoretical model of CR injection by diffusive shock acceleration followed by escape through the Galactic magnetic field with Kolmogorov turbulence [8].

Thus, it is important to consider the normalization factors $\Phi_0^{D,\Gamma}$ for the IC signal for different models of cosmic fluxes $E^\Gamma \Phi^{D,\Gamma}(E) \equiv \Phi_0^{D,\Gamma}(E)$ (in units: GeV cm^{-2} s^{-1} sr^{-1}. The D stands for $diffuse$.) The same number N_{IC} of events for different $\Phi^{D,\Gamma}(E)$ is obtained using the effective area $A_{IC}(E) \equiv [A_{IC}^{\nu_e} + A_{IC}^{\nu_\mu} + A_{IC}^{\nu_\tau}]$ and detector livetime T:

$$N_{IC} = T \cdot \int \Phi^{D,\Gamma}(E) \cdot A_{IC}(E) \cdot dE \cdot d\Omega = 4\pi T \cdot \Phi_0^{D,\Gamma}$$
$$\cdot \int E^{-\Gamma} \cdot A_{IC}(E) \cdot dE = 4\pi T \cdot \Phi_0^{D,\Gamma} \cdot \mathcal{D}_\Gamma \,. \tag{1}$$

The integral $\mathcal{D}_\Gamma$ (the *detector response*) extends over the energy range where $A_{IC}(E)$ is not null, and is computed numerically.

Let us assume that n_p events out of N_{IC} are produced by a point-like source with generic spectrum: $E^\Gamma \Phi^{p,\Gamma}(E) = \Phi_0^{p,\Gamma}$ (units: GeV cm^{-2} s^{-1}. The p stands for $point-like$.) The normalization factor $\Phi_0^{p,\Gamma}$ necessary to produce n_p events is obtained by requiring that:

$$n_p = T \cdot \int \Phi^{p,\Gamma}(E) \cdot A_{IC}(E) \cdot dE = T \cdot \Phi_0^{p,\Gamma} \int E^{-\Gamma} \cdot A_{IC}(E) \cdot dE = T \cdot \Phi_0^{p,\Gamma} \cdot \mathcal{D}_\Gamma \tag{2}$$

where T, $A_{IC}(E)$ and, consequently, the detector response $\mathcal{D}_\Gamma$ are the same as in Eq. (1). Then, the normalization factor for a point-like source flux of a given spectral index Γ is given by:

$$\Phi_0^{p,\Gamma} = 4\pi \cdot \left(\frac{n_p}{N_{IC}}\right) \cdot \Phi_0^{D,\Gamma} \,. \tag{3}$$

If a fraction $n_{\Delta\Omega}$ of the IceCube signal is produced in a region of the Southern sky of angular extension $\Delta\Omega \ll 4\pi$ sr, and flux $E^\Gamma \Phi^{D',\Gamma}(E) = \Phi_0^{D',\Gamma}$, the signal can be observed as an *enhanced diffuse flux*. Similarly to Eq. (3), using the detector response derived in (1), we obtain

$$\Phi_0^{D',\Gamma} = \left(\frac{n_{\Delta\Omega}}{N_{IC}}\right) \cdot \left(\frac{4\pi}{\Delta\Omega}\right) \cdot \Phi_0^{D,\Gamma} \,. \tag{4}$$

4. ANTARES constraints for the IC signal from the Southern sky

Point-like sources. Table 2 reports the normalization factor $\Phi_0^{p,\Gamma}$ for a point-like source necessary to produce $n_p = 1 \div 5$ HESE, as derived from Eq. (3). Four different values of Γ are considered. Point-like sources in the Galactic central region were searched for by ANTARES [9] and upper limits as a function of the source declination were derived assuming a spectral index $\Gamma = 2.0$. Following the procedure defined in [5], the the 90% C.L. upper limit for a point-like source has been translated to upper limits for softer spectral indexes. The ANTARES results for $\Gamma = 2.0$ in the Galactic Center region and the derived valued for Γ =2.2, 2.3 and 2.4 are reported in the last column of Table 2. The ANTARES 90% C.L. upper limit excludes a single point-like source with $\Gamma = 2.0$ producing more than 5 HESE. The derived limit excludes a single point-like source yielding a cluster of more than 2 events for $\Gamma = 2.3$, while the presence of a cluster made of two or more events is excluded for $\Gamma > 2.3$.

Table 2. Column 2 to 6: normalization factors $\Phi_0^{p,\Gamma}$ yielding $n_p = 1, ..., 5$ HESE in IceCube vs. Γ. The last column shows the 90% C.L. upper limits for a $\Gamma = 2.0$ point-like source derived from ANTARES [9]. The values for $\Gamma > 2.0$ were derived in [5]. The first value in each row excluded by these limits is underlined.

	units: (GeV cm^{-2} s^{-1}) $\Phi_0^{p,\Gamma}$ (from HESE)					ANTARES
$\Gamma =$	$n_p = 1$	$n_p = 2$	$n_p = 3$	$n_p = 4$	$n_p = 5$	90% C.L. limit
2.0	6.9 10^{-9}	1.4 10^{-8}	2.1 10^{-8}	2.8 10^{-8}	3.5 10^{-8}	4.0 10^{-8}
2.2	9.0 10^{-8}	1.8 10^{-7}	2.7 10^{-7}	<u>3.6 10^{-7}</u>	-	3.2 10^{-7}
2.3	3.3 10^{-7}	6.6 10^{-7}	<u>9.9 10^{-7}</u>	-	-	8.4 10^{-7}
2.4	1.2 10^{-6}	<u>2.3 10^{-6}</u>	-	-	-	2.2 10^{-6}

Table 3. Column 3 to 6: normalization factors $\Phi^{D'}, \Gamma_0$ for an enhanced diffuse flux, obtained assuming $n_{\Delta\Omega} = 3$ to 6 HESE in a circular window of 8° ($\Delta\Omega = 0.06$ sr). In the last column, the value for $\Gamma = 2.0$ corresponds to the ANTARES sensitivities from the FB regions [11]. The sensitivities for $\Gamma > 2.0$ are obtained in [5].

$\Delta\Omega$		units: (GeV cm^{-2} s^{-1} sr^{-1}) $\Phi_0^{D',\Gamma}$ (from HESE)				ANTARES
(sr)	$\Gamma =$	$n_{\Delta\Omega} = 3$	$n_{\Delta\Omega} = 4$	$n_{\Delta\Omega} = 5$	$n_{\Delta\Omega} = 6$	sensitivity
0.06	2.0	3.5 10^{-7}	4.6 10^{-7}	5.8 10^{-7}	7.0 10^{-7}	3.1 10^{-7}
	2.2	4.5 10^{-6}	6.0 10^{-6}	7.5 10^{-6}	9.0 10^{-6}	3.6 10^{-6}
	2.3	1.7 10^{-5}	2.2 10^{-5}	2.8 10^{-5}	3.3 10^{-5}	1.1 10^{-5}
	2.4	5.9 10^{-5}	7.8 10^{-5}	9.8 10^{-5}	1.2 10^{-4}	3.4 10^{-5}

Enhanced diffuse flux. Table 3 (columns from 3 to 6) shows the normalization factors from Eq. (4), assuming $n_{\Delta\Omega} = 3 \div 6$ HESE within a solid angle region $\Delta\Omega = 2\pi(1 - \cos\theta)$ corresponding to a circular windows of $\theta = 8°$. The ANTARES strategy for the study of an enhanced diffuse flux is different with respect to that for the search for point-like sources. This latter relies mainly on the pointing accuracy of the telescope. The expected background due to atmospheric neutrinos within a circular windows of $\theta \lesssim 1°$ is small and this is not anymore true for larger values of θ. As the energy spectrum from a cosmic signal (either point-like or diffuse) is expected to be harder than that of atmospheric neutrinos, the signal should exceed the background above a certain threshold of the reconstructed energy. Thus, the discrimination between signal and background needs the use of the estimated energy of the event, similarly to the case of the search for a diffuse flux of high energy ν_μ [10].

ANTARES has used an Artificial Neural Network to estimate the energy of the muons entering the detector for studying the Fermi bubbles (FB) [11]. The reported ANTARES sensitivity in terms of an enhanced diffuse flux from the FB region, assuming a $E^2\Phi(E)$ spectrum without cutoff up to the PeV energies is 3.1×10^{-7} GeV cm^{-2} s^{-1} sr^{-1}. In the analysis, using 806 days livetime, 16 events were found, with an expected background of 11 events. For $\Gamma = 2.0$ the background corresponds to 7.5 events/(sr · y). The derived 90% C.L. upper limit is $E^2\Phi^{FB}(E) = 5.4 \times 10^{-7}$ GeV cm^{-2} s^{-1} sr^{-1}. As the sensitivity depends on the background rate, different optimizations must be deduced for different spectral indexes; in general, it can be assumed that the background level slightly increases for softer spectral indexes [5]. The sensitivities extrapolated from the ANTARES FB analysis for $\Gamma > 2.0$ are reported in the last column of Table 3. According to these values, a dedicated search for a directional neutrino flux, for instance around the IC hot spot, would produce a positive result for any spectral indexes $\Gamma \geq 2.0$, if $\Delta\Omega \leq 0.06$ sr (or circular window of $\theta < 8°$) and $n_{\Delta\Omega} > 2$. For a signal spread out on a larger circular window, the minimum sensitivity would correspond to a higher $n_{\Delta\Omega}$.

Regions of large angular size (Fermi bubbles, Galactic plane). Recent predictions of the neutrino flux from the FB regions [12] allow to estimate the expected number of events for the IC detector assuming the effective area of the HESE. By folding the predicted ν spectra with the ANTARES effective area, the number of ν_μ induced events in ANTARES would correspond to $\sim 30\%, 50\%$ and 100% of the $\nu_e + \nu_\mu + \nu_\tau$ HESE in the same livetime, for $\Gamma = 2.0, 2.1$ and 2.3, respectively. However, this corresponds to a smaller ANTARES sensitivity with respect to IC, due to the larger background induced by atmospheric neutrinos from the wide FB region ($\Delta\Omega \sim 0.8$ sr).

The excess of HESE events from the Galactic region could finally be produced by interaction during propagation of freshly injected CRs with spectral index $\Gamma \simeq 2.4 - 2.5$ [8]. The preliminary ANTARES upper limits from the Galactic plane are reported in [13].

5. Conclusions

The ANTARES detector has sufficient sensitivity to test many models that explain a fraction of the HESE sample in IceCube in terms of a Galactic component. Models in which more than 2 HESE are originated from a point-like and steady source in the Southern hemisphere are excluded for spectral indexes $\Gamma \geq 2.3$. The possibility that a clustering of events is produced in a region of small angular size ($\Delta\Omega \simeq 0.1 - 0.2$ sr) in (or near) the Galactic Plane is under investigation in ANTARES. As reported in Table 3, the estimated ANTARES sensitivity is below the signal level (allowing a positive detection) for any spectral indexes $\Gamma \geq 2.0$, if $\Delta\Omega \leq 0.06$ sr (i.e. a circular window of $\theta < 8°$) and $n_{\Delta\Omega} > 2$. For a signal spread out on a larger solod angle, the minimum sensitivity would correspond to a higher $n_{\Delta\Omega}$. For very large regions (the FB, the Galactic plane) the present sensitivities using the ν_μ channel alone are above the model predictions. The inclusion of showering events, with a relatively looser angular precision, would significantly increase the ANTARES sensitivities for the study of extended regions.

I would like to thank many members of the ANTARES and KM3NeT Collaborations for comments and in particular J. Brunner and A. Kouchner.

References

[1] M. G. Aartsen et al. [IceCube Coll.], Science **342**, 1242856 (2013)
[2] M. G. Aartsen et al. [IceCube Coll.], Phys. Rev. Lett. **113** (2014) 101101
[3] M. Spurio. *Particles and Astrophysics*. Springer (2014) ISBN 978-3-319-08050-5
[4] M. G. Aartsen et al. [IceCube Coll.], Phys. Rev. D **91**, 022001 (2015); arXiv:1410.1749
[5] M. Spurio. Phys. Rev. D **90** (2014) 10, 103004
[6] M. Ageron et al. [ANTARES Coll.], Nucl. Instrum. Meth. A **656** (2011) 11
[7] S. Adrián-Martínez et al. [ANTARES Coll.], The Astrophysical Journal **760** (2012) 53
[8] A. Neronov, D. Semikoz. arXiv:1412.1690 [astro-ph.HE]
[9] S. Adrián-Martínez et al. [ANTARES Coll.], The Astrophysical Journal Letters **786**, L5 (2014)
[10] J. A. Aguilar et al. [ANTARES Coll.], Phys. Lett. B **696** (2011) 16
[11] S. Adrián-Martínez et al. [ANTARES Coll.], Eur. Phys. J. C **74** (2014) 2701
[12] C. Lunardini et al., Phys. Rev. D **90**, 023016 (2014)
[13] L.A. Fusco. These proceedings; E. Visser. To be publ. on the proceed. of ECRS 2014

EPJ Web of Conferences **121**, 05008 (2016) DOI: 10.1051/epjconf/201612105008
RICAP-2014

The optical detection unit for Baikal-GVD neutrino telescope

A.D. Avrorin[1], A.V. Avrorin[1], V.M. Aynutdinov[1,a], R. Bannash[7], I.A. Belolaptikov[2], D.Yu. Bogorodsky[3], V.B. Brudanin[2], N.M. Budnev[3], I.A. Danilchenko[1], G.V. Domogatsky[1], A.A. Doroshenko[1], A.N. Dyachok[3], Zh.-A.M. Dzhilkibaev[1], S.V. Fialkovsky[5], A.R. Gafarov[3], O.N. Gaponenko[1], K.V. Golubkov[1], T.I. Gress[3], Z. Honz[2], K.G. Kebkal[7], O.G. Kebkal[7], K.V. Konischev[2], E.N. Konstantinov[3], A.V. Korobchenko[3], A.P. Koshechkin[1], F.K. Koshel[1], A.V. Kozhin[4], V.F. Kulepov[5], D.A. Kuleshov[1], V.I. Ljashuk[1], M.B. Milenin[5], R.A. Mirgazov[3], E.R. Osipova[4], A.I. Panfilov[1], L.V. Pan'kov[3], A.A. Perevalov[3], E.N. Pliskovsky[2], M.I. Rozanov[6], V.Yu. Rubtzov[3], E.V. Rjabov[3], B.A. Shaybonov[2], A.A. Sheifler[1], A.V. Skurihin[4], A.A. Smagina[2], O.V. Suvorova[1], V.A. Tabolenko[3], B.A. Tarashansky[3], S.A. Yakovlev[7], A.V. Zagorodnikov[3], V.A. Zhukov[1], and V.L. Zurbanov[3]

[1] Institute for Nuclear Research, 60th October Anniversary pr. 7A, Moscow 117312, Russia
[2] Joint Institute for Nuclear Research, Dubna 141980, Russia
[3] Irkutsk State University, Irkutsk 664003, Russia
[4] Skobeltsyn Institute of Nuclear Physics MSU, Moscow 119991, Russia
[5] Nizhni Novgorod State Technical University, Nizhni Novgorod 603950, Russia
[6] St. Petersburg State Marine University, St. Petersburg 190008, Russia
[7] EvoLogics GmbH, Berlin, Germany

Abstract. The first stage of the GVD-cluster composed of five strings was deployed in April 2014. Each string consists of two sections with 12 optical modules per section. A section is the basic detection unit of the Baikal neutrino telescope. We will describe the section design, review its basic elements – optical modules, FADC readout units, slow control and calibration systems, and present selected results for section in-situ tests in Lake Baikal.

1. Introduction

The objective of the Baikal Project is the creation of a kilometer-scale high-energy neutrino observatory: the Gigaton Volume Detector (GVD) in Lake Baikal [1–3]. The first phase of GVD will consist of 12 clusters of strings – functionally independent subarrays connected to the shore by individual electro-optical cables. Each cluster comprises eight strings of optical modules – seven peripheral strings are uniformly arranged at 60 m distance around a central one. Each string minimally comprises 24 OMs. The possibility of installing 36 OM strings is to be studied at 2015–2016.

The present configuration of the first GVD-cluster compris five strings and was deployed in 2014. within the framework of BAIKAL-GVD project. Each string consists of two *sections* with 12 optical modules each. A section is the basic detection unit of the Baikal neutrino telescope. We describe the section design, review the basic section elements – optical

[a]Corresponding author: aynutdin@yandex.ru

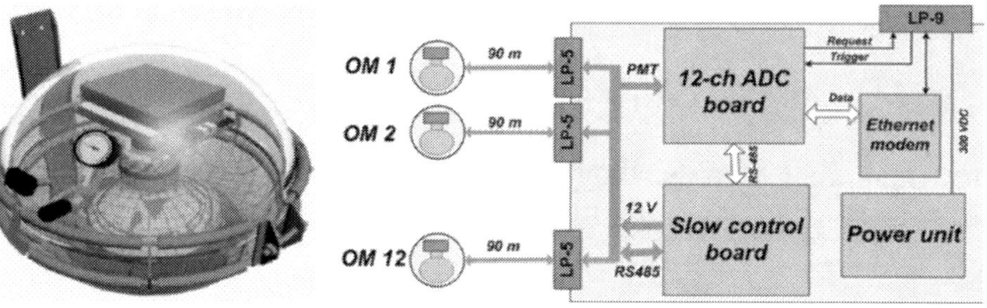

Figure 1. A sketch of a GVD optical module and a block diagram of the GVD section.

modules, FADC readout units, slow control and calibration systems and present selected results of section element in-situ testing in Lake Baikal.

2. GVD section

A section is the basic detection unit of the GVD neutrino telescope [4]. It comprises 12 optical modules (OM) and a central module (CeM). The block diagram of a section is presented in Fig. 1. PMT signals from all OMs are transmitted to the CeM via 90 meters of coaxial cables, where they are digitized by custom-made 12-channel ADC boards with 200 MHz sampling rate. The slow-control board located in the CeM provides data communication between OM and CeM via an underwater RS-485 bus. Also, this unit is intended for OM power control (to switch power on/off for each optical module independently). The ADC board provides trigger logic, data readout and digital processing, and connection via local Ethernet to the cluster DAQ center.

OM and CeM electronics are placed in a pressure-resistant glass sphere VETROVEX with 42 cm diameter. One connector SubConn Low Profile 5-contacts is installed on each optical module. It is used for analog pulse transmission, slow control (2-wire RS-485) and OM power supply (12 VDC). CeM has twelve 5-contact connectors to connect with OMs and one 9-contact SubConn connector for communication with the DAQ centre of the cluster.

3. Optical module

The optical modules [5] contain the photo-sensors of GVD: a photomultiplier tube (PMT), which detects the Cherenkov light produced by relativistic charged particles passing through the water. After testing different options, the photomultiplier Hamamatsu R7081-100 was selected as a light sensor for the OM. This PMT has a hemispherical photocathode with 10 inch diameter and a typical quantum efficiency of 35%. The PMT is placed in a pressure-resistant glass sphere. A high permittivity alloy cage surrounds the PMT, shielding it against the Earth's magnetic field. The OM electronics unit is mounted directly onto the PMT base. The OM electronics includes a controller, a high voltage (HV) power supply unit, a fast two-channel amplifier, and a LED flasher. The block diagram and a picture of the optical module electronics are presented in Fig. 2.

The OM controller is intended for HV regulation and monitoring, for PMT noise measurements, and for time and amplitude calibration. It is designed on the basis of SiLabs C8051F121 microcontroller. Slow control data to and from the OMs are transferred via an underwater RS-485 bus. The unit has an instruction set for the control of OM electronics: HV control (to switch HV on/off, to set PMT voltage and readout HV value), LED flasher

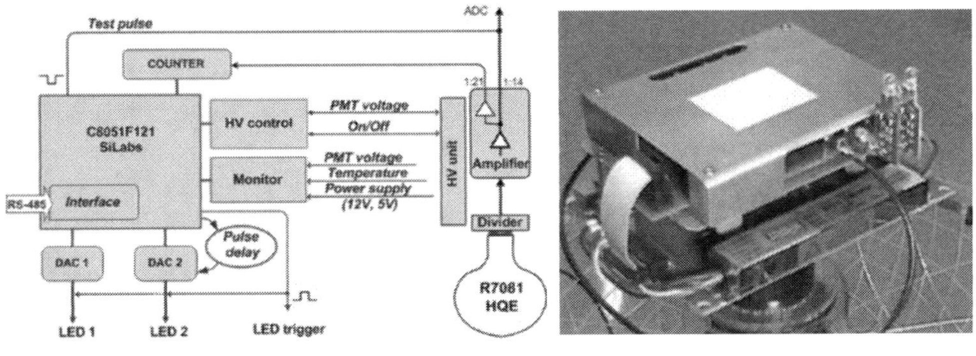

Figure 2. Block diagram and a picture of the OM electronics.

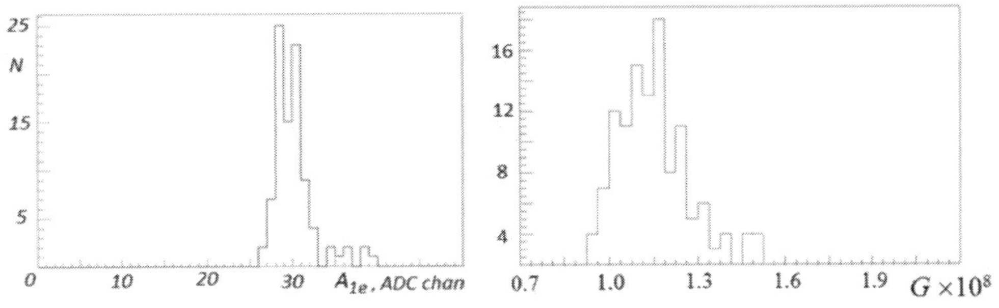

Figure 3. Distributions of the single electron amplitude $A_{1e}$and the gains G (12 May 2014).

control (to set LED intensities, the delay between LED pulses, and the period of flashes), the counter control (setting counter threshold, time window, off-duty factor and the size of circular memory data buffer). Also a set of procedures is foreseen for the laboratory calibration of the counter threshold and PMT voltage.

PMT power supply is provided by a passive HV divider circuit with the resistance of 18 MΩ and HV unit Traco Power SHV 12–2.0 K 1000 P with positive polarity. The tube gains have been adjusted to about 10^7. This gain is provided by divider voltages between 1150 V and 1800 V, depending on the individual tube (average value is about 1400 V). An additional signal amplification is provided by the amplifier. The PMT amplifier is comprised of two channels. The first channel with an amplification factor of 14 forms a signal with negative polarity that is transmitted to the ADC board. The second channel with an amplification factor of 21 produces positive pulses that are intended for PMT noise monitoring. The outer cascades are implemented as emitter-follower amplifiers and provide operation with 50Ω load. The maximum amplitude for both channels is limited to 4 Volts. Assuming an attenuation factor of 0.7 over 90 meter coaxial cable, the total channel amplification will be about 10^8. This corresponds to a single photoelectron value at 30 ADC channels (ADC scale is 1.25 mV per channel). During detector operation in 2014 several special runs were conducted to check the PMT gains. Distributions of single photoelectron amplitudes and the channel gains for May 2014 are presented in Fig. 3 (the average gain is 1.12×10^8). The same measurements conducted in July and August have shown a stability of the average channel gain: 1.17×10^8 and 1.14×10^8, respectively.

The LED flasher is intended for time and amplitude calibration of OM channels during long-term exposition. It includes two LEDs Kingbright L7113 and electronic circuits: LED

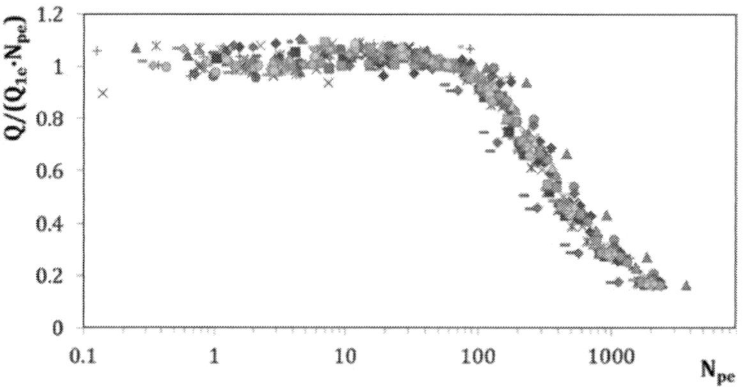

Figure 4. The channel nonlinearity curves (marked by symbols) for 24 OM as a function of photoelectron numbers N_{pe}.

drivers. The dominant wavelength of the LED is 470 nm, the LED pulse has a width of ~ 6 ns (FWHM). The control system of the flasher has to provide independent smooth regulation of two LED luminosities in a wide dynamic range (up to about 10^8 photons per LED flash), has to have minimal level of cross-talks between two LED channels (less than 1%). Independent regulation of light intensities of each flasher channel and absence of cross-talks provide a possibility to perform a linearity test of spectrometric channels. Results of linearity test conducted in a laboratory for 24 optical modules (one string) are presented in Fig. 4. The nonlinearity was calculated as $Q/(Q_{1e} \times N_{pe})$, where Q is a measured PMT pulse charge, Q_{1e} is single electron charge, and N_{pe} is a number of produced photoelectrons. The estimated linearity range is about 100 photoelectrons.

Another application of the LED flasher is a time calibration of the sections and channels. Time calibration means measurements of relative delays of signals from the individual channels. The channel delay accumulates two main components: cable delays and delays from the PMTs. Cable delays are measured once in the laboratory and do not change with time. PMT delays depend on power voltage and thus require regular calibration in situ. There is a specialized test pulse generated by the OM controller that is delivered to the point of signal creation in the PMT preamplifier. The difference between arrival times of the LED signal and the test pulse gives the delay of the PMT signal.

Due to the sufficiently large intensity of LED bursts, the signals generated by LEDs are detected by PMTs of neighboring sections and strings. The calibration coefficients are derived from the known positions of OMs (which are determined by acoustic methods). Figure 5 (left) shows an example of the detection of LED flashes with 8 OMs of one string. LED flasher on the first OM (R=0) is switched on with a maximum light intensity. Each point on the plot shows the OM position on the string versus flashes detection time. The straight line corresponds to the velocity of light in the Baikal water. The average pulse amplitude for the most distant OM (R=105 m) is about 1 photoelectron.

The OM counter with programmable threshold is intended for PMT count rate monitoring. Count rate data is accumulated in the circular buffer and transmitted to the shore with 10 minute interval for each OM. Figure 5 (right) shows the count rate data for all channels accumulated from April to August 2014 for all OMs. The average channel rate is 23 kHz for about 0.5 p.e. counter thresholds.

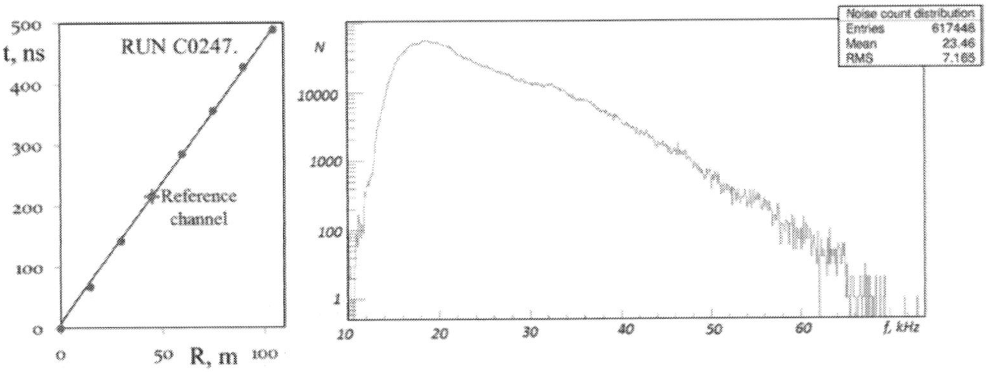

Figure 5. The PMT detection time of the LED flashes vs OM position on the string (left), and the count rates for all channels accumulated from April to August 2014 (right).

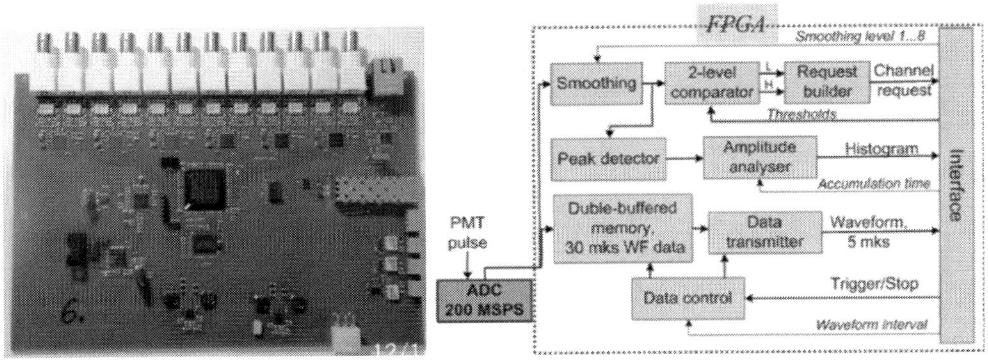

Figure 6. A picture of the custom made ADC board and a block diagram of the ADC channel.

4. Central electronics module

The basic elements of the central electronics module [6–9] are ADC board and slow control board (see Fig. 1). The ADC board is a 12-bit 200MS/s flash ADC waveform digitizer with 12 input channels. It provides trigger logic, data readout and connection via local Ethernet to the cluster DAQ center. The ADC board is implemented on the basis of microcircuits AD9613 and FPGA Xilinx Spartan 6. AD9613 is a dual 12-bit analog-to-digital converter with sampling speeds up to 250 MSPS (see Fig. 6). Data streams are continuously written into circular memory buffers of each ADC channel. When the trigger occurs, the digitizer freezes the buffer that can be read by readout links. The acquisition can continue without any dead time in the second buffer. A memory buffer of 12kB allows waveform data accumulation from the ADC for 30 μs. A common acquisition trigger signal can be provided externally as well as via the software, but it can also be generated internally thanks to self-trigger capability. Thanks to the FPGAs housed in the digitizer, ADC board provides possibilities of on-line signal processing rejecting samples smaller than programmable threshold. It allows for reducing the data flow more than 50 times.

Each ADC data channel is connected to a peak detector and an amplitude analyzer, which accumulates monitor histograms in a programmable time interval. An ADC trigger channel includes a smoothing unit for electronics noise reduction, a two-level adjustable digital comparator (low threshold L and high threshold H), and a request builder, that builds the trigger logic request. The basic trigger modes are (A) coincidences of >N L-requests within a selectable time window or (B) coincidences of L and H requests from any neighbouring OMs within a section.

The section data is transferred through Ethernet channel. Two Ethernet outputs are foreseen: DSL-modem (MOXA IEX-402-SHDSL) and optical output on ADC board for future detector extension.

5. Conclusion

The present configuration of the first GVD-cluster consists of five strings and has been deployed in 2014. The basic detection units of GVD are sections of optical modules (OM), with each OM housing a photomultiplier of the type R7081–100. In-situ tests of the installation in Lake Baikal show good performance of all section elements. The prototyping phase will be concluded in 2015 with the completion of the first GVD demonstration cluster with 8 strings.

This work was supported by the Russian Found for Basic Research (grants 13-02-12221, 14-02-00175, 14-02-00972).

References

[1] A. Avrorin, et al., Astronomy Letters **37**, 7 (2011)
[2] A. Avrorin et al., Nucl. Instr. and Meth. **A626**, 6 (2011)
[3] A. Avrorin et al., Nucl. Instr. and Meth. **A630**, 4 (2011)
[4] A. Avrorin et al., Nucl. Instr. and Meth. **A639**, 3 (2011)
[5] A. Avrorin et al., Instruments and Experimental Techniques **54**, 649–659 (2011)
[6] A. Avrorin et al., Nucl. Instr. and Meth. **A692**, 7 (2012)
[7] A. Avrorin et al., Nucl. Instr. and Meth. **A693**, 10 (2012)
[8] A. Avrorin et al., Nucl. Instr. and Meth. **A725**, 4 (2013)
[9] A. Avrorin et al., Instruments and Experimental Techniques **3**, 12 (2014)

EPJ Web of Conferences **121**, 05009 (2016) DOI: 10.1051/epjconf/201612105009
RICAP-2014

Muon and neutrino energy reconstruction for KM3NeT

Evangelia Drakopoulou[1,2,a] for the KM3NeT Collaboration

[1] N.C.S.R. Demokritos, Patriarchou Grigoriou and Neapoleos, Agia Paraskevi, Greece
[2] National Technical University of Athens, Heroon Polytechniou 9, Zografou Campus, Greece

Abstract. KM3NeT/ARCA is a European deep-sea research infrastructure that will host a neutrino telescope with a volume of several cubic kilometers at the bottom of the Mediterranean Sea. The telescope will search for galactic and extragalactic neutrinos from astrophysical sources like gamma ray bursts, super-novae or colliding stars. The analyses performed in large water Cherenkov detectors rely upon the reconstruction of the muon direction and energy, and consequently, those of the neutrino. The estimation of the muon energy is also critical for the differentiation of muons from neutrinos originating from astrophysical sources from muons and neutrinos that have been generated in the atmosphere and constitute the detector background. The energy is derived from the detection of the Cherenkov light produced by the muons that are created during the charged current interactions of neutrinos in or in the vicinity of the detector. We describe a method to determine the muon and neutrino energy employing a Neural Network. An energy resolution of about 0.29 has been achieved for muons at the TeV range.

1. Introduction

One of the aims of the KM3NeT collaboration is to deploy a neutrino telescope of several cubic kilometers at the bottom of the Mediterranean Sea in order to search for neutrinos of galactic and extragalactic origin. This project is called KM3NeT/ARCA. This detector consists of two building blocks of about one hundred detection units that are linked each other and to shore by means of electro-optical cables. Each detection unit hosts 18 optical modules (OMs) vertically aligned on a string. Each optical module consists of a 17 inches glass sphere with 31 photomultipliers (PMTs). These PMTs will detect the Cherenkov light emitted by charged particles which are produced during the neutrino interactions with the Earth or the sea water. The detector background consists of atmospheric ν_μ and μ events which are induced by cosmic rays interactions in the atmosphere and travel through the detector volume leaving similar signatures of astrophysical ν_μ and μ. Moreover, photons due to bioluminescence and ^{40}K decays in sea water are detected by the PMTs thus originating an optical background.

Two different samples of simulated Charged Current ν_μ events were used for this analysis to prevent from adding bias in our results. The first event sample was used to train and test the Neural Network while the second event sample was used during the evaluation procedure. Isotropic atmospheric ν_μ flux (by Bartol) [1] was assumed, for both samples, with neutrino

[a]e-mail: `liliadrak@inp.demokritos.gr`

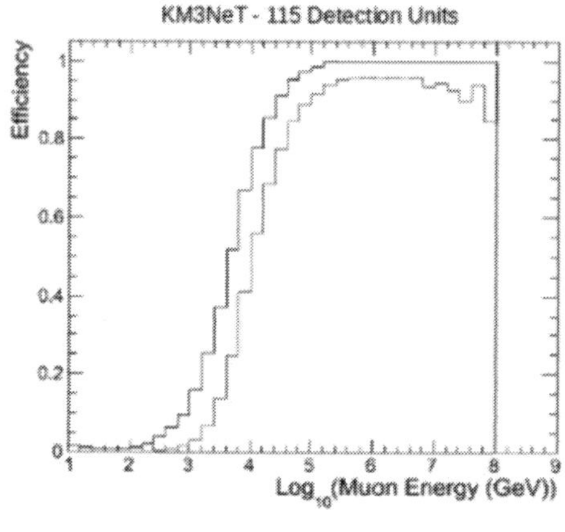

Figure 1. The efficiency of the reconstruction as a function of $\log_{10} E_\mu$. The black line corresponds to all events that cross the detector and the red line refers to events satisfying the selection criteria.

energy ranging from 10^2 GeV to 10^8 GeV. The first event sample assumes that neutrinos are generated with an energy spectrum $E^{-1.4}$ while the second event sample corresponds to a neutrino energy spectrum of E^{-2}. A random optical background rate of 5 kHz was assumed for each PMT, including dark current, ^{40}K decays and bioluminescence. In addition to random coincidences, a rate of 500 Hz due to genuine coincidences from ^{40}K decay has also taken into account.

2. Track direction reconstruction

The simulated events were reconstructed using the Chameleon reconstruction package [2]. This package consists of two main parts. The pattern recognition which includes algorithms for the selection and grouping of hits in track candidates and the fitting algorithm that uses a χ^2 minimizer in order to reconstruct the direction of the track.

A quality selection process is applied: A minimum number of OMs with hits, and a minimum path inside the instrumented volume are required in order to get a reliable reconstruction of the track direction. The efficiency and the purity of the track direction reconstruction is reported in Figs. 1 and 2, respectively, where the black line corresponds to all events that cross the detector volume and the red line refers to events that survive from the quality selection process. The purity of the reconstruction refers to the ratio of the well reconstructed events (events for which the angle difference ($\Delta\Omega$) between the simulated and the reconstructed muon track direction is less 1°) to all reconstructed events that survive from the quality selection process. A good purity ($\geq 80\%$) of the reconstructed events especially in high energies ($E_\mu \geq 10$ TeV) can be established, accompanied by an efficiency of $\simeq 55\%$ for $E_\mu \geq 10$ TeV and $\geq 80\%$ for $E_\mu \geq 40$ TeV rising with energy to approximately $\simeq 100\%$ for $E_\mu \geq 100$ TeV.

The median of the angle difference ($\Delta\Omega$) between the simulated and the reconstructed track direction is shown in Fig. 3. The median is less than 0.5° for $E_\mu > 1$ TeV while it reaches 0.35° for $E_\mu = 10$ TeV. A very good angular resolution is achieved in the high energy regime with the median of $\Delta\Omega = 0.15°$ for $E_\mu \simeq 3$ PeV. For muon ener-gies above 10 PeV the calculation of the median is domi-nated by limited statistics.

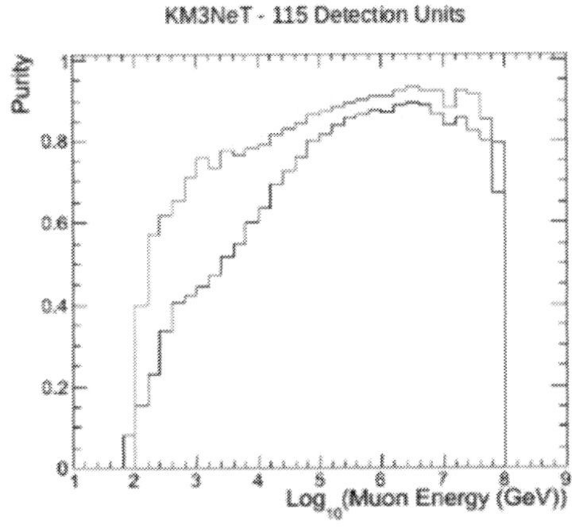

Figure 2. The purity of the reconstruction that is defined as the ratio of well reconstructed events ($\Delta\Omega \leq 1°$) to the number of reconstructed events as a function of $\log_{10} E_\mu$. The black line corresponds to all events that cross the detector and the red line refers to events satisfying the selection criteria.

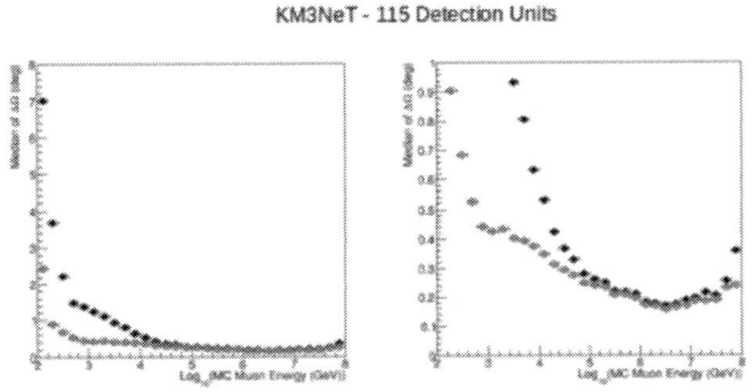

Figure 3. The median of the angle difference ($\Delta\Omega$) between the simulated and the reconstructed muon track direction as a function of $\log_{10} E_\mu$. The right plot depicts the distribution of the median of this angle difference for $\Delta\Omega < 1°$ while the left plot contains all values of median with respect to $\log_{10} E_\mu$. The black dots correspond to all reconstructed events and the red dots refer to events satisfying the selection criteria.

3. Energy estimation

Muons lose energy via ionization and by stochastic processes, such as bremsstrahlung, pair production, and photonuclear interactions. The total average energy loss of the muon is:

$$\frac{-dE}{dx} = a(E) + b(E)E_\mu$$

where $a \simeq 0.274\,\mathrm{GeV\,m^{-1}}$ accounts for the energy loss due to ionization and $b \simeq 0.000349\,\mathrm{m^{-1}}$ is due to the stochastic energy loss. Muons with $E_\mu > 1$ TeV lose energy stochastically, while for lower energies ionization dominates. We describe a method to derive muon (and neutrino) energy from the light collected along its passage through the detector volume.

When the neutrino interacts inside the instrumented volume, both photons from the hadronic shower and Cherenkov photons from the muon passage through sea water can be detected, leading to a reliable estimate of the neutrino energy. If the neutrino has interacted to muon far outside the instrumented volume, the majority of the light produced by the muon (and/or hadronic shower) never reaches the PMT sensors, leading to an underestimation of the neutrino energy. In this case the muon energy is used as a measure of the neutrino energy. In order to ensure that photons from the hadronic part can be detected, the neutrino energy is estimated for events with the reconstructed neutrino interaction vertex inside a cylindrical fiducial volume, which is slightly smaller than the detector volume. For events that cross the detector volume but the reconstructed neutrino interaction vertex lies outside the fiducial volume, only the muon energy is reconstructed. The MC muon energy that is considered for this study refers to the energy of the muon at the neutrino interaction vertex.

3.1 Neural network

Artificial neural networks (NNs) are gaining ground as a multipurpose, robust computational methodology which performs effectively in many analyses. The response of an artificial neural network is determined by the layout of the neurons, the weights of the inter-neuron connections and the neuron response function which describes the response of the neurons to the input variable. This analysis is based on the employment of a Multi-Layer Percepton (MLP) Neural Network which is part of the TMVA package included in ROOT [3]. The neurons in such neural network are organized in layers and only direct connections from a given layer to the following layer are allowed resulting in a less complex system. The first layer of a MLP is the input layer that hosts the neurons. Each neuron holds an input variable. The last layer is the output layer that holds the output variable, the neural network estimator. All intermediate layers are the hidden layers of the NN. A weight is associated to each directional connection between the output of one neuron and the input of another neuron. During the calculation of the input value to the response function of a neuron these weights are multiplied with the output values of all neurons connected to this given neuron [3]. The NN used for the current analysis consists of three layers. The first layer holds four input variables, the hidden layer consists of 10 nodes and the output layer returns the network result. The neuron activation function used was the hyperbolic tangent while the weights were adjusted with the use of the Broyden-Fletcher-Goldfarb-Shannon (BFGS) method [4].

3.2 Description of the neural network input variables for the energy estimation

Since the determination of neutrino and muon energy is based on the collection of light in PMTs, muons should have traveled an adequate distance inside the volume of the detector before an attempt to evaluate their energy is made. Furthermore, the optical modules (OMs) in the detector configuration are not homogeneously distributed in space. The distance of OMs on a string is less than half of the distance between neighboring strings. As a consequence the minimum expected path of a muon through the detector should differ with the reconstructed muon zenith angle. We define the minimum expected distance by the formula:

$$\text{Minimum Expected Path} = 0.5 * h + (R - 0.5 * h) * \sin(\theta_{rec})$$

where: h is the string height, R is the detector radius and θ_{rec} the reconstructed muon angle ($\theta_{rec} = \pi - \theta_{zenith}$).

According to this formula the "minimum expected path" for horizontal muons is the detector radius, while for vertical muons it is half of the height of the string. We calculate the distance between the first and the last PMT position along the track (the reconstructed track direction is not used in order to avoid bias due to poorly reconstructed events) and we accept events that fulfill the requirement:

$$\frac{Distance\ between\ PMTs}{Minimum\ Expected\ Path} > 0.5. \tag{1}$$

The quantities that have a strong dependence on the muon and neutrino energy and are used as input variables to feed the Neural Network are:

i. The number of OMs used in the reconstruction divided by the calculated muon length. A weight is used in order to also accept events with low energies that stop traveling inside the detector.

$$\frac{Number\ of\ OMs}{muon\ length} * \frac{muon\ length}{muon\ potential\ length}$$

ii. The number of PMTs that have L1 pulses (coincidence of two PMTs in the same OM in a time interval of 10 ns) and were used in the reconstruction weighted according to the vertical distance from the track.

$$\sum_{i=1}^{Number\ of\ PMTs} \frac{D_i}{D_{max}}$$

where D_i is the vertical distance between the PMT and the reconstructed track and D_{max} =500 m.

iii. The ratio of the total number of PMTs that have L1 pulses and were used in the reconstruction to the number of PMTs that could be hit according to the track and the PMT direction but have not recorded any pulses.

$$\frac{Number\ of\ PMTs\ with\ hits}{Number\ of\ PMTs\ with\ no\ hits}$$

iv. The Total Time over Threshold (ToT) of all PMTs used in the reconstruction. The Time over Threshold is used as a measure of charge in PMTs.

$$\sum_{i=1}^{Number\ of\ PMTs} ToT.$$

3.3 Results on muon and neutrino energy reconstruction

Once the weights of the inter-neuron connections of the NN have been produced, a different sample of ν_μ events is used for the evaluation of the Neural Network and for the estimation of muon and neutrino energy. The input variables are created for this new sample and are inserted in the NN which returns the logarithm of the reconstructed energy in GeV.

The relation between the estimated and the MC energy for muons and neutrinos is presented in Figs. 4 and 5 respectively. The energy range of interest for this experiment refers to muon and neutrino energy above 1 TeV and this is the energy range used in the figures above. An overestimation of the muon energy for $E_\mu < 10$ TeV can be observed which recedes as the energy increases resulting to a linear correlation for $E_\mu \geq 10$ TeV. This overestimation is expected as the track direction reconstruction applies a strict hit selection in order to reduce hits coming from background sources, reconstructing mainly events with $E_\mu \geq 10$ TeV (see Fig. 1). A very good correlation between the reconstructed and simulated energy can still be observed for $E_\nu \geq 10$ TeV while the deviation from the line for $E_\nu > 600$ TeV is due to low statistics.

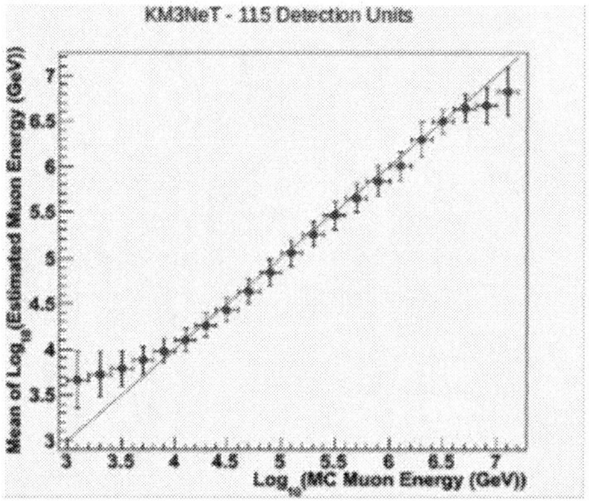

Figure 4. The mean (with RMS error) of the distribution of the logarithm of the reconstructed muon energy with respect to the logarithm of the MC muon energy.

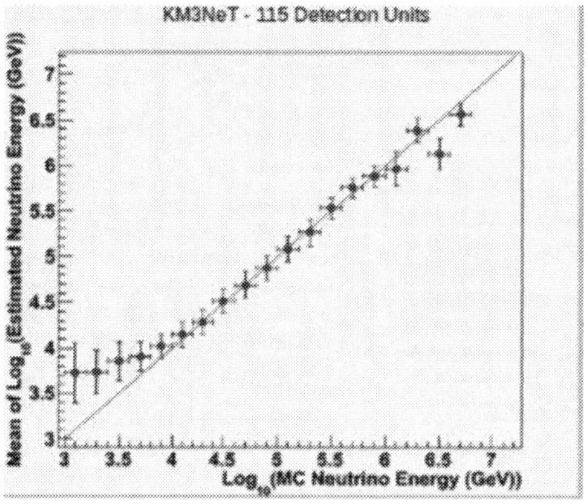

Figure 5. The mean (with RMS error) of the distribution of the logarithm of the reconstructed neutrino energy with respect to the logarithm of the MC neutrino energy for events with the reconstructed neutrino vertex inside the fiducial volume.

Figure 6 shows the distribution of Log_{10} (Estimated Muon Energy/MC Muon Energy) for all events for which the energy was estimated (100 GeV $\leq E_\mu \leq$ 20 PeV). The distribution is fit with a Gaussian. An offset of the mean at 0.14 can be observed which depicts the overestimation of energy for $E_\mu <$ 1 TeV due to the fact that in these energies ionization dominates, while the $\sigma \simeq$0.32. This overestimation of muon energy is absent at high muon energies ($E_\mu \geq$ 10 TeV) for which a mean at $\simeq$-0.03 and a $\sigma \simeq$0.26 is achieved (Fig. 7).

Figure 8 shows the distribution of Log_{10} (Estimated Neutrino Energy/MC Neutrino Energy) for events with the reconstructed neutrino vertex inside the fiducial volume. The distribution is fit with a Gaussian. An offset of the mean at 0.10 can be observed which depicts the overestimation of energy for $E_\nu <$ 1 TeV due to the fact that in these energies ionization dominates, while the $\sigma \simeq$0.29. This overestimation of neutrino energy is absent at high neutrino energies ($E_\nu \geq$ 10 TeV) for which a mean at $\simeq$0.014 and a $\sigma \simeq$0.26 is achieved (Fig. 9).

The muon energy resolution for all events for which the energy was estimated and the neutrino energy resolution for events with the reconstructed neutrino vertex inside the fiducial volume for three different energy intervals is reported in Fig. 10. The energy intervals were

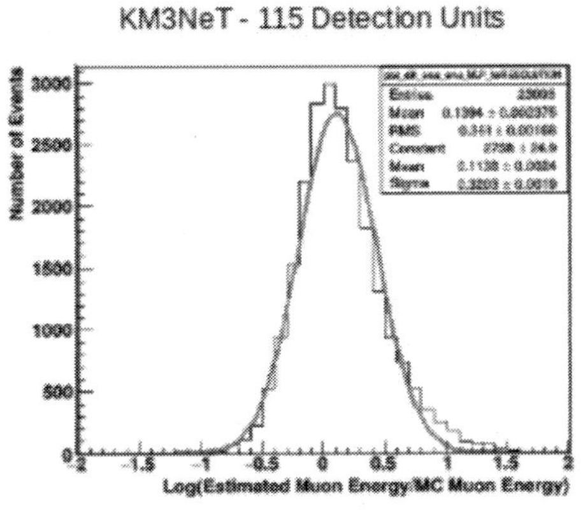

Figure 6. The distribution of Log_{10} (Estimated Muon Energy/MC Muon Energy) for all events for which the energy was estimated.

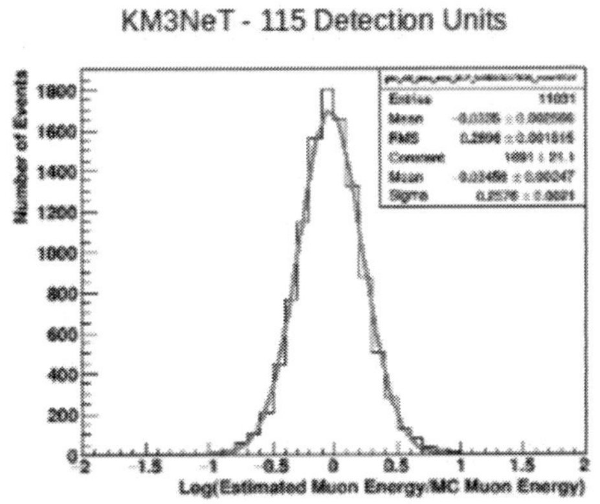

Figure 7. The distribution of Log_{10} (Estimated Muon Energy/MC Muon Energy) for events with $E_\mu \geq 10\,\text{TeV}$ for which the energy was estimated.

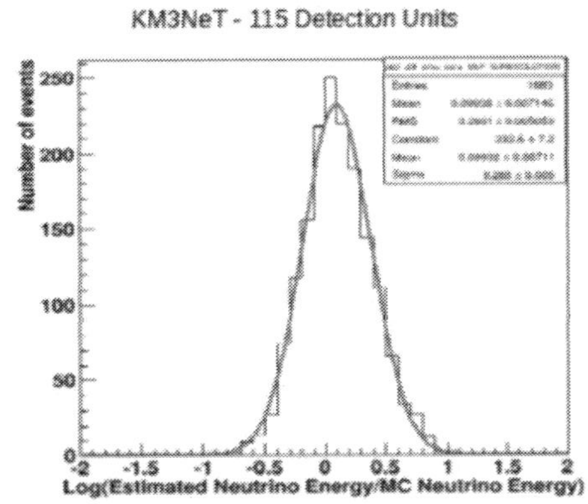

Figure 8. The distribution of Log_{10} (Estimated Neutrino Energy/MC Neutrino Energy) for events with the reconstructed neutrino vertex inside the fiducial volume.

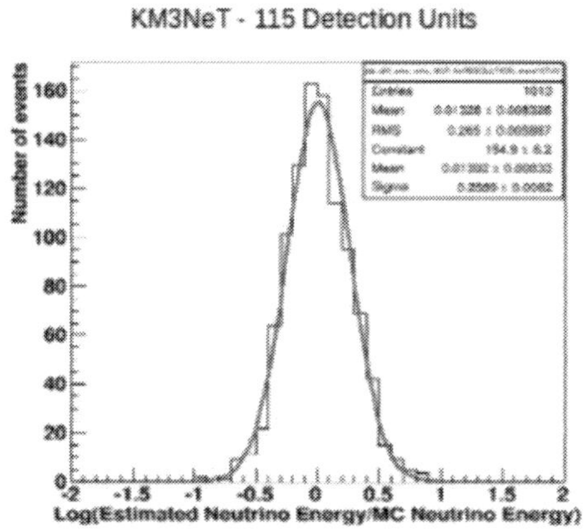

Figure 9. The distribution of Log_{10} (Estimated Neutrino Energy/MC Neutrino Energy) for events with $E_\mu \geq 10\,\text{TeV}$ for which the reconstructed neutrino vertex is inside the fiducial volume.

Energy Resolution

	$1\,TeV \leq E_\mu < 10\,TeV$	$10\,TeV \leq E_\mu < 30\,TeV$	$30\,TeV \leq E_\mu < 20\,\text{PeV}$
Muons crossing the detector volume	0.29 in $\log_{10} E_\mu$	0.25 in $\log_{10} E_\mu$	0.26 in $\log_{10} E_\mu$
	$1\,TeV \leq E_\nu < 10\,TeV$	$10\,TeV \leq E_\nu < 30\,TeV$	$30\,TeV \leq E_\nu < 20\,\text{PeV}$
Events with the reconstructed neutrino vertex inside a fiducial volume	0.27 in $\log_{10} E_\nu$	0.26 in $\log_{10} E_\nu$	0.25 in $\log_{10} E_\nu$

Figure 10. The energy resolution for three different energy intervals.

chosen in order to have comparable number of events in each energy range. The energy resolution is determined by the standard deviation of the distribution of Log_{10} (Estimated Muon Energy/MC Muon Energy) and Log_{10} (Estimated Neutrino Energy/MC Neutrino Energy) respectively. The muon energy resolution for $1\,\text{TeV} \leq E_\mu < 10\,\text{TeV}$ is 0.29 and it decreases to 0.25 for $10\,\text{TeV} \leq E_\mu < 30\,\text{TeV}$ and 0.26 for $30\,\text{TeV} \leq E_\mu < 20\,\text{PeV}$. The neutrino energy resolution corresponds to 0.27 for $1\,\text{TeV} \leq E_\nu < 10\,\text{TeV}$ and it decreases with the rising of the neutrino energy to 0.26 for $10\,\text{TeV} \leq E_\nu < 30\,\text{TeV}$ and 0.25 for $30\,\text{TeV} \leq E_\nu < 20\,\text{PeV}$.

Conclusions

A new method for the muon and neutrino energy estimation using a Multi-Layer Percepton Neural Network with appropriate input variables was presented. The results achieved lead to an energy resolution $\simeq 0.26$ in $\log_{10} E_\mu$ for muon energy at $10\,\text{TeV} \leq E_\mu \leq 20\,\text{PeV}$, while for muons over all energies ($100\,\text{GeV} \leq E_\mu \leq 20\,\text{PeV}$) the energy resolution is $\simeq 0.32$ in $\log_{10} E_\mu$. For the events with the neutrino vertex inside the fiducial volume the energy resolution is $\simeq 0.26$ in $\log_{10} E_\nu$ for neutrino energy at $10\,\text{TeV} \leq E_\nu \leq 20\,\text{PeV}$, while for neutrinos over all energies ($100\,\text{GeV} \leq E_\nu \leq 20\,\text{PeV}$) the energy resolution is $\simeq 0.29$ in $\log_{10} E_\nu$.

The presented work has been partially financed by the European Social Fund (NSRF 2007-2013) under the Thalis initiative GRBNeT project-MIS 360381. I would like to thank Katerina Tzamariudaki and

Christos Markou for their invaluable assistance during the supervision of the present work. I also owe many thanks to Konstantia Balasi for converting the initial event samples to the appropriate format used by the chameleon package and to Kostantinos Pikounis for providing his help whenever it was necessary.

References

[1] V. Agrawal, T. K. Gaisser, P. Lipari, T. Satnev, Phys Rev D. 53.1314 **v1**, 18 (1995)

[2] D. Lenis, G. Stavropoulos, astro-ph.IM, **v1**, 37 (2012)

[3] A. Hoecker, P. Speckmayer, J. Stelzer, J. Therhaag, E. von Toerne, H. Voss, *TMVAUsersGuide*, 152 (2013)

[4] C. G. Broyden, *The Convergence of a Class of Double-rank Minimization Algorithms*, J.Inst. of Math. and App. 6, 76 (1970); R. Fletcher, *A New Approach to Variable Metric Algorithms*, Computer J. 13, 317 (1970); D. Goldfarb, *A Family of Variable Metric Updates Derived by Variational Means*, Math. Comp. 24, 23 (1970); D.F. Shannon, *Conditioning of Quasi-Newton Methods for Function Minimization*, Math. Comp. **24**, 647 (1970)

EPJ Web of Conferences **121**, 05010 (2016) DOI: 10.1051/epjconf/201612105010
RICAP-2014

Operation and results of the prototype KM3NeT detection unit

Simone Biagi[1,2,a] on behalf of the KM3NeT Collaboration

[1] Dipartimento di Fisica e Astronomia dell'Università di Bologna, Viale Berti Pichat 6/2, 40127 Bologna, Italy
[2] INFN Sezione di Bologna, Viale Berti Pichat 6/2, 40127 Bologna, Italy

Abstract. KM3NeT will be a km^3-scale neutrino telescope in the Mediterranean Sea. The detector will consist of blocks of about one hundred detection units. Each detection unit will host 18 digital optical modules, connected along a 700 m-long vertical structure. Electro-optical cables allow for data transmission and power supply to the optical modules. The optical module comprises 31 photomultiplier tubes of 3″, instruments to monitor environmental variables and electronic boards to communicate onshore and operate the photomultipliers. A prototype detection unit has been deployed in May 2014 at the KM3NeT-It installation site 80 km SE offshore of Capo Passero, Sicily. This prototype allowed to test the deployment procedures, the mechanics and the electronic of the apparatus, the data taking and analysis procedures. A general description of the detector and some results of the prototype are presented. The first detection unit of the KM3NeT neutrino telescope will be deployed and become operative by the end of 2015.

1. Introduction

KM3NeT [1] is a deep-sea research infrastructure hosting a neutrino telescope with a volume of several cubic kilometres currently under construction in the Mediterranean Sea. The pilot projects ANTARES [2], NEMO [3] and NESTOR [4] paved the way to the design, construction and operation of the KM3NeT telescope. It will be composed by thousands of digital optical modules (DOMs), consisting of a 17-inch diameter glass pressure sphere containing 31 3-inch photomultiplier tubes (PMTs) and the front-end electronics (Fig. 1 left); the total photocathode surface is $1260\,cm^2$ [5]. The multi-PMT design provides a large photocathode area, very good separation between single-photon and multiple-photon hits and information on the photon direction. In addition, a single PMT loss minimally degrades the performance. These features improve the track reconstruction performance in a cost-effective way with respect to the use of single PMTs.

The detection principle exploits the observation of the Cherenkov light induced by charged particles produced by the interaction of neutrinos with matter inside or in the vicinity of the KM3NeT detector. A group of 18 DOMs distributed over a 700 m mooring line constitutes a detection unit (DU) of the telescope. The DUs are connected to shore via a sea-bottom network of electro-optical cables and junction boxes. Following the strategy adopted

[a]Now at INFN Laboratori Nazionali del Sud, via S. Sofia 62, 95123 Catania, Italy; e-mail: simone.biagi@bo.infn.it

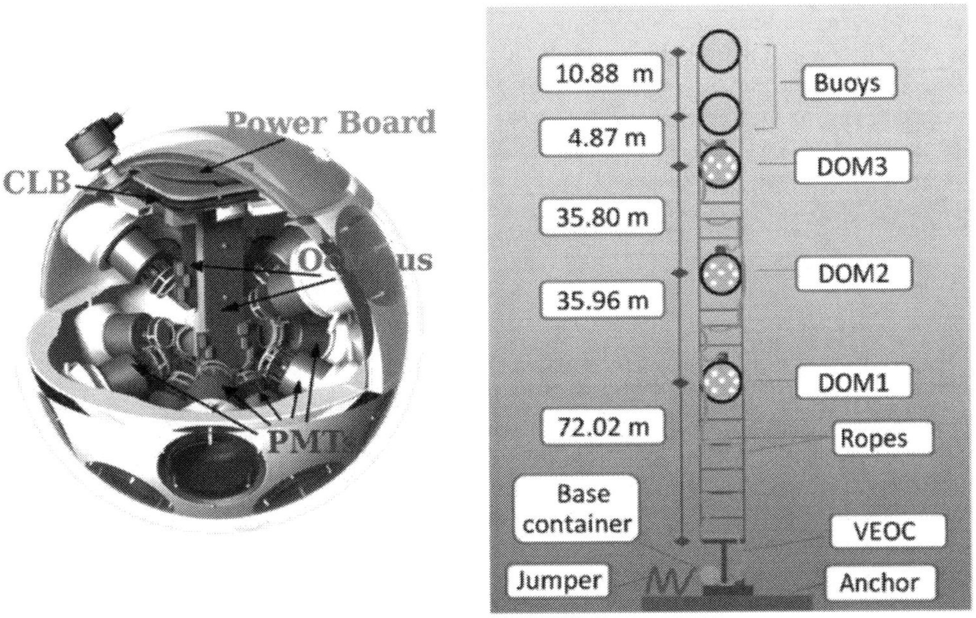

Figure 1. Left: a drawing of the digital optical module. 31 PMTs and the electronics are contained in a glass sphere – see text for details. Right: graphical representation of the PPM-DU installed in the deep water (vertical and horizontal dimensions not in scale). An electro-optical cable, represented with the cyan line marked VEOC, connects the DOMs to the base container; the latter is connected to the onshore station through a jumper cable.

by ANTARES, all data collected by PMTs are sent to shore where they are filtered according to trigger algorithms; no underwater data filtering is implemented.

A prototype DOM has been installed in the ANTARES site, 40 km offshore the French coasts close to Toulon, on one ANTARES line [6]. The Pre Production Model Digital Optical Module (PPM-DOM) was deployed in April 2013 and it is operating since its deployment. This qualification project validated the DOM technology and demonstrated the capability to select hits from muons looking for coincidences in time between PMTs inside the DOM.

2. The detector

The Pre Production Model Detection Unit (PPM-DU) was deployed during the night between May 6 and 7 2014 at the site located 80 km offshore Capo Passero, Sicily (latitude 36° 17′ 50″ N, longitude 15° 58′ 45″ E), at 3457 m depth. It is composed by three DOMs spaced by 36 m and kept vertical by two empty glass spheres that provide buoyancy, see Fig. 1 right. The PPM-DU is 160 m long; the full DU with 18 DOM will be 700 m long. The PPM-DU base is connected with the onshore station through an electro-optical cable of about 100 km length. DOM 1 and DOM 2 are equipped with ETEL D783FLA PMTs, while DOM 3 contains Hamamatsu R12199-02 PMTs. There are small differences in the global efficiency of the two kinds of PMTs that result in a slightly higher counting rate for Hamamatsu PMTs. In addition to the PMTs, each DOM also contains (Fig. 1 left):

- a power board to feed the electronics providing the needed voltage levels;
- a Central Logic Board (CLB) to operate the PMTs and transmit data to the onshore system;

- two "Octopus boards" to power the PMTs and connect them with the CLB;
- a LED beacon pointing upward for time calibrations between PMTs and DOMs;
- a piezo sensor for acoustic measurements;
- a compass and tiltmeter for relative positioning;
- a temperature and relative humidity sensor to monitor the environmental conditions inside the DOMs.

The PPM-DU is installed on the sea bed using a mechanical spherical structure that houses the full line. Once on the sea bottom, an acoustic release allows the unfurling of the line, and the mechanical spherical structure can be recovered at the sea surface. With the help of a Remotely Operated Vehicle (ROV), the electro-optical cable is then connected with the base of the PPM-DU, providing power supply and allowing communication with the onshore station. The detector was successfully operated from its deployment until its decommissioning in July 2015.

3. Calibration and data analysis

The PPM-DU provides useful information to characterise the marine site and to better understand the properties of the DOM. An interesting parameter that is taken under control is the counting rates measured on every PMT. Mean rates show whether the PMTs are well calibrated and give information on the total PMT efficiency. Single hit rates for the PMTs of the three DOMs are shown in Fig. 2 left. For every PMT, the mean rate is evaluated from the distribution of time differences between consecutive hits. The tail of the time difference distribution is fitted with an exponential:

$$f(\Delta t) = p_0 \cdot \mathrm{e}^{-p_1 \cdot \Delta t}$$

where p_0 is a scaling factor, p_1 is the single hit rate and Δt is the time difference. The values in the plot are evaluated from one run of 30′. Two PMT channels are switched off: channel 14 in DOM 2 due to hardware problems, channel 27 in DOM 3 due to anomalous high counting rates. All PMTs appear to be well calibrated since the mean rates are all about 5–6 kHz. This value is compatible with the expected hit rate due to the radioactive decay of ^{40}K present in sea water [7]. It was checked that these values are stable over time scales of months.

The relative time calibration of the detector is done through the evaluation of the time offsets between PMTs and between DOMs. The time offsets can be classified in:

- Intra-DOM time offsets (between PMTs) – they depend on the electronics inside the DOM and the PMT properties;
- Inter-DOM time offsets (between DOMs) – they mainly depend on the cable lengths and slightly on the electronics. The travel time of light in sea water must be taken into account in this calibration procedure.

The intra-DOM time offsets are evaluated studying the local coincidences between DOMs inside the same DOM. These local coincidences are correlated with the signal from ^{40}K decay. A χ^2 minimisation procedure is applied to coincidences between all PMT pairs in a DOM, in order to obtain the relative time offsets, the PMT detection efficiencies and the intrinsic time-spreads of all PMTs.

The inter-DOM time calibration is performed using the DOM with the flashing LED beacon and the DOM directly on top of it. The LED beacon installed in each DOM has a wavelength of 470 nm and is positioned in the top half of the DOM pointing upwards. LED beacons are controlled with adjustable frequency and intensity in order to illuminate the next DOM of the line without invoking a High Rate Veto of the PMTs. The time accuracy

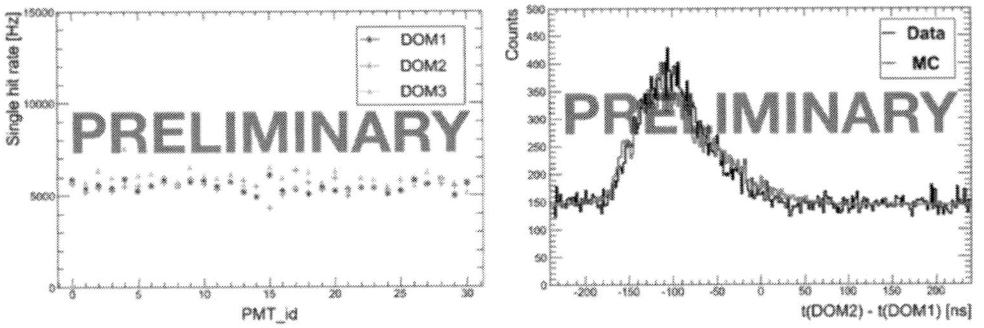

Figure 2. Left: single hit rates for the 31 PMTs of the DOMs. Rates are evaluated from a 30' run – see text. DOM 3 has slightly higher vales because it contains Hamamatsu PMTs that show a higher global efficiency respect to ETEL ones, mounted in DOM 1 and DOM 2. Right: distribution of differences in time between correlated hits in DOM 1 and DOM 2 for data and MC.

achieved is of the order of ~ 2 ns. Some tests have been done with PPM-DU LED beacons in coincidence with the laser beacon located in the base of the prototypal tower detection unit, installed in the site at a distance of about 60 m from PPM-DU [8].

A dedicated MC simulation framework, based on the ANTARES software [9], has been modified for the PPM-DU geometry. The simulation chain consists of the generation of atmospheric muons and their propagation in rock and sea water, the generation of Cherenkov light and the digitisation of the PMT signals. The PMT characteristics and the optical water properties measured at the Capo Passero site [10] are taken into account in the simulation.

Data taken in June-July 2014, corresponding to 80 hours of equivalent livetime, are compared to Monte Carlo (MC) simulations to double check the results of the time calibrations. The L1 trigger is defined as a time coincidence between two hits from different PMTs in the same DOM within 10 ns. Data and MC are triggered in order to have at least one L1. The time differences between DOM 1 and DOM 2 for events that trigger both corresponding DOMs are shown in Fig. 2 right. The MC is normalised to data. The time of the data is corrected with the relative time offsets obtained through the intra-DOM and inter-DOM calibration procedures. A good agreement is found showing the accuracy of the entire procedure.

A search for coincidences between PMTs and DOMs is optimised to select atmospheric muon events. A multiple coincidence is defined as all hits inside a DOM taken in time window of 130 ns around L1 triggers. Figure 3 left shows the rate of multiple coincidences for the three DOMs in the period June-July 2014, corresponding to 52 hours of equivalent livetime. The first filled bin is for multiple coincidences equal to two, i.e. events with only one L1. According to MC simulations, the blue region identifies the contribution due to muons. The clear change of shape at ≥ 8 identifies the region in which muons are dominant over the optical background due to ^{40}K.

Events selected with at least one L1 trigger in each DOM are essentially background free, as they are all muons. The data-MC comparison shown in Fig. 3 right demonstrates the good agreement with the muon simulations, triggering at least one L1 coincidence on each DOMs. The histogram is filled with hits of the selected events as a function of the orientation of the PMTs inside the three DOMs. 158 runs taken in June–July 2014 are used, corresponding to 80 hours of equivalent livetime. MC is normalised to data. MC simulations contain only muon events. Events due to the ^{40}K decay only are not simulated. As 3-DOM-triggered events, they are only muons; the contribution from the optical background is negligible. Furthermore, the

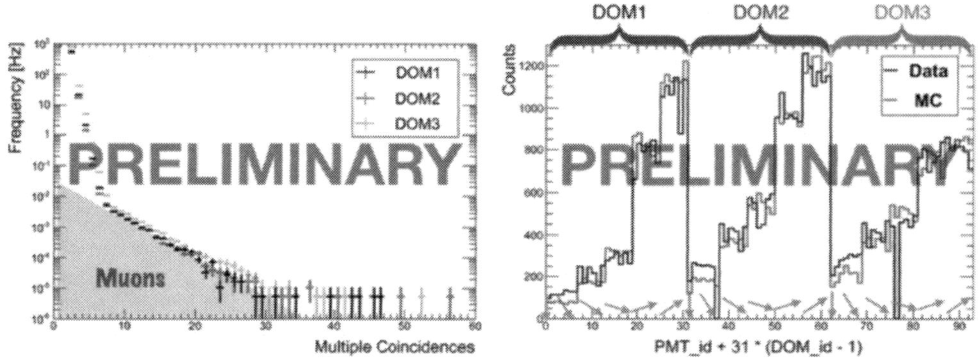

Figure 3. Left: frequency of multiple hit coincidences for the three DOMs from data taken in June-July 2014. The blue area represents events from muons. Right: distribution of hits as a function of the PMT id for events with one L1 trigger on each DOM. The cyan arrows are oriented as the zenith angle of the PMTs inside the DOM.

asymmetry between the upper and lower part of the DOMs seen in Fig. 3 right is a clear evidence of the downward direction of the atmospheric muons.

4. Conclusions

The Pre Production Model Detection Unit was active and took data since its deployment in the Mediterranean Sea until its decommissioning. The complete marine operations for the installation of the DUs (DU deployment on the sea bed, submarine connections with ROV, unfurling procedures) were tested successfully and proved reliable. This prototype project validated the DU structure and provided a test bench for the operation and data handling tools. The calibration procedures developed for the km^3-scale detector were tested and optimised. The PPM-DU helped to improve our knowledge of the marine site, through a long term evaluation of the optical background.

This qualification project leaded the way to the forthcoming installation of 24 detection units and 8 towers in the Capo Passero area (KM3NeT-IT) and 7 detection units at the Toulon site (KM3NeT-FR). The electronics inside the DOM for the complete DU has been redesigned in order to improve general performances and achieve an absolute timing of < 1 ns [11]. The detector construction is starting with the installation of the fist DU by the end of 2015.

References

[1] `http://www.km3net.org`
[2] M. Ageron et al. (ANTARES Collaboration), Nucl. Instrum. Meth. A **656**, 11 (2011)
[3] A. Capone et al. (NEMO Collaboration), Nucl. Instrum. Meth. A **602**, 47 (2009)
[4] P. A. Rapidis, Nucl. Instrum. Meth. A **602**, 54 (2009)
[5] S. Adrián-Martínez et al. (KM3NeT Collaboration), JINST **8**, T03006 (2013)
[6] S. Adrián-Martínez et al. (KM3NeT Collaboration), Eur. Phys. J. C **74**, 3056 (2014)
[7] C. M. F. Hugon, Nucl. Instrum. Meth. A **787**, 189 (2015)
[8] S. Aiello et al. (NEMO Collaboration), Astropart. Phys. **66**, 1 (2015)
[9] J. Brunner et al., Proceedings of the First VLVnT Conference, Amsterdam (2003)
[10] KM3NeT Technical Design Report, `http://www.km3net.org/TDR/TDRKM3NeT.pdf`.
[11] S. Biagi and A. Orzelli, JINST **9**, C12033 (2014)

EPJ Web of Conferences **121**, 05011 (2016) DOI: 10.1051/epjconf/201612105011
RICAP-2014

Potential of KM3NeT to observe galactic neutrino point-like sources

Agata Trovato[a] on behalf of the KM3NeT Collaboration

LNS-INFN, via S. Sofia 62, 95123 Catania, Italy

Abstract. KM3NeT (`http://www.km3net.org`) will be the next-generation cubic-kilometre-scale neutrino telescope to be installed in the depths of the Mediterranean Sea. This location will allow for surveying the Galactic Centre, most of the Galactic Plane as well as a large part of the sky. We report KM3NeT discovery potential for the SNR RXJ1713.7-3946 and the PWN Vela X and its sensitivity to point-like sources with an E^{-2} spectrum.

1. Introduction

In the last decades, also thanks to a multi-messenger approach, our knowledge of the high-energy Universe has been widely extended. Nevertheless many open questions remain. Neutrinos, being uncharged and weakly interacting, can provide unique information on astrophysical objects. Their measurement will allow for new insights into the acceleration mechanisms, clarifying the role of the hadronic component.

In this context, the first evidence for a high-energy neutrino flux of extraterrestrial origin [1, 2] recently reported by the IceCube neutrino telescope, opens a new observational window on our Universe. The origin of this signal is however not yet clear, due to the scarceness of the detected neutrinos and mostly to the limited angular resolution of the IceCube detector. A confirmation, possibly accompanied by the identification of the sources is therefore necessary.

Candidate neutrino sources in the cosmos are numerous, but the estimate of their extragalactic neutrino fluxes has large uncertainties due to model assumptions and to the intergalactic absorption of VHE γ-rays that strongly modifies the spectra measured at Earth. On the other hand, in the hypothesis of hadronic gamma emission, models for galactic neutrino sources are constrained by TeV γ-ray observations and allow to obtain realistic expectations on the detection perspectives [3], as in the case of SuperNova Remnants (SNR) and Pulsar Wind Nebulae (PWN) that are the among most intense sources.

Neutrino telescopes detect the charged leptons produced in the neutrino weak interaction. In transparent media, relativistic particles can be detected through the light produced via Cherenkov effect, with 3D arrays of optical sensors. The "golden channel" for neutrino astronomy is the ν_μ charged current (CC) interaction because the muon range in water is, at $E_\mu \sim$ TeV, of the order of kilometres and the muon track is almost co-linear to the ν_μ permitting to point back to the neutrino cosmic source. This is the channel considered in this work. High energy neutrino astronomy requires detector volumes of the km^3 scale hosted in deep water or in deep Antarctic ice, where several thousands of metres of water (or ice) reduce the flux of atmospheric muons by several orders of magnitude. Since neutrinos are the only particles that can pass the whole Earth, neutrino telescopes look mainly at the up-going

[a]e-mail: `atrovato@lns.infn.it`

neutrinos coming from the opposite hemisphere. A detector in the northern hemisphere is therefore necessary to complement the IceCube sky coverage.

KM3NeT [4] is an international collaboration with the aim to build a research infrastructure in the Mediterranean Sea hosting an underwater multi-km^3 high-energy neutrino telescope. Thanks to its geographical location in the Northern hemisphere, KM3NeT can observe most of the Galactic Plane, including the galactic centre, so our Galaxy is its prime field of investigation. The KM3NeT neutrino telescope will consist of a 3D-array of Digital Optical Modules (DOMs) made of pressure resistant glass spheres each containing 31 photomultiplier tubes (PMTs) with a diameter of about 3 inch and their electronics. The first prototype of a DOM, called also "multi-PMT", has been deployed at a depth of 2500 m in the Antares site. The analysis of first months of data taking highlights the capabilities of the new module design in terms of background suppression and signal recognition [5]. The DOMs are hosted in vertical string-like structures, approximately 1 km in height, called Detection Units (DUs) and described in the Technical Design Report [6] (TDR). The detector will be made of building blocks of 115 DUs each, arranged uniformly in a circular area at an average distance of 90 m. Each block has a volume of about 0.5 km^3. Two of these building blocks will be installed in Italian site of Capo Passero for the so called KM3NeT/ARCA. An extension to 6 building blocks is also envisaged to fully exploit the scientific potential of the detector especially in the search of Galactic point sources.

2. Simulation codes and analysis

The simulation of detector response to astrophysical neutrino fluxes provides a guideline for detector design and optimisation. The software used in this work has been developed by the ANTARES Collaboration [7] and adapted to km^3-scale detectors. The code provides a complete simulation of the incident muon neutrinos with energy in the range 10^2–10^8 GeV, including their interaction in the medium and the propagation of the resulting secondary particles, the light generation and propagation in water and the detector response. The depth and the optical water properties measured at the Sicilian Capo Passero site have been used [8]. Background light due to the presence of ^{40}K in salt water and bioluminescence has been simulated adding an uncorrelated hit rate of 5 kHz per PMT and a time-correlated hit rate of 500 Hz per DOM (two coincident hits in different PMTs inside the same DOM) due to the genuine coincidences from ^{40}K decays.

After the event generation, a track reconstruction algorithm is employed to estimate muon (and consequently neutrino) direction from the arrival times of the photons on the PMTs. After an initial hit selection requiring space-time coincidences between hits, the reconstruction proceeds through four consecutive fitting procedures, each using the result of the previous fit as starting point, except for the first one, called prefit, that is a linear fit through the positions of the hits with the hit time as independent variable [9, 10]. The detector resolution, i.e. the median angle between the reconstructed track and the generated neutrino track, reaches about 0.2° at 10 TeV.

The simulated events are analysed through statistical technique to look for a statistical excess originating in narrow regions of the sky around the sources. Indeed, the weak neutrino signal from a cosmic source stems on the large background of atmospheric muons and neutrinos, both produced by the interaction of primary cosmic rays with the atmosphere. The atmospheric neutrino event rate is calculated assuming the Honda et al. [11] conventional atmospheric model, the Enberg et al. [12] prompt component and adding the correction due to the cosmic rays knee on the neutrino spectrum as described in [13]. Atmospheric muons are simulated using the MUPAGE event generator [14]. The optimisation of the signal to background ratio is far from being trivial and it is strongly dependent on the source features (e.g. energy spectrum, angular extension). As figure of merit of the telescope performance we have taken the discovery potential, that is the signal flux required to obtain an observation at a given significance level (e.g. 5σ or 3σ) with 50% probability [15]. The method used in this work to calculate the discovery potential is the "unbinned" method [16] that relies on the maximisation of a likelihood ratio to evaluate the probability that a set of events is compatible with the hypothesis of signal+background instead of the hypothesis of background only.

3. Results

Amongst the galactic objects, SNRs are probably the most promising ones. In particular, the SNR RXJ1713.7-3946 and the PWN VelaX are at present between the most intense known galactic objects in the high-energy gamma-ray band and are used here to evaluate the KM3NeT performances. In addition the discovery flux is evaluated also for a generic E^{-2} point-source.

The young shell-type SNR RX J1713.7-3946 has been observed by HESS in several campaigns [17] and its energy gamma spectrum is measured up to about 100 TeV. The source has large intensity and a relatively large size with a complex morphology. Moreover, it is visible for about the 80% of the time by KM3NeT therefore is a good reference case.

This SNR has been simulated as an extended neutrino source of 0.6° with an energy spectrum calculated under the hypothesis of a transparent source and 100% hadronic emission and parametrised as [18]:

$$\Phi(E) = 16.8 \times 10^{-15} \left[\frac{E}{\text{TeV}}\right]^{-1.72} e^{-\sqrt{E/2.1\text{TeV}}}\text{GeV}^{-1}\text{s}^{-1}\text{cm}^{-2}. \quad (1)$$

Vela X is one of the nearest pulsar wind nebulae (PWN) and it is associated with the energetic Vela pulsar PSR B0833-45. Even if PWNs are generally treated as leptonic sources (γ-ray emitters through inverse Compton), interpretation of TeV γ-ray emission from Vela X in terms of hadronic interaction is discussed by some authors (see i.e. [19, 20]). The first VHE γ-ray emission from Vela X was reported by the H.E.S.S. Collaboration [21] has been recently updated [22] with data from the 2005–2007 and 2008–2009 observation campaigns and using a more accurate method for the background subtraction. The new data are characterised by a higher gamma flux and a harder energy spectrum. From the differential energy spectrum extracted from an integration radius of 0.8° around the centre at RA = $08^h\ 35^m\ 00^s$, Dec = $-45°\ 36'\ 00''$, the corresponding neutrino emission spectrum has been derived using the Vissani prescription [3] based on the hypothesis of a transparent source and 100% hadronic emission. The neutrino spectrum is parametrised as:

$$\Phi(E) = 7.2 \times 10^{-15} \left[\frac{E}{\text{TeV}}\right]^{-1.36} e^{-(E/7\text{TeV})}\,\text{GeV}^{-1}\text{s}^{-1}\text{cm}^{-2}. \quad (2)$$

Also in this case the source extension has been simulated as a flat spatial distribution within a disk with 0.8° radius.

Under the assumed hypotheses, the significance of the sources observation as a function of the observation years for KM3NeT/ARCA and the full KM3NeT with 6 building blocks has been calculated at a confidence level (CL) of 50% and is shown in Fig. 1. An observation with 5σ significance is expected with the full KM3NeT after 2.5 and 4 years for the VelaX and the RXJ1713.7-3946 respectively. Anyway, even with KM3NeT/ARCA a 3σ observation is possible after a reasonable amount of time (2.5 years for the VelaX and 4.5 years for the RXJ1713.7-3946). Figure 1 shows also how the expected significance can change due

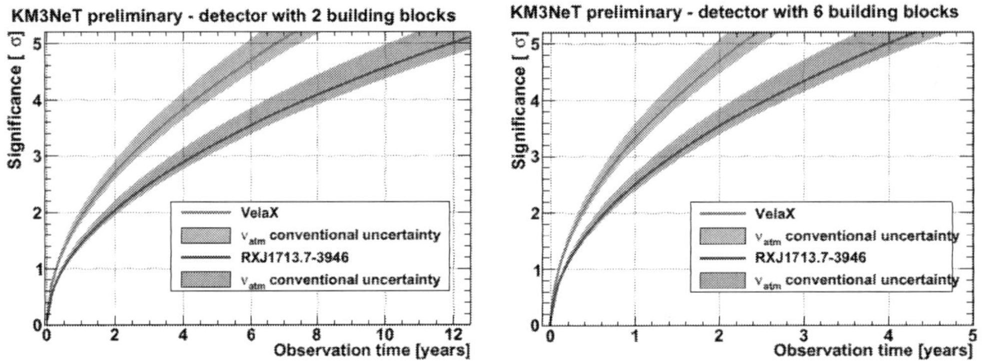

Figure 1. Significance of the RXJ1713.7-3946 and Vela X observation as a function of the years of data taking for the KM3NeT with 2 (left) and 6 (right) building blocks. The shaded bands show the effect of the uncertainties on the conventional component of the atmospheric neutrino flux.

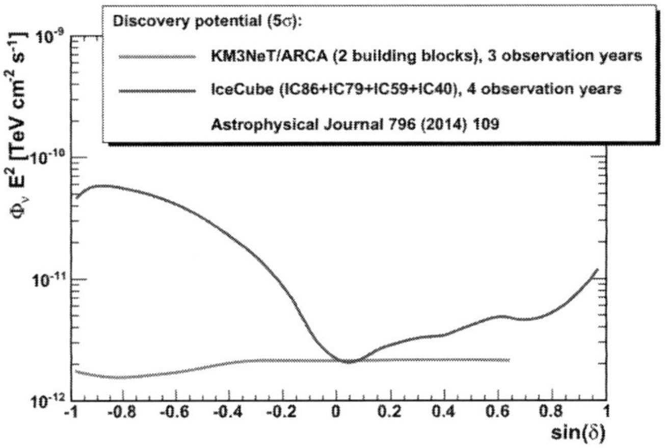

Figure 2. KM3NeT 5σ discovery potential for point sources emitting a neutrino flux with a E^{-2} spectrum, for 3 years of data taking with KM3NeT/ARCA. The flux values are shown as function of the source declination. For comparison the corresponding discovery potential to sources with an E^{-2} spectrum is also shown for the IceCube detector [23].

to the major uncertainty in this calculation, that is the conventional component of the conventional neutrino flux. This effect is taken into account assuming a variation of $\pm 25\%$ in the normalization of the Honda et al. model [11].

In order to compare the KM3NeT performances with IceCube, the 5σ discovery flux $\Phi_{5\sigma}$ has been calculated for generic point-sources with an E^{-2} spectrum as a function of their declination (see Fig. 2). The 3 years observation time has been chosen in such a way that KM3NeT/ARCA has a comparable exposure w.r.t. the IceCube results in [23]. The different trend of the KM3NeT and IceCube curves is mainly due to the geographical location of the detectors.

4. Conclusions and perspectives

The results presented in this paper represent the status of the galactic point-like sources analysis and show that at least the more intense ones are at reach for KM3NeT. In particular we report expectation for RXJ1713.7-3946 and Vela X. The inclusion in our simulation of a realistic source morphology extrapolated from the high energy γ-ray maps measured by HESS is in progress. This will provide a more realistic description of the spatial extension of the sources. The estimate of discovery potential for other galactic sources, as well as a staking analysis of several candidate galactic sources, will be investigated in the near future.

The discovery potential for point sources with E^{-2} spectrum show that KM3NeT has a very large field of view thus not only complementing, but also overlapping to large extent to the IceCube field of view. KM3NeT can thus provide important contributions to the new born field of neutrino astronomy.

References

[1] M. G. Aartsen et al. (IceCube Collaboration), Science **342**, 1242856 (2013)
[2] M. G. Aartsen et al. (IceCube Collaboration), Phys. Rev. Lett. **113**, 101101 (2014)
[3] F. L. Villante and F. Vissani, Phys. Rev. D **78**, 103007 (2008); F. Vissani and F. L. Villante, NIM A **588**, 123 (2008) and F. Vissani, Astr. Ph. **26**, 310 (2006)
[4] `http://www.km3net.org/home.php`.
[5] S. Adrian-Martinez et al. (KM3NeT Collaboration), Eur. Phys. J. **C74**, 3056 (2014)
[6] KM3NeT Technical Design Report, available on `http://www.km3net.org/TDR/TDRKM3NeT.pdf`
[7] D. Bailey, Ph.D. Thesis, available on `http://antares.in2p3.fr/Publications/index.html`
[8] G. Riccobene et al., Astropart. Phys. **27**, 1 (2006)
[9] A. Heijboer, Ph.D. Thesis, available on `http://antares.in2p3.fr/Publications/index.html`
[10] A. Trovato, *Development of reconstruction algorithms for large volume neutrino telescopes and their application to the km3net detector*, Ph.D. Thesis (2013)
[11] M. Honda, T. Kajita, K. Kasahara, S. Midorikawa, and T. Sanuki, Phys. Rev. D **75**, 043006 (2007)
[12] R. Enberg, M. H. Reno, and I. Sarcevic, Phys. Rev. D **78**, 043005 (2008)
[13] M. G. Aartsen et al. (IceCube Collaboration), Phys. Rev. D **89**, 062007 (2014)
[14] Y. Becherini et al., Astropart. Phys. **25**, 1 (2006)
[15] G.C. Hill et al., World Scientific eProceedings, Statistical problems in particle physics, astrophysics and cosmology, 108 (2005)
[16] J. Braun et al., Astropart. Phys. **29**, 299 (2008)
[17] F. Aharonian et al., Astronomy & Astrophysics **464**, 235–243 (2007)
[18] S.R. Kelner et al., Phys. Rev. D **74**, 034018 (2006)
[19] D. Horns et al. Astronomy & Astrophysics **451**, L1 (2006)
[20] L. Zhang and X. C. Yang, ApJ **699**, L153 (2009)
[21] F. Aharonian et al., Astronomy & Astrophysics **448**, L43 (2006)
[22] F. Aharonian et al., Astronomy & Astrophysics **548**, A38 (2012)
[23] M. G. Aartsen et al. (IceCube Collaboration), Astroph. J. **796**, 109 (2014)

EPJ Web of Conferences **121**, 05012 (2016) DOI: 10.1051/epjconf/201612105012
RICAP-2014

Cascade sensitivity studies for KM3NeT

Luigi Antonio Fusco[1,2,a] on behalf of the KM3NeT Collaboration

[1] Dipartimento di Fisica e Astronomia dell'Universitá, Viale Berti-Pichat 6/2, 40127 Bologna, Italy
[2] INFN – Sezione di Bologna, Viale Berti-Pichat 6/2, 40127 Bologna, Italy

Abstract. KM3NeT is a future research infrastructure in the deep seas of the Mediterranean housing a large scale neutrino telescope. The first phase of construction of the telescope has started. Next step is an intermediate phase realising a detector volume of about one-third of the final detector volume. We report on calculations of the sensitivity of the KM3NeT detector to showering neutrino events, the strategy to optimise the detector to a cosmic neutrino flux analogous to the one reported by the IceCube Collaboration and the results of this strategy applied to the intermediate phase detector.

1. Introduction

KM3NeT [1] is the next generation neutrino telescope being built in the Mediterranean Sea. It will consist of several thousand Digital Optical Modules (DOMs) [2], pressure resistant glass spheres housing 31 3″ photomultiplier tubes (PMTs). DOMs will be distributed along vertical strings, each holding 18 of them vertically spaced by 36 m. Strings will be anchored to the seabed at an average distance of 90 m from each other. 115 strings arranged together will constitute a detector building block.

The final detector will consist of 6 building blocks, for an instrumented volume of about 3–5 km^3, distributed at different locations in the Mediterranean Sea. The main aim for the KM3NeT Collaboration will be to search for cosmic neutrino sources, especially in the Galactic Region. The Phase 1 of the project is currently in its final development steps [3]. The first phase of construction of the detector has started.

The next phase of construction is an intermediate step to construct a detector of two building blocks. This phase of the detector ("Phase 1.5") will be dedicated to the confirmation of the isotropic and all flavor cosmic neutrino signal recently reported by the IceCube Collaboration [4]. This signal can be, for example, parametrised as

$$\mathrm{E}_{\nu}^{2}\Phi_{\nu} = 1.2 \cdot 10^{-8} e^{-\mathrm{E}_{\nu}/3\,\mathrm{PeV}}\,\mathrm{GeV s^{-1} sr^{-1} cm^{-2}}. \quad (1)$$

The observation of such a signal by an independent experiment, with different systematics using a different neutrino converting and light detection medium, would constitute an extremely important discovery.

2. Event selection procedure

Many of the neutrino events reported by IceCube look like shower events (cascades), generated by NC and CC ν_e and ν_τ interactions and ν_μ NC interactions. Cascades release

[a] e-mail: luigiantonio.fusco@bo.infn.it

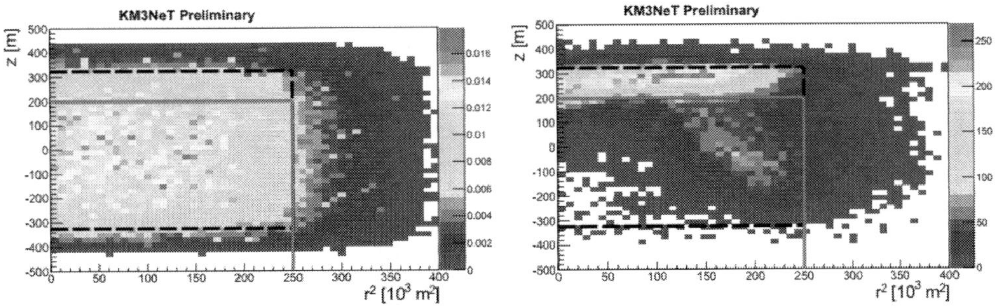

Figure 1. Output of the vertex reconstruction for signal cosmic neutrinos (left) and background atmospheric muons (right). The z and r^2 coordinates (see text) of the reconstructed vertex are plotted. The black dashed line represents the instrumented volume of a KM3NeT detector building block, while the red solid line is the containment requirement for the analysis. This selection allows for a good muon rejection.

a large amount of light starting from the interaction vertex region and this signature can be properly reconstructed.

A large background of atmospheric events is expected when searching for cosmic neutrinos. Atmospheric neutrinos and muon bundles from air showers can reach large sea depths and mimic the signal. Background neutrinos can be rejected with a proper energy cut, as their steep spectrum lowers their contribution at high energy. Atmospheric muon rejection requires different strategies.

Four selection cuts for signal events have been defined using a set of simulated events reconstructed as cascades.

1. N_{hit}: more than 2000 PMT pulses are required. Below this threshold the signal to noise ratio is always negligible because of atmospheric events.
2. Containment: the distribution of reconstructed vertices for atmospheric muons shows an accumulation in the uppermost part of the instrumented volume. On the other hand, cosmic showers vertices are expected to be uniformly distributed in the detector (Fig. 1). As a consequence, cutting on the reconstructed vertex position improves the rejection of atmospheric muon bundles.
3. Time over Threshold: the Time over Threshold (ToT) of each PMTs, summed over all hits that are causally correlated with the reconstructed vertex, is used to discriminate events both by their topology and by their energy. A shower event is expected to have more selected hits than a track, leading to a larger sum of ToT; high energy events would be selected because of higher hit-wise ToT.
4. Boosted Decision Tree: a machine learning technique has been used as shower selector, trained with different quantities defined in the reconstruction algorithms for KM3NeT events. The output of this method is then used, together with the energy estimator ρ, in a two dimensional Model Rejection Factor minimization (MRF) [5], to obtain the final event sample (Fig. 2).

The resulting chain of cuts, maximising the sensitivity, is:

$$N_{hit} > 2000; r < 500 \text{ m and } z < 200 \text{ m}; \text{ ToT} > 15\ \mu s; \text{ BDT} > 0.35 \text{ and } \rho > 10^6, \quad (2)$$

being r and z the cylindric coordinates of the reconstructed vertex.

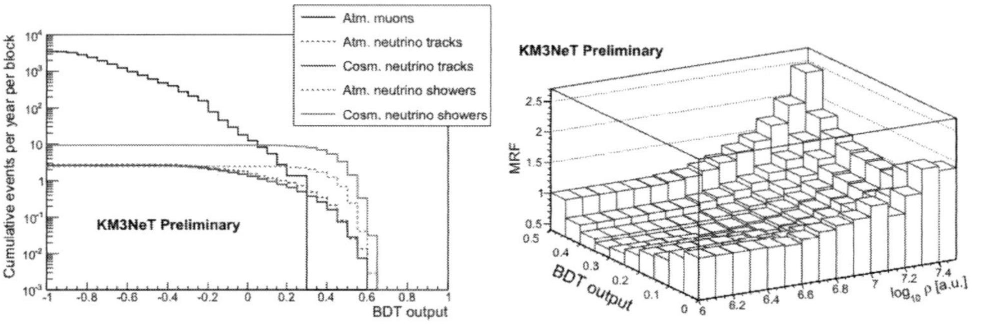

Figure 2. Output of the BDT application to every event type (left) – color code as in the legend. Application of the 2D MRF minimization with the BDT output and the energy estimator ρ (right). The minimum is found at $\rho > 10^6$ and BDT > 0.35, the selection cut maximising the sensitivity [5].

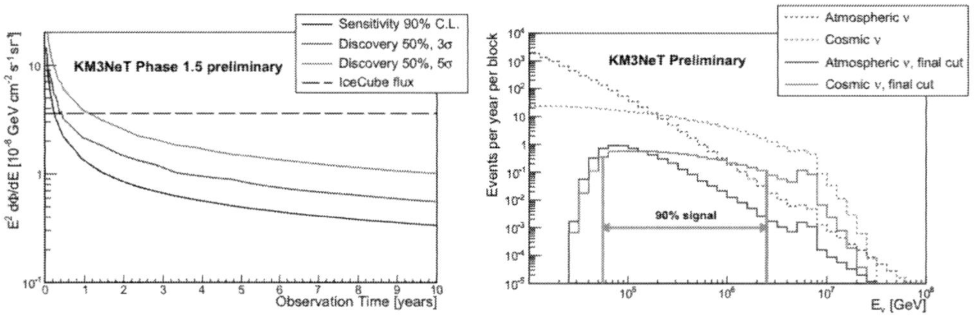

Figure 3. Left: 90% C.L. upper limit sensitivity and 3 and 5σ discovery potential for the detector of KM3NeT Phase 1.5 to the flux of Eq.1, color code as in the legend. Right: energy distribution of signal and background events after selection cuts; the central 90% of the signal lies between 56 TeV and 2.5 PeV.

3. Results

The background is extremely reduced after the cut optimization, as 6.7 signal events are expected to be detected on top of a background of 2.2 per year per detector building block: the performance of the detector is consequently very promising.

After the MRF procedure, the 90% C.L. sensitivity flux for the cosmic neutrino flux reported in [4] is obtained. The KM3NeT Phase 1.5 detector would have the possibility to probe an IceCube-like signal in a short time. As shown in Fig. 3, this limit is valid over and energy range between 56 TeV to 2.5 PeV. Using the optimized selection cut, the 3σ and 5σ significance for a 50% discovery can be calculated, and the evolution with observation time of the detector performance is shown in Fig. 3; with the neutrino detector of KM3NeT Phase 1.5 it will be possible to discover a cosmic neutrino flux similar to the IceCube signal with a 5σ significance in less than 1.5 years.

References

[1] U. Katz for the KM3NeT Collaboration, AIP Conf. Proc. **1630**, 38–43 (2014)
[2] S. Adriáąn-Martinez et al. (KM3NeT Collaboration), EPJC **74**, 3056 (2014)
[3] S. Biagi (for the KM3NeT Collaboration), these proceedings
[4] M.G. Aartsen et al. (IceCube Collaboration), Phys. Rev. Lett. **113**, 101101 (2014)
[5] G.C. Hill & K. Rawlins, Astrop. Phys. **19**, 393 (2003)

EPJ Web of Conferences **121**, 05013 (2016) DOI: 10.1051/epjconf/201612105013
RICAP-2014

Searches for diffuse fluxes of cosmic neutrinos with the ANTARES telescope

Luigi Antonio Fusco[1,2,a] on behalf of the ANTARES Collaboration

[1] Dipartimento di Fisica e Astronomia dell'Università, Viale Berti-Pichat 6/2, 40127 Bologna, Italy
[2] INFN – Sezione di Bologna, Viale Berti-Pichat 6/2, 40127 Bologna, Italy

Abstract. In this proceedings we report on the status of searches for diffuse fluxes of cosmic neutrinos with the ANTARES neutrino telescope data. A complete overview of full sky searches will be given, together with the analysis of possible diffuse neutrino emission from regions such as the Fermi Bubbles or the Galactic Plane.

1. Introduction

Neutrinos are predicted to be produced nearby the expected Cosmic Rays (CR) accelerators, such as Supernova Remnants, Active Galactic Nuclei or Gamma Ray Bursts. A diffuse flux of cosmic neutrinos is expected from unresolved individual sources. The energy spectrum of these neutrinos should be similar to that of primary CRs, produced by Fermi shock acceleration, and flatter than the observed atmospheric neutrino background.

The IceCube Collaboration has reported [1] the observation of an flavour diffuse and isotropic excess of high energy neutrinos, not compatible with atmospheric expectations. This observation has opened the path to high energy neutrino astronomy.

The ANTARES detector [2] is currently the largest neutrino telescope in the Northern Hemisphere, located at a depth of 2475 m in the Mediterranean Sea, 40 km from Toulon, France and continuously operated since 2007. It consists of almost 900 photomultiplier tubes (PMT), distributed on 12 vertical strings anchored to the sea bed. It detects neutrinos using the Cherenkov light emitted by particles produced after neutrino interactions in the surroundings of the detector. Charged current (CC) ν_μ interactions, where a muon straight track is produced, can be properly reconstructed in direction with a good pointing accuracy [3]. All flavours neutral current (NC) interactions, as well as CC ν_e and ν_τ interactions can be reconstructed by dedicated shower algorithms.

The main background in searches for diffuse fluxes of a cosmic signal is given by atmospheric neutrinos and muon bundles produced in CR air showers. To reduce the background from down-going muons only tracks reconstructed as upward-goinh are selected, as only neutrinos can traverse the Earth. Nonetheless many wrongly reconstructed atmospheric muons can survive this directional selection, but a proper cut on the reconstruction quality parameter can help to reduce this background.

[a]e-mail: `lfusco@bo.infn.it`

2. Full sky searches

In this section full sky searches of cosmic neutrinos will be presented, both in the CC ν_μ channel (Sect. 2.1) and for all flavours neutrino interactions reconstructed as shower events (Sect. 2.2). Along with this, an estimation of the atmospheric ν_μ energy spectrum – the neutrino background for cosmic searches – will be given in Sect. 2.3. Blinded analysis procedures have been followed, with only 10% of data available for tests and data/Monte Carlo comparison. The good agreement between data and Monte Carlo allowed for a final cut choice based on simulations.

2.1 Muon neutrinos diffuse flux analysis

An update of the analysis on diffuse ν_μ fluxes reported in [4], with two additional years of ANTARES data (2007–2011), is presented. The equivalent livetime is 885 days, about a factor three larger than the previous analysis.

Most of the down-going atmospheric muon events are rejected applying a preliminary cuts on the reconstructed zenith angle ($\theta_{zen} > 90°$) and on the angular error estimation $\beta(\beta < 0.5°)$. To further reduce the atmospheric background combined cut on reconstruction quality parameter, Λ, and on the number of hit related to the reconstructed track, N_{hit}, has been applied thus reducing the muon background level well below 1%.

Atmospheric neutrinos can be rejected by applying an energy related cut, based on the estimation of the muon energy loss in the detector [5]. The optimal cut on this variable is chosen through a Model Rejection Factor (MRF) procedure [6], determining the best sensitivity flux $E^2\hat{\Phi}^{90\%}_{track} = 4.7 \cdot 10^{-8}$ GeV cm^{-2} s^{-1} sr^{-1}. After the final cut 8.4 events are expected from the background and 1.4 signal events should be observed from the signal reported in [1]. The 90% of the expected signal after cuts has energy in the range from 45 TeV to 10 PeV.

After unblinding 8 events are found in data. Using the method from [7] the upper limit at 90% confidence level is $E^2\Phi^{90\%}_{track} = 5.1 \cdot 10^{-8}$ GeV cm^{-2} s^{-1} sr^{-1} (Fig. 1). The systematic error on the background is taken from the observed normalization of atmospheric neutrino flux measurement shown in section Sect. 2.3 with respect to the Bartol flux [8] model adopted in this work. Systematics on the signal are taken into account by varying water properties, PMT efficiency and angular acceptance in the simulation.

2.2 All flavour diffuse flux analysis with showers

All events reconstructed with the shower algorithm of [9] are considered in this analysis. A vertex likelihood fit, followed by an energy and direction fit are performed by the reconstruction method. This leads to an energy resolution of 0.2–0.3 for the $\log_{10}E_{shower}$ and a median angular resolution of 6° in the hundred TeV region.

The analysis collects all data from 2007 to 2012, with a total livetime of 1247 days. An event pre-selection is done using a cut on the vertex likelihood and requiring signal hits to be present on at least three detector lines. This selection significantly reduces the the contribution from track events, including atmospheric muon background. Additionally the reconstructed vertex position is required not to be closer than 15 m to any PMT, to exclude spurious events from sparks in optical modules.

The MRF optimisation is done on the fitted shower energy and zenith angle. The optimal cut is found to be $E_{shower} > 10$ TeV and $\theta > 94°$, leading to a sensitivity with $E^2\hat{\Phi}^{90\%}_{showers} = 2.2^{+0.9}_{-0.7} \cdot 10^{-8}$ GeV cm^{-2} s^{-1} sr^{-1}. 5±3 background events are expected, and the IceCube flux would correspond to $2.1^{+0.5}_{-0.7}$ cosmic events added to the atmospheric expectations. 8 events

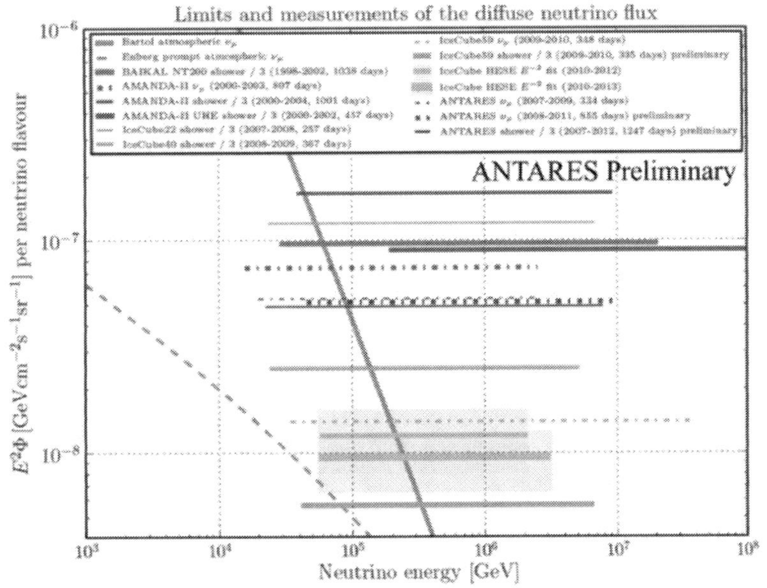

Figure 1. Summary of diffuse flux limits from various experiments in the different detection channels for neutrinos, as reported in the legend.

are observed in data after unblinding. The excess over background has a significance of 1.5σ. Considering an E^{-2} spectrum, the 90% C.L. upper limit on the cosmic neutrino diffuse flux is $E^2\Phi^{90\%}_{showers} = 4.9 \cdot 10^{-8}$ GeV cm^{-2} s^{-1} sr^{-1}, including systematics on signal and background. The validity energy range for this limit is 23 TeV – 7.8 PeV (Fig. 1).

2.3 Atmospheric neutrino energy spectrum

Atmospheric neutrinos are sources of background in neutrino telescopes, but their observation can give valuable information on the detector performances as well as on the systematics affecting the energy measurement. Moreover great uncertainties are present at high energies, both in conventional fluxes – from charged π and K meson decays – and in prompt fluxes – from decaying charmed particles. Measuring their flux can thus improve the knowledge of the neutrino detection itself.

Using the same data sample of Sect. 2.1 two different analysis have been performed, with different selection cuts and unfolding methods and with a complete evaluation of systematic errors. Analogous results in the evaluation of the atmospheric neutrino spectrum from 100 GeV to 200 TeV [5] are achieved in both analyses. The observed spectrum (Fig. 2) is compatible with the expectations from conventional fluxes, being 25% higher in the overall normalization than the flux of [8] but within reported uncertainties. No limit on the prompt contribution can be given as large uncertainties are present at the highest energies.

3. Special regions

In this section searches devoted to possible extended neutrino emission regions will be shown. In particular the so called Fermi Bubbles and the Galactic Plane regions have been analysed, as ANTARES has an excellent coverage of these areas of the sky. For these analyses an on/off-zone procedure has been followed. Considering regions of the sky with the same

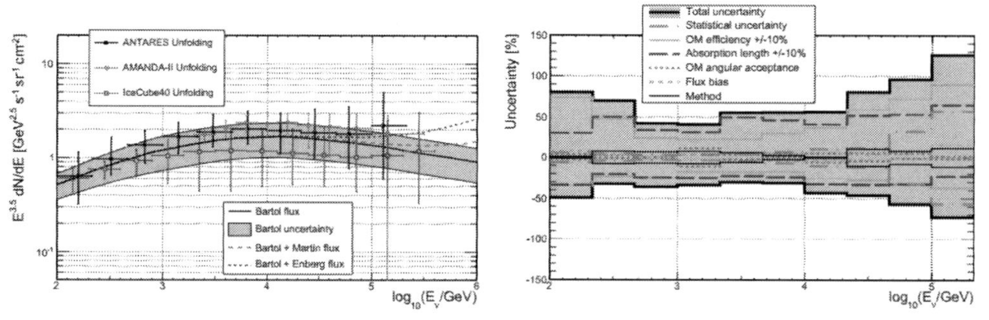

Figure 2. Left: measurement of the atmospheric muon neutrino energy spectrum compared with theoretical predictions (solid line for conventional, dashed for prompt) and experimental results from AMANDA and IceCube. Right: uncertainties on the atmospheric neutrino energy spectrum measurement, considering different effects on water and OM properties, as well as the uncertainties on the underlying spectrum and deriving from the unfolding methods. Color code in the legend.

shape, detector efficiency and visibility of the expected signal regions the background has been estimated and then compared to the actual observation in the signal region. In absence of a signal, the observed number of events in on/off-zones should be statistically compatible.

3.1 Fermi bubbles

Fermi telescope data [10, 11] have revealed the presence of two large γ-ray emission regions above and below the Galactic plane. If hadronic mechanisms are responsible for the production of such a signal, diffuse neutrino emissions are expected from these regions with various possible energy cut-offs, from few to some hundreds of TeV [12].

Data collected in the ν_μ CC channel with the ANTARES telescope from 2007 to 2011 are considered in the analysis [14]. The event selection is based on cuts on the quality of up-going reconstructed tracks and on the energy estimation through an Artificial Neural Network (ANN) [13]. The final selection cut, obtained minimising the MRF, is $\Lambda > -5.14$ and $E_{ANN} >$ 11 TeV, when optimizing for a neutrino flux having a cut-off at 100 TeV. After the unblinding of the on-zone, 16 events are observed, while 11 are expected, on average, from the off-zones. The significance of this excess can be estimated, following the prescription of [15], as 1.2σ and upper limits are calculated (Fig. 3).

3.2 Galactic plane

A diffuse neutrino flux is expected from the decays of charged mesons produced in CR interactions in the interstellar medium in the Galactic Plane. The corresponding emission from neutral mesons is clearly visible in γ-ray observation of the sky [16]. Different models for the CR propagation are proposed (e.g. [17–19]) as well as direct computations from Fermi data [16], each leading to a different neutrino flux expectation but in agreement within a factor ~2. *NoDrift* models [17, 18] do not take into account magnetic fields while the *Drift* model [19], doing so, produces an enhancement of the neutrino flux in the Galactic Centre region.

Muon neutrinos collected by ANTARES from 2007 to 2011 have been analysed [20] following the MRF procedure to set an optimal cut on the area of the on-zone (78° size in longitude and 9° size in latitude centered at the Galactic Centre) and on the quality and energy of reconstructed tracks ($\Lambda > -5.6, \beta < 8°$ and $E_{ANN} > 250$ GeV). The 8 selected

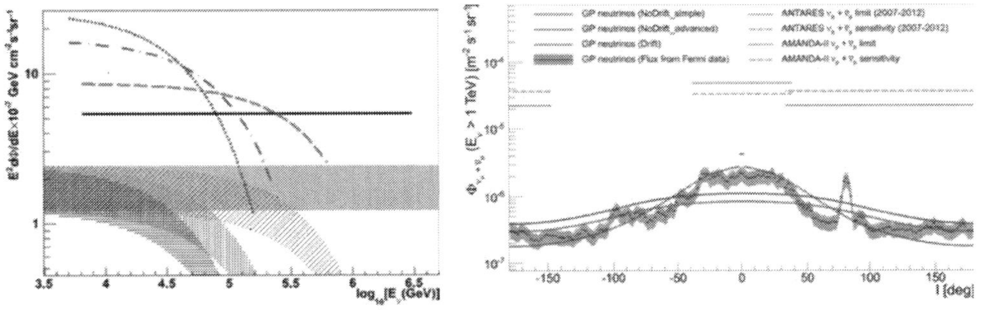

Figure 3. Left: upper limit calculation for neutrino fluxes from the Fermi Bubbles for different energy cut-offs (∞, 100 TeV, 50 TeV, 10 TeV – black, red, green and blue lines). Shaded areas are for theoretical predictions with the various cuf-offs. Right: upper limits and sensitivity for Galactic Plane analyses from ANTARES and AMANDA, compared to theoretical models as reported in the legend – see text for explanations.

off-zones show on average 166 events, and 177 are observed in the on-zone after unblinding. The significance of this excess is 0.8 σ, and the calculated upper limits are reported in Fig. 3.

4. Conclusions

The ANTARES neutrino telescope is in its 7th year of operation. Despite its moderate size, but thanks to its location and good event reconstruction performances, it yields good diffuse flux sensitivity in the relevant range and the best limits for the Galactic Plane and the Fermi bubble regions. In the near future more data will be added and joint track and shower analyses will be performed. The next generation KM3NeT neutrino telescope will further improve the results for diffuse flux sensitivities.

References

[1] M. G. Aartsen et al., Phys. Rev. Lett. **113**, 101101 (2013)
[2] M. Ageron et al., Nucl. Instrum. Methods A **656**, 11 (2011)
[3] S. Adrián-Martínez et al. Astrophys. J **760**, 53 (2012)
[4] J.A. Aguilar et al., Phys. Lett. B **696**, 16 (2011)
[5] S. Adrián-Martínez et al. Eur. Phys. J. C **73**, 2606 (2013)
[6] G.C. Hill and K. Rawlins, Astropart. Phys. **19**, 393 (2003)
[7] J. Conrad et al., Phys. Rev D **67**, 012002 (2003)
[8] G.D. Barr et al, Phys. Rev. D **74**, 094009 (2006)
[9] F. Folger, PhD thesis, FAU (2014)
[10] M. Su et al., Astrophys. J. **724**, 1044 (2010)
[11] M. Ackermann et al., Astrophys. J. **793**, 64 (2014)
[12] R. M. Crocker and F. Aharonian, Phys. Rev. Lett. **106**, 101102 (2011)
[13] J. Schnabel, Nucl. Instrum. Methods A **725**, 106 (2013)
[14] S. Adrián-Martínez et al. Eur. Phys. J. C **74**, 2701 (2014)
[15] T. P. Li and Y. Q. Ma, Astrophys. J. **272**, 317 (1983)
[16] M. Ackermann et al., Astrophys. J. **750**, 3 (2012)
[17] G. Ingelman and M. Thunman, Arxiv preprint hep-ph/9604286 (1996)
[18] J. Candia and E. Roulet, JCAP **09**, 005 (2003)
[19] J. Candia, JCAP **11**, 002 (2005)
[20] E. Visser, PhD thesis, University of Leiden, to be published

EPJ Web of Conferences **121**, 05014 (2016) DOI: 10.1051/epjconf/201612105014
RICAP-2014

The multi-messenger search programme and results of the ANTARES neutrino telescope

Giulia De Bonis[a] on behalf of the ANTARES Collaboration[b]

Istituto Nazionale di Fisica Nucleare (INFN), Piazzale Aldo Moro, 2, 00185 Roma, Italy

Abstract. The key-word of modern astronomy and astrophysics is multi-messenger: not only photons used as probes for the investigation of the Universe, but also cosmic-rays, neutrinos and gravitational waves. The multi-messenger approach is important in particular for neutrino detectors: potential astrophysical sources are predicted to emit a very faint neutrino signal and the presence of an isotropic flux of atmospheric background requires the development of effective search strategies. The multi-messenger approach can increase the discovery potential, the statistical significance of the observations and the efficiency of the detection. The advantages of the multi-messenger approach are evident, in particular, when looking at transient or flaring sources. In ANTARES, a wide programme of multi-messenger searches is active; the most relevant results will be presented in this contribution.

1. Introduction

In the last decades the community of astronomers and astrophysicists has been witness to an important change in the modalities of observing the Universe. More than an evolution, the revolution of modern astronomy has consisted in moving from traditional astronomy to *multi-wavelength astronomy*, in which observations of light in the visible band are complemented by the information collected in the radio, X-ray and gamma-ray regions of the electromagnetic spectrum. The next step in this mission of exploration of the Cosmos is the progress from *multi-wavelength astronomy* to *multi-messenger astronomy*: no more "only photons", but new "actors" are coming into play in the scene of astronomy and astrophysics. New messengers from the Universe can be added in the list: cosmic rays, gravitational waves and neutrinos, to enrich and complement the information provided by photons at any wavelength. The multi-messenger approach connects astronomy with astrophysics and particle physics, it opens new windows in cosmology and in the theory of gravity and can give some hints of new physics beyond the standard model. The manifesto is completed bearing in mind that it promotes collaborations between different experiments: researchers involved in shared projects join efforts and skills, and can make available additional tools for data analysis.

The multi-messenger approach assumes that the different messengers (all of them or some of them, depending on the model and on the type of source) are expected to be produced in the same astrophysical site. This allows to increase the discovery potential by observing the same source with different probes. Moreover, the statistical significance of the observations can be

[a]e-mail: `giulia.debonis@roma1.infn.it`
[b]`http://antares.in2p3.fr`

improved by coincident detection (sustained by the development of alert systems between the experiments) and the efficiency of the detection can be refined profiting from relaxed cuts and exploiting the advantages of time-dependent analysis. The advantages of the multi-messenger approach are evident, in particular, when looking at transient or flaring sources.

Dealing with the data analysis in a multi-messenger perspective is crucial for neutrino detectors, since potential astrophysical sources of neutrinos are predicted to emit faint signals and the presence of an isotropic flux of atmospheric background makes the detection challenging and requires the development of effective search strategies. The strongest connection is with gamma-ray and X-ray astronomy, since both high energy photons and neutrinos are expected from cosmic emitters if hadronic processes take place. In the standard picture, known as Fermi acceleration [1], hadrons, confined by magnetic fields inside the astrophysical sites, are accelerated through repeated scattering by plasma shock fronts; collisions of hadrons with ambient plasma produce photons and neutrinos through pion photoproduction mechanisms. In this model, both gammas and neutrinos are linked to cosmic rays; the flux predictions for neutrinos can be computed given the observations of the gamma-ray spectra. Confirmed gamma-ray sources are thus the first target of observations for neutrino telescopes, see for instance the ANTARES results for point-like sources [2], in which the list of candidate sources is a list of gamma-ray emitters, and the ANTARES results for the diffuse neutrino emission from Fermi Bubbles and from the Galactic Plane [3], in which the "on-region" is delimited by gamma-ray observations. The searches included in the ANTARES multi-messenger programme, presented in this contribution after a brief account on the ANTARES detector (Sect. 2), are those that regard specifically the time-dependent analysis: flaring sources, both extragalactic (blazars, Sect. 3) and galactic (micro-quasars, Sect. 4); transient sources, like Gamma-Ray Bursts (GRBs) (Sect. 5); the TAToO project for the optical follow-up (Sect. 6). As examples of looking for coincidences with messengers other than photons, the search for a correlation with Ultra-High-Energy Cosmic Rays (UHECRs) (Sect. 7) and with gravitational waves (GWs) (Sect. 8) is presented. The ANTARES programme includes activities for which other experiments trigger neutrino observations, but also the opposite case, with neutrinos triggering others; the TAToO project and the collaboration with the network of gravitational waves interferometers are examples of this kind.

2. The ANTARES detector

ANTARES (Astronomy with a Neutrino Telescope and Abyss environmental RESearch) [4], deployed in the Mediterranean Sea, 40 km offshore Toulon, France, at a depth of about 2500 m, is the largest neutrino telescope in operation in the Northern hemisphere. It consists of 12 strings composed of 25 triplets of PMTs (storey), for a total of 885 PMTs; the distance between strings is about 70 m; the distance between storeys along the string is 14.5 m; the total length of the string is 450 m, including the length of 100 m between the bottommost storey and the seabed. The deployment of the telescope started in 2007 with the first 5 lines; the detector has been completed in May 2008. Data taking is continuously on-going since 2007 and it will be on up to end of 2016.

3. Neutrino signal from flaring blazars

The first study of a possible correlation of the ANTARES signal with the enhanced photon emission during AGN flares has been carried on with data collected in 2008 (61 days live time) [5].Ten blazars have been selected from the Fermi/LAT AGN catalog, among those that showed a large variability (flaring state) in the time interval under investigation. A dedicated algorithm has been developed for the identification of flares in the light curve. The idea behind

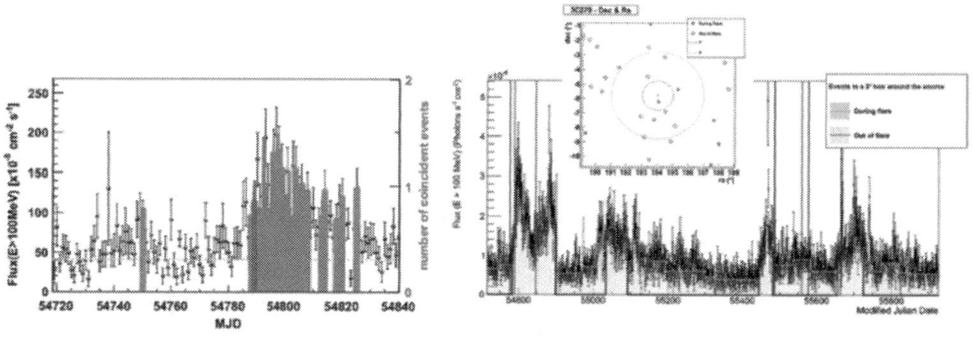

Figure 1. Left: light curve (dots) of the blazar 3C279 measured by Fermi/LAT; the blue (dark grey) shaded histogram indicates the flaring periods selected for the analysis (the dashed line, in green, is the fitted baseline); the red (dark grey) histogram displays the time of the associated ANTARES neutrino event. Right: light curve of 3C279 considering the time interval of the improved analysis; flaring periods are in evidence (in yellow, in the coloured version of this article), the vertical lines identify the time of detected neutrinos around the source; the insert report the position of the candidate neutrino events in a map centred at the source location.

this study is that selecting a narrow time window during which the neutrino signal is expected, in coincidence with the gamma flare, the atmospheric background is cut down. The estimated performance of the method is that the flare timing information yields an improvement of the discovery potential by about a factor 2 (1.3) with respect to a standard time-integrated point source search when looking for a 1 (10) day flare. The outcome of the analysis was one neutrino candidate event compatible in space with source 3C279 and coincident in time with a flaring period (Fig. 1, left); the p-value of the observation was 1% but post-trial probability was 10%; the occurrence is thus not statistically significant and consistent with a background fluctuation. Fluence upper limits have been set for any of the selected sources, with fluence defined as the integral in energy and time of the flux upper limit. Even in absence of a discovery, the method looked very promising and the period under study was extended, considering the interval 2008–2011. The improved analysis is currently in progress and has already selected 40 sources with 86 flaring periods and 6 specially significant flares, among which source 3C279 is still one of the most promising (Fig. 1, right – preliminary). A further extension consists in searching for coincidences not only with FERMI/LAT data, but also with the information provided by IACT telescopes (HESS, MAGIC, VERITAS).

4. Neutrino signal from galactic micro-quasars

Micro-quasars are galactic X-ray binary systems with relativistic jets; several models indicate micro-quasars as possible sources of high-energy neutrinos, with flux expectations depending on the baryonic content of the jets [6]. The analysis has been carried out using the data set collected in the time interval 2007–2010 [7], selecting six sources in the ANTARES visibility and showing an outburst in the period under study. For each source, the data analysis has been restricted to the flaring time periods, selected with a dedicated algorithm in a multi-wavelength approach (X-rays/gamma-rays), using data from RXTE/ASM, Swift/BAT and Fermi/LAT. No statistically significant excess has been found above the expectation of the atmospheric background, thus upper limits have been derived and compared to theoretical predictions, assuming a power-law spectrum (E^{-2}) and considering also the case of an exponential energy cut-off at 100 TeV. The results can be used also to infer information on the jet composition; for some sources, the obtained upper limits already constrain some emission models concerning the ratio of proton to electron luminosity in the jets.

5. Neutrino emission in coincidence with GRBs

The most recent analysis published by ANTARES on the search for neutrino emission in coincidence with detected GRBs has considered data collected in the time interval 2007–2011 and a list of 296 GRBs [8]. The requirement of temporal and spatial coincidence (within 10°) with a GRB reduces the background to $\sim \mathcal{O}(10^{-4})$ per GRB; as a consequence, selection cuts can be loosened. The novelty of the analysis consists also in considering the predictions of the spectra given by the NeuCosmA code [9], based on numerical calculations of the full photohadronic interaction processes. The neutrino flux expected from this model is an order of magnitude below that predicted by previous analytic approaches and can be a key element in the light of the non-observation by IceCube of GRB neutrinos. The analysis has been carried on looking for muon neutrinos; no coincident event was found, 90% C.L. upper limits have been set.

6. TAToO

The TAToO (Telescopes-ANTARES Target of Opportunity) project has established an alert system for sending triggers to a network of optical telescopes, aiming at the optical follow-up of the signal of neutrino candidate events [10]. The opportunity offered by the ANTARES detector is related to the large sky coverage and the high duty cycle of the ANTARES observations; in addition, the pointing accuracy for muon tracks is about 0.5° for energies above the TeV scale. Coincident observations of neutrinos and optical signals are expected from transient sources, like GRBs. A fast online muon track reconstruction algorithm is used to trigger the optical telescopes; the alert criteria include the identification of a multiplet of events within given time and angular window (interpreted as a burst), or the occurrence of an high energy event (signature of energetic processes). The information from the online reconstruction is used for the starting of the follow-up; offline reconstruction is then used for the computation of the refined direction, that is taken into account in the subsequent observations. Several improvements have been introduced since the beginning of the project, the most important being related to the latency of the alert system, that is now reduced to about 15 seconds; data analysis is in progress.

7. Search for correlation with the arrival directions of Auger UHECRs

The arrival directions of 2190 neutrino candidate events (detected by ANTARES in the period 2007–2008, corresponding to an effective live time of 304 days) and 69 UHECRs (detected by the Pierre Auger Observatory in the period 2004–2009, with energy $E > 10^{19.74}$ eV) have been compared, in search for correlations [11]. All the Auger events are in the ANTARES field of view; a light composition has been assumed for this study, so that the magnetic deflection angle could be set at the reasonable value of 3°. A source stacking method has been applied; no significant correlation was observed; the obtained 90% C.L. upper limit on the neutrino flux per source is 5×10^{-8} GeV cm^{-2} s^{-1} (for a E^{-2} energy spectrum).

8. Search for coincidences with the signal of gravitational waves

The GWHEN (Gravitational Wawes + High Energy Neutrinos) working group has been established between the ANTARES and the VIRGO-LIGO Collaborations to set up a common data analysis in search of coincident signals from common astrophysical sources, among which: GRBs, bursting magnetars, topological defects, and also sources with no electro-magnetic counterpart ("hidden sources"). The first study has considered data collected in 2007, when ANTARES construction was in progress and the detector was in a 5-line configuration [12]. The live time of the search was 91 days, during which 216 neutrinos were reconstructed, but only 158 occurred when at least 2 GW detectors were in operation; events reconstructed with less than 2 lines show an azimuthal degeneracy that produces mirror tracks; only 14 candidate neutrinos fulfilled the requirement of at least 3 lines. The result of the search was no gravitational wave burst found associated with any selected neutrinos; all the computed p-values were compatible with the null hypothesis. This outcome can be translated in 90% C.L. limits on the exclusion distances of the sources and on the rate density of common sources (GW+HEN). Work in progress considers an extended data set, collected in 2009 with ANTARES in its final configuration (12 lines), for a total of 129 days of common data taking and an amount of 1989 neutrino candidates. Improved reconstruction algorithms are applied for both neutrinos and gravitational waves, resulting in a factor 7 of improvement in sensitivity and discovery potential.

9. Summary and conclusions

Neutrino astronomy is always carried on in a multi-messenger framework, and given the peculiarity of neutrino interaction properties neutrinos have the ambition of assuming non only a "supporting" role (as complementary source of information for photon astronomy), but a leading role. Astronomy can be done even without photons, with neutrinos offering unique opportunities to look further away and deeper inside astrophysical objects: indeed, in contrast to photons and charged cosmic rays, they can find their way out, without scattering or absorption, from the very inner core of dense astrophysical objects, providing information about sites that are opaque to electromagnetic radiation, revealing the existence of sources so far undetected ("hidden sources") and releasing triggers for other "penetrating" probes (like gravitational waves). Results obtained so far by the ANTARES detector using a multi-messenger approach in data analysis are very competitive; data taking is ongoing and updated results will be soon released.

References

[1] R. Blandford, J. Ostriker, The Astrophysical Journal **221**, L29 (1978)
[2] F. Schüssler, this Conference
[3] L. Fusco, this Conference
[4] J. Aguilar et al., Nucl. Instr. and Meth.A **656**, 11 (2011)
[5] S. Adrián-Martínez et al., Astropart. Phys. **36**, 204 (2012)
[6] H. Christiansen et al., Phys. Rev. D **73**, 063012 (2006)
[7] S. Adrián-Martínez et al., Journal of High Energy Astrophysics **3–4**, 9 (2014)
[8] S. Adrián-Martínez et al., A&A **559**, A9, (2013)
[9] S. Hümmer et al., Phys. Rev. Lett. **108**, 231101 (2012)
[10] M. Ageron et al., Astropart. Phys **35**, 530 (2012)
[11] S. Adrián-Martínez et al., ApJ **774**, 19 (2013)
[12] S. Adrián-Martínez et al., JCAP **06**, 008 (2013)

EPJ Web of Conferences **121**, 05015 (2016) DOI: 10.1051/epjconf/201612105015
RICAP-2014

Measurement of the atmospheric muon flux at 3500 m depth with the NEMO Phase-2 detector

C. Distefano[3,a], S. Aiello[4], F. Ameli[9], M. Anghinolfi[6], G. Barbarino[7,15], E. Barbarito[1,11], F. Barbato[7,15], N. Beverini[8,16], S. Biagi[3], B. Bouhadef[8], C. Bozza[10,18], G. Cacopardo[3], M. Calamai[8,16], C. Calì [3], A. Capone[9,17], F. Caruso[3], A. Ceres[1], T. Chiarusi[2], M. Circella[1], R. Cocimano[3], R. Coniglione[3], M. Costa[3], G. Cuttone[3], C. D'Amato[3], A. D'Amico[3,b], G. De Bonis[9], V. De Luca[3], N. Deniskina[7], G. De Rosa[7,15], F. Di Capua[15], P. Fermani[9,16], V. Flaminio[8], L.A. Fusco[2,12], F. Garufi[7,15], V. Giordano[4], A. Gmerk[3], R. Grasso[3], G. Grella[10,18], C. Hugon[6], M. Imbesi[3], V. Kulikovskiy[3], G. Larosa[3], D. Lattuada[3], K.P. Leismueller[3], E. Leonora[4], P. Litrico[3], A. Lonardo[9], F. Longhitano[4], D. Lo Presti[4,13], E. Maccioni[8,16], A. Margiotta[2,12], A. Martini[5], R. Masullo[9,17], P. Migliozzi[7], E. Migneco[3], A. Miraglia[3], C.M. Mollo[7], M. Mongelli[1], M. Morganti[8,16,19], P. Musico[6], M. Musumeci[3], C.A. Nicolau[9], A. Orlando[3], R. Papaleo[3], C. Pellegrino[2,12], M.G. Pellegriti[3], C. Perrina[9,17], P. Piattelli[3], C. Pugliatti[4,12], S. Pulvirenti[3], A. Orselli[6], F. Raffaelli[8,16], N. Randazzo[4], G. Riccobene[3], A. Rovelli[3], M. Sanguineti[6,14], P. Sapienza[3], V. Sciacca[3], I. Sgura[1], F. Simeone[9], V. Sipala[4], F. Speziale[3], M. Spina[3], A. Spitaleri[3], M. Spurio[2,12], S.M. Stellacci[10,18], M. Taiuti[6,14], G. Terreni[8,16], L. Trasatti[5], A. Trovato[3], C. Ventura[4], P. Vicini[9], S. Viola[3], and D. Vivolo[7,15]

[1] INFN Sezione Bari, via E. Orabona 4, 70126 Bari, Italy
[2] INFN Sezione Bologna, V. le Berti Pichat 6/2, 40127 Bologna, Italy
[3] INFN Laboratori Nazionali del Sud, via S. Sofia 62, 95123 Catania, Italy
[4] INFN Sezione Catania, via S. Sofia 64, 95123 Catania, Italy
[5] INFN Laboratori Nazionali di Frascati, via Enrico Fermi 40, 00044 Frascati (RM), Italy
[6] INFN Sezione Genova, via Dodecaneso 33, 16146 Genova, Italy
[7] INFN Sezione Napoli, via Cintia, 80126 Napoli, Italy
[8] INFN Sezione Pisa, Polo Fibonacci, Largo Bruno Pontecorvo 3, 56127 Pisa, Italy
[9] INFN Sezione Roma, P.le A. Moro 2, 00185 Roma, Italy
[10] INFN Gruppo Collegato di Salerno, via Giovanni Paolo II 132, 84084 Fisciano, Italy
[11] Dipartimento Interateneo di Fisica Università di Bari, via E. Orabona 4, 70126 Bari, Italy
[12] Dipartimento di Fisica Università di Bologna, V.le Berti Pichat 6/2, 40127 Bologna, Italy
[13] Dipartimento di Fisica e Astronomia Università di Catania, via S. Sofia 64, 95123 Catania, Italy
[14] Dipartimento di Fisica Università di Genova, via Dodecaneso 33, 16146 Genova, Italy
[15] Dipartimento di Scienze Fisiche Università di Napoli, via Cintia, 80126 Napoli, Italy
[16] Dipartimento di Fisica Università di Pisa, Polo Fibonacci, Largo Bruno Pontecorvo 3, 56127 Pisa, Italy
[17] Dipartimento di Fisica Università "Sapienza", P.le A. Moro 2, 00185 Roma, Italy
[18] Dipartimento di Fisica Università di Salerno, Italy
[19] Accademia Navale di Livorno, viale Italia 72, 57100 Livorno, Italy

[a] e-mail: distefano_c@lns.infn.it
[b] *Present address*: Nikhef, Science Park, Amsterdam, The Netherlands

Abstract. In March 2013, the Nemo Phase-2 tower was successfully deployed at 80 km off-shore Capo Passero (Italy) at 3500 m depth. The tower operated continuously until August 2014. We present the results of the atmospheric muon analysis from the data collected in 411 days of live time. The zenith-angle distribution of atmospheric muons was measured and results compared with Monte Carlo simulations. The associated depth intensity relation was then measured and compared with previous measurements and theoretical predictions.

1. Introduction

High energy neutrinos are considered optimal probes to identify the sources of high energy cosmic rays. Their detection is therefore one of the most interesting topics of astroparticle physics. High energy neutrino astronomy requires km^3-scale detectors because of the low expected neutrino fluxes [1]. After the first generation of neutrino telescopes, such as BAIKAL [2], AMANDA [3] and ANTARES [4], the IceCube telescope at the South Pole recently showed the first evidence for high energy extraterrestrial neutrinos [5]. The next step will be the construction in the Mediterranean Sea of a deep-sea km^3-scale detector: the KM3NeT telescope [6], result of a joined effort of the ANTARES, NEMO and NESTOR Collaborations, which conducted intense R&D activities. KM3NeT will be implemented as a distributed infrastructure in three sites: Toulon (France) [7], Capo Passero (Italy) [8] and Pylos (Greece) [9].

A first small-scale prototype, a tower-like detection unit (NEMO Phase-1 tower), was operated for five months during 2007 at 2080 m depth offshore Catania (Italy) [10]. In March 2013, a larger scale prototype (NEMO Phase-2 tower) was deployed in the Capo Passero site at a depth of 3458 m. This detector was continuously operated for more than one year. In this paper, after a description of the detector, we report on the measurement of the atmospheric muon angular distribution and on the comparison with Monte Carlo simulations. The corresponding muon depth intensity relation is evaluated and compared with theoretical predictions and with previous results. The present analysis extends the muon depth intensity relation measured in water up to 13 km.

2. The NEMO Phase-2 tower

The NEMO Phase-2 tower was deployed on March 23 2013 at the site located about 80 km offshore Capo Passero (latitude: 36° 17′ 48″ N, longitude: 15° 58′ 57″ E) at a depth of 3458 m. The tower operated continuously until August 4 2014, when it was disconnected to allow for an upgrade of the infrastructure. The tower is a three-dimensional flexible structure composed of eight horizontal *floors* interlinked by a system of tensioning ropes. The structure is anchored at the seabed and kept taut by a system of buoys at the top. The tower floor, a 8 m long structure, is connected to the next ones with four ropes such that each floor is perpendicular to its vertical neighbours. The floors are vertically spaced by 40 m, with the lowermost one located 100 m above the sea bottom [11]. Each floor supports four Optical Modules (OMs), two at each end, one down-looking and one looking horizontally. An OM is composed by a photo-multiplier tube (PMT) enclosed in a 13″ pressure-resistant glass sphere [12]. The PMT is a 10" Hamamatsu[1] R7081Sel with 10 dynode stages. A Front-End Module board (FEM) is also placed inside the OM. It applies a quasi-logarithmic compression on the analog signal, which is then sampled at 200 MHz by means of two 100 MHz Flash ADCs, staggered by 5 ns [13].

[1] Hamamatsu Photonics, 812 joko-cho, Hamamatsu city, 431-31 Japan, web-site: `www.hamamatsu.com`

 DOI: 10.1051/epjconf/201612105015

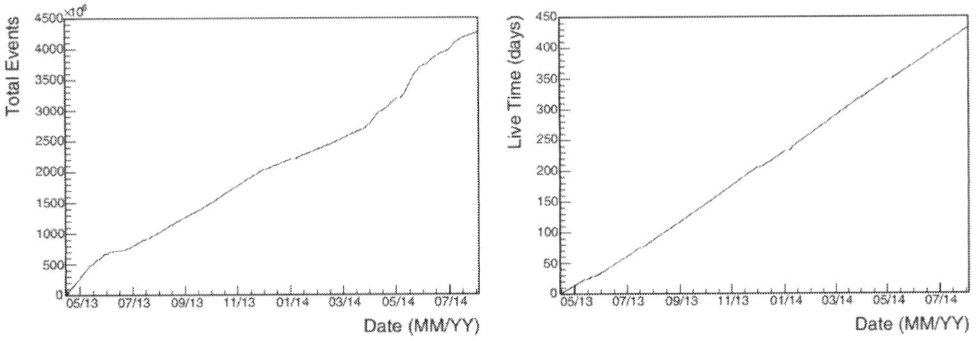

Figure 1. Cumulative number of triggered events and the corresponding live time as a function of time.

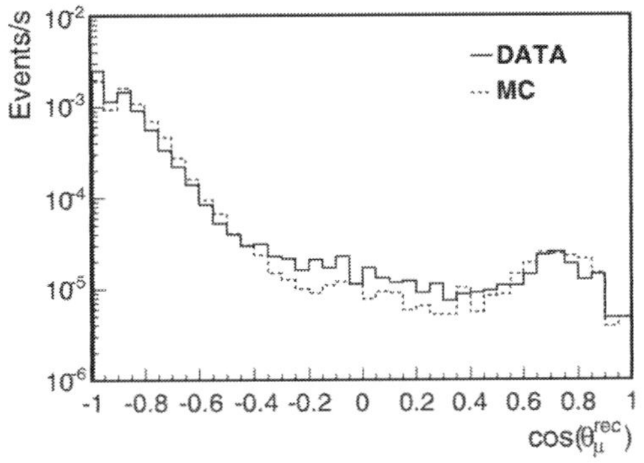

Figure 2. Zenith angular distributions of reconstructed atmospheric muon tracks. Solid histogram is data, dashed histogram is Monte Carlo simulation.

3. Muon data taking

The presence of an optical background, due to ^{40}K decays and bioluminescence, gives an average rate of about 50 kHz of uncorrelated single photo-electron signals on each PMT [14]. An on-line muon selection algorithm is used to reduce the amount of PMT data due to the optical background. This algorithm is based on searching *trigger seeds* among the PMT signals (*hits*), such as Simple Coincidences (SCs): coincidences within 20 ns between two hits occurring on two PMTs located at the same extremity of a given floor; Floor Coincidences (FCs): coincidences within 100 ns between two hits recorded at the opposite ends of the same floor; Charge Shootings (CSs): hits exceeding a charge threshold of about 10 p.e. [15]. The cumulative number of triggered events and the correspond live time as a function of time are plotted in Fig. 1.

4. Muon depth intensity relation

Muon data recorded between April 2013 and August 2014 were analysed [16]. A total of 606,546 atmospheric muon tracks were reconstructed during 411.1 days of live time. The zenith-angle distribution of reconstructed events and its comparison with Monte Carlo simulations are shown in Fig. 2. The true number of muon events $N_\mu(\theta_\mu)$ as a function of zenith-angle is obtained applying an unfolding procedure on the reconstructed angular

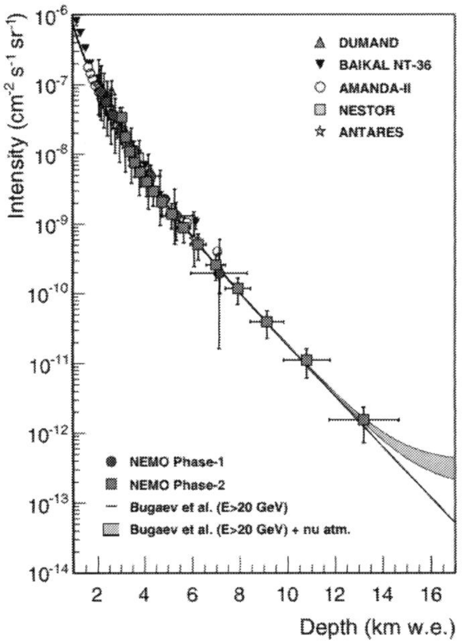

Figure 3. Vertical muon intensity, $I(\theta_Z = 0, h)$, versus depth measured with the NEMO Phase-2 tower [16]. For comparison, results from other experiments are shown: DUMAND [22], BAIKAL [23], NESTOR [24], AMANDA [25], ANTARES [26] and NEMO Phase-1 [10]. The solid line is the prediction of Bugaev et al. [27]. The shaded area at large depths includes atmospheric neutrino-induced muons. Error bars include statistical and systematic uncertainties.

distribution. The muon intensity $I(\theta_\mu)$ is then calculated using the relation

$$I(\theta_\mu) = \frac{N_\mu(\theta_\mu) \cdot m(\theta_\mu)}{A_{eff}(\theta_\mu) \cdot T \cdot d\Omega}, \tag{1}$$

where $m(\theta_\mu)$ and $A_{eff}(\theta_\mu)$ are respectively the average muon multiplicity and the detector effective area as determined from the Monte Carlo simulations; T is the detector live time corresponding to the selected data sample and $d\Omega$ is the detection solid angle.

The final step of the analysis was the evaluation of the Depth Intensity Relation (DIR). The DIR describes the vertical muon flux intensity as a function of the muon slant. An atmospheric muon, reaching the detector located at a vertical depth D from a zenith angle $\theta_Z = 180° - \theta_\mu$, propagates through a water slant depth $h = D/\cos\theta_Z$. The measured flux $I(\theta_\mu)$ was therefore transformed into the DIR using:

$$I(\theta_Z = 0, h) = I(\theta_Z) \cdot \cos(\theta_Z) \cdot c_{corr}(\theta_Z), \tag{2}$$

where the term $c_{corr}(\theta_Z)$ is a geometrical correction factor which takes into account the curvature of the Earth [21]. Results are plotted in Fig. 3. For details about the analysis and discussion on results see [16].

5. Conclusions

The NEMO Collaboration has achieved a major milestone with the installation and operation of a tower-like prototype at 3500 m depth. The NEMO Phase-2 tower, composed by 8 floors for a total height of 380 m, equipped with 32 PMTs, was deployed in 2013 about 80 km offshore Capo Passero (Italy). It was continuously operated for more than one year.

Atmospheric muon tracks have been reconstructed and their measured angular distribution has been compared with Monte Carlo simulations. The muon depth intensity relation has been evaluated and compared with previous data and predictions, showing a good agreement. With the present analysis, the muon depth intensity relation has been measured in water for the first time up to an equivalent depth of 13 km.

References

[1] J.G. Learned, K. Mannheim, Ann. Rev. Nucl. Part. Sci. **50** (2000) 679
[2] V. Aynutdinov et al., Nucl. Instr. Meth. A **602** (2009) 14
[3] R. Abbasi et al., Phys. Rev. D **79** (2009) 062001
[4] J.A. Aguilar et al., Nucl. Instr. Meth. A **656** (2011) 11
[5] M.G. Aartsen et al., Science **342** (2013) 1242856
[6] KM3NeT web site: `http://www.km3net.org`
[7] H. Yepes-Ramírez, for the ANTARES Coll., Nucl. Instr. Meth. A **725** (2013) 203
[8] G. Riccobene et al., Astropart. Phys. **27** (2007) 1
[9] E.G. Anassontzis et al., Astropart. Phys. **34** (2010) 187
[10] S. Aiello et al., Astropart. Phys. **33** (2010) 263
[11] T. Chiarusi et al., JINST **9** (2014) C03045
[12] S. Aiello et al., JINST **8** (2013) P07001 (erratum published in JINST **8**, E08001 (2013))
[13] C.A. Nicolau, for the NEMO Coll., Nucl. Instr. Meth. A **567** (2006) 552
[14] M.G. Pellegriti et al., Proc. of the VLVnT13, Stockholm, Sweden, August, 2013
[15] C. Pellegrino et al., AIP Conf. Proc. **1630** (2014) 158
[16] S. Aiello et al., Astropart. Phys. **66** (2015) 1
[17] S. Adrián-Martínez et al., ApJ **760** (2012) 53
[18] S. Aiello et al., Astropart. Phys. **28** (2007) 1
[19] S. Viola et al., Proc. of the VLVnT13, Stockholm, Sweden, August, 2013
[20] G. D'Agostini, Nucl. Instr. Meth. A **362** (1995) 487
[21] P. Lipari, Astropart. Phys. **1** (1993) 195
[22] J. Babson et al., Phys. Rev. D **42** (1990) 3613
[23] I.A. Belolaptikov et al., Astropart. Phys. **7** (1997) 263
[24] E. G. Anassontzis et al., in Proc. of the 23rd ICRC, Calgary, Canada, August, 1993; G. Aggouras et al., Astropart. Phys. **23** (2005) 377
[25] P. Desiati for the AMANDA Coll. and K. Bland, in Proc. of the 28th ICRC, Tsukuba, Japan, July 31–August 7, 2003
[26] J.A. Aguilar et al. Astropart. Phys. **34** (2010) 179
[27] E. Bugaev et al., Phys. Rev. D **58** (1998) 05401

EPJ Web of Conferences **121**, 05016 (2016) DOI: 10.1051/epjconf/201612105016
RICAP-2014

The design of the optical modules of the KM3NeT-Italia project towers

Emanuele Leonora[a], Sebastiano Aiello, and Valentina Giordano on behalf of the KM3NeT-Italia project

INFN, Sezione di Catania, via S. Sofia 64, 95125 Catania, Italy

Abstract. The KM3NeT-Italia project aims to construct a large volume underwater neutrino telescope, to be installed in the depths of the Mediterranean Sea. The R&D and mass production phases of the detection elements of the telescope, the optical modules, were entirely performed in the INFN-LNS site in the harbour of Catania. In November 2014 a first tower of 14 storeys equipped with 84 optical modules was successfully deployed in the Mediterranean Sea site. The design of the optical modules and their main components are described in this paper.

1. Introduction

The main goal of an high-energy neutrino telescopes is the study of the Universe by observing extremely high-energy neutrinos (10^{11}–10^{16} eV). The most suitable technique to perform such measurement is to instrument large volumes of water in order to detect the Cherenkov radiation emitted by the secondary muons and hadrons produced in the neutrino interactions with the matter surrounding the detector. This faint light can be detected by means of a three-dimensional array of optical modules (OMs), systems composed by high-sensitive photodetectors integrated into transparent pressure-resistant vessels. The KM3NeT-Italia project, PON 2007–2013 by the Italian Ministry of Education, University and Research (MIUR), aims to deploy a deep sea neutrino detector off shore the Mediterranean coast of Sicily at about 3500 m of depth, connected to the ground station in the harbour of Capo Passero through a 100-km electro-optical cable. The detector will be composed of 8 towers spaced by a mean distance of 120 m. Each tower has 14 storeys, 8m-long, interleaved vertically by 20 m plus a tower base module. In each storey will be hosted 6 optical modules, tilted vertically and horizontally at each extremity and in its centre 45° downwards. In November 2014 a first tower of 14 storeys equipped with 84 OMs was successfully deployed in the Mediterranean Sea site, at a depth of over 3500 meter, and now it is running in data taking. The optical modules R&D and production phases were entirely performed in the INFN-LNS site in the harbour of Catania. The design and the main components of the optical modules are described in this paper.

2. The optical module main components

The key detection element of an underwater Cherenkov detector is then the “optical module”. The mean intensity of the Cherenkov light signal, which arrives at each OM in water, could

[a]Corresponding author: emanuele.leonora@ct.infn.it

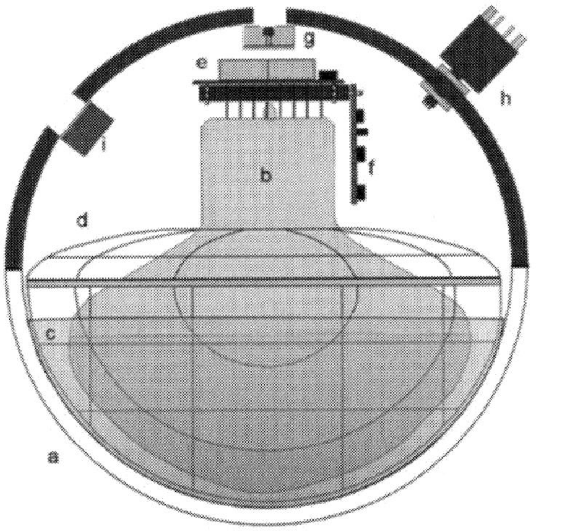

Figure 1. A schematic drawing of the optical module components: (a) 13-inch glass sphere; (b) 10-inch photomultiplier; (c)optical gel; (d) mu-metal cage; (e) base voltage supply; (f) Front End Module (FEM) board; (g) LED beacon system; (h) 8-pin connector; (i) manometer.

be low as few photons, and so optimization of the photon sensitive is therefore mandatory. Among the main requirements, the optical coupling between the water and the photodetector has to be optimized, the influence of the Earth's magnetic field must be minimized, the electronic power supply board and the electronic front-end must be installed inside the OM vessel and all the components must be highly reliable. Indeed, the neutrino telescope lifetime must be higher than 10 years.

Figure 1 shows schematically the main components of the optical module, designed following the experience of the successfully phase 1 and phase 2 of the NEMO project [1, 2].

Each optical module is hence composed by a transparent pressure-resistant vessel, which works as water and pressure protection ensuring a good light transmission. The vessel must have capability to withstand hydrostatic pressure, up to 400 atm, and a transparency to photons in the 350–600 nm wavelength range, with a refractive index that ensures a good optical matching between the sea water and the photodetector hosted inside. A 13-inch spherical deep-sea vessel made in borosilicate glass, produced by Benthos® [3] was selected. Each glass sphere is composed by one transparent hemi-spherical halve that hosts the photodetector, and by a second halve painted in black to stop the photons impinging from the back. Into the back black hemisphere a pressure gauge is glued to monitor the pressure inside the OM and a 8-pin SEACON® feedthrough is also mounted, which provides electrical connections (see Fig. 1). The selected photodetector was a Hamamatsu [4] large area photomultiplier (PMT) type R7081. It has 10-stages and a 10-inch bialkali photocathode with a typical surface around 500 cm^2 and a quantum efficiency of about 25% at 400 nm wavelength. This photomultipler was selected for its low dark count rate, typically less than 3 kHz at a threshold 1/3 of single photoelectron threshold, good time resolution, with a transit time of about 3 ns as FWHM and for its good charge properties, such as the high gain, over 5×10^6 with a voltage less than 1800 volt, a peak to valley ratio over 2 and a charge resolution of about 40% as sigma.

The high voltage system is a PHQ7081-i-2m integrated active board produced by ISEG® [5]. Damping resistors properly designed were added to the last output stages to minimize the ringing effect on the anode signal. The Cockroft-Walton scheme requires only a low voltage supply (+ 5 Volt), with the advantages of reducing power consumption and making very compact the design.

As confirmed by several studies [6], the performance of a large area PMT are significantly affected by its orientation with respect to the Earth's magnetic field. Therefore, the use of a magnetic shield is mandatory. The passive magnetic shield adopted is a cage made of 1 mm-diameter mu-metal wire, a nickel-iron alloy with high magnetic permeability. The cage, produced by ITEP [7], is composed of two parts. A hemispherical part of 30 cm-diameter and 14 cm-height that surrounds the entire photocathode area, and a flat part of a 30 cm-diameter disk with a hole in its centre through which the PMT neck could fit. The scale of the grid is 68×68 mm^2, giving a shadow effect on the photocathode smaller than 4% and a reduction factor of the magnetic field inside the cage of about 4.

Into the transparent glass hemisphere, an optical silicon gel creates the optical coupling between the photocathode of the photomultiplier and the glass sphere, while ensuring the mechanical assembly of the PMT with the glass sphere and the mu-metal cage. The selected material is a two-component (A and B) silicone gel, WACKER® [8] SilGel 612. Taking into account the results obtained from measurements of the optical and the mechanical properties, the 60B/100A mixture was selected.

Inside the optical module, close to the PMT, is placed the Front End Module board (FEM). This is a 10×10 cm^2 square printed circuit board that digitizes the PMT analog signals, encodes and transmits the data to the Floor Control Module that is located on each floor of the tower.

The time calibration of the detector is performed by means of a system of optical beacons, i.e. a series of pulsed light sources placed inside the optical modules and distributed throughout the detector to illuminate large groups of optical modules in the sea water. These well-controlled light pulses emitted in water allow the time response calibration of the optical modules as well as studying the sea water properties. To this aim, each of the two vertically downward-oriented optical modules on each storey of the towers hosts a so-called LED beacon system, which emits a series of light pulses outside the optical modules. The 470 nm LED and the pulse generator circuit, driven by the FEM board, are mounted vertically on a small black PVC structure, which is glued at the centre of the upper black hemisphere, directly on the inner surface of the glass, into a proper small window where the black paint was removed.

3. Conclusions

A 13-inch diameter optical module with a large 10-inch PMT was designed as the detection element of a large volume underwater neutrino telescope, to be installed in the depths of the Mediterranean Sea in the frame of the KM3NeT-Italia project. The whole optical module design, as well as each single component, was chosen after an intense R&D work on photomultipliers, high voltage supply circuit, optical gel, mechanical supports and effects of the Earth's magnetic field. Following a well-established production procedure, almost 700 optical modules were assembled at the INFN-LNS integration workshop located at the Harbour of Catania, and their functionality tested from electrical and mechanical point of view by using properly test-benches and a hyperbaric chamber.

References

[1] S. Aiello, E. Leonora et al., JINST **8**, P07001 (2013)
[2] S. Aiello et al., Astroparticle Physics **66**, 1–7 (2015)

[3] Teledyne Benthos Glass Sphere, `https://teledynebenthos.com`
[4] ISEG Bautzer Landstr. 23, 01454 Radeberg / OT Rossendorf, Germany
[5] Hamamatsu-Photonics, `http://www.hamamatsu.com`
[6] S. Aiello, E. Leonora, et al., IEEE Trans. Nucl. Sci. **59** (2012) 1259
[7] Institute for Theoretical and Experimental Physics, Moscow, `http://www.itep.ru`
[8] Wacker Chemie AG Hanns-Platz 4, 81737 München, Germany

EPJ Web of Conferences **121**, 05017 (2016) DOI: 10.1051/epjconf/201612105017
RICAP-2014

Study on 3-inch Hamamatsu photomultipliers

Valentina Giordano[a], Sebastiano Aiello, and Emanuele Leonora on behalf of the KM3NeT Collaboration

INFN Sezione di Catania, via Santa Sofia 64, 95125 Catania, Italy

Abstract. Several kinds of photomultipliers are widely used in astroparticle physics detectors to measure Cherenkov light in media like water or ice. In neutrino telescopes the key element of the detector is the optical module, which consists of one or more photodetectors inside a transparent pressure-resistant glass sphere. It serves as mechanical protection while ensuring good light transmission. The KM3NeT collaboration has developed an innovative design of an optical module composed by 31 photomultipliers (PMTs) of 3-inch diameter housed in a 17-inch glass shpere. The performance of the telescope is largely dependent on the presence on noise pulses present on the anode of the photomultipliers. A study was conducted of noise pulses of Hamamatsu 3-inch diameter photomultipliers measuring time and charge distributions of dark pulses, pre-pulses, delayed pulses and after-pulses, focusing in particular on analysis on multiple afterpulses. Effects of the Earth's magnetic field on 3-inch PMTs were also studied.

1. Introduction

The KM3NeT Observatory [1] is a large scale neutrino telescope to be built in the deep waters of the Mediterranean Sea. With several cubic kilometres instrumented with thousands of optical sensors, KM3NeT will be the largest and most sensitive high energy neutrino detector. In neutrino telescopes the key element of the detector is the optical module (OM), which consists of one or more photomultipliers (PMTs) stored inside a transparent pressure-resistant glass sphere that serves as mechanical protection while ensuring good light transmission. The final configuration of KM3NeT neutrino telescope in Italian site of Capopassero will be hybrid, with 8 towers instrumented with large area PMTs of 10-inch diameter and 24 strings instrumented with 3-inch diameter PMTs. Concerning the strings; the KM3NeT collaboration has developed an innovative design of an optical module composed by 31 PMTs of 3-inch diameter and the readout electronics housed together in a 17-inch glass sphere looking downwards and upwards. This is called Digital Optical Module (DOM). After the successful experience of the NEMO project, in November 2014 the first KM3NeT tower with large area PMTs has been deployed in the test-site at 3500 m depth offshore the coast of Portopalo di Capopassero in Sicily (Italy) and this is now in taking data. The performance of the telescope is largely dependent on the main properties of the PMTs and on the presence of noise pulses on their anode. In order to investigate the DOM performance, a complete study was conducted on noise pulses of 3-inch diameter 10-stages R12199-02 bialkali photocathode PMTs produced by Hamamatsu. We have measured time and charge distributions of dark pulses, pre-pulses, delayed pulses and afterpulses, focusing also on the analysis of multiple afterpulses. For all projects where the orientation of the PMTs is critical, the variations in

[a] e-mail: `valentina.giordano@ct.infn.it`

their performance due to Earth's magnetic field must be investigated. Indeed the performance of a PMT could be subject to significant variation due to magnetic fields. Detection efficiency, Transit time spread (TTS) and gain were measured simultaneously while varying the PMT's orientations with respect to the Earth's magnetic field, both with and without a mu-metal cage as magnetic shielding.

2. Noise and spurious pulses measurements

Noise pulses can be defined as anode output pulses that are not the response of the photomultiplier to a light event detected by the photocathode. They often disturb accurate measurements of low-level signals, degrade energy measurements and causes errors in pulse counting applications. We will distinguish dark pulses from spurious pulses. Dark pulses are random pulses that can be measured at the anode of a photomultiplier even in total darkness.

Spurious pulses are instead noise pulses time-correlated with the main PMT response to detected light events, because they can be early or delayed by a characteristic time with respect to the electron transit time through the PMT. Spurious pulses are the pre-pulses (a pulses in time that ranges between 80 ns and 10 ns before the main temporal position), the delayed pulses (pulses delayed in time from 10 ns to 80 ns after the main temporal position) and the afterpulses (noise pulses that appear following the main PMT response to a detected light event). Pre-pulses and delayed are present in place of the main pulse while afterpulses follow it.

The PMT under test was located in a light-tight dark box in order to shield it from the environmental light. This box was made in opaque black plexiglass, to avoid internal reflection of the light coming from the calibrated light source. This was a pulsed laser with a laser head of 410 nm and a width of 60 ps, attenuated to a single photoelectron (spe) condition, 3 photoelectron (pe) condition and 5 pe condition. The laser source was located outside the box, and the light pulses were guided inside the box by means of a multimode optical fiber coupled to the laser head. All measurements were carried out at room temperature and atmospheric pressure, with the PMT powered by a passive base made by the ECAP-Erlangen team of KM3NeT collaboration. In order to set the signals and acquire the data, NIM electronics, time to amplitude converters and analog charge amplitude converter were used. A digital oscilloscope was also used to digitize the PMT signals during multiple after pulses measurements. The dark count rate was measured using a NIM counter, with a threshold set to 1/3 of the spe amplitude. The first batch of 20 PMTs under study shows the following characteristics. Without laser, the dark rate counts found was about 1 kHz after 12 hours in the dark. In Fig. 1(a) the typical charge spectrum distribution for a 3-inch PMT in spe condition at a gain of $5 \cdot 10^6$ is shown. The first peak is the pedestal due to electronic noise without signal, while the second one is the spe peak. The peak to valley ratio P/V was measured as the ratio of the number of events of the spe peak and the number of events of the region of the valley in each spe charge spectrum and has a mean value of about 3 for the first batch of PMTs tested. The time distributions show a median value of 4 ns (full width half maximum FWHM) in TTS in spe (blue line on Fig. 1(b)) and typical fractions on spurious pulses of 0.1% for pre-pulses, 5% for delayed and 6% of afterpulses. Increasing the laser intensity in order to change the PMT response condition from 1 pe to 3 pe and 5 pe the TTS decreased from 4 ns to 3 ns and 2.8 ns as seen in Fig. 1(b). Moreover fractions of pre-pulses and delayed pulses became lower as shown on Fig. 1(b): indeed pre-pulses vary from 0.3% to 0.01% until 0.002% going from 1 pe condition to 3 pe and 5 pe condition. The delayed pulses vary from 6.40% to 1.14% until 0.47% going from 1 pe condition to 3 pe and 5 pe condition. Concerning afterpulses (AP), we have to distinguish the first type of

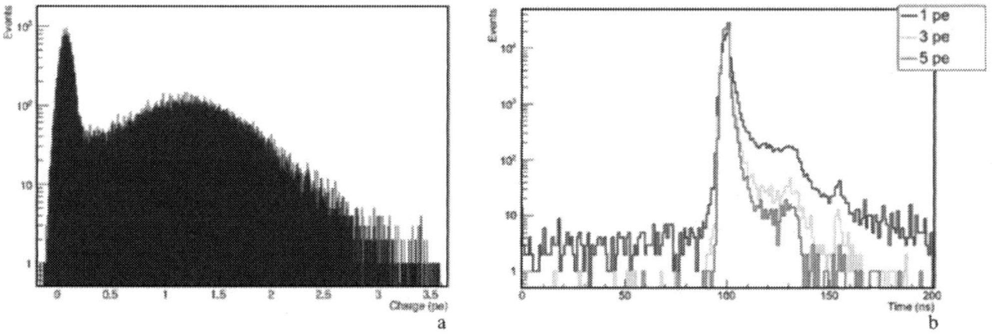

Figure 1. (a) Charge spectrum of a 3-inch PMT at a gain of $5 \cdot 10^6$. (b) Transit Time Distribution at different PMT response condition: blue line for spe, green line for 3 pe and red line for 5 pe. Main pulse is at about 100 ns; pre-pulses ranges from 20 ns until 90 ns a nd delayed from 110 ns to 190 ns.

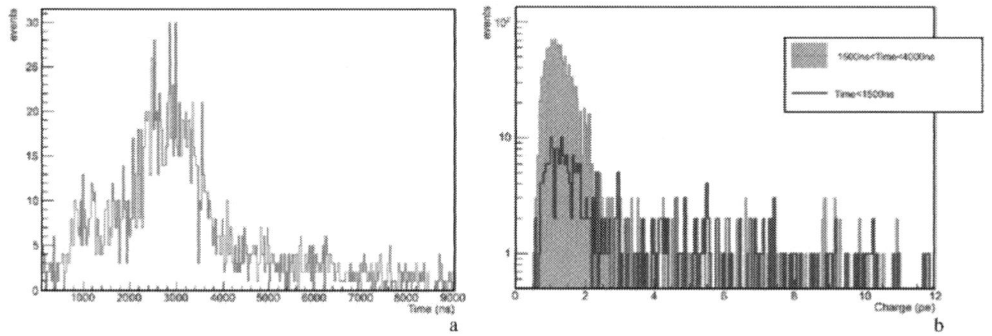

Figure 2. (a) Typical time distribution of type 2 after pulses. (b) Charge distribution of after pulses for two different time ranges. In magenta area there is CH_4 (arrival time lower than 1.5μs) while in blue there is Cs (with arrival time between 1.5 μs and 4 μs).

afterpulses (AP1) that ranges from 10 to 80 ns after the main pulses and the afterpulses type 2 (AP2) that ranges from 80 ns until 5 μs after the main pulse. For the first batch of PMTs, we measured a mean value of 0.1% of AP1 and 4% of AP2. AP1 are mainly of spe in the charge spectrum. AP2 can consist of several pulses. The majority of AP2 events consist of single pulse with a fraction of 3.81% and a mean charge per pulse of 1.2 pe. We measured 2 AP2 with a percentage of 0.25%, 3 AP2 with 0.07% and 4 AP2 with 0.02%. For multiple AP2 the mean charge per pulse is about 1.2 pe. The mean charge per AP2 is lower than that found for large area 10-inch PMTs used for KM3NeT tower construction [2]. The AP2 are due mainly to residual gases during the construction of the PMT. Looking at the time distribution in Fig. 2(a) we can see that there are two major components that are associated to CH_4^+ ions (at 1 μs) and Cs^+ ions (at 3 μs). In 10-inch PMTs we pointed out that the CH_4^+ peak had more charge per AP2 in confront of the Cs^+ peak [2]. For 3-inch PMTs this effect is present, even if less highlighted, as shown on bidimensional plot on Fig. 2(b) where Cs peak is mainly of spe instead of CH_4 peak that has an higher charge component.

3. Influence of the Earth's magnetic field

Laboratory tests have shown that an ambient magnetic field with a value of 44 μTesla (the Mediterranean Sea zone) could degrade significantly timing and charge characteristics on

Table 1. Gain Measurements.

Gain	$\Phi = 90°$		$\Phi = -50°$		$\Phi = 50°$	
	naked	shielded	naked	shielded	naked	shielded
Minimum Value [10^6]	4.59	4.75	4.67	4.77	4.63	4.77
Maximum Value [10^6]	4.72	4.83	4.91	4.86	4.80	4.90
Average Value [10^6]	4.67	4.78	4.83	4.81	4.72	4.82
Maximum Variation [%]	2.98	1.62	5.10	1.79	4.10	2.83

Table 2. Relative Detection Efficiency Measurements.

Rel. Detection Eff.	$\Phi = 90°$		$\Phi = -50°$		$\Phi = 50°$	
	naked	shielded	naked	shielded	naked	shielded
Minimum Value [10^{-3}]	11.86	12.34	12.34	12.34	12.34	11.90
Maximum Value [10^{-3}]	13.28	13.17	15.22	13.17	13.11	13.20
Average Value [10^{-3}]	12.58	12.36	14.17	12.67	12.35	12.48
Maximum Variation [%]	11.93	6.76	23.30	6.75	11.90	11.10

Table 3. Transit Time Spread Measurements.

TTS FWHM (ns)	$\Phi = 90°$		$\Phi = -50°$		$\Phi = 50°$	
	naked	shielded	naked	shielded	naked	shielded
Minimum Value (ns)	3.638	3.550	3.632	3.525	3.578	3.492
Maximum Value (ns)	3.785	3.610	3.753	3.577	3.866	3.667
Average Value (ns)	3.725	3.576	3.713	3.550	3.708	3.591
Maximum Variation [%]	4.04	1.69	3.30	1.49	8.10	5.01

large area PMTs [3]. Magnetic shielding is thus largely used to reduce magnetic effects and make the response of the PMT sufficiently orientation independent. Since PMTs are installed into an underwater neutrino telescope, they can change their orientation because of the movements of the detector structure due to sea currents so the influence of Earth's magnetic field should be investigated. We tested two 3-inch PMTs R12199-02. In order to characterize the PMT response to an injected light source while varying its orientation relative to the Earth's magnetic field, a light-tight dark box able to rotate with respect to vertical axes (step of 1°) and to change its inclination (step of 10°) was constructed.It is possible to shield a 3-inch PMTs against the magnetic field using a passive magnetic shield. The shield studied in this work was a wire cage, made of 1 mm diameter wire of mu-metal, a nickel-iron alloy with very high magnetic permeability 10^5. The used cage was composed of two parts. A hemispherical part which surrounds the entire PMT and a flat part. The pitch of the grid was $(68 \cdot 68)$ mm, giving a shadow effect on the photocathode less than 1%. The reduction factor of the magnetic field inside the cage is about 4. In order to compare results obtained in the different photomultipliers, each PMT started its rotation from the same position with respect to the box and to the Earth's magnetic field.

Both PMTs were measured in three inclinations of the box: horizontally oriented ($\Phi = 90°$), 50° downwards and 50° upwards. For each inclination, the PMT under test was rotated of 360° in step of 45°. All measurements were made first with the PMTs un-shieleded and then repeated with the mu-metal magnetic shield. The Tables 1, 2 and 3 summarize the main measurements. For all sets of measured parameters the minimum, maximum, average values and maximum variation $((max.value - minvalue)/min.value)$ are given for the unshielded and shielded condition of the PMT. Gain and TTS measurements in Tables 1 and 3 don't show high variations in the two PMTs while the detection efficiency (defined as the ratio between the number of detected pulses and those emitted by the laser) in Table 2 seems to be a parameter with higher variation between shielded and un-shielded condition.

The same measurements were performed on 10-inch PMTs used for KM3NeT tower and the larger nvariations from un-shielded to shielded condition, required to use the mu-metal cage in the optical module construction [3]. For the DOM this constraint does not seem to be so crucial since the magnetic effects are lower.

4. Conclusion

A large study was conducted on 3-inch bialkali photocathode R12209-02 PMTs produced by Hamamatsu in order to measure time and charge distributions of noise pulses and the variations of the PMTs performances with the Eart's magnetic field. In the study of spurious pulses the laser was regulated at different response of the PMT: spe, 3 pe and 5 pe. Concerning dark pulses, they correspond mainly to single photoelectron pulses with typical rates of 1 kHz. Pre-pulses have a random distribution before the main pulses, in a range of 10–80 ns, and they have an average charge per pulse lower than that of the main PMT response. Their fraction is usually very small, less than 1% in spe and this fraction decreases when the number of photoelectrons per light pulse increases. The average charge per pulse measured corresponds mainly to single photoelectrons in 1 pe condition and about 2 pe and 4 pe in 3 pe and 5 pe PMT conditions. Their typical fraction is about 6%, and this fraction decreases when the number of photoelectrons per light pulse increases. The measurements of afterpulses show fractions of about 0.11% for type 1 after pulses and of 4% for type 2. The time distribution of type 2 after pulses shows a peak at around 1 μs presumably produced by CH_4 ions and another peak around 3 μs presumably produced by Cs ions. The last measurements done on 3-inch PMTs deal with the influences of the Earth's magnetic field. The maximum variations are not so high for TTS (8.1%) and gain (4.1%) changing from shielded to un-shielded measurements, but cause an higher effect on detection efficiency it causes an higher effect. Given the moderate size of performance variation, KM3NeT has chosen to not use the mu-metal shielding.

References

[1] KM3NeT Technical Design Report ISBN 978-90-6488-033-9
[2] E. Leonora et al., TNS-IEEE doi: 10.1109/TNS.2014.2322655 (2014)
[3] S. Aiello et al., TNS-IEEE doi: 10.1109/TNS.2012.2189245 (2012)

EPJ Web of Conferences **121**, 06001 (2016) DOI: 10.1051/epjconf/201612106001
RICAP-2014

Direct detection of Dark Matter

P. Belli[1,2]

[1] Dip. di Fisica, Università di Roma "Tor Vergata", 00133 Rome, Italy
[2] INFN, sez. Roma "Tor Vergata", 00133 Rome, Italy

Abstract. An overview of the latest results of Dark Matter direct detection will be summarized, with particular care to the DAMA/LIBRA-phase1 results and the evidence with high confidence level obtained by exploiting the model independent Dark Matter annual modulation signature for the presence of Dark Matter particles in the galactic halo. Results from other experiments using different procedures, different techniques and different target-materials will be shortly discussed. Results, implications and experimental perspectives will be addressed.

1. Introduction

Nearly a century of experimental observations and theoretical arguments has pointed out that a large fraction of the Universe is composed by Dark Matter Particles and Dark Energy. Considering all the observational data coming from the study of the CMB, of the Supernovae Ia, of the Baryonic Acoustic Oscillations (BAO) and of the large-scale structures, the following contributions to Ω are obtained[1, 2]: i) $\Omega_r \approx 5 \times 10^{-5}$ for the radiation density; ii) $\Omega_b \approx 0.05$ for the baryonic matter; iii) $\Omega_{dm} \approx 0.27$ for the non baryonic Dark Matter; iv) $\Omega_\Lambda \approx 0.68$ for the Dark Energy. A value $\Omega_b \approx 0.05$ for the baryonic matter is also supported by the Big-Bang nucleosynthesis (BBN), that is based on the predictions of the abundances of the light elements. This value is much larger than the cosmic density of the luminous matter, $\Omega_{lum} \approx 0.004$ [3], crediting that most baryons are dark and, probably, in the form of diffuse intergalactic medium [4].

The values of the cosmological parameters support that most of the matter in the Universe has a non baryonic nature.

As regards the light massive neutrino DM candidate, its density can be strongly constrained by cosmology; in fact, a value above the limit $\Omega_\nu \approx 0.03$ gives an unacceptable lack of small-scale structure [5, 6]. In addition, a pure light massive neutrino scenario is ruled out by the measurements of the CMB radiation, which does not show sufficiently large inhomogeneity. Therefore, a significant role should be played by non-baryonic relic particles from the Big Bang, outside the Standard Model of particle physics; they have to be stable or with a lifetime comparable with the age of the Universe to survive up to now in a significant amount.

In theories extending the Standard Model of particle physics, many candidates as Dark Matter particles have been proposed having different nature and interaction types. Among the many DM candidates there are: SUSY particles (as e.g. neutralino or sneutrino in various scenarios), inelastic Dark Matter in various scenarios, electron interacting Dark Matter, a heavy neutrino of the 4-th family, sterile neutrino, Kaluza-Klein particles, self-interacting Dark Matter, axion-like (light pseudoscalar and scalar candidate), mirror Dark Matter in various scenarios, Resonant Dark Matter, DM from exotic 4th generation quarks,

Elementary Black holes, Planckian objects, Daemons, Composite DM, Light scalar *WIMP* through Higgs portal, Complex Scalar Dark Matter, specific two Higgs doublet models, exothermic DM, Secluded *WIMPs*, Asymmetric DM, Isospin-Violating Dark Matter, Singlet DM, Specific GU, SuperWIMPs, WIMPzilla, etc.. Moreover, even a suitable particle not yet foreseen by theories could be the solution or one of the solutions; in fact, considering the richness in particles of the visible matter which is less than 1% of the Universe density, one could also expect that the particle component of the Dark Matter in the Universe may be multicomponent.

Depending on the DM candidate, the interaction processes can be various, as e.g.: 1) elastic scatterings on target nuclei with either spin-independent or spin-dependent or mixed coupling; moreover, an additional electromagnetic contribution can arise, in case of few GeV candidates, from the excitation of bound electrons by the recoiling nuclei [7]; 2) inelastic scatterings on target nuclei with either spin-independent or spin-dependent or mixed coupling in various scenarios [8–13]; 3) interaction of light DM (LDM) either on electrons or on nuclei with production of a lighter particle [14]; 4) preferred interaction with electrons [15]; 5) conversion of DM particles into electromagnetic radiation [16]; 6) etc.. Often, the elastic scattering on target nuclei is the considered interaction process, but other processes are possible and considered in literature, as those aforementioned where also electromagnetic radiation is produced. Hence, considering the richness of particle possibilities and the existing uncertainties on related astrophysical (e.g. halo model and related parameters, etc.), nuclear (e.g. form factors, spin factors, scaling laws, etc.) and particle physics (e.g. particle nature and interaction types, etc.), a widely-sensitive model independent approach is mandatory. Most of the activities in the field are instead based on a particular *a priori* assumption on the nature of the DM particle and of its interaction, in order to try to overcome the limitation arising from their generally large originally measured counting rate.

The expected energy distribution for the interactions of Dark Matter particles in a terrestrial detector depends – among others – on their density and velocity distribution at Earth's position. However, the experimental observations regarding the dark halo of our Galaxy do not allow one to get information on this crucial aspect without introducing a model for the Galaxy matter density. Because of its simplicity, the isothermal sphere model (which consists in a spherical infinite system with a flat rotational curve) is a widely used assumption for the DM density distribution, and thus in the evaluation of Dark Matter expected rates. However many of its underlying assumptions (sphericity of the halo, absence of rotation, isotropy of the dispersion tensor, flatness of the rotational curve) are not strongly constrained by astrophysical observations. Moreover, the isothermal sphere is strictly unphysical and may only represent the behavior of the inner part of physical systems, since it has a total infinite mass and needs some cutoff at large radii. Thus, the use of more realistic halo models is mandatory in the interpretation and comparison procedures, since the model dependent results can significantly vary.

Generally, the halo models can be grouped in the following classes: i) spherically symmetric matter density with isotropic velocity dispersion; ii) spherically symmetric matter density with non-isotropic velocity dispersion; iii) axisymmetric models; iv) triaxial models; v) axisymmetric models either with a halo co-rotation or a halo counter-rotation. The possible rotation can be considered also for the other kinds of halo. A parameterization of these classes of halo models has been given in Ref. [17, 18], taking into account the available observational data; anyhow, information on the galactic dark halo can be obtained only in indirect way considering some hypotheses on its form and characteristic.

In conclusion, the uncertainties still present on the shape of the DM halo and on the density and velocity distribution prevent the definition of a "standard" halo and illustrate how the comparisons among the experiments of direct detection of DM particles (see later) can be consistent even just considering this particular aspect (also see Ref. [19]). Moreover, many other experimental and theoretical uncertainties exist and must be considered in a suitable comparison among the experiments of direct detection of DM particles.

2. The Dark Matter particles detection

It is worth noting that experiments at accelerators may prove – when they can state a solid model independent result – the existence of some possible DM candidate particles, but they could never credit by themselves that a certain particle is a/the only solution for Dark Matter particle(s). Moreover, DM candidate particles and scenarios (even e.g. in the case of the neutralino candidate) exist which cannot be investigated at accelerators. Thus, a model independent approach, a ultra-low-background suitable target material, a very large exposure and the full control of running conditions are mandatory to pursue a widely sensitive direct detection of DM particles in the galactic halo.

Considering the many available candidate particles and scenarios and the existing uncertainties on the astrophysical (e.g. halo model and related parameters, etc.), nuclear (e.g. form factors, spin factors, scaling laws, etc.) and particle physics (e.g. particle nature and interaction types, etc.), a widely-sensitive model independent approach is mandatory as well as a suitable exposure, and full control of the running condition over the whole data taking.

Indeed, most activities in the field release marginal exposures even after many years underground; they do not offer suitable information e.g. about operational stability and procedures during the running periods, and base their analysis on a particular *a priori* assumption on the nature of the DM particle and its interaction, and of all the used parameters.

In particular, the applied rejection and subtraction procedures to reduce the experimental counting rate, in order to derive a set of recoil-like candidates, is pursued by experiments as CDMS, EDELWEISS, CRESST, XENON, LUX, etc.. It is worth noting that the applied subtraction procedures are statistical and cannot offer an unambiguous identification of the presence of DM particle elastic scatterings because of the known existing recoil-like indistinguishable background; tails of the subtracted populations can play a role as well. Finally, the electromagnetic component of the counting rate, statistically "rejected" in this approach, can contain either the signal or part of it, and it will be lost. In the following few experimental activities are mentioned as examples.

In the double read-out bolometric technique, the heat signal and the ionization signal are used in order to discriminate between electromagnetic and recoil-like events. This technique is used by CDMS and EDELWEISS collaborations. In particular, the CDMS-II detector consists of 19 Ge bolometers of about 230 g each one and of 11 Si bolometers of about 100 g each one. The experiment released data for an exposure of about 194 kg $\times$ day [20] using only 10 Ge detectors in the analysis (discarding all the data collected with the other ones) and considering selected time periods for each detector. EDELWEISS employs a target fiducial mass of about 2 kg of Ge and has released data for an exposure of 384 kg $\times$ day collected in two different periods (July-Nov 08 and April 09-May 10) [21] with a 17% reduction of exposure due to run selection. These two experiments claim to have an "event by event" discrimination between noise + electromagnetic background and recoil + recoil-like (neutrons, end-range alphas, fission fragments,...) events by comparing the bolometer and the ionizing signals for each event. Thus, their results are, actually, largely based on huge data selections, as for example, the time cut analysis used to remove the so-called surface electrons that are distributed in both the electromagnetic and recoil bands. The stability, the unlinear response and the robustness of the reconstruction procedure are key points, as well as the associated systematical errors. In these experiments few recoil-like events survive the many selections/subtractions cuts applied in the data analysis; these events are generally interpreted in terms of background. As regards, in particular, their application to the search for time dependence of the data (such as the annual modulation signature), it would require – among other – to face the objective difficulty to control all the operating conditions – at the needed level ($< 1\%$) – despite of the required periodical procedures e.g. for cooling and for radiation source introduction for calibration as well as of the limitation arising from the reachable duty cycle. For example, the attempt by CDMS-II to search for annual modulation in Ge target has been performed with a marginal exposure by using only 8 detectors over 30 and using – among others – data that are not continuous over the whole annual periods considered in the analysis [22]; the use of non-overlapping time periods, collected with detectors having different background rate within the signal box does not allow one to get any reliable result in the investigation of an effect at few percent level (see e.g. arguments in [23]).

Recently, the results of CDMS-II with the Si detectors were published in two close-in-time data releases [24, 25]; while no events in six detectors (corresponding exposure of only 55.9 kg$\times$day before analysis cuts) were reported in the former [24], three events in eight detectors (corresponding raw exposure of 140.2 kg$\times$day) were reported over the residual background, estimated after subtraction: $\simeq 0.4$ in the second one [25]. The latter result could be interpreted – under certain assumptions – in terms of a DM candidate with spin-independent interaction and a mass around 10 GeV, which is compatible with interpretations of the model independent DM annual modulation result already reported by DAMA in terms of this kind of DM candidate and with some other hints reported by CRESST and CoGeNT (see later).

The XENON project uses instead dual phase liquid/gas detectors. Experiments exploiting such technique (as also LUX, DARKSIDE) perform statistical discrimination between nuclear recoil-like candidates and electromagnetic component of the measured counting rate through the ratio of the prompt scintillation signal ($S1$) and the delayed signal ($S2$) due to drifted electrons in the gaseous phase. The XENON100 experiment has released data for an exposure of 224.6 days, using a fiducial volume of just 34 kg of Xenon target mass [26]. See related discussions in literature about the detector response of such devices, in particular, to low energy recoils [27, 28]. The technical performance of the apparata, confirmed also by similar experiments, has shown e.g. that: i) the detectors are affected by large non-uniformity; some kind of corrections may be estimated and applied, but significant systematics has to be accounted for; ii) the response of these detectors is not linear, i.e. the number of photoelectrons/keV depends on the energy scale and depends also on the applied electric field; iii) the physical energy threshold is not suitably proved iv) the use of energy calibration lines from Xe activated by neutrons cannot be applied as routine and the studies on a possible calibration with internal sources in the same running conditions have not been realized so far; v) despite of the small light response (2.28 photoelectron/keVee), an energy threshold at 1.3 keVee is claimed; vi) the energy resolution is poor; vii) in the scale-up of the detectors the performances deteriorate; viii) the behaviour of the light yield for recoils at low energy is uncertain in every case. Recently, LUX reports the first results corresponding to an exposure of 85.3 days, using a fiducial volume of 118 kg [29]; similar considerations, as above, hold.

The CRESST experiment exploits the double read-out bolometric technique, using the heat signal due to an interacting particle in the $CaWO_4$ crystals and the scintillation light produced. A statistical discrimination of nuclear recoil-like events from electromagnetic radiation is performed. As regards the cuts and selection procedures applied, most of the above discussion also holds. A previous run (8 detectors of 300 g each one, for an exposure of about 730 kg $\times$ day [30]) showed, after selections, 67 nuclear recoil-like events have been observed in the Oxygen band. The background contribution estimated by authors ranges from 37 to 43 events, and does not account for all the observed events. The remaining number of events and their energy distribution could be interpreted – under certain assumptions – in terms of a DM candidate with spin-independent interaction and a mass in the range of 10–30 GeV. This result has been not confirmed in the last run [31], where a more marginal exposure has been used, confirming the difficulties to manage the systematics in such experiments.

Another positive hint for a signal of light Dark Matter candidates inducing just nuclear elastic scatterings have been reported by the CoGeNT experiment [32, 33]. The set-up is composed by a 440 g, p-type point contact (PPC) Ge diode, with a very low energy threshold at 0.4 keVee. It is located in the Soudan Underground Laboratory. In the data analysis no discrimination between electromagnetic radiation and nuclear recoils is applied; only noise events are rejected. The experiment observes more events than they expect from an estimate of the background in the energy range 0.4–3.2 keVee. The energy spectrum of these events is compatible – under certain assumptions – with a signal produced by the interaction of a DM particle with a mass around 10 GeV. In addition, considering an exposure of 146 kg $\times$ days CoGeNT experiment also reports an evidence at about 2.2σ C.L. of an annual modulation of the counting rate in (0.5–2) keVee with phase and period compatible – although the small confidence level – with a Dark Matter signal [33]. This result is compatible with interpretations of the DM model-independent annual modulation result already reported by DAMA in terms of this kind of DM candidate and with the possible hints reported above.

In conclusion, suitable experiments offering a model independent signature for the presence of Dark Matter particles in the galactic halo are mandatory.

2.1 DM model independent signature and DAMA results

To obtain a reliable signature for the presence of DM particles in the galactic halo, it is necessary to exploit a suitable model independent signature. With the present technology, one feasible and able to test a large range of cross sections and of DM particle halo densities, is the so-called DM annual modulation signature [34]. The annual modulation of the signal rate originates from the Earth revolution around the Sun. In fact, as a consequence of its annual revolution around the Sun, which is moving in the Galaxy traveling with respect to the Local Standard of Rest towards the star Vega near the constellation of Hercules, the Earth should be crossed by a larger flux of Dark Matter particles around $\sim$2 June (when the Earth orbital velocity is summed to the one of the solar system with respect to the Galaxy) and by a smaller one around $\sim$2 December (when the two velocities are subtracted). Thus, this signature has a different origin and peculiarities than the seasons on the Earth and than effects correlated with seasons (consider the expected value of the phase as well as the other requirements listed below). This DM annual modulation signature is very distinctive since the effect induced by DM particles must simultaneously satisfy all the following requirements: (1) the rate must contain a component modulated according to a cosine function; (2) with one year period; (3) with a phase that peaks roughly around $\sim$ 2nd June; (4) this modulation must be present only in a well-defined low energy range, where DM particles can induce signals; (5) it must be present only in those events where just a single detector, among all the available ones in the used set-up, actually "fires" (*single-hit* events), since the probability that DM particles experience multiple interactions is negligible; (6) the modulation amplitude in the region of maximal sensitivity has to be $\lesssim 7\%$ in case of usually adopted halo distributions, but it may be significantly larger in case of some particular scenarios such as e.g. those in refs. [12, 35]

At present status of technology it is the only DM model independent signature available in direct Dark Matter investigation that can be effectively exploited.

This signature has been exploited with large exposure – using highly radiopure NaI(Tl) as target material – by the former DAMA/NaI ($\simeq$ 100 kg sensitive mass) experiment and the present DAMA/LIBRA ($\simeq$ 250 kg sensitive mass), within the DAMA project [14–16, 18, 36–55].

In particular, the experimental observable in DAMA experiments is the modulated component of the signal in NaI(Tl) target and not the constant part of it, as done in the other approaches aforementioned.

The whole DAMA/LIBRA-phase1 results correspond to seven annual cycles for an exposure of 1.04 ton$\times$yr [44–46]. Considering these data together with those previously collected by DAMA/NaI over 7 annual cycles (0.29 ton$\times$yr), the total exposure collected over 14 annual cycles is 1.33 ton$\times$yr; this is orders of magnitude larger than the exposures typically collected in the field.

The DAMA/NaI set up and its performances are described in [18, 37–39], while the DAMA/LIBRA set-up and its performances are described in Ref. [43, 45]. In particular, the sensitive part of the DAMA/LIBRA set-up is made of 25 highly radiopure NaI(Tl) crystal scintillators placed in a 5-rows by 5-columns matrix; each crystal is coupled to two low background photomultipliers working in coincidence at single photoelectron level. For details see Ref. [36, 43].

The DAMA/LIBRA–phase1 results give evidence for the presence of Dark Matter particles in the galactic halo, on the basis of the exploited model independent DM annual modulation signature by using highly radio-pure NaI(Tl) target, at 7.5 σ C.L. Including also the first generation DAMA/NaI experiment (cumulative exposure 1.33 ton $\times$ yr, corresponding to 14 annual cycles), the C.L. is 9.3 σ and the modulation amplitude of the *single-hit* events in the (2–6) keV energy interval in NaI(Tl) target is: (0.0112 ± 0.0012) cpd/kg/keV; the measured phase is (144 ± 7) days and the measured period is (0.998 ± 0.002) yr, values well in agreement with those expected for DM particles. No systematic or side reaction able to mimic the exploited DM signature has been found or suggested by anyone over more than a decade.

After a first upgrade in 2008, a further upgrade of DAMA/LIBRA has been performed at the end of 2010 when all the PMTs have been replaced with new ones having higher quantum efficiency [48]. Since then, after a test and optimization period, the DAMA/LIBRA–phase2 is continuously running in order: (1) to increase the experimental sensitivity lowering the software energy threshold of the experiment; (2) to improve the corollary investigation on the nature of the DM particle and related astrophysical, nuclear and particle physics arguments; (3) to investigate other signal features and second order effects. DAMA/LIBRA also continue its study on several other rare processes [51–55].

2.1.1 *Implications and comparisons*

The DM model independent evidence obtained by DAMA is compatible with a wide set of scenarios regarding the nature of the DM candidate and related astrophysical, nuclear and particle Physics. For example some given scenarios and parameters are discussed e.g. in refs. [14–16, 18, 39–42, 44, 50]. Further large literature is available on the topics [56]; other possibilities are open.

It is worth noting that no other experiment exists, whose result can be directly compared in a model-independent way with those by DAMA/NaI and DAMA/LIBRA. Some activities (e.g. [20, 21, 26]) claim a model-dependent exclusion under many largely arbitrary assumptions (see for example discussions in Refs. [18, 27, 28, 39, 44]). However, often some critical points exist in their experimental aspects, as mentioned above, and the existing experimental and theoretical uncertainties are generally not considered in their presented single model dependent result; moreover, implications of the DAMA results are often presented in incorrect/partial/unupdated way. Both the accounting of the existing uncertainties and the existence of alternative scenarios (see literature) allow one to note that model dependent reults by indirect and direct experiments actually are not in conflict with the DAMA model independent result.

3. Prospects for the DM directionality approach

The directionality approach – based on the study of the correlation between the recoil direction of the target nuclei and the Earth motion in the galactic rest frame – can offer a good approach to study those DM candidate particles able to induce just nuclear recoils. In fact, the dynamics of the rotation of the Milky Way galactic disc through the halo of DM causes the Earth to experience a wind of DM particles apparently flowing along a direction opposite to that of the solar motion relative to the DM halo. Hence, in the case of DM candidate particles interacting with nuclei the induced nuclear recoils are expected to be strongly correlated with the impinging direction of DM, while the background events are not; therefore, the study of the nuclear recoils direction can offer a way for pointing out the presence of these DM candidate particles.

In the practice, this approach has some technical difficulties because it is arduous to detect the short recoil track. Different techniques are under consideration but, up to now, they are at R&D stage and have not produced yet competitive results in the field (see e.g. the DRIFT project or the DM-TPC experiment). In fact, they are generally limited by the difficulty of detecting very short tracks and of achieving high stability, large sensitive volume and very good spatial resolution.

To overcome such a difficulty, it has been suggested the use of anisotropic scintillator detectors [57–59]; their use was proposed for the first time in [57] and revisited in [58]. In particular, low background $ZnWO_4$ crystal scintillators have been recently proposed since their features and performances are very promising [60]. In fact, both the light output and the scintillation pulse shape depend on the impinging direction of heavy particles (p, alpha, nuclear recoils, etc.) with respect to the crystal axes and can supply two independent ways to study the directionality and to discriminate the electromagnetic background (that does not give rise to any anisotropic effects).

Other advantages offered by $ZnWO_4$ detectors are very good radio-purity levels (about 0.1 cpd/kg/keV at low energy) and the possibility to reach energy thresholds at keV level. Both these features can also be improved (e.g., the light yield shows a significant enhancement when working at low temperatures – about 100 K – and better radiopurity levels can be reached with dedicated R&D). A detailed discussion can be found in [60].

It is also worth to mention the new idea of using DNA for nanometer tracking [61].

4. Conclusions and perspectives

Large efforts are in progress with different approaches and target materials to investigate various kinds of DM candidates and scenarios. Due to the difficulty of measuring at very low energy, several techniques still would require further work for results' qualifications before enlarging their target mass.

As regards the possibility to exploit the directionality for some DM candidates, new efforts have to be encouraged towards a first realistic exploitation.

The DM model independent annual modulation signature with widely sensitive target materials still remains a major approach, offering an unique possibility for detection; it requires well known techniques, full proved detector stability, well known and proved detector response in all the aspects, etc. At present the DAMA positive model independent evidence for the presence of DM particles in the galactic halo is supported at 9.3σ C.L. (on a cumulative exposure of 1.33 ton$\times$yr i.e. 14 annual cycles of DAMA/NaI and DAMA/LIBRA). It has been shown in literature that this is compatible with many DM scenarios.

I have been also recalled the recent possible positive hints exploiting different approaches and different target materials, and the existing uncertainties in the model dependent results and comparisons.

Finally very useful complementary results can arise from experiment exploiting other target detectors and approaches adopting adequately safe experimental procedures. For completeness, it is also worth to mention the efforts to investigate the axion candidate and in particular the present efforts of the ADMX experiment.

References

[1] J. Beringer et al., Phys. Rev. D **86** (2012) 010001
[2] Planck coll., arXiv:1303.5076v1
[3] M. Fukugita and P.J.E. Peebles, Astrophys. J. **616** (2004) 643
[4] R. Cen and J.P. Ostriker, Astrophys. J. **514** (1999) 1
[5] R. A. C. Croft et al., Phys. Rev. Lett. **83** (1999) 1092
[6] O. Elgaroy et al., astroph/0204152
[7] R. Bernabei et al., Int. J. Mod. Phys. A **22** (2007) 3155
[8] R. Bernabei et al., Eur. Phys. J. C **23** (2002) 61
[9] S. Chang et al., Phys. Rev. D **79** (2009) 043513
[10] S. Chang et al., Phys. Rev. Lett. **106** (2011) 011301
[11] S. Chang et al., J. Cosm. Astrop. Phys. **08** (2010) 018
[12] D. Tucker-Smith and N. Weiner, Phys. Rev. D **64** (2001) 043502
[13] D. Tucker-Smith and N. Weiner, Phys. Rev. D **72** (2005) 063509
[14] R. Bernabei et al., Mod. Phys. Lett. A **23** (2008) 2125
[15] R. Bernabei et al., Phys. Rev. D **77** (2008) 023506
[16] R. Bernabei et al., Int. J. Mod. Phys. A **21** (2006) 1445
[17] P. Belli et al., Phys. Rev. D **66** (2002) 043503
[18] R. Bernabei el al., La Rivista del Nuovo Cimento **26**(1) (2003) 1-73
[19] Y. Y. Mao, L. E. Strigari and R. H. Wechsler, arXiv:1304.6401
[20] Z. Ahmed et al., Science **327** (2010) 1619
[21] E. Armengaud et al., Phys. Lett. B **702** (2011) 329
[22] Z. Ahmed et al., arxiv:1203.1309
[23] J.I. Collar and N.E. Fields, arxiv:1204.3559
[24] R. Agnese et al., arXiv:1304.3706
[25] R. Agnese et al., arXiv:1304.4279
[26] E. Aprile et al., Phys. Rev. Lett. **107** (2011) 131302
[27] R. Bernabei et al., "Liquid Noble gases for Dark Matter searches: a synoptic survey", Exorma Ed., Roma, ISBN 978-88-95688-12-1, 2009, pp. 1–53, arXiv:0806.0011v2
[28] J.I. Collar and D.N. McKinsey, arXiv:1005.0838; arXiv:1005.3723; J.I. Collar, arXiv:1006.2031; arXiv:1010.5187; arXiv:1103.3481; arXiv:1106.0653; arXiv:1106.3559

[29] D.S. Akerib et al., Phys. Rev. Lett. **112** (2014) 091303
[30] G. Angloher et al., Eur. Phys. J. C **72** (2012) 1971
[31] G. Angloher et al., arXiv:1407.3146
[32] C.E. Aalseth et al., Phys. Rev. Lett. **106** (2011) 131301
[33] C.E. Aalseth et al., arXiv: 1401.3295
[34] A.K. Drukier, K. Freese, and D.N. Spergel, Phys. Rev. D **33** (1986) 3495; K. Freese, J. A. Frieman and A. Gould, Phys. Rev. D **37** (1988) 3388
[35] K. Freese et al., Phys. Rev. D **71** (2005) 043516; K. Freese et al., Phys. Rev. Lett. **92** (2004) 11301
[36] R. Cerulli, at this Conference
[37] R. Bernabei et al., Il Nuovo Cim. A **112** (1999) 545
[38] R. Bernabei et al., Eur. Phys. J. C **18** (2000) 283
[39] R. Bernabei et al., Int. J. Mod. Phys. D **13** (2004) 2127
[40] R. Bernabei et al., Eur. Phys. J. C **47** (2006) 263
[41] R. Bernabei et al., Int. J. Mod. Phys. A **22** (2007) 3155
[42] R. Bernabei et al., Eur. Phys. J. C **53** (2008) 205
[43] R. Bernabei et al., Nucl. Instr. and Meth. A **592** (2008) 297
[44] R. Bernabei et al., Eur. Phys. J. C **56** (2008) 333
[45] R. Bernabei et al., Eur. Phys. J. C **67** (2010) 39
[46] R. Bernabei et al., Eur. Phys. J. C **73** (2013) 2648
[47] P. Belli et al., Phys. Rev. D **84** (2011) 055014
[48] R. Bernabei et al., J. of Instr. **7** (2012) P03009
[49] R. Bernabei et al., Eur. Phys. J. C **72** (2012) 2064
[50] R. Bernabei et al., Int. J. of Mod. Phys. A **28** (2013) 1330022
[51] R. Bernabei et al., Eur. Phys. J. C **62** (2009) 327
[52] R. Bernabei et al., Eur. Phys. J. C **72** (2012) 1920
[53] R. Bernabei et al., Eur. Phys. J. A **49** (2013) 64
[54] R. Bernabei et al., Eur. Phys. J. C **74** (2014) 2827
[55] R. Bernabei et al., Eur. Phys. J. C **74** (2014) 3196
[56] A. Bottino et al., Phys. Rev. D **81** (2010) 107302; N. Fornengo et al., Phys. Rev. D **83** (2011) 015001; A.L. Fitzpatrick et al., Phys. Rev. D **81** (2010) 115005; D. Hooper et al., Phys. Rev. D **82** (2010) 123509; A.V. Belikov et al., Phys. Lett. B **705** (2011) 82; E. Kuflik et al., Phys. Rev. D **81** (2010) 111701; S. Chang et al., Phys. Rev. D **79** (2009) 043513; S. Chang et al., Phys. Rev. Lett. **106** (2011) 011301; R. Foot, Phys. Rev. D **81** (2010) 087302; Y. Bai, P.J. Fox, JHEP **0911** (2009) 052; J. Alwall et al., Phys. Rev. D **81** (2010) 114027; M.Yu. Khlopov et al., arXiv:1003.1144; S. Andreas et al., Phys. Rev. D **82** (2010) 043522; J. Kopp et al., JCAP **1002** (2010) 014; V. Barger et al., Phys. Rev. D **82** (2010) 035019; P. Belli et al., Phys. Rev. D **84** (2011) 055014; J. L. Feng et al., Phys. Lett. B **703** (2011) 124; A. Bottino et al., Phys. Rev. D **85** (2012) 095013; C.E. Aalseth et al., arXiv:1401.3295; R. Foot, arXiv:1401.3965; Q. Wallemacq, arXiv:1401.5243; S. Scopel et al., arXiv:1405.0364; C. Arina et al., arXiv:1406.5542; R. Foot, arXiv:1407.4213; G. Barello et al., arXiv:1409.0536; Q. Wallemacq, arXiv:1411.3178
[57] P. Belli et al., Il Nuovo Cim. C **15** (1992) 475
[58] R. Bernabei et al., Eur. Phys. J. C **28** (2003) 203
[59] N.J.C. Spooner et al., IDM Workshop 481 (1997); Y. Shimizu et al., Nucl. Instrum. and Meth. A **496** (2003) 347
[60] F. Cappella et al., Eur. Phys. J. C **73** (2013) 2276
[61] A. Drukier et al., arXiv:1206.6809

EPJ Web of Conferences **121**, 06002 (2016) DOI: 10.1051/epjconf/201612106002
RICAP-2014

Dark Matter indirect detection:
Some anomalies and many constraints

Marco Cirelli[a]

Institut de Physique Théorique, CNRS, UMR 3681 & CEA/Saclay, 91191 Gif-sur-Yvette, France

Abstract. I discuss four recent anomalies in Dark Matter Indirect Detection (the positron excess, the 130 GeV line, the GeV GC excess and the 3.5 KeV line) and some relevant constraints.
[SACLAY-T15-003]

1. Introduction

Indirect searches for Dark Matter (DM) aim at detecting the signatures of the annihilations or decays of DM particles in the fluxes of Cosmic Rays (CRs), intended in a broad sense: charged particles (electrons and positrons, protons and antiprotons, deuterium and antideuterium), photons (gamma rays, X-rays, synchrotron radiation), neutrinos. In general, a key point of all these searches is to look for channels and ranges of energy where it is possible to beat the background from ordinary astrophysical processes. This is for instance the basic reason why searches for charged particles focus on fluxes of antiparticles (positrons, antiprotons, antideuterons), much less abundant in the Universe than the corresponding particles. Similarly, searches for photons or neutrinos have to look at areas where the DM-signal to astro-noise ratio can be maximised (typically the Galactic Centre and DM-dominated structures such as dwarf satellite galaxies).

Pioneering work has explored indirect detection (ID) as a promising avenue of discovery since the late-70's. Since then, innumerable papers have explored the predicted signatures of countless particle physics DM models. In the past 6 years or so, however, the field has experienced a significant burst of activity, mainly due to the results presented by a few very well performing experiments, above all the PAMELA satellite, the FERMI satellite and the HESS telescope. It is fair to say that the field has passed, for better or for worse, from a theory-driven state to a data-driven phase.

In this presentation I intend to briefly review the current status of the field, using the pretext of discussing four recent experimental "anomalies" and the ensuing phenomenological activity. The four anomalies are: 1) the positron and electron excesses, first soundly detected by PAMELA in 2008 in the positron fraction and then corroborated by many results from FERMI, HESS and recently AMS-02; 2) the "130 GeV line" from the Galactic Centre (GC), first identified in 2012 by Christoph Weniger and collaborators in FERMI data; 3) the "GeV Galactic Centre γ-ray excess", promoted since 2010 most notably by Dan Hooper; 4) the 3.5 KeV X-ray line, supposedly detected in March 2014 in data from the XMM-NEWTON satellite from several galaxy clusters and the Andromeda galaxy (M31).

---

[a] e-mail: marco.cirelli@cea.fr

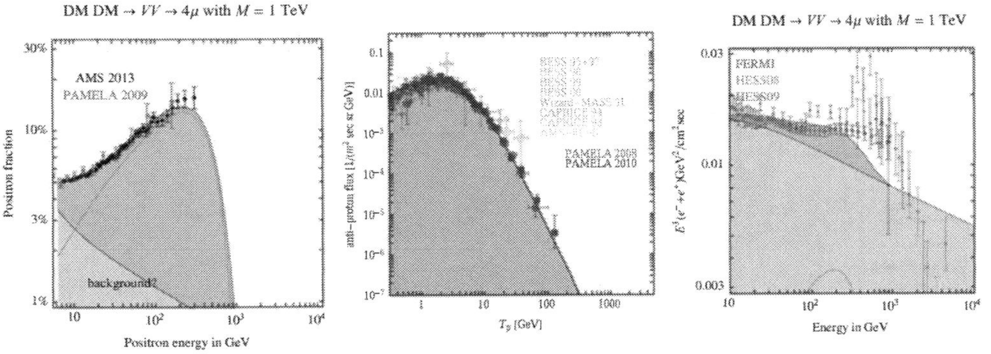

Figure 1. A compilation of recent and less recent data in charged cosmic rays, superimposed on plausible but uncertain astrophysical backgrounds from secondary production and on the flux produced by Dark Matter annihilations for a specific model. Left: positron fraction. Center: antiproton flux. Right: sum of electrons and positrons. Figures from ref. [9].

2. The positron and electron excesses

There has been a flurry of positive results from a few indirect detection experiments looking at the fluxes of charged cosmic rays. In particular, the signals pointed to an excess of electrons and positrons at the TeV and sub-TeV scale:

- Renowned data from the PAMELA satellite [1] showed, back in 2008, a steep increase in the energy spectrum of the positron fraction $e^+/(e^+ + e^-)$ above 10 GeV up to 100 GeV, compatibly with previous hints from HEAT [2] and AMS-01 [3]. These findings have later been confirmed with independent measurements by the FERMI satellite [4] and, recently, by the AMS-02 experiment [5] and extended to about 430 GeV.
- Data from PAMELA [6] also showed no excess in the $\bar{p}$ energy spectrum compared with the predicted background.
- In the $e^+ + e^-$ energy spectrum, the results of the FERMI satellite [7], combined with the results from the HESS telescope [8], hint to an excess (with respect to the expected background) reproduced by a simple power law up to about 1 TeV and eventually a steepening at energies of a few TeVs.

The data are displayed in Fig. 1, together with the expected astrophysical "backgrounds" and with the contribution from an annihilating DM particle which fits them reasonably well (see below). The properties of such a particle are pin-pointed quite precisely by the data. The DM has to be:

- With a mass of *1 to few TeV*, in order to reproduce the feature in the $e^+ + e^-$ spectrum. Actually, the hint of a flattening in the positron fraction suggested by AMS-02 favours a DM mass below about 1 TeV with about 3σ statistical significance, depending on the DM annihilation channel, so that a little bit of a tension is present with the $e^+ + e^-$ spectrum, which requires a slightly larger value.
- *Leptophilic*, i.e. annihilating almost exclusively into leptonic channels, otherwise the antiproton measurements would be exceeded.
- With a *very large annihilation cross section*, of the order of 10^{-23} cm^3/sec or more (for the masses under consideration), much larger than the thermal one, in order to produce a large enough flux that can fit the positron rise and the $e^+ + e^-$ bump.

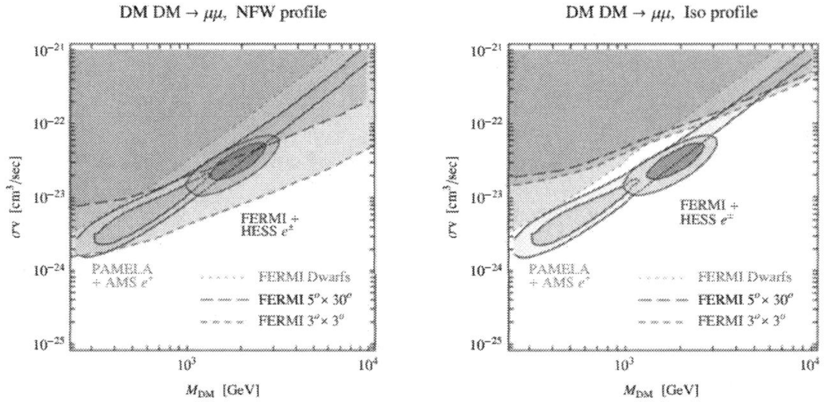

Figure 2. Best fit regions for the positron and electron excesses, together with some representative γ-ray constraints. The left (right) plot assumes a NFW (Isothermal) DM profile. Figure from Ref. [9].

As tantalising as these hints of DM can be, they have to be confronted with associated constraints. Many possible constraints can be considered, but here I will focus on two classes only. The first one is observations of γ-rays. In Fig. 2 we show representative γ-ray bounds (the constraints are taken from [10, 11], more recent analyses find similar or slightly more stringent bounds). In the left panel, we see that, if we have chosen a benchmark NFW galactic Dark Matter profile, the fit region is either in some tension or clearly ruled out (depending on at which FERMI constraint one looks). Choosing instead the shallower isothermal profile (right panel) makes all the constraints looser. It is therefore difficult to get a final answer from γ-rays. The second class of constraints comes from observations of the cosmic microwave background (CMB), which imposes bounds on DM annihilations (based on the fact that they would have re-ionised the primordial universe) that disfavour at various degrees and for most channels the DM interpretation of the positron excess [12].

3. The 130 GeV line

The "130 GeV line" claim has gathered a lot of attention in the past two years (for a more thorough review see [13]). Originally spotted by [14] and, above all, by [15] in the publicly available FERMI data from an extended region including the GC (Fig. 3 left reports the most evocative of the original analysis' figures), it has later found support in other analyses [16–19], with varying degrees of accuracy and claimed significance. [16, 19] have seen it in what could possibly be DM sub-haloes of the MW, and there might be two lines, at 111 GeV and 129 GeV [17, 20]. [18] has seen it in galaxy clusters too. For a response, [21–23] challenged the analyses in a number of ways, suggesting that the line(s) could be due to unidentified instrumental, statistical or astrophysical origin. Although it is probably too early for a final conclusion on this claim, it is fair to say that the current consensus seems to be that the line has been a rather unfortunate combination of an instrumental effect and a statistical fluctuation. The right panel of Fig. 3 illustrates that, as more data are accumulated, the significance of the signal lowers, hence pointing at something which is probably not an actual signal. C. Weniger, at this conference, presented recent work that supports these conclusions [25].

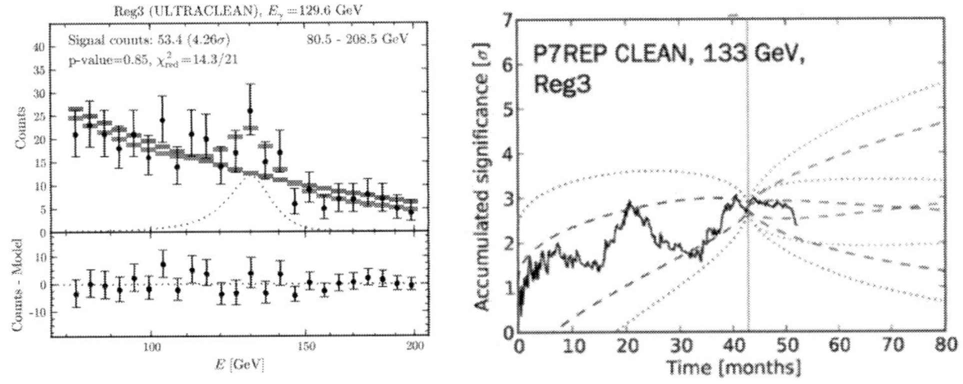

Figure 3. Left: FERMI γ-ray data and fits pointing to a line at about 130 GeV. Right: behaviour with time of the accumulated significance for this signal. Figures from Ref. [15] and Ref. [24].

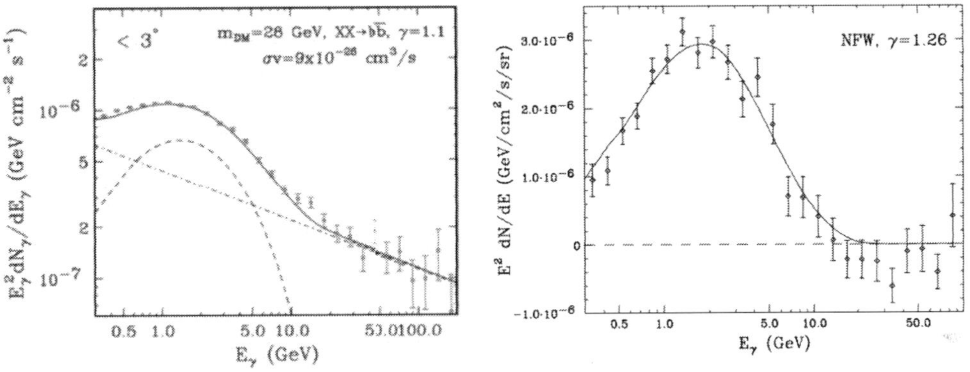

Figure 4. Earliest and latest fits to the GeV excess at the GC. From Ref. [26] and Ref. [28].

4. The GeV Galactic Center excess

Several authors have reported since 2009 the detection of a gamma-ray signal from the inner few degrees around the GC [26, 27], with the most notable early claims by Dan Hooper. Its spectrum and morphology are found to be compatible with those expected from annihilating DM particles: to fix the ideas, the results of one of the most recent analysis [28] confirm the presence of this excess at an incredibly high level of significance (if taken at face value) and find this signal to be best fit by 31–40 GeV DM particles distributed according to a (contracted) NFW profile and annihilating into $b\bar{b}$ with $\langle \sigma v \rangle = 1.4 \div 2 \times 10^{-26}$ cm^3/s. Figure 4 displays the earliest fit to the data (from [26]) and one of the most recent ones (from [28]).

Of course, one should not forget that, in very general terms, the identification of an "excess" strongly relies on the capability of carefully assessing the background over which the excess is supposed to emerge. The claim under scrutiny constitutes no exception, quite the contrary. The extraction of the residuals strongly relies on the modelling of the diffuse gamma-ray background (in particular the one publicly made available by the FERMI collaboration) as well as on additional modelling of astrophysical emissions, e.g. from FERMI bubbles, isotropic component, unresolved point sources, molecular gas... For instance, at this Conference, C' Weniger has presented the results of a thorough analysis of the background models [25]. While this is probably the best that can be done, it is not guaranteed to be (and

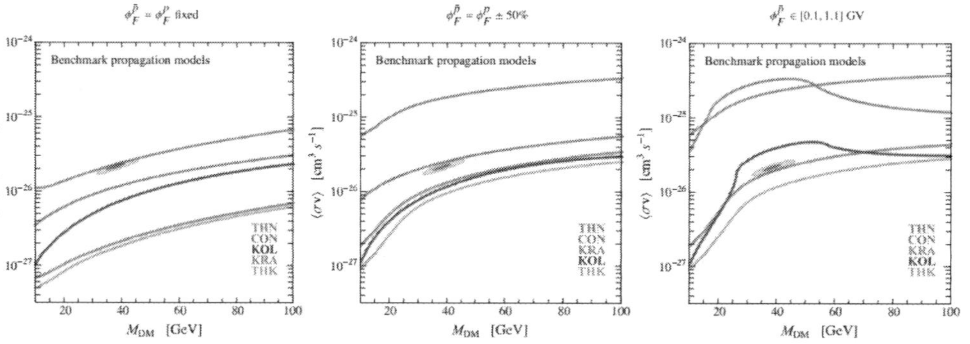

Figure 5. 3-σ exclusion contours on $\langle \sigma v \rangle$ for 100% DM annihilation into $b\bar{b}$, for the three approaches to solar modulation briefly discussed in the text. The grey area is the best-fit region. Figure from Ref. [33].

in general is not expected to be) the optimal strategy. Also, one should not forget that there might be alternative astrophysical explanations for the excess. A population of milli-second pulsars has been extensively discussed since the beginning [29], as well as the possibility of a spectral break in the emission of the central Black Hole [30]. More recently, the possibility has been suggested that isolated injections of charged particles (electrons [31] or protons [32]) sometime in the past, possibly connected with the activity of the central Black Hole, can produce secondary radiation able to account for the anomalous signal. While reproducing with these models all the details of the observed emission might be not easy, they represent plausible and useful counterexamples to the DM interpretation.

Still, it is interesting to insist on the tantalising DM hypothesis and to explore ways to confirm or disprove the result within the DM framework. In particular, given the alleged hadronic origin of the signal, it is very useful to analyse the antiproton channel to put constraints on the DM interpretation of such excess. Reference [33] delved precisely into this issue, and the condensed results are displayed in Fig. 5. It considered several galactic propagation models for antiprotons (THN, CON, KOL, KRA, THK, roughly distinguished by the thickness of the diffusive halo, the diffusion properties and the presence of side effects such as convection) and several assumptions for the so-called solar modulation, i.e. the complicated effect of the magnetic field and solar cosmic ray wind of the heliosphere on the last segment of the antiproton journey. More precisely, it considered a solar force field for $\bar{p}$ fixed and equal to p one (left panel of Fig. 5), variable within 50% (central panel) or free within a wide range (right panel). The overall conclusions are the following: adopting the most realistic propagation models and well motivated choices for the solar modulation potential, the hadronic ($b\bar{b}$) DM interpretation for the GeV excess is definitely in strong tension with the antiproton data. Nevertheless, given that our knowledge of CR diffusion both in the Galaxy and in the heliosphere is far from being accurate and complete, there are still conservative choices of the parameters involved that do not result in ruling it out, namely thin halo models and large solar modulation potentials. The authors of Ref. [34] have also discussed the antiproton bounds, reaching somewhat different conclusions.

5. The 3.5 KeV X-ray line

One of the latest claims in the field of indirect detection comes from a different range of energies: X-rays. In datasets from the XMM-NEWTON satellite, two independent groups [35, 36] have found evidence for an unexplained line at 3.5 KeV. The former group found it in observations of a set of 73 galaxy clusters with redshift between 0.01 and 0.35. The latter one

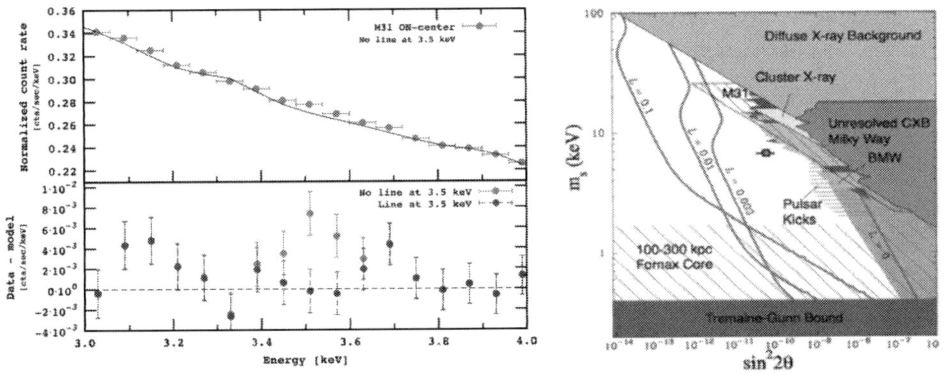

Figure 6. Identification of the 3.5 KeV line in XMM-NEWTON data (left) and the parameter space of its interpretation in terms of a decaying sterile neutrino. From Ref. [36] and Ref. [35].

in observations both of the Perseus cluster and of Andromeda, with no detection in "blank sky" measurements. Figure 6, left, displays an extraction of the spectrum showing the line, from [36].

The complication is that the X-ray spectrum in this range of energies is crowded with atomic de-excitation lines from elements such as Cr, Mn, K, Fe, Ni, Ca, Cu... Reference [37] has indeed very recently argued that previously-unaccounted-for potassium lines can well explain the signal. Reference [38] reiterates, however, that data from Andromeda are instead solid and make the potassium interpretation problematic. On another side, Ref. [39] has argued that no line is seen in Chandra data from the GC, although this conclusion depends on how one models the local background. The discussion is currently unfolding and probably more data from independent instruments will be needed.

If confirmed, however, the most straightforward explanation of the line in terms of new physics is of great interest for the field of DM indirect detection as it consists of a sterile neutrino of mass 7 KeV decaying into an ordinary ν and a photon (the detected X-ray). The decay rate turns out to be $\mathcal{O}(10^{-29})$ sec^{-1}. This, translated in terms of particle physics parameters by the effective mixing angle, lies in a region of parameter space still allowed by other constraints, as illustrated by the right panel of Fig. 6. The production mechanism of a population of sterile neutrinos in the early universe would involve active-sterile oscillations helped by the presence of a sizeable leptonic asymmetry, quite uncompelling, but possible.

6. Conclusions

There are arguably no firm conclusions in this field at this moment in time. There are tantalising hints (the positron and electron excess, the gamma-ray line, the GeV GC excess and the X-ray line) and there are stringent constraints. Such constraints, however, are often relaxed by appropriate assumptions, which can be extreme or not (the illustration with the antiproton constraints on the GeV excess in Sect. 4 is exemplar). The only firm albeit generic conclusions seem to be that:

- ◇ current experiments are clearly reaching (and in some cases have already reached) the sensitivities for which they were designed, and hence they probe very promising regions of the parameter space;
- ◇ astrophysics, in different manifestations, is the main killjoy, introducing alternative compelling explanation, irreducible uncertainties, unbeatable background noise...;

◇ hence, it is important to pursue a multi-messenger approach in all instances, investigating associated signals in other channels, cross-checking constraints and confirmations from independent targets etc.;

◇ in any case, the profusion of data from the recent experiments have spurred a remarkable proliferation of DM models, so that "traditional" DM models (such as SuSy DM) have, for better or for worse, been joined by many other possibilities.

I thank my collaborators on the papers which led to the results presented here and who helped me with many useful discussions. I acknowledge the hospitality of the Institut d'Astrophysique de Paris, where part of this work was done. Funding and research infrastructure acknowledgements: European Research Council (ERC) under the EU Seventh Framework Programme (FP7/2007-2013)/ERC Starting Grant (agreement n. 278234 – "NEWDARK" project); French national research agency ANR under contract ANR 2010 BLANC 041301.

References

[1] O. Adriani et al. [PAMELA Collaboration], Nature **458**, 607–609, 2009, arXiv:0810.4995

[2] S. W. Barwick et al. [HEAT Coll.], Astrophys. J. **482** (1997) L191 [astro-ph/9703192]

[3] M. Aguilar et al. [AMS-01 Coll.], Phys. Lett. B **646** (2007) 145 [astro-ph/0703154]

[4] M. Ackermann et al. [The Fermi LAT Collaboration], arXiv:1109.0521 [astro-ph.HE]

[5] M. Aguilar et al. [AMS Collaboration], Phys. Rev. Lett. **110** (2013) 141102. (An updated measurement has only been presented in a talk by S. Ting (and subsequent talks elsewhere).)

[6] O. Adriani et al. [PAMELA Coll.], Phys. Rev. Lett. **105** (2010) 121101 [arXiv:1007.0821]

[7] A. Abdo et al. [Fermi-LAT Coll.], Phys. Rev. Lett. **102** (2009) 181101, arXiv:0905.0025
M. Ackermann et al. [Fermi LAT Coll.], Phys. Rev. D **82** (2010) 092004, arXiv:1008.3999

[8] F. Aharonian et al. [H.E.S.S. Coll.], Phys. Rev. Lett. **101** (2008) 261104 [arXiv:0811.3894]
F. Aharonian et al. [H.E.S.S. Coll.], Astron. Astrophys. **508** (2009) 561 [arXiv:0905.0105]

[9] M. Cirelli, M. Kadastik, M. Raidal and A. Strumia, Nucl. Phys. B **813** (2009) 1 [Addendum-ibid. B **873** (2013) 530] [arXiv:0809.2409 [hep-ph]]

[10] M. Cirelli, P. Panci and P. D. Serpico, Nucl. Phys. B **840** (2010) 284 [arXiv:0912.0663]
M. Papucci and A. Strumia, JCAP **1003** (2010) 014 [arXiv:0912.0742]

[11] Fermi-LAT collaboration, Phys. Rev. Lett. **107** (2011) 241302 [arXiv:1108.3546]
Fermi-LAT collaboration, Astrophys. J. **761** (2012) 91 [arXiv:1205.6474] For this figure we used the dwarfs data presented by A. Drlica-Wagner at the 2012 Fermi symposium

[12] S. Galli, F. Iocco, G. Bertone and A. Melchiorri, Phys. Rev. D **80** (2009) 023505 [arXiv:0905.0003]. T.R. Slatyer, N. Padmanabhan, D.P. Finkbeiner, Phys. Rev. D **80** (2009) 043526, 0906.1197. G. Huetsi, A. Hektor, M. Raidal, Astron. Astrophys. **505** (2009) 999, arXiv:0906.4550. M. Cirelli, F. Iocco and P. Panci, JCAP **0910** (2009) 009, arXiv:0907.0719. See also: T. Kanzaki, M. Kawasaki and K. Nakayama, Prog. Theor. Phys. **123** (2010) 853, arXiv:0907.3985. Q. Yuan, B. Yue, X.-J. Bi, X. Chen, X. Zhang, JCAP **1010** (2010) 023, arXiv:0912.2504. G. Hutsi, J. Chluba, A.

Hektor and M. Raidal, arXiv:1103.2766. S. Galli, F. Iocco, G. Bertone, A. Melchiorri, Phys. Rev. D **84** (2011) 027302, arXiv:1106.1528. A. Natarajan, arXiv:1201.3939. G. Giesen, J. Lesgourgues, B. Audren and Y. Ali-Haimoud, JCAP **1212** (2012) 008 [arXiv:1209.0247]. J. M. Cline and P. Scott, JCAP **1303** (2013) 044 [arXiv:1301.5908]
[13] T. Bringmann and C. Weniger, arXiv:1208.5481 [hep-ph]
[14] T. Bringmann, X. Huang, A. Ibarra, S. Vogl and C. Weniger, arXiv:1203.1312 [hep-ph]
[15] C. Weniger, JCAP **1208** (2012) 007 [arXiv:1204.2797 [hep-ph]]
[16] E. Tempel, A. Hektor and M. Raidal, arXiv:1205.1045 [hep-ph]
[17] M. Su and D. P. Finkbeiner, arXiv:1206.1616 [astro-ph.HE]
[18] A. Hektor, M. Raidal and E. Tempel, arXiv:1207.4466 [astro-ph.HE]
[19] M. Su and D. P. Finkbeiner, arXiv:1207.7060 [astro-ph.HE]
[20] A. Rajaraman, T. M. P. Tait and D. Whiteson, arXiv:1205.4723 [hep-ph]
[21] A. Boyarsky, D. Malyshev and O. Ruchayskiy, arXiv:1205.4700 [astro-ph.HE]
[22] N. Mirabal, arXiv:1208.1693 [astro-ph.HE]
[23] A. Hektor, M. Raidal and E. Tempel, arXiv:1208.1996 [astro-ph.HE]
[24] C. Weniger, presentation at the FERMI meeting for alternative observations, 25 Jul. 2013
[25] C. Weniger, presentation at this RICAP 2014 Conference
[26] L. Goodenough and D. Hooper, arXiv:0910.2998 [hep-ph]
[27] V. Vitale et al. [Fermi-LAT Collaboration], arXiv:0912.3828 [astro-ph.HE]. D. Hooper and L. Goodenough, Phys. Lett. B **697** (2011) 412 [arXiv:1010.2752 [hep-ph]]. D. Hooper and T. Linden, Phys. Rev. D **84** (2011) 123005 [arXiv:1110.0006 [astro-ph.HE]]. D. Hooper, Phys. Dark Univ. **1** (2012) 1 [arXiv:1201.1303 [astro-ph.CO]]. K. Abazajian, M. Kaplinghat, Phys. Rev. D **86** (2012) 083511 [arXiv:1207.6047. D. Hooper, T. Slatyer, Phys. Dark Univ. **2** (2013) 118 [arXiv:1302.6589 [astro-ph.HE]]. C. Gordon and O. Macias, Phys. Rev. D **88** (2013) 083521 [arXiv:1306.5725 [astro-ph.HE]]. W.-C. Huang, A. Urbano and W. Xue, arXiv:1307.6862 [hep-ph]. K. N. Abazajian, N. Canac, S. Horiuchi and M. Kaplinghat, arXiv:1402.4090 [astro-ph.HE]
[28] T. Daylan, D. P. Finkbeiner, D. Hooper, T. Linden, S. K. N. Portillo, N. L. Rodd and T. R. Slatyer, arXiv:1402.6703 [astro-ph.HE]
[29] K. N. Abazajian, JCAP **1103** (2011) 010 [arXiv:1011.4275 [astro-ph.HE]]. D. Hooper, I. Cholis, T. Linden, J. Siegal-Gaskins and T. Slatyer, Phys. Rev. D **88** (2013) 083009 [arXiv:1305.0830 [astro-ph.HE]]. Q. Yuan and B. Zhang, arXiv:1404.2318 [astro-ph.HE]
[30] A. Boyarsky, D. Malyshev and O. Ruchayskiy, Phys. Lett. B **705** (2011) 165 [arXiv:1012.5839 [hep-ph]]
[31] J. Petrovic, P. D. Serpico and G. Zaharijas, arXiv:1405.7928 [astro-ph.HE]
[32] E. Carlson and S. Profumo, arXiv:1405.7685 [astro-ph.HE]
[33] M. Cirelli, D. Gaggero, G. Giesen, M. Taoso and A. Urbano, arXiv:1407.2173 [hep-ph]
[34] T. Bringmann, M. Vollmann and C. Weniger, arXiv:1406.6027 [astro-ph.HE]
[35] E. Bulbul, M. Markevitch, A. Foster, R. K. Smith, M. Loewenstein and S. W. Randall, Astrophys. J. **789** (2014) 13 [arXiv:1402.2301 [astro-ph.CO]]
[36] A. Boyarsky, O. Ruchayskiy, D. Iakubovskyi and J. Franse, arXiv:1402.4119 [astro-ph.CO]
[37] T. E. Jeltema and S. Profumo, arXiv:1408.1699 [astro-ph.HE]
[38] A. Boyarsky, J. Franse, D. Iakubovskyi and O. Ruchayskiy, arXiv:1408.4388 [astro-ph.CO]
[39] S. Riemer-Sorensen, arXiv:1405.7943 [astro-ph.CO]

EPJ Web of Conferences **121**, 06003 (2016) DOI: 10.1051/epjconf/201612106003
RICAP-2014

Highlights on gamma rays, neutrinos and antiprotons from TeV Dark Matter

Viviana Gammaldi[a]
University Complutense Madrid, Spain

Abstract. It has been shown that the gamma-ray flux observed by HESS from the J1745-290 Galactic Center source is well fitted as the secondary gamma-rays photons generated from Dark Matter annihilating into Standard Model particles in combination with a simple power law background. The neutrino flux expected from such Dark Matter source has been also analyzed. The main results of such analyses for 50 TeV Dark Matter annihilating into W^+W^- gauge boson and preliminary results for antiprotons are presented.

1. Introduction

Multimessenger astroparticle study is fundamental for Dark Matter (DM) indirect search. Signatures of DM annihilating or decay in astrophysical sources may be observed in cosmic-ray fluxes by Cerenkov telescopes such as VERITAS, HESS, MAGIC and CTA; neutrino telescopes such as ANTARES or IceCube; satellites such as PAMELA, AMS and Fermi or ballon experiments like CAPRICE or BESS [1–3]. The secondary products of annihilation and decay of DM particles contribute to the cosmic-rays differential flux at the Earth as

$$\frac{d\Phi_{\text{cr-DM}}}{dE} = \eta_{\text{cr}} \cdot \sum_{a=1}^{2} \sum_{i}^{\text{SM channels}} \frac{\zeta_i^{(a)}}{a} \frac{dN_i^{(\text{cr})}}{dE} \cdot \frac{\kappa_{\text{cr}}^{(a)}}{4\pi M^a}, \tag{1}$$

where:

- η_{cr} depends on the secondary particles of interest (observed cosmic-ray flux) and its propagation.
- the total flux is given by decay ($a = 1$) or annihilation ($a = 2$) events of DM particles into the i-th Standard Model (SM) particle (annihilation/decay channel). The ζ factor discerns between these two cases: $\zeta_i^{(1)} = 1/\tau_i^{\text{decay}}$ and $\zeta_i^{(2)} = \langle\sigma_i v\rangle$ are respectively the inverse of the decay time and and thermal averaged annihilation cross section times velocity. The probability that DM annihilates or decays into the i-th channel depends on the nature of DM.
- the differential number $dN_i^{(\text{cr})}/dE$ of cosmic-rays produced at the source by subsequent events of annihilation or decay of SM particles is simulated by means of Monte Carlo events generator software, such as `PYTHIA` or `HERWIG`. Some uncertainties may be introduced in the evaluation of both the $\zeta_i^{(a)}$ and κ_{cr} factor due to the choice of the Fortran or C++ versions of `PYTHIA` or `HERWIG` software, as discussed in [4] for gamma-rays.

[a] e-mail: `vivigamm@ucm.es`

- the κ_{cr} factor depends on the astrophysics of DM distribution as well as on the cosmic-rays propagation. For *neutral* cosmic-rays (n-cr) it is the astrophysical factor

$$\kappa^{(a)}_{\rm n-cr} \equiv \langle J \rangle_{\Delta\Omega} = \frac{1}{\Delta\Omega}\int_{\Delta\Omega} {\rm d}\Omega \int_0^{l_{max}(\Psi)} \rho^{(a)}[r(l)]dl(\Psi)\,. \quad (2)$$

Here, $\rho(r)$ is the DM density in the halo and l is the distance from the Sun to any point in the Galaxy disk and halo. The radial distance r of any point in the halo is measured from the Galactic Center (GC), and is related to l by $r^2 = l^2 + D_\odot^2 - 2D_\odot l \cos\Psi$, where $D_\odot \simeq 8.5$ kpc is the distance from the Sun to the center of the Galaxy. The distance from the Sun to the edge of the halo in the direction θ from the GC is $l_{max} = D_\odot \cos\theta + \sqrt{r^2 - D_\odot^2 \sin\theta}$. For *neutral* particles, *directional* observations are achievable. In this case, the flux must be averaged over the solid angle of the detector, that is typically of order of $\Delta\Omega = 2\pi(1 - \cos\Psi)$, being Ψ the angular resolution of the telescope.

For *charged* cosmic-rays (c-cr), *directional* observations are *not* feasible. In fact, charged particles observed in a given direction might have been originated everywhere in the sky. In this case the $\kappa_{\rm c-cr}$ factor is proportional to a diffusion term

$$\kappa^{(a)}_{\rm c-cr} \equiv \left(\frac{\rho_\odot}{M}\right)^{(a)} R_{\rm c-cr}(r_\odot, E). \quad (3)$$

The diffusion factor at the position of the sun $R_{\rm c-cr}(r_\odot, E)$ for *charged* cosmic-ray ($e^\pm$, $p^\pm$ etc.) is the solution of a diffusion equation that depends on the particle of interest and DM distribution. It describes the diffusion of particles in the Galaxy and the production of secondary cosmic-rays due to the interaction with the Interstellar Medium (ISM). The final flux at the position of the Earth also includes Solar magnetic field effect [5, 6].

In Sect. 2, I present the main results of the cosmic-rays analysis of the prospective DM source at the GC: the gamma-rays cut-off detected by HESS at the GC and the prospective neutrino flux are reviewed. New constraints obtained from the antiprotons study are also presented in Sect. 2, while the conclusions are in Sect. 3.

2. Cosmic rays from TeV Dark Matter at the Galactic Center

In previous works it has been shown that the collection of data from HESS telescope from the J1749-290 source at the GC is well fitted by model independent DM annihilating into SM channels [7]. Such study provides gamma-rays fits with DM masses between 1–100 TeV. While annihilation in lepton channel is excluded at the 99% confidence level, gamma-rays secondary flux produced by DM annihilating into quark and bosons channels well reproduce the data. Here, Eq. (1) for the cosmic-rays differential flux at the Earth is discussed for heavy DM particle of ≈ 50 TeV annihilating into W^+W^- channel. Extra-dimensional particles as branons may justify TeV DM annihilating in boson channels [8, 9].

Gamma-rays. The differential secondary gamma-ray flux emitted by DM annihilating 100% in W^+W^- SM channel can be easily written as the Eq. (1) with $\eta_\gamma = 1$, and $\kappa_\gamma = \langle J \rangle_{\Delta\Omega}$. Such gamma-ray flux from DM combined with a power-law background is able to fit HESS data from the GC and Fermi-LAT data [7]:

$$\frac{d\Phi_{\gamma-Tot}}{dE} = B^2 \cdot \left(\frac{E}{\rm GeV}\right)^{-\Gamma} + A_W^2 \cdot \frac{dN_W^{(\gamma)}}{dE}\,. \quad (4)$$

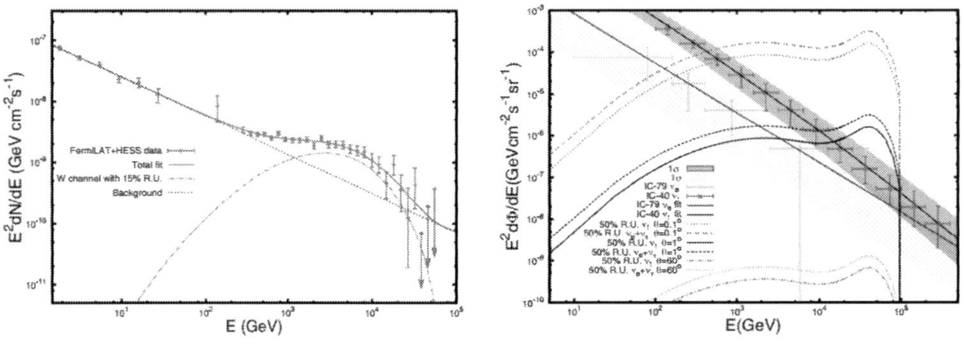

Figure 1. Left panel: the combination of the gamma-ray flux originated by DM ≈ 50 TeV annihilating into W^+W^- channel and a power-law background well fits HESS [2] and Fermi-LAT [3] data from the GC. Above 10 GeV the Fermi-LAT has an angular resolution comparable to the angular resolution of HESS at TeV energies, which makes significant the joint analysis of the Fermi and HESS data [3]. An enhancement in the secondary gamma-rays originated by annihilation of DM is needed. An estimation of such enhancement is given throughout the boost factor $b \equiv \langle J_{(2)}\rangle / \langle J_{(2)}^{\rm NFW}\rangle$ with $\langle J_{(2)}^{\rm NFW}\rangle \simeq 280 \cdot 10^{23}$ GeV^{2}cm^{-5}. The background spectral index $\Gamma \simeq 2.63 \pm 0.02$ is compatible with Fermi-LAT observations [7]. Right panel: expected neutrino fluxes corresponding to muon neutrinos and electron plus tau neutrinos from ≈ 50 TeV DM annihilating into W^+W^- bosons for an angular field of view of $\theta = 60°$, $1°$ and $0.1°$. The observed atmospheric muon by the IceCube telescope in the 40-string configuration (IC-40) and electron neutrinos by the 79-string configuration (IC-79) are also shown together with the corresponding shared regions at 1σ confidence level [12].

where $A_W = \frac{\zeta_W^{(2)}}{2} \cdot \frac{\Delta\Omega^{\rm HESS} \langle J_{(2)}\rangle_{\Delta\Omega}}{4\pi M^2}$. The cosmological constraints for thermal DM annihilation cross section is given by CMB observation by PLANCK and WMAP satellite experiments among other measurements [10, 11]. In the model independent analysis developed here, we assume such value $\sum_i^{\rm SM} \zeta_i^{(2)} \equiv \zeta_W^{(2)} \equiv \langle \sigma v\rangle \simeq 3 \times 10^{-26}\,{\rm cm}^3{\rm s}^{-1}$ and $\Delta\Omega^{\rm HESS} \approx 10^{-5}$. The left panel of Fig. 1 shows the model independent fit of HESS + Fermi-LAT gamma-ray data.

Neutrinos. The differential flux of neutrinos of flavor ν_f observed on the Earth can be computed as Eq. (1) with $\eta_\nu \equiv \sum_{p=1}^3 P_{fp}$. This parameter depends on the dimensions of the source and its distance from the observer:

$$\frac{d\Phi_{\nu_f}}{dE} = \sum_{p=1}^{3} P_{fp} \cdot \frac{\zeta_W^{(2)}}{2} \frac{dN_W^{(\nu_p)}}{dE} \cdot \frac{\Delta\Omega \langle J_{(2)}\rangle_{\Delta\Omega}}{4\pi M^2} . \tag{5}$$

Neutrinos allow *directional* observation of the GC, although the angular resolution is worse than that for gamma rays. Moreover, it depends on the kind of signal (track or shower) and the position of the source with respect to the horizon of the telescope. All this affects the effective area of the telescope and the possibility to detect a neutrino signal above the background at a given confidence level [12]. The right panel of Fig. 1 shows the detectability of the neutrino signal from ≈ 50 TeV at the GC up the background component, depending on the resolution angle of the telescope. More details on the detectability with 5, 3 and 2σ confidence level may be found in [12].

Antiprotons. As a difference with *neutral* cosmic-rays that preserve information about the emission direction and spectra, *charged* cosmic-rays propagation in the Galaxy is a complex process affected by many different physical phenomena: energy losses, convection

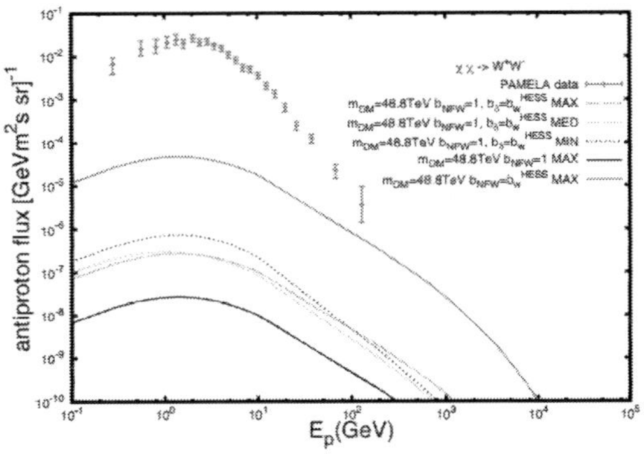

Figure 2. Antiproton differential flux at the TOA after propagation for ≈ 50 TeV DM annihilating into W^+W^- pairs for different diffusion models and distribution profiles: The lower signal (black solid line) corresponds to the non-boosted NFW profile by employing the maximum diffusion model. On the other hand, the same assumptions give raise to the highest flux at high energies (violet solid line) but with a boost factor of $b^{WW}_{\rm NFW} = b^{WW}_{\rm HESS} = 1767$. We show the antiproton flux at TOA for the medium diffusion model for non-boosted NFW profile plus enhanced δ-DM distribution ($b^{WW}_{\delta-\rm NFW} = b^{\rm HESS}_{WW}$) at the GC for three diffusion models (blue big-dotted, green rushed and red little-dotted line) togehter with PAMELA data [5].

and reacceleration affect the flux at the Top of Atmosphere (TOA). Due to these effects, the direct observation of antiprotons produced at the GC is unlikely. The available antiproton data at TOA are in agreement with secondary antiprotons diffusion in the Galaxy [1]. In any case, antiprotons generated at the GC might affect the antiproton flux at the TOA. The prospective flux is given by Eq. (1) with $\eta_{\bar{p}} = v_{\bar{p}}/4\pi$ and $\kappa^{(a)}_{\bar{p}} = \rho^{(a)} R(r_\odot, E_{\bar{p}})/M^{(a)}$. For DM totally annihilating into W^+W^- channel the antiproton differential flux at the Solar system is:

$$\frac{d\Phi_{(\bar{p},\odot)}}{dE_{(\bar{p},\odot)}} = \frac{v_{\bar{p}}}{4\pi} \cdot \frac{\zeta^{(2)}_W}{2} \frac{dN^{(\bar{p})}_W}{dE_{\bar{p}}} \cdot \left(\frac{\rho_\odot}{M}\right)^2 R(r_\odot, E_{(\bar{p},\odot)}) \cdot \tag{6}$$

The diffusion parameter accounts for two different contribution from the Navarro-Frank-White DM halo and the δ-like DM distribution at the GC:

$$R(r_\odot, E_{\bar{p}}) = b^W_{\rm NFW} \cdot R^{\rm NFW}(E_{\bar{p}}) + b^W_{\delta-\rm NFW} \cdot R^{\delta-\rm NFW}(E_{\bar{p}}) \cdot \tag{7}$$

The equivalent boost factor normalization constant is given by $\langle J \rangle^{NFW}_{\Delta\Omega} \Delta\Omega_{\rm HESS} \left(\frac{D_\odot}{\rho_\odot}\right)^2 \simeq 2.13 \cdot 10^{60}\, m^3$ sr [5]. We include in the analysis the solar modulation effect on the antiprotons propagation in the Solar system [6]. It can be neglected for antiprotons production from heavy DM, because the solar modulation affects energies between 1–10 GeV [5]. More important is the effect of Galactic wind, that is described as a convective velocity V_c. Such velocity at the Galactic disk is $V_c \simeq 0$, while at the GC there is evidence of a strong convective velocity of $V_c \simeq 100$–1000 km/s. Such effect reduces the antiproton flux at low energy. In any case, a detailed model of antiproton diffusion at the GC has not been developed so far.

3. Conclusion

I have reviewed the main results for the study of heavy DM sources at the GC through the observation of different cosmic-rays. The gamma-rays analysis show that HESS and Fermi-LAT data are well fitted with a ≈ 50 TeV DM candidate annihilating into W^+W^-, among other channels. The expected neutrino flux from the same source may be detected after an improvement of the resolution angle of current generation of neutrino detectors. Antiprotons observation at higher energies may also set more stringent constraints on the model. Heavy DM candidate with mass of tens of TeV scale is unconstrained so far. Future investigations will be developed in the direction of both indirect and direct searches, DM model building and deeper understanding of the astrophysics at the GC.

This work was supported by UCM FPI grants G/640/400/8000 (2011 Program), the Spanish MINECO projects numbers FIS2011-23000, FPA2011-27853-C02-01 and MULTIDARK CDS2009-00064 (Consolider-Ingenio 2010 Programme). The author thanks José A. R. Cembranos and Antonio L. Maroto for their supervision and participation in the research reviewed in this contribution.

References

[1] A.V. Belikov, G. Zaharijas, J. Silk, Phys. Rev. D **86**, 083516 (2012). A. A. Abdo et al. [arXiv:astro-ph.CO/1001.4531v1] (2010). T. Linden, E. Lovegrove and S. Profumo, arXiv:1203.3539 [astro-ph.HE]. K. Kosak, H. M. Badran, I. H. Bond et al., ApJ **608**, L97 (2004). F. Aharonian, A. G. Akhperjanian, K.M. Aye et al. A&A **425**, L13 (2004b). J. Albert, E. Aliu, H. Anderhub et al., ApJ **638**, L101 (2006). MAGIC collaboration, [arXiv:astro-ph/1103.0477v1] (2011). Maier G. [arXiv:astro-ph.IM/0907.5118v1]. The CTA consortium [arXiv:astro-ph.IM/1003.3703v2], `http://www.cta-observatory.org/`
[2] F. Aharonian, A. G. Akhperjanian, K.M. Aye et al., A&A **503**, 817 (2009)
[3] M. Chernyakova et al., ApJ **726**, 60 (2011), [arXiv:1009:2630]
[4] J. A. R. Cembranos, A. de la Cruz-Dombriz, V. Gammaldi, R. A. Lineros, A. L. Maroto, JHEP **1309** (2013) 077
[5] J. A. R. Cembranos, V. Gammaldi, A. L. Maroto [arXiv:1410.6689], Accepted in JCAP
[6] J.S. Perko, A&A **184**, 119 (1987)
[7] J. A. R. Cembranos, V. Gammaldi, A. L. Maroto, JCAP **1304** (2013) 051; Phys. Rev. D **86**, 103506 (2012)
[8] A. Dobado and A. L. Maroto, Nucl. Phys. B **592**, 203 (2001); J. A. R. Cembranos, A. Dobado and A. L. Maroto, Phys. Rev. Lett. **90**, 241301 (2003); Phys. Rev. D **68**, 103505 (2003); A. L. Maroto, Phys. Rev. D **69**, 043509 (2004); Phys. Rev. D **69**, 101304 (2004); Int. J. Mod. Phys. **D13**, 2275 (2004). J. A. R. Cembranos et al., JCAP **0810**, 039 (2008)
[9] J. A. R. Cembranos, A. de la Cruz-Dombriz, V. Gammaldi and A. L. Maroto, Phys. Rev. D **85**, 043505 (2012)
[10] E. Komatsu et al. [WMAP Collaboration], ApJ. Suppl. **192**, 18 (2011)
[11] Planck Collaboration (Ade, P.A.R. et al.) A&A (2014) arXiv:1303.5076 [astro-ph.CO]
[12] J. A. R. Cembranos, V. Gammaldi, A. L. Maroto, Phys. Rev. D **90**, 043004 (2014)

EPJ Web of Conferences **121**, 06004 (2016) DOI: 10.1051/epjconf/201612106004
RICAP-2014

Secluded Dark Matter search in the Sun with the ANTARES neutrino telescope

Silvia Adrián-Martínez[a] on behalf of the ANTARES Collaboration

Institut d'Investigació per a la Gestió Integrada de les Zones Costaneres (IGIC), Universitat Politècnica de València, C/Paranimf, nº1, 46730 Grao de Gandia, València, Spain

Abstract. Models where Dark Matter (DM) is secluded from the Standard Model (SM) via a mediator have increased their presence during the last decade to explain some experimental observations. This is a special scenario where DM, which would gravitationally accumulate in sources like the Sun, the Earth or the Galactic Centre, is annihilated into a pair of non-standard Model mediators which subsequently decay into SM particles, two co-linear muons for example. As the lifetime of the mediator could be large enough, its decay may occur in the vicinity of the Earth and the resulting SM particles could be detected. In this work we will describe the analysis for Secluded Dark Matter (SDM) annihilation from the Sun with ANTARES in three different cases: a) detection of di-muons that result of the mediator decay, or neutrino detection from: b) mediator that decays into di-muon and, in turn, into neutrinos, and c) mediator that directly decays into neutrinos. The ANTARES limits for these kinds of SDM case will be presented.

1. Introduction

The possibilities of DM detection have been motivated by its gravitational capture, in massive objects like the Sun, and subsequent annihilation. If as expected DM self-annihilates, the capture is balanced by the annihilation of DM particles. The intensity of the annihilation signal would be a probe of the DM scattering cross-section on nucleons. Neutrinos could be produced in the annihilations, also SM particles which interact strongly with the interior of the Sun being largely absorbed but, during this process, producing high-energy neutrinos which could scape and can be potentially seen by neutrino detectors as ANTARES. Limits from ANTARES on WIMP DM annihilation in the Sun have been reported in [1], also from Baksan [2], Super-Kamiokande [3] and IceCube [4]. Another popular explanation is based in the idea that DM annihilates first into metastable mediators (ϕ), which subsequently decay into SM states, [5–9]. In all of these models, the thermal relic WIMP DM scenario is considered as usual while there is also the potential to explain some astrophysical observations as the positron excess observed by PAMELA [10] or FERMI [11]. In the Secluded Dark Matter scenario, the presence of a mediator, as a communication way between DM and SM, can dramatically change the annihilation signature of DM captured in the Sun. If the mediators are long-lived enough to escape the Sun before decaying, they can produce detectable charged-particle or γ-ray fluxes [12, 13] and also neutrinos that could reach the Earth and be detected. In many of the secluded dark matter models, ϕ can decay into leptons near the Earth. Some differences appear in the leptons created by the neutrino interaction

[a]e-mail: siladmar@upv.es

and the leptons arising from ϕ decays. In the latter case, for kinematics, as the DM mass is greater than the ϕ mass the leptons may be boosted and parallel. If these leptons are muons the signature in the vicinity of the detector would be two muon tracks almost parallel. The di-muon signature could be interpreted as a single muon and it could be discriminated (or at least the selection could be optimized for these cases) from the atmospheric neutrino signal by their energy deposition topology. Even being short-lived and decaying inside the Sun energetic neutrinos would remain the only signature. Also in these situations the neutrino signal could be enhanced compared to the standard scenario where high energy neutrinos can interact with nuclei and be absorbed before escaping the Sun. The fact that the solar density decreases exponentially with radius facilitates that the neutrinos injected by ϕ at larger radii propagate out of the Sun because they are subjected to much less absorption. In this work an indirect search for SDM using the 2007–2012 data recorded by the ANTARES neutrino telescope is reported by looking at the different mediator decay products: a) di-muons (a good discrimination between di-muons and single muon is not the priority, but to obtain the best efficiency to detect di-muons from secluded dark matter in the Sun), b) di-muons which in turn, decay into neutrinos and c) neutrinos.

1.1 The ANTARES Neutrino Telescope

The ANTARES [14] neutrino detector was completed in 2008 and is presently the largest neutrino telescope in the northern hemisphere. Its scientific scope is very broad, but the two main goals are the observation of astrophysical sources and the indirect detection of dark matter. The latter is possible through neutrinos produced after the annihilation of WIMPs, which would accumulate in sources like the Sun, the Earth or the Galactic Centre. The operation principle is as follows: when a high energy neutrino interacts via charged current inside the detector or close to it, it produces a relativistic muon which, in the water, induces Cherenkov light observable by the photomultipliers (PMTs). The information of the time, position and amplitude of the photon signals in the PMTs is used in order to reconstruct the muon track and therefore the direction of the original neutrino. The muon track is reconstructed with the BBFit algorithm [15], which provides an angular resolution of about two degrees at energies of tens of GeVs for the selected tracks (less than two degrees for higher energies).

2. Signal and background simulation

Two main sources of background are present in ANTARES: 1) Down-going atmospheric muons resulting from the interaction of cosmic rays in the atmosphere. Almost all of this is reduced by the deep sea location and by the reconstruction algorithms that are tuned to up-going events. Cuts on the quality of the tracks are also applied to reject down-going muons wrongly reconstructed as up-going. 2) Atmospheric neutrinos produced by cosmic rays. These neutrinos can traverse the Earth, so they can be detected as upgoing tracks. This is an irreducible background. For background estimation scrambled data (randomizing the time) during the period under study has been used. This allows to reduce the effect of systematic uncertainties (efficiency of the detector, assumed flux, etc.). The data used correspond to the period from 2007–2012. During almost all 2007 only 5 lines were installed. The number of operative lines was increasing until arriving to 12 in 2008. In order to test the SDM model, a new tool for Di-Muon signal generation (DiMugen) has been developed. DiMugen generates and propagates di-muons coming from decay of mediators resulting of dark matter annihilation. For this analysis, the mediator arrives from Sun direction following the zenith and azimuth info about the Sun position during the period under study with respect to the

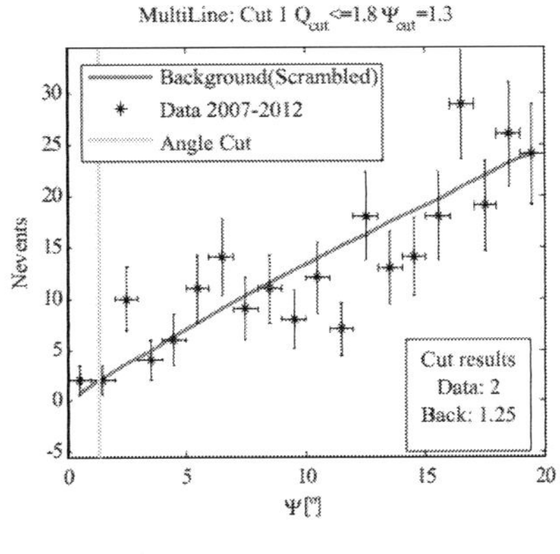

Figure 1. Differential distribution of the angular separation of the even tracks with respect to the Sun's direction for the expected background (red line) compared to the data (black).

ANTARES position. Different DM masses in the range between 30 GeV to 10 TeV have been simulated and a typical ϕ mass of 1 GeV. Di-muon flux sensitivity would be given by Eq. (1). For the cases where the neutrino is the final decay product that arrives to the Earth we have used the ANTARES effective areas for neutrinos in this energy range. In b) neutrino spectra have been obtained by Michel's spectra [16] taking into account the boost. For c) long lifetime mediators have been considered, for these cases the neutrino spectra is almost flat in the energy region under study [17]. Neutrino flux sensitivity would be given by Eq. (2). We have considered the more conservative assumption that after oscillations all neutrino flavours arrive to the Earth with the same ratio 1:1:1.

$$\Phi_{di-\mu} = \frac{\left(\frac{N_{sim}}{T_{live}}\right)\frac{\bar{\mu}_{90}}{n_s}}{\left(\left(\frac{4}{3}\right)\pi R_{sph}^{3}\right)^{2/3}} \quad (1) \qquad \Phi_{\upsilon+\bar{\upsilon}} + \frac{\bar{\mu}_{90}/Aeff_{\upsilon+\bar{\upsilon}}}{T_{live}}. \quad (2)$$

3. Optimization of the event selection criteria

In this analysis a blinding policy in order to avoid biases in the event selection has been followed. The values of the cuts have been chosen before looking at the region where the signal is expected. Best sensitivities for di-muon flux and cross-sections are extracted with the Model Rejection factor (MRF) method [18]. It consists of finding the set of cuts which provide, in average, the best flux upper limit. MRF is used to optimize the half-con angle around the sun and the track quality cut parameters (Q).

4. Results and discussion

After a detailed study of three different SDM scenarios the cut optimization procedure provided a pair of optimised values Q and ψ for each DM mass and for each SDM scenario. Finally it was decided, for the unblinding, selecting a total of 6 cuts (on quality parameter (Q_{cut}) and angular distance to the source (ψ_{cut}) for all considered masses and scenarios.

Figure 1 shows the distribution of the angular separation between the events and the Sun's direction obtained after applying the selection criteria on the zenith angle, the minimum number of hits and lines, and a Q_{cut} =1.8. No statistically significant excess has been

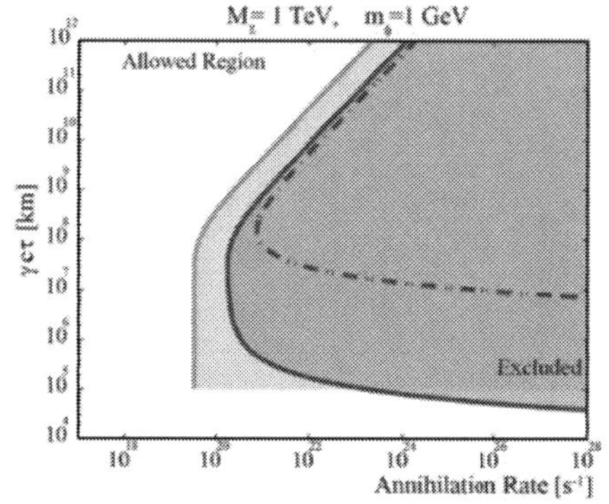

Figure 2. ANTARES exclusion limits for the three SDM scenarios (products of DM annihilation in the Sun through mediators decaying into: dimuons (dash-dotted blue), neutrinos from dimuons (solid blue), directly into neutrinos (green)) as a function of the annihilation rate and the decay length.

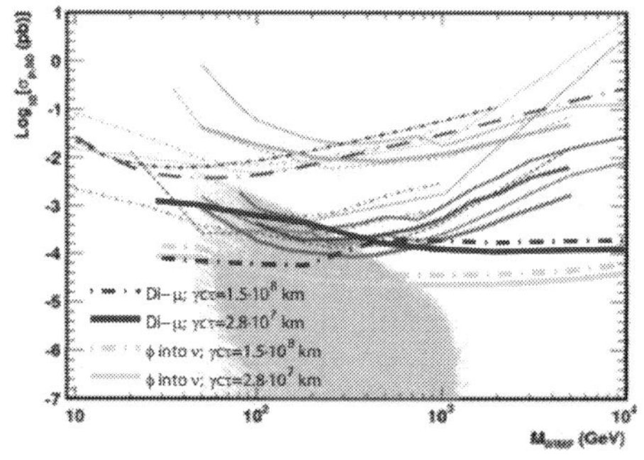

Figure 3. Sensitivities collection of different experiments of search for dark matter in the Sun. ANTARES 2007–2008 (thin solid lines: W^+W^-(blue), $\tau^+\tau^-$(red), $b\bar{b}$ (green)), ANTARES 2007–2012 for the three annihilation channels (thick solid lines) [1], IceCube-79 (dashed lines) [4], SuperKamiokande (colored dash-dotted lines) [3], SIMPLE (black short dash-dotted line) [22], COUPP (black long dash-dotted line) [23], SDM ANTARES limits (super-thick lines. For colour identification see legend).

observed above the expected background in the Sun's direction in neither of the cuts. It has been proceed to put limits for this model. Taking into account the number of events observed and the expected background, the upper limit for di-muon or neutrino flux can be obtained as the 90% confidence interval in the Feldman-Cousins approach [19]. Considering this, limits for the annihilation rates [20] have been obtained for each scenario. Figure 2 shows the ANTARES exclusion limits for the three Secluded DM scenarios for DM mass of 1 TeV and the typical ϕ mass of 1 GeV. Assuming that the annihilation balances the DM capture in the Sun,$\Gamma_{ann} = C_{\odot}/2$, the capture is approximately [21]:

$$C_{\odot} \simeq 10^{20}\mathrm{s}^{-1}\left(\frac{1TeV}{M_{\chi}}\right)^2 \frac{2.77\sigma_{SD} + 4.27\cdot 10^3\sigma_{SI}}{10^{-40}\mathrm{cm}^2} \tag{3}$$

where, σ_{sc} and σ_{SI} are the spin-dependent (SD) and spin-independent (SI) cross sections respectively. The limits on the SD and SI WIMP-proton scattering cross-sections are derived for the case in which one or the other is dominant. Sensitivity in terms of the annihilation rates depends on the lifetime of the ϕ. For cross-section evaluation it will be necessary fixing its lifetime. To see the maximum potential for these models, lifetime values for which the sensitivities are the best have been assumed. For the dimuon case, the lifetime has to be long enough to assure that the mediator reach the vicinity of the Earth. Following this, the best sensitivity would be obtained for mediators with decay length about Sun-Earth distance. In both neutrino cases the lifetime of the mediator for best sensitivity has to be long enough to ensure that the mediator escapes of the solar dense core, but not too long so that it decays before reaching the Earth. The lifetime of the mediator for the best sensitivity must be enough to travel an approximately distance of forty times the solar radius (value obtained by minimization of the decay probability equation). To finalize the cross section limits are presented and compared with other experiments. Figure 3 shows Antares proton-WIMP cross-section limits for SDM scenario (products of DM annihilation in the Sun through mediators decaying into: di-muons (blue) and directly into neutrinos (green)) for favourable mediator life time values.

We acknowledge the financial support of the Spanish Ministerio de Ciencia e Innovación (MICINN) and Ministerio de Economía y Competitividad (MINECO), Grants Consolider MultiDark CSD2009-00064 and FPA2012-37528-C02-02, and of the Generalitat Valenciana, Grants ACOMP/2014/153, ACOMP/2015/175 and PrometeoII/2014/079.

References

[1] S. Adrián-Martínez et al., JCAP **2013**, 032 (2013)
[2] M. M. Boliev et al., JCAP **1309**, 019 (2013)
[3] T. Tanaka et al., Astrophys. J **742**, 78 (2011)
[4] IceCube. collaboration, Phys. Rev. Lett. **110** (2013)
[5] M. Pospelov, A. Ritz, M. B. Voloshin, Phys. Lett. B **662**, 53–61 (2008)
[6] N. Arkani-Hamed, D. P. Finkbeiner, T. R. Slatyer, N. Weiner, Phys. Rev. D **79**, 015014 (2009)
[7] M. Pospelov, A. Ritz, Phys. Lett. B **671**, 391–397 (2009)
[8] I.Z. Rothstein, T. Schwetz, J. Zupan, JCAP **07**, 18 (2009)
[9] F. Chen, J. M. Cline, A. R. Frey, Phys. Rev. D **80**, 083516 (2009)
[10] O. Adriani et al., Phys. Rev. Lett. **105**, 121101 (2010)
[11] A. A. Abdo et al., Phys. Lett. **102**, 181101 (2009)
[12] M. Batell, A. Pospelov, Y. Ritz, Shang, Phys. Rev. D **81**, 075004 (2010)
[13] P. Schuster, N. Toro, N. Weiner, I. Yavin, Phys. Rev. D **82**, 115012 (2010)
[14] M, Aegon et al., ANTARES, Nucl. Instr. Meth. Phys. Res. A **656**, 11–38 (2011)
[15] J. A. Aguilar et al., Astropart. Phys. **34**, 652–662 (2011)
[16] P. Divari, S. Galanopoulos, G. Souliotis, J. Phys. G. **39**, 095204 (2012)
[17] N. F. Bell, K. Petraki, JCAP, **2011**, 003 (2011)
[18] G. C. Hill, K. Rawlins, Astropart, Phys. **19**, 393 (2003)
[19] G. J. Feldman, R. D. Cousins, Phys. Rev. D. **57**, 3873–3889 (1998)
[20] P. Meade, S. Nussinov, M. Papucci, T. Volansky, J. High Energy Phys. **2010**, 1–25 (2010)
[21] A. Ibarra, M. Totzauer, S. Wild, JCAP **1404**, 012 (2014)
[22] M. Felizardo et al. Phys. Rev. Lett. **108**, 201302 (2012)
[23] E. Behnke et al. Phys. Rev. D **86**, 052001 (2012)

EPJ Web of Conferences **121**, 06005 (2016) DOI: 10.1051/epjconf/201612106005
RICAP-2014

DAMA/LIBRA-phase1 results and perspectives of the phase2

R. Bernabei[1,2], P. Belli[1,2], F. Cappella[3], V. Caracciolo[3], R. Cerulli[3], C.J. Dai[4], A. d'Angelo[5,6], S. d'Angelo[1,2], A. Di Marco[1,2], H.L. He[4], A. Incicchitti[5,6], H.H. Kuang[4], X.H. Ma[4], F. Montecchia[2,7], X.D. Sheng[4], R.G. Wang[4], and Z.P. Ye[4,8]

[1] Dip. di Fisica, Università di Roma "Tor Vergata", 00133 Rome, Italy
[2] INFN, sez. Roma "Tor Vergata", 00133 Rome, Italy
[3] Laboratori Nazionali del Gran Sasso, I.N.F.N., Assergi, Italy
[4] Key Laboratory of Particle Astrophysics, Institute of High Energy Physics, Chinese Academy of Sciences, PO Box 918/3, 100049 Beijing, China
[5] Dip. di Fisica, Università di Roma "La Sapienza", 00185 Rome, Italy
[6] INFN, sez. Roma, 00185 Rome, Italy
[7] Dip. di Ingegneria Civile e Ingegneria Informatica, Università di Roma "Tor Vergata", 00133 Rome, Italy
[8] University of Jing Gangshan, Jiangxi, China

Abstract. The DAMA/LIBRA experiment ($\sim$ 250 kg of highly radio-pure NaI(Tl)) is running deep underground at the Gran Sasso National Laboratory (LNGS) of the I.N.F.N. Here we briefly recall the results obtained in its first phase of measurements (DAMA/LIBRA–phase1; total exposure: 1.04 ton $\times$ yr). DAMA/LIBRA–phase1 and the former DAMA/NaI (cumulative exposure: 1.33 ton $\times$ yr) give evidence at 9.3 σ C.L. for the presence of DM particles in the galactic halo by exploiting the model-independent DM annual modulation signature. No systematic or side reaction able to mimic the exploited DM signature has been found or suggested by anyone over more than a decade. At present DAMA/LIBRA–phase2 is running with increased sensitivity.

1. Introduction

The DAMA project is based on the development and use of low background scintillators. In particular, the second generation DAMA/LIBRA apparatus [1–13], as the former DAMA/NaI (see for example Ref. [8, 14, 15] and references therein), is further investigating the presence of DM particles in the galactic halo by exploiting the model independent DM annual modulation signature, originally suggested in the mid 80's [16]. At present DAMA/LIBRA is running in its phase2 with increased sensitivity. The detailed description of the DAMA/LIBRA set-up during the phase1 has been discussed in details in Ref. [1–4, 8].

The signature exploited by DAMA/LIBRA (the model independent DM annual modulation) is a consequence of the Earth's revolution around the Sun; in fact, the Earth should be crossed by a larger flux of DM particles around $\simeq$ 2 June (when the projection of the Earth orbital velocity on the Sun velocity with respect to the Galaxy is maximum) and by a smaller one around $\simeq$ 2 December (when the two velocities are opposite). This DM annual modulation signature is very effective since the effect induced by DM particles must simultaneously satisfy many requirements: the rate must contain a component modulated according to a cosine function (1) with one year period (2) and a phase peaked

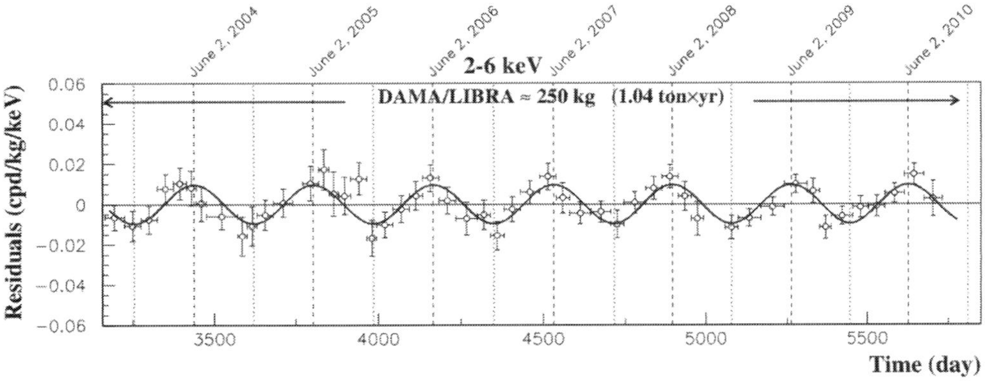

Figure 1. Experimental residual rate of the *single-hit* scintillation events measured by DAMA/NaI and DAMA/LIBRA–phase1 in the (2–6) keV energy interval as a function of the time. The data points present the experimental errors as vertical bars and the associated time bin width as horizontal bars. The superimposed curves are the cosinusoidal functions behaviors $A\cos\omega(t-t_0)$ with a period $T=\frac{2\pi}{\omega}=$ 1 yr, a phase $t_0=152.5$ day (June 2nd) and modulation amplitudes, A, equal to the central values obtained by best fit on the data points. The dashed vertical lines correspond to the maximum expected for the DM signal (June 2nd), while the dotted vertical lines correspond to the minimum.

roughly $\simeq$ 2 June (3); this modulation must only be found in a well-defined low energy range, where DM particle induced events can be present (4); it must apply only to those events in which just one detector of many actually "fires" (*single-hit* events), since the DM particle multi-interaction probability is negligible (5); the modulation amplitude in the region of maximal sensitivity must be $\simeq$ 7% for usually adopted halo distributions (6), but it can be larger (even up to $\simeq$ 30%) in case of some possible scenarios such as e.g. those in Ref. [17, 18]. Thus this signature is model independent, very discriminating and, in addition, it allows the test of a large range of cross sections and of halo densities. This DM signature might be mimicked only by systematic effects or side reactions able to account for the whole observed modulation amplitude and to simultaneously satisfy all the requirements given above. No one is available [1–4, 7, 8, 12–15, 19].

2. The results of DAMA/LIBRA–phase1 and DAMA/NaI

The total exposure of DAMA/LIBRA–phase1 is 1.04 ton $\times$ yr in seven annual cycles; when including also that of the first generation DAMA/NaI experiment it is 1.33 ton $\times$ yr, corresponding to 14 annual cycles [2–4, 8].

To point out the presence of the signal the time behaviour of the experimental residual rates of the *single-hit* scintillation events for DAMA/NaI and DAMA/LIBRA–phase1 in the (2–6) keV energy interval are plotted in Fig. 1. The χ^2 test excludes the hypothesis of absence of modulation in the data: χ^2/d.o.f. = 83.1/50 for the (2–6) keV energy interval (P-value = 2.2 $\times$ 10^{-3}). When fitting the *single-hit* residual rate of DAMA/LIBRA–phase1 together with the DAMA/NaI ones, with the function: $A\cos\omega(t-t_0)$, considering a period $T=\frac{2\pi}{\omega}=1$ yr and a phase $t_0=152.5$ day (June 2nd) as expected by the DM annual modulation signature, the following modulation amplitude is obtained: $A=(0.0110\pm 0.0012)$ cpd/kg/keV corresponding to 9.2 σ C.L.. When the period, and the phase are kept free in the fitting procedure, the modulation amplitude is (0.0112 ± 0.0012) cpd/kg/keV (9.3 σ C.L.), the period $T=(0.998\pm0.002)$ year and the phase $t_0=(144\pm7)$ day, values well in agreement with expectations for a DM annual modulation signal. In particular, the phase is consistent with about June 2nd and is fully consistent with the value independently determined by Maximum Likelihood analysis [4][1]. The run test and the χ^2 test on the data have shown that the modulation amplitudes singularly calculated for each annual cycle of DAMA/NaI and DAMA/LIBRA–phase1 are normally fluctuating around their best fit values [2–4].

We have also performed a power spectrum analysis of the *single-hit* residuals of DAMA/LIBRA–phase1 and DAMA/NaI [8], obtaining a clear principal mode in the (2–6) keV energy interval at a frequency of 2.737×10^{-3} d^{-1}, corresponding to a period of $\simeq 1$ year, while only aliasing peaks are present just above.

Absence of any other significant background modulation in the energy spectrum has been verified in energy regions not of interest for DM [4]; it is worth noting that the obtained results account of whatever kind of background and, in addition, no background process able to mimic the DM annual modulation signature (that is able to simultaneously satisfy all the peculiarities of the signature and to account for the whole measured modulation amplitude) is available (see also discussions e.g. in Ref. [1–4, 7, 8, 12, 13]).

A further relevant investigation in the DAMA/LIBRA–phase1 data has been performed by applying the same hardware and software procedures, used to acquire and to analyse the *single-hit* residual rate, to the *multiple-hit* one. In fact, since the probability that a DM particle interacts in more than one detector is negligible, a DM signal can be present just in the *single-hit* residual rate. Thus, the comparison of the results of the *single-hit* events with those of the *multiple-hit* ones corresponds practically to compare between them the cases of DM particles beam-on and beam-off. This procedure also allows an additional test of the background behaviour in the same energy interval where the positive effect is observed. In particular, the residual rates of the *single-hit* events measured over the DAMA/LIBRA–phase1 annual cycles are reported in Ref. [4], as collected in a single cycle, together with the residual rates of the *multiple-hit* events, in the (2–6) keV energy interval. A clear modulation satisfying all the peculiarities of the DM annual modulation signature is present in the *single-hit* events, while the fitted modulation amplitude for the *multiple-hit* residual rate is well compatible with zero: $-(0.0005 \pm 0.0004)$ cpd/kg/keV in the same energy region (2–6) keV. Thus, again evidence of annual modulation with the features required by the DM annual modulation signature is present in the *single-hit* residuals (events class to which the DM particle induced events belong), while it is absent in the *multiple-hit* residual rate (event class to which only background events belong). Similar results were also obtained for the last two annual cycles of the DAMA/NaI experiment [15], when the electronics allowed it. Since the same identical hardware and the same identical software procedures have been used to analyse the two classes of events, the obtained result offers an additional strong support for the presence of a DM particle component in the galactic halo.

By performing a maximum-likelihood analysis of the *single-hit* scintillation events, it is possible to extract from the data the modulation amplitude, S_m, as a function of the energy considering $T = 1$ yr and $t_0 = 152.5$ day. Again the results have shown that positive signal is present in the (2–6) keV energy interval, while S_m values compatible with zero are present just above; for details see Ref. [4]. Moreover, as described in Ref. [2–4, 8], the observed annual modulation effect is well distributed in all the 25 detectors, the annual cycles and the energy bins at 95% C.L. Further analyses have been performed. All of them confirm the evidence for the presence of an annual modulation in the data satisfying all the requirements for a DM signal.

[1] For completeness, we recall that a slight energy dependence of the phase could be expected in case of possible contributions of non-thermalized DM components to the galactic halo, such as e.g. the SagDEG stream [20–22] and the caustics [23].

Sometimes naive statements were put forwards as the fact that in nature several phenomena may show some kind of periodicity. The point is whether they might mimic the annual modulation signature in DAMA/LIBRA (and former DAMA/NaI), i.e. whether they might be not only quantitatively able to account for the observed modulation amplitude but also able to contemporaneously satisfy all the requirements of the DM annual modulation signature. The same is also for side reactions. This has already been deeply investigated in Ref. [1–4] and references therein; the arguments and the quantitative conclusions, presented there, also apply to the entire DAMA/LIBRA–phase1 data. Additional arguments can be found in Ref. [7, 8, 12, 13].

No modulation has been found in any possible source of systematics or side reactions; thus, cautious upper limits on possible contributions to the DAMA/LIBRA measured modulation amplitude are summarized in Ref. [2–4]. It is worth noting that they do not quantitatively account for the measured modulation amplitudes, and also are not able to simultaneously satisfy all the many requirements of the signature. Similar analyses have also been done for the DAMA/NaI data [14, 15]. In particular, in Ref. [13] a simple and intuitive way why the neutrons, the muons and the solar neutrinos cannot give any significant contribution to the DAMA annual modulation results is outlined.

In conclusion, DAMA give model-independent evidence (at 9.3σ C.L. over 14 independent annual cycles) for the presence of DM particles in the galactic halo.

As regards comparisons, we recall that no direct model independent comparison is possible in the field when different target materials and/or approaches are used; the same is for the strongly model dependent indirect searches. In particular, the DAMA model independent evidence is compatible with a wide set of scenarios regarding the nature of the DM candidate and related astrophysical, nuclear and particle Physics; for examples some given scenarios and parameters are discussed e.g. in Ref. [2, 8, 14] and references therein. Further large literature is available on the topics. In conclusion, both negative results and possible positive hints are compatible with the DAMA model-independent DM annual modulation results in various scenarios considering also the existing experimental and theoretical uncertainties; the same holds for the strongly model dependent indirect approaches (see e.g. arguments in Ref. [8] and references therein).

Recently the *single-hit* low energy scintillation events collected by DAMA/LIBRA–phase1 have also been investigated in terms of possible diurnal effects. In particular, a diurnal effect with the sidereal time is expected for DM because of Earth rotation; this DM second-order effect is model-independent and has several peculiar requirements as the DM annual modulation effect does. At the present level of sensitivity the presence of any significant diurnal variation and of diurnal time structures in the data can be excluded for both the cases of solar and sidereal time; in particular, the DM diurnal modulation amplitude expected, because of the Earth diurnal motion, on the basis of the DAMA DM annual modulation results is below the present sensitivity [12]. It will be possible to investigate such a diurnal effect with adequate sensitivity only when a much larger exposure will be available; moreover better sensitivities can also be achieved by lowering the software energy threshold as in the presently running DAMA/LIBRA–phase2.

For completeness we recall that other rare processes have also been searched for by DAMA/LIBRA-phase1; see for details Refs. [9–11].

3. DAMA/LIBRA–phase2 and perspectives

After a first upgrade of the DAMA/LIBRA set-up in September 2008, a more important upgrade has been performed at the end of 2010 when all the PMTs have been replaced with new ones having higher Quantum Efficiency (Q.E.), realized with a special dedicated development by HAMAMATSU co.. Details on the developments and on the reached performances in the operative conditions are reported in Ref. [6] where the feasibility to decrease the software energy threshold below 2 keV in the new configuration has been also demonstrated.

Since the fulfillment of this upgrade, DAMA/LIBRA–phase2 – after optimization periods – is continuously running in order: (1) to increase the experimental sensitivity lowering the software energy threshold of the experiment; (2) to improve the corollary investigation on the nature of the DM particle and related astrophysical, nuclear and particle physics arguments; (3) to investigate other signal features. This requires long and heavy full time dedicated work for reliable collection and analysis of very large exposures.

Another upgrade at the end of 2012 was successfully concluded: new-concept preamplifiers were installed, with suitable operative and electronic features. Moreover, further improvements are planned; in particular, new trigger modules have been prepared and ready to be installed.

Finally, further improvements to increase the sensitivity of the set-up can be considered; in particular, the use of high Q.E. and ultra-low background PMTs directly coupled to the NaI(Tl) crystals is an interesting possibility. This possible configuration can allow a further large improvement in the light collection and a further lowering of the software energy threshold. Moreover, efforts towards a possible highly radiopure NaI(Tl) "general purpose" experiment (DAMA/1ton) having full sensitive mass of 1 ton (we already proposed in 1996 as a general purpose set-up) have been continued in various aspects.

References

[1] R. Bernabei et al., Nucl. Instr. and Meth. A **592** (2008) 297
[2] R. Bernabei et al., Eur. Phys. J. C **56** (2008) 333
[3] R. Bernabei et al., Eur. Phys. J. C **67** (2010) 39
[4] R. Bernabei et al., Eur. Phys. J. C **73** (2013) 2648
[5] P. Belli et al., Phys. Rev. D **84** (2011) 055014
[6] R. Bernabei et al., J. of Instr. **7** (2012) P03009
[7] R. Bernabei et al., Eur. Phys. J. C **72** (2012) 2064
[8] R. Bernabei et al., Int. J. of Mod. Phys. A **28** (2013) 1330022
[9] R. Bernabei et al., Eur. Phys. J. C **62** (2009) 327
[10] R. Bernabei et al., Eur. Phys. J. C **72** (2012) 1920
[11] R. Bernabei et al., Eur. Phys. J. A **49** (2013) 64
[12] R. Bernabei et al., Eur. Phys. J. C **74** (2014) 2827
[13] R. Bernabei et al., Eur. Phys. J. C **74** (2014) 3196
[14] R. Bernabei el al., La Rivista del Nuovo Cimento **26** (1) (2003) 1–73
[15] R. Bernabei et al., Int. J. Mod. Phys. D **13** (2004) 2127
[16] K.A. Drukier et al., Phys. Rev. D **33**, 3495 (1986); K. Freese et al., Phys. Rev. D **37**, 3388 (1988)
[17] D. Smith and N. Weiner, Phys. Rev. D **64** (2001) 043502; D. Tucker-Smith and N. Weiner, Phys. Rev. D **72** (2005) 063509; D. P. Finkbeiner et al, Phys. Rev. D **80** (2009) 115008
[18] K. Freese et al., Phys. Rev. D **71** (2005) 043516; K. Freese et al., Phys. Rev. Lett. **92** (2004) 11301
[19] R. Bernabei et al., Eur. Phys. J. C **18** (2000) 283
[20] R. Bernabei et al., Eur. Phys. J. C **47** (2006) 263
[21] K. Freese et al., Phys. Rev. D **71** (2005) 043516; New Astr. Rev. **49** (2005) 193; astro-ph/0310334; astro-ph/0309279
[22] G. Gelmini, P. Gondolo, Phys. Rev. D **64** (2001) 023504
[23] F.S. Ling, P. Sikivie and S. Wick, Phys. Rev. D **70** (2004) 123503

EPJ Web of Conferences **121**, 06006 (2016) DOI: 10.1051/epjconf/201612106006
RICAP-2014

Direct Dark Matter search with XENON100

S.E.A. Orrigo[a,b] on behalf of the XENON Collaboration

Department of Physics, University of Coimbra, Coimbra, Portugal

Abstract. The XENON100 experiment is the second phase of the XENON program for the direct detection of the dark matter in the universe. The XENON100 detector is a two-phase Time Projection Chamber filled with 161 kg of ultra pure liquid xenon. The results from 224.6 live days of dark matter search with XENON100 are presented. No evidence for dark matter in the form of WIMPs is found, excluding spin-independent WIMP-nucleon scattering cross sections above $2 \times 10^{-45}\,\mathrm{cm}^2$ for a $55\,\mathrm{GeV/c}^2$ WIMP at 90% confidence level (C.L.). The most stringent limit is established on the spin-dependent WIMP-neutron interaction for WIMP masses above $6\,\mathrm{GeV/c}^2$, with a minimum cross section of $3.5 \times 10^{-40}\,\mathrm{cm}^2$ (90% C.L.) for a $45\,\mathrm{GeV/c}^2$ WIMP. The same dataset is used to search for axions and axion-like-particles. The best limits to date are set on the axion-electron coupling constant for solar axions, $g_{Ae} < 7.7 \times 10^{-12}$ (90% C.L.), and for axion-like-particles, $g_{Ae} < 1 \times 10^{-12}$ (90% C.L.) for masses between 5 and $10\,\mathrm{keV/c}^2$.

1. Introduction

Non-Barionic dark matter constitutes the 26.8% of the total energy and the 84.5% of the total matter of the known universe. The nature of the dark matter is among the fundamental open questions in modern physics. One of the most favorite dark matter candidates are the Weakly Interacting Massive Particles (WIMPs), cold and non-charged exotic relics of the Big Bang that interact with the ordinary matter only through gravity and the weak interaction. Axions and Axion-Like Particles (ALPs) are other well-motivated candidates for cold dark matter, introduced by many extensions of the Standard Model of particle physics.

The XENON program aims at the direct detection of the dark matter in the universe using dual-phase Time Projection Chambers (TPCs) of increasing sensitivity [1–3] filled with ultra pure liquid xenon (LXe). All the XENON experiments are located underground at the Laboratori Nazionali del Gran Sasso (LNGS) in Italy. The XENON100 experiment [2], having a total mass of 161 kg of LXe, is the second phase of the program and has set the most stringent limits on the WIMP-nucleon spin-independent [4] and spin-dependent [5] cross sections and on the axion-electron coupling [6]. The successor, XENON1T [3, 7], is a ton-scale TPC that will be commissioned in 2015 and it is expected to improve the sensitivity by two orders of magnitude.

2. The XENON100 experiment

The XENON100 experiment [2], aimed primarily at the detection of WIMPs, is operated at LNGS since 2009 with outstanding stability and performance. The XENON100 detector is a

[a] e-mail: Sonja.Orrigo@ific.uv.es
[b] *Present address*: IFIC, CSIC-Universidad de Valencia, 46071 Valencia, Spain

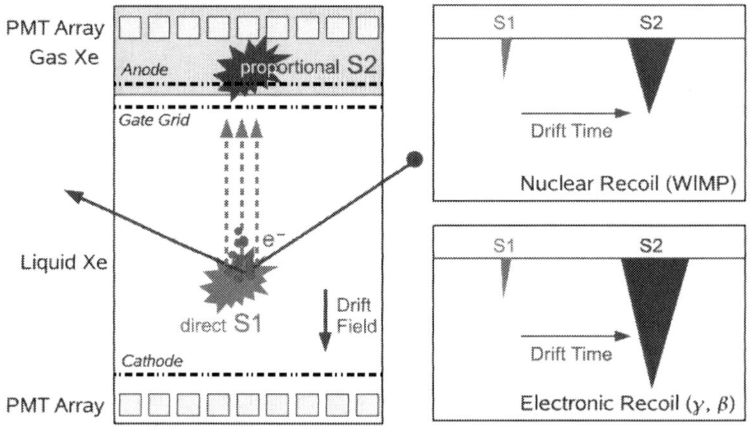

Figure 1. Left: working principle of the XENON TPC. Right: the signal ratio, $S2/S1$, allows discrimination between nuclear and electronic recoils.

position-sensitive two-phase (liquid-gas) TPC with a total mass of 161 kg of LXe, divided into a sensitive volume of 62 kg and 99 kg of active veto. Ultra pure LXe acts both as a target and detector medium, with the advantage of combining a high WIMP sensitivity with excellent self-shielding capability crucial for background suppression. The TPC is mounted in a double-walled vacuum-insulated ^{316}Ti stainless steel cryostat, embedded in a passive radiation shield. Thanks to the careful material screening and selection [8] and the detector design [2], XENON100 has achieved its design goal of reducing the electromagnetic background of two orders of magnitude in comparison to XENON10 [1] while increasing the target mass by a factor of ten. The experimentally verified electromagnetic background is $< 5 \times 10^{-3}$ events/(keV$_{ee}$[1] kg day) in the low-energy region of interest for the dark matter search before signal discrimination [9].

A scheme of XENON100 is shown in Fig. 1 (left). The TPC is equipped with low-radioactivity Hamamatsu R8520-06-Al 1" square photomultiplier tubes (PMTs), 98 of which are located in the gas phase (*top* array). A second array of 80 PMTs is immersed in LXe below the cathode (*bottom* array). Finally, 64 PMTs watch the active veto volume. A particle interaction in LXe produces direct scintillation light ($S1$ signal) detected by the bottom array. Ionization electrons are also produced; they are drifted towards the liquid-gas interface by the applied electric field and extracted into the Xe gas. A second signal, $S2$, is generated via proportional scintillation in the gas. The x- and y-coordinates of the interaction vertex are inferred from the proportional scintillation hit pattern on the top array. The time difference between the $S1$ and $S2$ signals, caused by the finite electron drift velocity in LXe, is proportional to the z-coordinate. Therefore the XENON100 TPC provides full 3-dimensional vertex reconstruction on an event-by-event basis. This property allows the definition of an ultra-low-background *fiducial volume* inside the target. Moreover, the $S2/S1$ ratio provides additional background discrimination because, as shown in Fig. 1 (right), it differs for electronic recoil signals (ER, from the γ and β backgrounds) and nuclear recoils (NR, from WIMPs and neutrons).

[1] keV electron-equivalent.

3. Spin-independent and spin-dependent WIMP-nucleon couplings

The interaction of the WIMPs with the target nuclei can be spin-independent (scalar coupling), which is possible with all the nuclei, and spin-dependent (axial-vector coupling). The latter is only possible if the WIMP carries spin and with odd mass isotopes, which have a non-zero nuclear spin (^{129}Xe and ^{131}Xe in the case of XENON100). Recent results from the XENON100 experiment have put new constraints on the elastic, spin-independent [4] and spin-dependent [5] WIMP-nucleon cross sections.

224.6 live days of data were acquired between February 2011 and March 2012 [4]. The data analysis is described in details in [10]. The data were blinded below the 90% ER quantile in $log_{10}(S2/S1)$-space for $S1 < 160$ PE. All the data quality cuts and topology requirements were designed and fixed before unblinding the signal region, according to the expected characteristics of the WIMP NR signal. The latter was calibrated using a ^{241}AmBe source both before and after the science run. Calibrations of the low-energy ERs were taken regularly during the science run using ^{60}Co and ^{232}Th sources. A fit to the light yield L_y measured with all the available calibration lines [11] determined the light yield $L_y(122\,\text{keV}_{ee}) = (2.28 \pm 0.04)$ PE/keV$_{ee}$ at the applied drift field (530 V/cm). The detector response to single-electron charge signals has been characterized and used to infer crucial parameters for the experiment [12], showing an excellent sensitivity of XENON100 to small charge signals.

The NR energy scale has been inferred from the $S1$ signal as $E_{nr} = (S1/L_y)(1/L_{eff})$ (S_{ee}/S_{nr}) [13], where the relative scintillation efficiency L_{eff} was extracted using all the existing direct measurements including [14]. The NR energy scale has been verified by a detailed Monte Carlo simulation of the whole detector, including the shield [15]. The simulated $S1$ and $S2$ responses of XENON100 to NRs from an AmBe source have been compared to the experimental $S1$ and $S2$ spectra obtained by the AmBe calibration. Remarkable agreement has been found over the whole spectra (down to 2 PE in $S1$), reflecting the excellent understanding of the detector response to NRs at % level.

Possible sources-of background in the WIMP search region are NRs induced by neutrons and leaking ER events caused by the intrinsic radioactivity of LXe and the detector materials. The NR background was estimated by Monte Carlo simulations [16] and the ER background from the calibration data [9]. In the selected fiducial volume of 34 kg and for 99.75% ER rejection, the background expectation in the WIMP search region is (1.0 ± 0.2) events [4], prediction verified in the high energy sideband (30–130 PE). After unblinding the data, 2 events survived to all the requirements for WIMP-induced single-scatter NRs, which do not constitute evidence for dark matter when compared to the expected background. Moreover, the profile likelihood analysis [17] did not find any signal excess.

A limit on the spin-independent WIMP-nucleon elastic scattering cross section has been calculated [4] using the profile likelihood analysis [17], where the uncertainties in the background expectation and in the energy scale were profiled out. Poisson fluctuations in the generation of the PEs in the PMTs, which influence the energy resolution of the $S1$ signal, were also taken into account. It was assumed that the WIMPs are distributed in an isothermal halo with local density of $\rho_\chi = 0.3\,\text{GeV/cm}^3$, local circular velocity of $v_0 = 220$ km/s, and a Galactic escape velocity $v_{esc} = 544$ km/s. The 90% confidence level (C.L.) limit of Fig. 2 has a minimum at $2 \times 10^{-45}\,\text{cm}^2$ for a 55 GeV/c^2 WIMP.

The same data and event selection as the spin-independent WIMP search, 224.6 live days for 34 kg of exposure, have been used to set a limit on the spin-dependent coupling, assuming that WIMPs couple predominantly to protons or neutrons [5]. The nuclear model of [18] was employed, which yields a far superior agreement between the calculated and measured spectra of ^{129}Xe and ^{131}Xe both in energy and in the ordering of the nuclear levels. The most

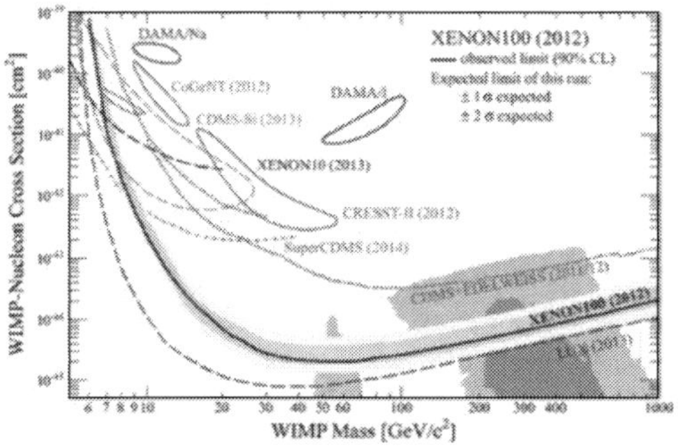

Figure 2. Spin-independent WIMP-nucleon elastic scattering cross section as a function of the WIMP mass. The XENON100 exclusion limit (90% C.L.) is shown in blue; the green/yellow bands show the expected sensitivity ($1\sigma/2\sigma$) of this science run. Other experimental limits (90% C.L.) and detection claims (2σ) are also shown for comparison.

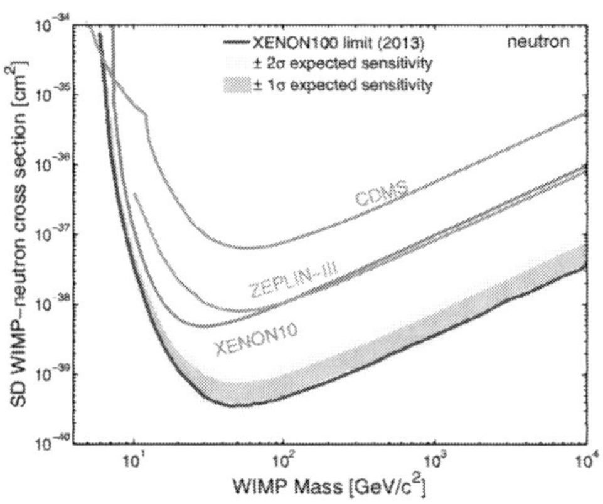

Figure 3. Spin-dependent WIMP-neutron elastic scattering cross section as a function of the WIMP mass. The XENON100 upper limit (90% C.L.) is shown by the blue line; the $1\sigma/2\sigma$ uncertainty on the expected sensitivity of the run is shown as a green/yellow band.

sensitive limit on the pure WIMP-neutron cross section has been obtained above $6\,GeV/c^2$ (Fig. 3), with a minimum at $3.5 \times 10^{-40}\,cm^2$ (90% C.L.) for a WIMP mass of $45\,GeV/c^2$. Even if the sensitivity on the pure WIMP-proton coupling is weaker because ^{129}Xe and ^{131}Xe have an unpaired neutron but even number of protons, the limit we get on the coupling to protons is also competitive [5].

4. Solar axions and axion-like particles coupling to electrons

QCD axions and ALPs are light cold DM candidates which interact predominantly with atomic electrons in the medium and can couple to photons, electrons and nuclei. The coupling to electrons, g_{Ae}, can be tested through the *axio-electric effect*. This process is analogous to the photoelectric effect, but with an axion/ALP playing the role of the photon: the axion ionizes an atom and it is absorbed, extracting an electron. Therefore axions and ALPs can scatter off the electrons of the LXe target, and so they can be searched for in XENON100 by looking at the ER events.

The same science run of 224.6 live days and 34 kg of exposure (see Sect. 3) has been used to perform the first axion searches with XENON100 [6]. The very low background and

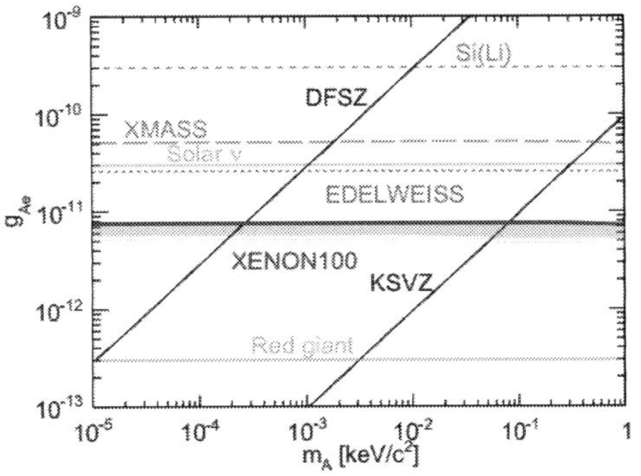

Figure 4. Axion-electron coupling constant for solar axions as a function of the axion mass. The XENON100 limit (90% C.L.) is indicated by the blue line. The expected sensitivity, based on the background hypothesis, is shown by the green/yellow bands ($1\sigma/2\sigma$). Recent experimental constraints, astrophysical bounds and theoretical benchmark models are also shown for comparison.

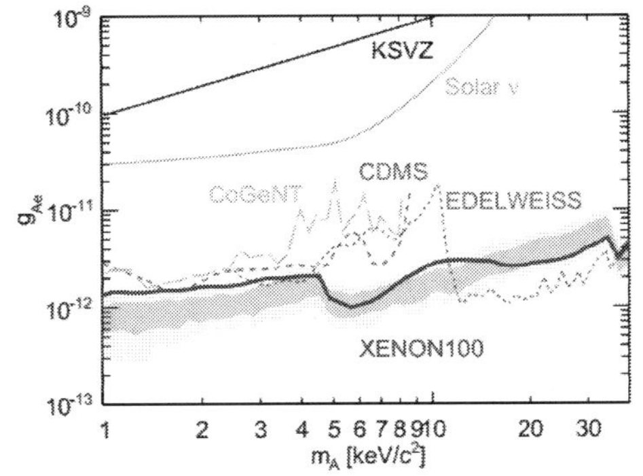

Figure 5. Axion-electron coupling constant for galactic axion-like particles as a function of their mass. The XENON100 limit (90% C.L.) is shown by the blue line. The green/yellow bands indicate the expected sensitivity. Constraints set by other experiments, the solar neutrinos astrophysical bound and the benchmark KSVZ model are also shown.

threshold of XENON100 are crucial for these analyses, leading to the first axion results from a dual-phase TPC.

Solar axions are postulated to be produced in the Sun via Bremsstrahlung, Compton scattering, atomic recombination and atomic de-excitation. The event distribution for solar axions in XENON100 is expected to be a continuum [6]. A profile likelihood analysis gives no signal excess for the background-only hypothesis, thus the data are compatible with the background model. The best limit to date, shown in Fig. 4, is set on the axion-electron coupling constant for solar axions with masses below 1 keV/c^2, $g_{Ae} < 7.7 \times 10^{-12}$ (90% C.L.). In the frame of the DFSZ and KSVZ models, we exclude QCD axions heavier than 0.3 eV/c^2 and 80 eV/c^2, respectively.

Galactic ALPs in the keV range may have been generated via a non-thermal production mechanism in the early universe. The axion-electron coupling for galactic ALPs can be tested in XENON100 [6], under the assumption that they constitute the whole abundance of dark matter in our galaxy. A monoenergetic peak at the axion mass is expected in the spectrum of galactic ALPs. The width of this signal is given by the energy resolution of the detector at the relevant $S1$ signal size. As for the solar axion search, the profile likelihood analysis shows no evidence for an ALP signal and so the data are consistent with the background hypothesis.

The most sensitive limit, $g_{Ae} < 1 \times 10^{-12}$ (90% C.L.), is obtained for galactic ALPs with masses between 5 and 10 keV/c^2 (see Fig. 5).

References

[1] J. Angle et al. (XENON10 Collaboration), Phys. Rev. Lett. **107**, 051301 (2011), 1104.3088
[2] E. Aprile et al. (XENON100 Collaboration), Astropart. Phys. **35**, 573–590 (2012), 1107.2155
[3] Elena Aprile. (XENON1T collaboration), Springer Proc. Phys. C **12-02-22**, 93–96 (2013), 1206.6288
[4] E. Aprile et al. (XENON100 Collaboration), Phys. Rev. Lett. **109**, 181301 (2012), 1207.5988
[5] E. Aprile et al. (XENON100 Collaboration), Phys. Rev. Lett. **111**, 021301 (2013), 1301.6620
[6] E. Aprile et al. (XENON100 Collaboration), Phys. Rev. D **90**, 062009 (2014), 1404.1455
[7] A. Rizzo. (on behalf of the XENON Collaboration), Proceedings of RICAP-14, EPJ Web of Conferences (2014)
[8] E. Aprile et al. (XENON100 Collaboration), Astropart. Phys. **35**, 43–49 (2011), 1103.5831
[9] E. Aprile et al. (XENON100 Collaboration), Phys. Rev. D **83**, 082001 (2011), 1101.3866
[10] E. Aprile et al. (XENON100 Collaboration), Astropart. Phys. **54**, 11–24 (2014), 1207.3458
[11] E. Aprile et al. (XENON100 Collaboration), Phys. Rev. Lett. **105**, 131302 (2010), 1005.0380
[12] E. Aprile et al. (XENON100 Collaboration), J. Phys. G **41**, 035201 (2014), 1311.1088
[13] E. Aprile et al. (XENON100 Collaboration), Phys. Rev. Lett. **107**, 131302 (2011), 1104.2549
[14] G. Plante, E. Aprile, R. Budnik, B. Choi, K.L. Giboni, et al., Phys. Rev. C **84**, 045805 (2011), 1104.2587
[15] E. Aprile et al. (XENON100 Collaboration), Phys. Rev. D **88**, 012006 (2013), 1304.1427
[16] E. Aprile et al. (XENON100 Collaboration), J. Phys. G **40**, 115201 (2013), 1306.2303
[17] E. Aprile et al. (XENON100 Collaboration), Phys. Rev. D **84**, 052003 (2011), 1103.0303
[18] J. Menendez, D. Gazit, and A. Schwenk. Phys. Rev. D **86**, 103511 (2012), 1208.1094

EPJ Web of Conferences **121**, 06007 (2016) DOI: 10.1051/epjconf/201612106007
RICAP-2014

Dark matter searches using superheated liquids

Bou-Cabo Manuel[2,a], Ardid Miguel[1], and Felis Ivan[1] on behalf of PICO Collaboration

[1] IUPV-IGIC, Universidad Politécnica de Valencia, Spain
[2] IFIC, Universidad de Valencia, Spain

Abstract. Direct detection of dark matter is one of the most important topics in modern physics. It is estimated that 22% of universe matter is composed by dark matter in front of 0.4% of ordinary matter like stars, galaxies planets and all kind of known astrophysical objects. Several kinds of experiments are nowadays involved in detection of one of the more accepted particle candidates to be dark matter: WIMPs (Weakly Interacting Massive Particles). These detectors, using several kinds of techniques: Cryogenic semiconductors, scintillation materials like I Na or noble gas chambers among others, are reporting very interesting but inconclusive results. In this paper a review of detectors that are using the superheated liquid technique in bubble chambers in order to detect WIMPs is reported. Basically, we will report about Coupp (Chicagoland observatory for underground particle physics), PICO that is composed by Coupp and Picasso researchers having the aim to build a ton experiment and also about a new detector named MOSCAB (Materia oscura a bolle) that recently published a first results of a test chamber that uses also superheated liquid technique but as a Geyser chamber.

1. Introduction

Direct detection of WIMPs is commonly accepted like the most used way to try to detect non baryonic dark matter of the Universe. This paper is related with the use of superheated liquid to detect nuclear recoil produced by WIMP interaction with nucleus present in target material. The most tested and succesful detector of dark matter that uses this technique is the COUPP/PICO bubble chamber detector. Basically, it consists of a bubble chamber filled with CF_3I or C_3F_8 that operates in superheated regime. Detection of nuclear recoil is based on bubble formation. When nuclear recoil is produced with enough energy and this energy is deposed in the medium in a space smaller than the critical radius for a given thermodynamic conditions a macroscopic bubble is generated. Several authors have described the growth of the bubble in a superheated liquid, the first ones were Rayleigh [1], Plesset & Zwick [2] or Foster and Zuber [3] that arrive to approximate solutions of bubble growth rate, ((1) Rayleigh – Plesset equation).

$$\frac{p_v(T_v) - P_\infty}{\rho_l} = R \cdot \frac{d^2R}{d^2t} + \frac{3}{2} \cdot \left(\frac{dR}{dt}\right)^2 + \frac{4 \cdot \eta_1}{R} \cdot \frac{dR}{dt} + \frac{2 \cdot \sigma}{\rho_1 \cdot R} \tag{1}$$

being $p_v(T_v)$ pressure within the bubble, P_∞ external pressure far away from the bubble, η_1 represents the kinematic viscosity of the fluid, σ surface tension of the bubble, ρ_1 density of fluid and R(t) is the radius of the bubble.

[a] e-mail: manuelbcabo@gmail.com

In the case that is being treated here, the liquid is in a superheated state that basically consists in having the liquid to a pressure such that is kept in liquid state at a temperature above its evaporation temperature.

The growth of the bubble is composed by three different processes attending to the nature of the bubble formation:

- Nucleation
- Regime dominated by inertia where (radius $\propto$ time)
- Regime dominated by heat diffusion where (radius $\propto$ time$^{1/2}$).

In order to determine the thermodynamic conditions of the superheated liquid in terms of pressure and temperature of the fluid that fix the desired energy threshold of the detector for nuclear recoils, the "hot spike" Seitz model [4] is applied. This model considers the necessary energy to overcome the surface energy and the latent heat, in first approximation the energy threshold can be parameterized as follows:

$$E_{th} = 4 \cdot \pi \cdot r_c^2 \cdot \left(\sigma - T \cdot \frac{\delta\sigma}{\delta T}\right) + \frac{4}{3} \cdot \pi \cdot r_c^3 \cdot \rho_v \cdot (h_b - h_l) \qquad (2)$$

where σ is the surface tension, T is the temperature of the fluid, ρ_v is the bubble vapour density, h_l, h_b are the specific enthalpies of the bubble vapor and superheated fluid and r_c represents the critical radius of the bubble.

Detection of WIMP nuclear recoils using superheated liquid technique presents important advantages with respect other techniques, such as strong rejection of minimum ionizing particles and gamma rays or electron background coming from gamma rays or beta decay, easy and cheap mechanical construction, possibility of use different target material being possible distinguish among spin dependent and spin independent interactions and measuring in a simple way multiple neutron interactions that can help to avoid later this kind of background. In next sections the operation mode of this kind detectors (Sect. 2) will be discribed and results obtained by COUPP [5, 6] collaboration will be discussed.The future plans for new merged PICO (Picasso [7] + COUPP) collaboration will be presented as well as the new european initiative, MOSCAB [8], to build a detector able to detect dark matter via WIMP nuclear recoil using superheated liquid using the "geyser" technique. Finally the conclusions of this communication will be summarized in Sect. 4.

2. Detectors and collaborations

The basic description of the bubble chamber detectors for dark matter detection consists of a fused silica quartz vessel immersed in a hydraulic fluid (propylene glycol commonly used), a pressure control unit to maintain the pressure inside and outside of the jar at a desired value to put the target material in a superheated state. All this equipment is contained inside of stainless steel pressure vessel. The inner part of the bubble jar is recorded by 2 VGA cameras with a typical sample frequency of ~100 frames/second allowing to reconstruct the spatial coordinates of bubbles inside of the vessel. The temperature is also monitored using temperature sensors mounted along the detector with the aim to control the thermodynamic conditions of target material. An important part of the instrumentation used is composed by several piezo transducers attached to the wall of the vessel. These piezo transducers can be used to register the acoustic signal produced by the bubble growth when nuclear recoil is produced. As it will be shown at next section the acoustic signal emitted by the bubble growth gives us the possibility to distinguish between neutron nuclear recoils, and alpha nuclear recoils, being very useful in terms of alpha background rejection.

Moving specifically to COUPP/PICO detectors, the superheated state of target material is achieved performing cycles of compression and decompression of the chamber. Typically the chamber is in a compressed mode at a pressure of ~215 psia, the superheated state is achieved performing a decompression going from 215 psia to tens of psia (depending of the energy threshold chosen) and after a stabilization time the detector is able to "look and listen" wimp interaction for a period of hundreds of seconds (typically around 500 s). During this time an algorithm compares consecutive frames and if some pixels change the trigger of the detector is activated and the event detected is recorded, typically 50 ms–100 ms of signals are recorded. After that, the bubble chamber is compressed and the cycle starts again.

The main difference between COUPP/PICO and MOSCAB detectors is that with the use of "geyser" technique, MOSCAB operates by controlling the temperatures of the superheated liquid and the colder vapour and therefore does not have cycles of compression and expansion. Once the bubble is created at MOSCAB detector it goes up passing through a gradient of temperature that collapses the bubble from vapour to liquid. This has the advantage of not needing to reset the bubble chamber, decreasing the dead time of the detector. During last 5 years, the COUPP collaboration developed few detectors of different size, performing several physics runs running at different levels of energy threshold.

One common system present in this kind of detectors is the acoustic system that plays an important role in background rejection. The major part of background rejection can be avoided considering the location of the detectors placed at deep underground laboratories like SNOLAB for COUPP/PICO detectors, but it is necessary to have an additional method to reject the alpha background. This rejection [9] has been achieved obtaining using the acoustic discrimination based on the calculation of an acoustic parameter that basically calculates the power spectrum density of the acoustic signal generated by the bubble growth. In COUPP/PICO detectors the acoustic parameter is obtained using Eq. (3).

$$\mathrm{AP} = A(\mathrm{T}) \cdot \sum_{\mathrm{j}} \mathrm{G_j} \cdot \sum_{\mathrm{n}} \mathrm{C_n}(\vec{\mathrm{x}}) \cdot \sum_{\mathrm{f^n_{min}}}^{\mathrm{f^n_{max}}} \mathrm{f} \cdot \mathrm{psd^j_f} \tag{3}$$

being A(T) a temperature dependent scale factor, $\mathrm{G_j}$ the gain of the acoustic transducer j, $\mathrm{C_n}(\vec{\mathrm{x}})$ a position correction factor, f is the frequency and $\mathrm{psd^j_f}$ represents the power spectrum density centered at frequency bin f for transducer j. In next section the results obtained for alpha rejection in COUPP/PICO family detectors will be shown.

The schedule of COUPP & PICO family detectors can be summarized as follows.

COUPP-4:

It was a 2–Litre bubble chamber filled with CF_3I. This detector was running at Snolab from 2010 to 2012, being stopped to implement and install some improvements in its instrumentation. Some of this changes involved the replacement of acoustic transducers, due to find some contamination at PZ 27 piezo electric ceramic, specifically were found 4 ppm of U and 2 ppm Th. New generation of piezo transducers, were developed using radio clean piezo ceramic salt. The new piezos were used during the second physics run.

COUPP-60:

Second detector implementing the superheated liquid technique was developed by the COUPP collaboration after the results of the first 4 kg bubble chamber built. The aim of this detector was increasing the effective volume of the target material filling the vessel up to 40 litres of CF_3I. In spite of all implemented systems works fine, some problems were found due to the appearance of some 'anomalous events' at the neutron recoil zone.

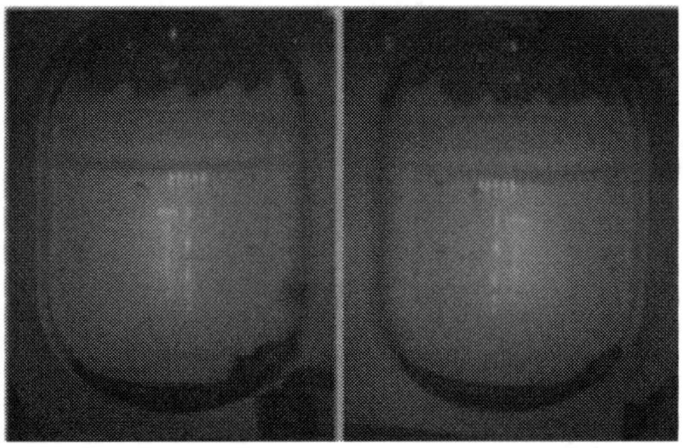

Figure 1. Detail of single bubble event at Coupp-4.

It is planned for 2015 to upgrade the system to solve this problem installing a microfiltering system to avoid micro impurities that could be the cause of the problem. Also the target material could be changed in order to use C_3F_8.

PICO-2L:

Since 2013 PICASO and COUPP collaborations merged in a new one named PICO. The PICO – 2L detector is the first detector built by the new collaboration. This 2L bubble chamber has been taken data during 2014 and the main difference with respect to the previous commented detectors, is the change of the target material, using l C_3F_8 that has twice fluorine density in comparison with CF_3I and a lower energy threshold can be achieved. After almost a year of data taking and observing arather good behaviour of all systems integrated at the detector, the detector has been stopped and probably a new physics run will be performed maybe using CF_3I.

PICO-250L:

The mean objective for PICO collaboration consists of building a ton scale detector based on 250L bubble chambers. The plan is starting to build the chamber in 2016, but the final timing will depend on the results obtained from previous COUPP/PICO detectors.

MOSCAB:

MOSCAB stands for (Materia Oscura A Bolle) and represents an european initiative promoted by U. Milano–Biccoca team leaded by professor A. Pullia, with the aim to build a bubble chamber able to detect wimp nuclear recoil. A first prototype has been tested using 0.5 kg of C_3F_8, at U. Milano–Biccoca laboratory. A 40 kg detector is now being commissioned and the future plans consider moving the detector to Gran Sasso laboratory [10].

The main results obtained for each detector are discussed in next section, giving an idea of the state of art of superheated liquid chamber detectors for dark matter searches.

3. Results

In 2010 begun to take data de first COUPP detector, that was bubble chamber filled with 2 liters of CF3I. An extensive information about results obtained can be found at [5], but can be summarized as follows.

The total exposure of the detector was 553 kg-day and were performed 3 levels of energy threshold. Multibuble events were recorded, but the counts observed were in good agreement

Figure 2. MOSCAB detector installed on Milano – Biccoca laboratory.

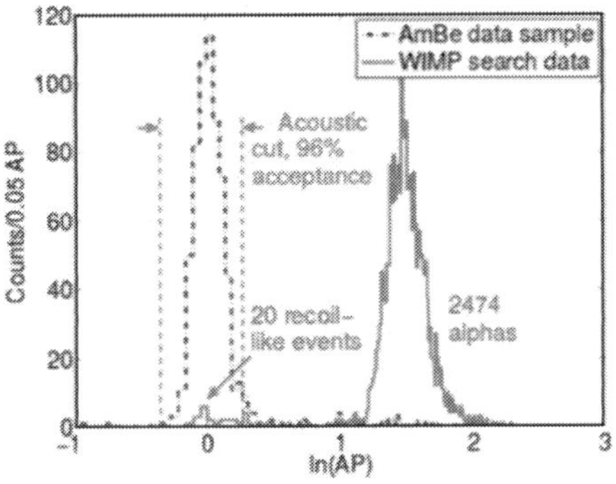

Figure 3. Distribution of acoustic parameter for 553 kg-day compilation data. The alpha background rejection presents a 96% of acceptance for neutron recoil zone. In this physics run considered 20 like WIMPs events were detected at the zone of neutron nuclear recoil.

with the total neutron background expected. The total exposure for single bubble events was 437 kg-day. Concerning about the alpha background rejection using acoustic techniques, results obtained were very good in terms of acoustic neutron acceptance ~96% and alpha background rejection ~99% as can be observed at Fig. 3.

A total of 20 like WIMP events were observed: 6 measured at 7,8 keV threshold level, 6 at 11 keV and 8 at 15,5 keV. Acoustic discrimination technique has been revealed a powerful alpha rejection method.

After obtaining these results, the next detector developed was COUPP-60. This detector has accumulated over 4,500 kg–day of exposure running 80%–95% of live time at 11keV threshold level. All systems that compose the detector worked properly during this time, and no major problems were observed during this time from technical point of view. As it can be observed in Fig. 4 the neutron AP distribution of calibration events presents a narrow shape and good defined region for neutron nuclear recoil events.

However, during the analysis of experimental data anomalous events overlapped with neutron calibration zone were found (Fig. 5). The nature of these anomalous events remains unclear yet but some hypotheses are under discussion, being considered the presence of micro impurities in the CF_3I the most probable cause.

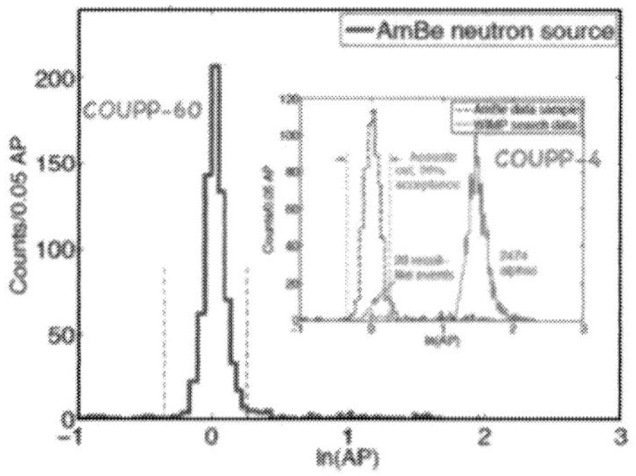

Figure 4. Acoustic parameter distribution for neutron recoil like WIMPs events (COUPP60).

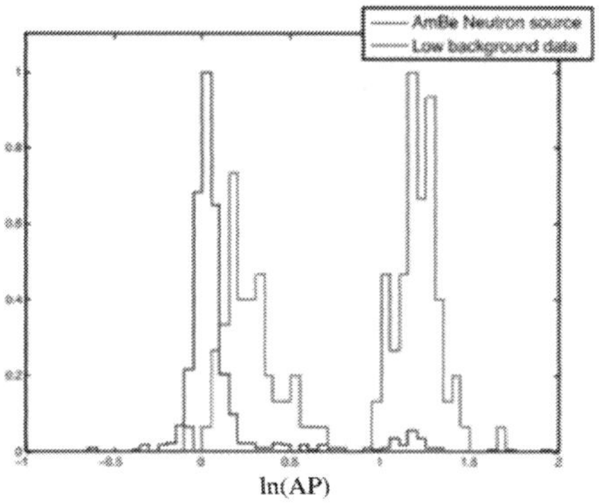

Figure 5. Overlap found in COUPP60 detector, probably due to the presence of micro impurities in the target material.

Hopefully, this data can still be used in spite of the presence of the anomalous events, going further in the acoustic analysis implementing new parameters that can discriminate as much as possible among the two kinds of events. For the near future, it is planned to install a micro-filtering system with the aim to mitigate this problem. The first detector built under the PICO collaboration is a bubble chamber of 2 liters with C_3F_8. The use of this new target material has the advantage to have twice density of fluorine with respect to previous target and being also possible to decrease the energy threshold of the detector. It is important to comment that during the physics run performed by this detector a level of 3 keV was successfully achieved. The acoustic techniques in this case have been also applied successfully and the presence of anomalous events was much more smaller, obtaining a good separation between alpha background and neutron event zone. Figure 6 presents preliminary results for this detector in relation with alpha background rejection and defined zone for WIMP-like nuclear recoil events.

With respect to the results obtained with PICO-2L, the final analysis is almost complete and soon they will be published. A second physics run is scheduled to be performed during next months.

Attending to results obtained for all COUPP/PICO family detectors, the limits (and sensitivity projections for the future PICO 250L) for spin dependent and spin independent interactions are shown in Fig. 7 and compared with other experiments.

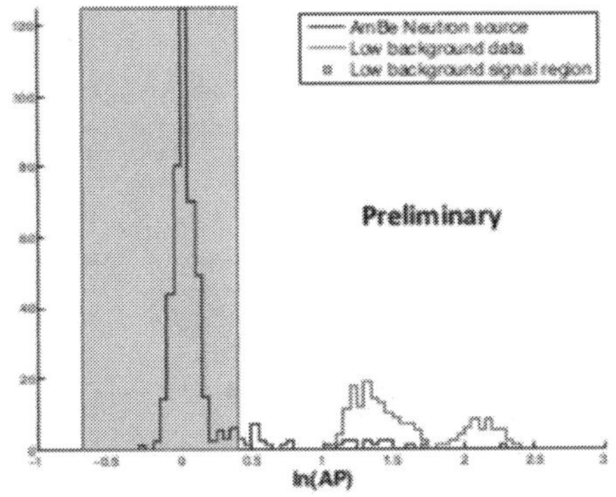

Figure 6. Acoustic parameter distribution for neutron recoil like WIMPs events (PICO2L).

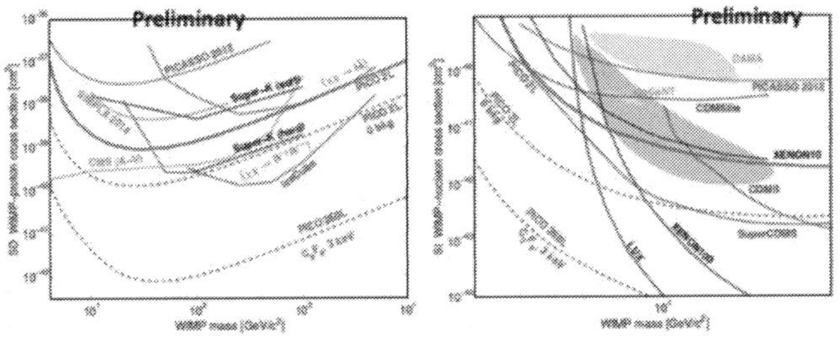

Figure 7. Limit projections for Pico 250 for Spin dependent and Spin independent interaction. Also is included in this plot preliminary results obtained for Pico 2L assuming data collected during its first physics run and also assuming 0 background.

With respect to the results obtained with PICO-2L, the final analysis is almost complete and soon they will be published. A second physics run is scheduled to be performed during next months. Attending to results obtained for all COUPP/PICO family detectors, the limits (and sensitivity projections for the future PICO 250L) for spin dependent and spin independent interactions are shown in Fig. 6 and compared with other experiments.

4. Conclusions

Summarizing all results obtained by detectors presented previously it seems clear that this kind of detectors will play an important role in direct dark matter detection. Following results obtained for PICO-2L detector the best world limit for proton-WIMP spin-dependent interaction has been achieved.

With respect to results for COUPP-60 bubble chamber, analysis of data is still ongoing trying to minimize the impact of anomalous events on physics run data. In any case, a new physics run with upgraded equipment is ongoing, so future cleaner data will be taken soon.

In addition, a new collaboration with the aim to build a dark matter detector using superheated liquids in a geyser detector has been appeared, obtaining in the first tests very good results in terms of functionality of all systems that compose the detector. News about the installation and tests of the detector in Gran Sasso facilities will come soon.

References

[1] L. Rayleigh, On the pressure developed in a liquid during the collapse of a spherical cavity (Philos. Mag **34**, 94, 1917)

[2] M.S. Plesset, The growth of vapor bubbles in superheated liquid (J. Appl. Phys. **25**, 493, 1954)

[3] H.K. Forster, Growth of vapor bubbles in superheated liquid (J. Appl. Phys. **05**, 474, 1954)

[4] F. Seitz (Phys. Fluids **1**, 2, 1958)

[5] E. Behnke et al., Coupp Collaboration, First dark matter search results from a 4-kg CF3I bubble chamber operated in a deep underground site (Phys. Rev. D **86**, 2012)

[6] Behnke et al., Coupp Collaboration, Direct measurement of the bubble-nucleation energy threshold in a CF3I bubble chamber (Phys. Rev. D **88**, 2013)

[7] S. Archambault, Picasso Collaboration, Dark Matter Spin-Dependent Limits for WIMPs Interactions on ^{19}*F by PICASSO* (Phys. Lett. B, 2009)

[8] R. Bertoni et al., A new technique for direct investigation of dark matter (Nuclear Instruments and Methods in Physics Research, A, 61–68, 2014)

[9] M.Bou-Cabo et al., LOW Radioactivity Techniques 2013 (LRT 2013): Proceedings of the IV International Workshop in Low Radioactivity Techniques, V: **1549**, 142 – 147

[10] M. Ardid et al., MOSCAB: Direct dark matter search using the Geyser technique, Proc. ICHEP 2014 Conf., Nucl. Phys. B: Proc. Supp. (in press)

EPJ Web of Conferences **121**, 06008 (2016) DOI: 10.1051/epjconf/201612106008
RICAP-2014

ANAIS: Status and prospects

J. Amaré[1,2], S. Cebrián[1,2], C. Cuesta[3], E. García[1,2], C. Ginestra[1,2], M. Martínez[1,2], M.A. Oliván[1,2,a], Y. Ortigoza[1,2], A. Ortiz de Solórzano[1,2], C. Pobes[4], J. Puimedón[1,2], M.L. Sarsa[1,2], J.A. Villar[1,2], and P. Villar[1,2]

[1] Laboratorio de Física Nuclear y Astropartículas, Universidad de Zaragoza, Pedro Cerbuna 12, 50009 Zaragoza, Spain
[2] Laboratorio Subterráneo de Canfranc, Paseo de los Ayerbe s/n, 22880 Canfranc Estación, Huesca, Spain
[3] CENPA and Department of Physics, University of Washington, Seattle, WA, USA
[4] Instituto de Ciencia de Materiales de Aragón, Universidad de Zaragoza – CSIC

Abstract. ANAIS experiment will look for dark matter annual modulation with large mass of ultra-pure NaI(Tl) scintillators at the Canfranc Underground Laboratory (LSC), aiming to confirm the DAMA/LIBRA positive signal in a model-independent way. Two 12.5 kg each NaI(Tl) crystals provided by Alpha Spectra are currently taking data at the LSC. Present status of ANAIS detectors background and general performance is summarized; in particular, thanks to the high light collection efficiency prospects of lowering the threshold down to 1 keVee are reachable. Crystal radiopurity goals are fulfilled for ^{232}Th and ^{238}U chains and ^{40}K activity, although higher than original goal, could be accepted; however, high ^{210}Pb contamination out-of-equilibrium has been identified. More radiopure detectors are being built by Alpha Spectra. The ongoing high quantum efficiency PMT tests and muon veto characterization are also presented. Finally, the sensitivity of the experiment for the annual modulation in the WIMP signal, assuming the already achieved threshold and background in ANAIS-25 is shown. Further improvement should be achieved by reducing both threshold and background, as expected.

1. Introduction

ANAIS [1, 2] (Annual modulation with NAI Scintillators) experiment will look for dark matter annual modulation with large mass of ultra-pure NaI(Tl) scintillators at the Canfranc Underground Laboratory (LSC), aiming to confirm the DAMA/LIBRA positive signal [3] in a model-independent way, using the same target and technique. The experimental goals for such an experiment have been set in an energy threshold equal or below 2 keVee and a background near the threshold below 2 counts/(keVee kg day).

2. ANAIS-25

ANAIS-25 prototype consists of two modules of 12.5 kg each provided by Alpha Spectra. The main goals for this prototype have been to measure the crystal internal contamination, determine light collection efficiency, fine tune the data acquisition and test the filtering and analysis protocols. The modules are cylindrical, 4.75″ diameter and 11.75″ length,

[a] e-mail: maolivan@unizar.es

Figure 1. The two modules of ANAIS-25 being introduced inside the shielding.

Table 1. Light collection for two ANAIS-25 modules.

Detector	PMT model	phe^-/keV
D0	Hamamatsu R12669SEL2	16.13 ± 0.66
D1	Hamamatsu R11065SEL	12.56 ± 0.13

with quartz windows for PMTs coupling. A Mylar window in the lateral face allows low energy calibration. Two types of photomultiplier have been tested: one module coupled to two Hamamatsu R12669SEL2 and the other coupled to Hamamatsu R11065SEL. The modules have been surrounded by 10 cm of archaeological lead plus 20 cm of low activity lead shielding at the Canfranc Underground Laboratory (see in Fig. 1).

This prototype has been taking data since December 2012. The first feature to be remarked is the excellent light collection as it can be seen in Table 1. This light collection has a good impact in both resolution and energy threshold. A preliminary study has been done with the coincident events in both prototype modules showing two low energy populations that have been attributed to internal ^{40}K and cosmogenic ^{22}Na as it is shown in Fig. 2. The K-shell electron binding energy following electron capture in ^{40}K (3.2 keV) and ^{22}Na (0.9 keV) can be tagged by the coincidence with a high energy γ ray (1461 keV and 1274 keV respectively). Hence, a threshold of the order of 1 keVee seems to be achievable. The trigger and filtering efficiencies at the threshold level are currently under study.

On the other hand, background contributions have been thoroughly analyzed. Figure 2 shows the low energy spectrum at the beginning of the data taking and fifteen months later, showing a high suppression of most of the lines except the corresponding to ^{210}Pb, highlighting their cosmogenic origin; a more detailed study of radionuclide production in NaI(Tl) derived from this data can be found at reference [4]. Table 2 shows the results of the activities determined for the main crystal contaminations: ^{40}K content has been measured performing coincidence analysis between 1461 keV and 3.2 keV lines [5] and the activities from ^{210}Pb and ^{232}Th and ^{238}U chains have been determined on the one hand, by quantifying Bi/Po sequences, and on the other, by comparing the total alpha rate with the low energy depositions attributable to ^{210}Pb, which are fully compatible. These results give a moderate contamination of ^{40}K, above the initial goal of ANAIS (20 ppb of K) but acceptable (see Sect. 3.5), a high suppression of ^{232}Th and ^{238}U chains but a high activity of ^{210}Pb at the mBq/kg level. The origin of such contamination was identified at crystal growing procedure and it is being solved by Alpha Spectra. A new module, grown following the new protocol, will be available for radiopurity checks at the beginning of 2015.

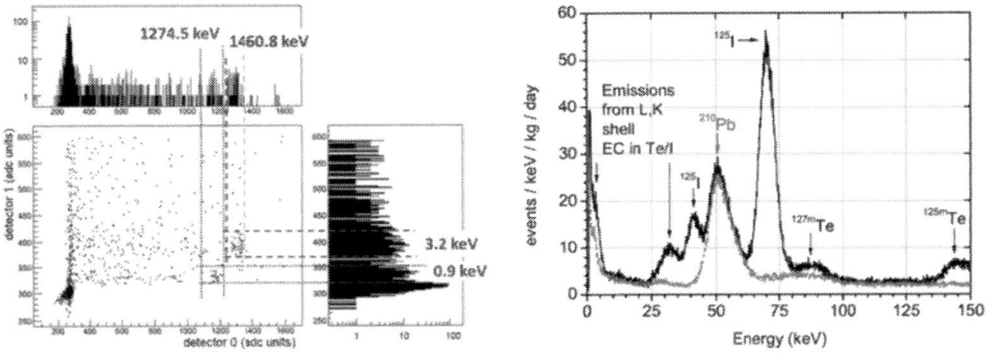

(a) (b)

Figure 2. Coincidence scatter plot showing events from ^{40}K and ^{22}Na (a) and cosmogenic radionuclide decay at low energy (b) showing the spectrum for the first month of data taking (black) and fifteen months later (red).

Table 2. Internal contamination measured in ANAIS-25 prototype.

^{40}K (mBq/kg)	^{238}U (mBq/kg)	^{210}Pb (mBq/kg)	^{232}Th (mBq/kg)
1.25 ± 0.11 (41 ppb K)	0.010 ± 0.002	~ 3.15	0.002 ± 0.001

3. Beyond ANAIS-25

The progress in the development and testing of several subsystems towards the full ANAIS experiment is reported in this section: the status of the muon veto system in Sect. 3.1, the low energy calibration system in Sect. 3.2, the PMT quality tests in Sect. 3.3 and other improvements in Sect. 3.4.

3.1 Muon veto

The muon flux at the Canfranc Underground Laboratory is approximately $5 \times 10^{-3}\ \mu s^{-1} m^{-2}$ being one order the magnitude above the flux at LNGS. For this reason, a plastic veto system to tag muon related events has been designed and tested in order to monitor possible systematic effects in the annual modulation analysis. The veto system will consist of sixteen plastic vetoes covering the top and all vertical faces of the shielding. The first underground data revealed a population faster than muon events and with a rate substantially higher than expected, which increased unnecessarily the experiment dead time if a *threshold only* strategy is chosen to tag muons. This fast population, which is also present at surface, is clearly dominating the underground rate as can be observed in Fig. 3.

A dedicated front-end setup and data acquisition chain are being developed in order to separate the fast population from the real muon events. This distinction can be performed via pulse shape analysis, using shaping amplifiers and analog-to-digital converters with different temporal windows. The muon peak and the fast population are identified at surface in order to set the hardware threshold and the conditions to select muons with the different ADC values, extracting the efficiency of the cut.

3.2 Low energy calibration

A new calibration system scalable to a larger number of modules was developed. This system consists of flexible wires with ^{57}Co and ^{109}Cd sources along them. The sources are outside

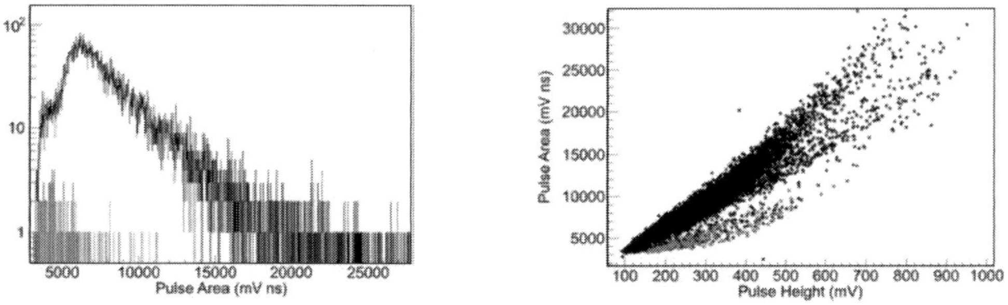

Figure 3. Spectra (left) and scatter plot (pulse height vs. pulse area, right) taken at surface (black) and underground (red) with fast events dominating underground rate.

the shielding and they are introduced inside the shielding facing the Mylar windows during calibration keeping the system closed and Radon free. This system has been tested with ANAIS-25 modules.

3.3 PMT quality tests

Forty two high quantum efficiency low background Hamamatsu R12669SEL2 PMTs have been received and are being tested. High purity germanium spectroscopy has been performed at LSC verifying the required radiopurity levels and the homogeneity among them. Operational parameters such as single electron response (SER), gain, relative quantum efficiency and dark rate are being tested at the University of Zaragoza test bench. First, a setup to expose PMTs to controlled ultraviolet LED light, data acquisition chain and algorithms to extract SER parameters have been developed and tested. The protocol to characterize all units is currently being developed. In particular, all tested units have a dark current below 500 Hz.

3.4 Front-end, DAQ, analysis software and slow control

The front-end setup and software chain have been redesigned in order to scale from ANAIS-25 to the full experiment. The front-end was modified with the introduction of a new high density NIM delay module and a new scalable design of linear preamplifiers. The DAQ software was designed to allow to configure the number of channels and the analysis software was modified to follow these changes. The full software stack has been tested with the low energy selection protocols from ANAIS-0 prototype [6] and it is being used in the following prototypes such as ANAIS-25, with configurations up to three detectors (six channels). A slow control system has also been developed and it is currently monitoring parameters such as temperatures inside the shielding and next to the electronic front-end, the external Rn activity, N_2 flux into the shielding and high voltage power supply voltage and current.

3.5 Prospects

We have evaluated the expected sensitivity of ANAIS in different background scenarios [7]: the measured background in ANAIS-25 (in blue in Fig. 4) and a model accounting for the measured contributions (PMTs, ^{40}K and ^{22}Na in the crystal, still without ^{210}Pb or ^{232}Th and ^{238}U internal contributions, which are being added to the model) and taking into account coincidence event rejection in a 5 × 4 module setup (in red). The considered exposures are 100 kg × 5 years (solid lines) and 250 × 5 years (dotted). The sensitivity shown in Fig. 4

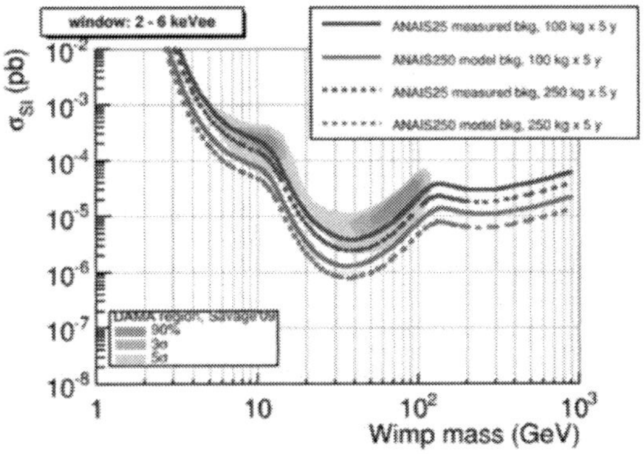

Figure 4. Sensitivity projections for 2–6 keVee energy window in four different scenarios (see text).

corresponds to a 2–6 keVee window, and even in the most pessimistic scenario allows testing a significant part of the DAMA/LIBRA singled-out parameter space regions [8]. Further sensitivity improvement should be achieved by reducing both, threshold and background, as expected.

This work has been supported by the Spanish Ministerio de Economía y Competitividad and the European Regional Development Fund (MINECO-FEDER) (FPA2011-23749), the Consolider-Ingenio 2010 Programme under grants MULTIDARK CSD2009-00064 and CPAN CSD2007-00042, and the Gobierno de Aragón (Group in Nuclear and Astroparticle Physics, ARAID Foundation and C. Cuesta predoctoral grant). C. Ginestra and P. Villar have been supported by the MINECO Subprograma de Formación de Personal Investigador. We also acknowledge LSC and GIFNA staff for their support.

References

[1] C. Cuesta et al., Nucl. Instrum. Meth. A **742** (2014) 187–190
[2] J. Amaré et al., `arXiv:1404.3564`, to appear in Physics Procedia
[3] R. Bernabei et al., Eur. Phys. J. C. **73** (2013) 2648
[4] J. Amaré et al., `arXiv:1411.0106`, submitted to JCAP
[5] C. Cuesta et al., Int. J. of Mod. Phys. A. **29** (2014) 144301
[6] C. Cuesta et al., Eur. Phys. J. C **74** (2014) 3150
[7] J. Amaré et al., `arXiv:1410.5949`, to appear in Nuclear Physics B - Proceedings Supplements
[8] C. Savage et al., JCAP **04** (2009) 010

The next generation dark matter hunter: XENON1T status and perspective

A. Rizzo[a] on behalf of the XENON Collaboration

Department of Astrophysics, Columbia University in the City of New York, USA

Abstract. The XENON Dark Matter Experiment has been ongoing at LNGS since 2005 with the goal of searching for dark matter WIMPs with liquid xenon as target and detector material. With detectors of increasing target mass and decreasing background, the XENON program has achieved competitive limits on WIMP-nucleon interaction couplings, but also on axions and axion like particles. With the start of the next generation experiment, XENON1T expected in 2015, XENON Dark Matter Experiment will continue to lead field of dark matter direct detection. XENON1T will be the first experiment to use multi-tons of liquid xenon in a time projection chamber and is designed to achieve two orders of magnitude higher sensitivity than the current best limits. I will review the status of construction and the scientific goals of XENON1T.

1. Introduction

There is an increasing number of astrophysical and astronomical observations pointing to the existence of a non-luminous, non-baryonic and cold (i.e. non-relativistic) matter component in the universe, called Cold Dark Matter (CDM) [1, 2]. The most appealing candidates for CDM are Weakly Interactive Massive Particles (WIMPs) predicted by supersymmetric theories (SUSY), models with extra dimensions and little Higgs models [3–5].

The XENON project is currently one the most promising experiments for the direct detection of dark matter. After the successful results of the first 10 and 100 kg scale liquid xenon detectors, XENON10 and XENON100 [6], the XENON project will continue to lead the field of worldwide direct dark matter experiments with its multi-ton liquid xenon detector called XENON1T, which will have an expected sensitivity to the spin-independent WIMP-nucleon cross section of 2×10^{-47} cm^2.

2. The XENON1T experiment

The construction of XENON1T started in 2013 at the Laboratori Nazionali del Gran Sasso (LNGS) in Italy, and it is expected to enter its first commissioning phase by Summer 2015. A drawing of the XENON1T facility is shown on Fig. 1.

The core of the detector (like its predecessors XENON10 and XENON100) is a dual-phase liquid xenon time projection chamber (TPC), which detection principle can be summarized as follows: if an incoming particle interacts with a nucleus in the target volume, it will produce both scintillation light, and ionization electrons. These last ones are drifted by an applied electric field towards the gas phase above the liquid, producing proportional

[a] e-mail: `ar3081@columbia.edu`

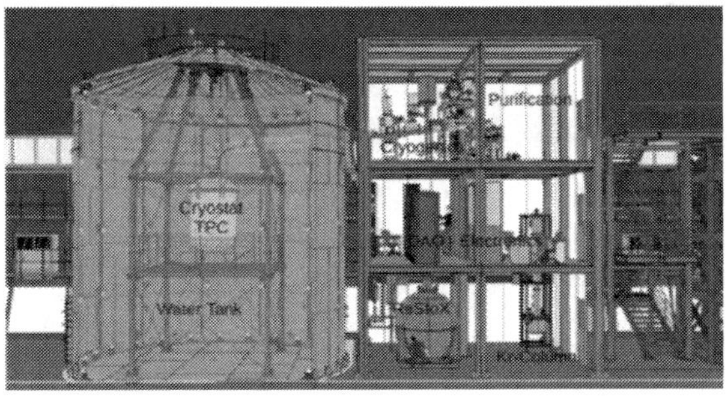

Figure 1. Sketch of the XENON1T facility. On the left the the water tank with inside the cryostat which holds the TPC; on the right the service building with the different detector subsystems: on the ground floor ReStoX and Kr-Column, on the first floor DAQ and Electronics and on the second floor Cryogenics and Purification.

light. The scintillation light (S1 signal) is promptly detected, while the proportional light (S2 signal) is delayed due to the drift time of the electrons. Being the S2 a localized signal, it allows the reconstruction of the (x,y) position of the interaction; the z position can be then derived from the time difference between the S1 and the S2 signals.

The discrimination principle between hypothetical signal (WIMP) and background particle starts from the different targets they are supposed to interact: nuclei for WIMPs and electrons that surround nuclei for most background. Nuclear and electron interactions have then different ratio of the S2 over S1 signal, thus allowing their separation via offline analysis. The study of the detector response to nuclear and electronic interactions will be achieved via calibration campaigns, with neutron and gamma sources respectively.

The XENON1T detector will contain ~3.3 tons of high-purity liquid xenon (~2 tons of which as active volume inside TPC), which provides excellent discrimination of electronic and nuclear interaction. Other remarkable property of liquid xenon is its self-shielding capability: the outer volume of xenon effectively shields the inner volume from background interactions occurring outside the TPC. This last one will be instrumented from both above and below with a total of 248 low-background photomultiplier tubes (PMT), and its volume will be kept under a homogeneous electric field $E_d > 500\,\mathrm{V/cm}$ over a drift distance of 1m.

To store and recovery the xenon a system called ReStoX will run in XENON1T, it was installed on August 2014 (see Fig. 2). It is a double-walled stainless steel sphere of 2.1m diameter capable to store xenon in different thermal conditions, even in different phases, depending on the needs, under high purity conditions. For this purpose it is well insulated and equipped with sophisticated cooling systems able to provide powerful cooling power with high response. The cooling system is based on (gaseous and liquid) nitrogen, so to make ReStoX functionalities totally independent from power failures. The stored xenon can be continuously purified thanks to a dedicated connection through a heat exchanger with the XENON1T purification system (which requires xenon in gaseous phase). As recovery system, ReStoX uses the simple principle of the communicating vases. ReStoX is kept at low pressure all the time (below a barg). In case of need of a recovery (voluntary or emergency-triggered) from the cryostat, the pressure in the cryostat (usually bigger than 2 barg) is enough to push liquid xenon through a dedicated recovery line in connection with ReStoX. Because of the siphon principle, all liquid xenon will be transferred in a very natural way once the transfer starts. Then ReStoX will be able to recover also most of the remaining gaseous xenon by subcooling the stored xenon and thanks to a direct gas-gas connection with the cryostat.

Figure 2. On the right the ReStoX stainless steel sphere to store and recovery the xenon which was put on site on August 2014. On the left it is shown its slow control system (PLC and touchscreen) enclosed in a 19-inches rack.

Figure 3. On the left the stainless steel water tank with a height of $\sim$10 m and a diameter of 9.6 m. It is equipped to work as Cherenkov detector; the implemented veto will reject the muon background. On the right the attractive service building with its 3 floors.

2.1 Expected background and sensitivity

In order to get the best sensitivity, a low-rate counting experiment like XENON1T has to reduce the sources of background to approximately zero level. One of this source is represented by cosmic muons which produce secondary neutrons as they interact near the detector, thus producing nuclear recoil background. Even if the muon flux is highly reduced at the site (LNGS stays under 3600 m.w.e.), the not much muon induced neutrons can be still nuisance in the signal region. To avoid this, the TPC and its cryostat will be immersed in 700 t of pure water contained in a stainless steel tank. The tank is equipped with an active muon veto with 84 PMTs, allowing a 99% efficiency in tagging crossing muons and 74% efficiency for showering events. The expected total background is expected to be around 0.01 event per year [7]. The water tank was completed in fall of 2013 and it is shown on the left of Fig. 3; on the right of the same picture it is shown the XENON1T service building which was completed at the beginning of 2014.

Other background from within the underground walls will be also effectively reduced by the passive shielding provided by the water tank and the outer volume of xenon.

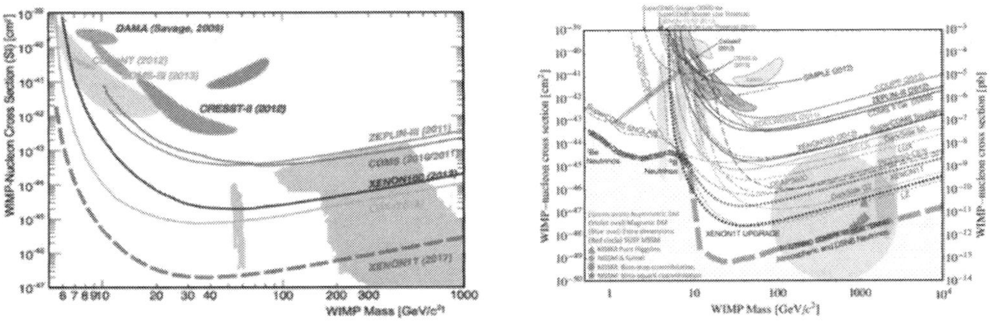

Figure 4. On the left the projected spin-independent WIMP-nucleon cross section sensitivity of XENON1T (red dashed line) after two years of operation, compared with current experimental results (adapted from [10]). On the right the projected sensitivity of the XENON1T Upgrade (blue dotted line) compared with present and future direct dark matter experiments (adapted from [11]).

The use of radio-pure raw material to build the different detector components is essential in order to reduce background contamination, and for this purpose several screening campaigns were performed and others are ongoing. The choices of the stainless steel used to build the cryostat and of the part of material which will be used for the TPC are indeed outcomes of these screening campaigns. The two arrays of low-background PMTs, 127 on top and 121 on bottom of the TPC, are 3-inches R11410-21 manufactured by Hamamatsu Photonics K.K., characterized by a high Q.E. = 32.5% @ 178 nm and a gain of a factor 3×10^6. The contamination measurements was performed with different techniques, such as high purity germanium screening, mass spectrometry and neutron activation [8, 9]. The estimate of the total background from the different material contamination results then to be about 0.25 events/year.

The xenon is commercially obtained from the fractional distillation of the air, thus containing krypton contamination at a certain level. This is then a source of intrinsic background since a factor of $\sim 2 \times 10^{-11}$ of krypton occurs as the unstable isotope ^{85}Kr, a β emitter. The only way to reduce such krypton impurities is via distillation. So, before to start the physical run all the xenon will pass once in a krypton distillation column with the aim to reduce the levels of contamination to below the ppt level (0.1×10^{-12} Kr/Xe). To filter the xenon from the electronegative impurities like O_2, N_2, and H_2O, it will keep circulating in a purification system with getters. The estimate of this intrinsic background results in about 0.15 events/year.

The solar neutrino and the double beta decay of ^{136}Xe (about 10% of natural xenon) represent the two sources of irreducible background. The contribution of both of these backgrounds is estimate to about 0.09 events/year, making the total background prediction from all the sources so far discussed to about 0.5 events/year.

The different background estimates so far discussed come from the assumption that the XENON1T analysis will have similar acceptance and background rejection with respect to the XENON100 analysis. Under this assumption, the increase in total mass (to 3 tonnes), coupled with the reduction in background levels to 6×10^{-5} events/kg/day/keVee, allows for two order of magnitude improvement in sensitivity with respect to XENON100. The analysis is based on a one ton fiducial volume, 5–50 keV search window for nuclear recoils, rejection of 99.5% of electronic recoil events, and acceptance of 50% of nuclear recoil events. The estimate sensitivity to the spin-independent WIMP-nucleon cross section is shown on the left of Fig. 4.

The XENON1T experiment has been designed to allow for the rapid deployment of a new detector with a total xenon mass of up to seven tonnes, named as XENON1T Upgrade. This new detector can be realized using the same cryogenic and purification infrastructure and will be housed in the existing XENON1T water tank (see next section for details).

3. XENON1T upgrade

Once completed in 2015, XENON1T will be proved to be the most sensitive worldwide dark matter experiments. The XENON1T will reach its design sensitivity after two years of stable operation.

The XENON1T Upgrade project will contain about 7 tons of xenon with an achievable goal of 20 ton-years of exposure. This will allow a sensitivity down to 3×10^{-48} cm^2 in the spin-independent WIMP-nucleon cross section.

The construction of the detector upgrade is foreseen to start as early as 2018, and will run in parallel with the operation of XENON1T. Most of the detector installation have been conceived to be reused to host a larger volume TPC without modifications: only a new inner vessel of the cryostat, which will fit with the existing outer cryostat, and the TPC itself will be needed. Signal and high voltage cables for the new PMTs have already been installed within the cable pipes connecting the detector to the outside electronics and the existing DAQ, slow control, and computing infrastructure will be reused, as well as the cryogenic and purification systems for the operation of the new detector. However the main cost of the experiment will be procurement of additional xenon and PMTs.

The projected sensitivity of the XENON1T Upgrade is shown on the right of Fig. 4.

4. Conclusion

We have presented the status of the XENON1T experiment which is under construction at the Laboratori Nazionali del Gran Sasso in Italy. It is currently under commissioning and the first results are foreseen for later 2015. With XENON1T and the planned upgrade, the XENON Collaboration plans to remain at the forefront of the exciting and challenging field of dark matter direct detection.

References

[1] W. Freeman and M. Turner, Rev. Mod. Phys. **75**, 1433 (2003)
[2] M. J. Jee et al., `http://xxx.lanl.gov/abs/0705.2171`
[3] A. Bottino et al., Phys. Rev. D **69**, 037302 (2004)
[4] J. Ellis et al., Phys. Rev. D **71**, 095007 (2005)
[5] A.Birkedal-Hansen and J. G. Wacker, Phys. Rev. G **69**, 065022 (2004)
[6] S. E. A. Orrigo (on behalf of the XENON Collaboration), This Proceeding (2015)
[7] E. Aprile et al., JINST **9**, P11006 (2014)
[8] L. Baudis et al., JINST **8**, P04026 (2013)
[9] K. Lung et al., Nucl. Inst. and Meth. **A 696**, 32-39 (2012)
[10] E. Aprile et al, Phys. Rev. Lett. **109**, 181301 (2012)
[11] P. Cushman et al., arXiv:1310.8327v2 (2013)

EPJ Web of Conferences **121**, 06010 (2016) DOI: 10.1051/epjconf/201612106010
RICAP-2014

The DarkSide Program

B. Rossi[15,6,a], P. Agnes [2], T. Alexander[3], A. Alton[4], K. Arisaka[5], H.O. Back[6], B. Baldin[7], K. Biery [7], G. Bonfini[8], M. Bossa[9], A. Brigatti[10], J. Brodsk y[6], F. Budano[11], F. Calaprice[6], N. Canci[5], A. Candela[8], M. Cariello[12], P. Cavalcante[8], S. Catalanotti[15], A. Chavarria[13], A. Chepurnov[14], A. G. Cocco[15], G. Covone[15], D. D'Angelo[10], M. D'Incecco[8], M. De Deo[8], A. Derbin[16], A. Devoto[17], F. Di Eusanio[6], E. Edkins[18], A. Empl[1], A. Fan[5], G. Fiorillo[15], K. Fomenko[19], D. Franco[2], F. Gabriele[8], C. Galbiati[6], A. Goretti[6], L. Grandi[13], M.Y. Guan[20], Y. Guardincerri[7], B. Hackett[18], K. Herner[7], E.V. Hungerford[1], Al. Ianni[8], An. Ianni[6], C. Kendziora[7], G. Koh[6], D. Korablev [19], G. Korga[1], A. Kurlej[3], P.X. Li[20], P. Lombardi[10], S. Luitz[21], I. Machulin[22], A. Mandarano[9], S. Mari[11], J. Maricic[18], L. Marini[11], C. J. Martoff[23], P.D. Meyers[6], D. Montanari[7], M. Montuschi[8], M.E. Monzani[21], P. Musico[12], S. Odrowski[8], M. Orsini[8], F. Ortica[24], L. Pagani[12], M. Pallavicini[12], E. Pantic[5], L. Papp[25], S. Parmeggiano[10], N. Pelliccia[24], S. Perasso[2], A. Pocar[3], S. Pordes[7], H. Qian[6], K. Randle[3], G. Ranucci[10], A. Razeto[8], B. Reinhold[18], A. Renshaw[5], A. Romani[24], N. Rossi[8], S.D. Rountree[25], D. Sablone[1], R. Saldanha[13], W. Sands[6], E. Segreto[8], E. Shields [6], O. Smirnov[19], A. Sotnikov[19], C. Stanford[6], Y. Suvorov[5], R. Tartaglia[8], J. Tatarowicz[23], G. Testera[12], A. Tonazzo[2], E. Unzhakov[16], R.B. Vogelaar[25], M. Wada[6], S. Walker[15], H. Wang[5], A. Watson[23], S. Westerdale[6], M. Wojcik[26], X. Xiang [6], J. Xu[6], C.G. Yang[20], J. Yoo[7], S. Zavatarelli[12], A. Zec[3], C. Zhu[6], and G. Zuzel[26]

[1] Department of Physics, University of Houston, Houston, TX 77204, USA
[2] APC, Université Paris Diderot, Sorbonne Paris Cité, Paris 75205, France
[3] Physics Department, University of Massachusetts, Amherst, MA 01003, USA
[4] Physics and Astronomy Department, Augustana College, Sioux Falls, SD 57197, USA
[5] Physics and Astronomy Department, University of California, Los Angeles, CA 90095, USA
[6] Physics Department, Princeton University, Princeton, NJ 08544, USA
[7] Fermi National Accelerator Laboratory, Batavia, IL 60510, USA
[8] Laboratori Nazionali del Gran Sasso, Assergi (AQ) 67010, Italy
[9] Gran Sasso Science Institute, L'Aquila 67100, Italy
[10] Physics Department, Università degli Studi and INFN, Milano 20133, Italy
[11] Physics Department, Università degli Studi Roma Tre and INFN, Roma 00146, Italy
[12] Physics Department, Università degli Studi and INFN, Genova 16146, Italy
[13] Kavli Institute, Enrico Fermi Institute and Dept. of Physics, University of Chicago, Chicago, IL 60637, USA
[14] Skobeltsyn Institute of Nuclear Physics, Lomonosov Moscow State University, Moscow 119991, Russia
[15] Physics Department, Università degli Studi Federico II and INFN, Napoli 80126, Italy
[16] Saint Petersburg Nuclear Physics Institute, Gatchina 188350, Russia
[17] Physics Department, Università degli Studi and INFN, Cagliari 09042, Italy
[18] Department of Physics and Astronomy, University of Hawai'i, Honolulu, HI 96822, USA
[19] Joint Institute for Nuclear Research, Dubna 141980, Russia
[20] Institute of High Energy Physics, Beijing 100049, China
[21] SLAC National Accelerator Laboratory, Menlo Park, CA 94025, USA

[a] e-mail: rossib@princeton.edu

[22] National Research Nuclear University Moscow Engineering Physics Institute, 115409 Moscow, Russia
[23] Physics Department, Temple University, Philadelphia, PA 19122, USA
[24] Chemistry, Biology and Biotechnology Department, Università degli Studi and INFN, Perugia 06123, Italy
[25] Physics Department, Virginia Tech, Blacksburg, VA 24061, USA
[26] Smoluchowski Institute of Physics, Jagiellonian University, Krakow 30059, Poland

Abstract. DarkSide-50 at Gran Sasso underground laboratory (LNGS), Italy, is a direct dark matter search experiment based on a liquid argon TPC. DS-50 has completed its first dark matter run using atmospheric argon as target. The detector performances and the results of the first physics run are presented in this proceeding.

1. Introduction

The majority of the matter and of the energy content of the universe seems to be dark. Cosmological and astronomical observations supported by the very recent results from the Plank satellite indicate that: the dark energy, responsible of the observed accelerated expansion of the cosmos, accounts for the 68% of the total budget; the dark matter, a yet-undetected form of matter which forms the observed large-scale structures and galaxies, accounts for the 27%, while the remaining 5% is composed by "ordinary" baryonic matter. Many theories beyond the standard model of particle physics predict possible candidates for dark matter particles and among those the Weakly Interacting Massive Particles (WIMPs). Although the strength of the interaction is supposed to be very small, the thermal motion of these WIMPs comprising the dark matter halo surrounding the galaxy and the earth is expected to result in WIMP-nuclear collisions of sufficient energy to be observable by very sensitive laboratory apparatus. The fact that dark matter exists is one of the strongest indications for physics beyond the standard model of particle physics. Its direct detection and characterization is a major experimental challenge of modern particle and astroparticle physics. The DarkSide project attempts to detect WIMP-induced nuclear recoils using two-phase Argon time projection chambers (TPCs) with scalable, zero-background technology. The largest challenge in searching for dark matter is the suppression of the rate of background events to below the very low WIMP interaction rates (a few events per ton-year) to which current dark matter experiments are sensitive. The prompt scintillation light (S1) and the delayed proportional light signal from the charge (S2) are simultaneously measured in two-phase TPCs and are used for a precise reconstruction of the event vertex and its multiplicity. Both signals are employed to suppress the backgrounds by target fiducialization and by rejection of multiple-scatter interactions (WIMPs are expected to only interact once). The charge-to-light ratio, S2/S1, is exploited to separate expected signal events, which are nuclear recoils (NR) from the dominant electronic recoil (ER) background and it allows discrimination up to a factor 10^2. One reason that makes Argon a promising medium for dark matter searches is that it has a unique powerful feature, the pulse shape discrimination (PSD). Particles interacting in noble liquids induce excitation and ionization of the medium, leading to the emission of scintillation light whose time structure is strongly correlated with the nature of the interaction. This time dependence of the scintillation signal provides an additional way to detect rare nuclear recoil events, possibly induced by WIMPs. With sufficient photon statistics, PSD can allow discrimination of nuclear recoil events from electron-induced background events at better than 10^8 [1, 2].

The ultimate goal of DarkSide-50 is to conduct a background-free dark matter search with its 50-kg TPC filled with argon derived from underground sources (UAr) [3, 4], to reduce the

Figure 1. The detector system of DarkSide-50. The outermost dark gray cylinder is the WCD, the sphere is the LSV, and the gray cylinder at the centre of the sphere is the LAr TPC.

rate of ^{39}Ar decays in the active volume. The present exposure amounts to (1422±67)kg d using an initial fill of atmospheric argon, obtained while the final purification of the UAr supply was still in progress. Atmospheric argon contains approximately 1 Bq/kg of cosmogenic 39Ar [5–7].

2. The DarkSide-50 detector

Figure 1 shows the experimental setup of the DarkSide-50 experiment composed by three detectors: the Liquid Argon Time Projection Chamber (LAr TPC) which is the dark matter detector, the organic Liquid Scintillator Veto (LSV), serving as shielding and as anti-coincidence for radiogenic and cosmogenic neutrons, γ-rays and cosmic muons; and the Water Cherenkov Detector (WCD), serving as a shield and as anti-coincidence for cosmic muons. The detector system is located in Hall C of LNGS at a depth of 3800 meter-water-equivalent.

The WCD is an 11 m-diameter, 10 m-high cylindrical tank filled with high purity water. The inside surface of the tank is covered with a laminated Tyvek-polyethylene-Tyvek reflector. An array of 80 ETL 9351 8″ PMTs, with 27% average quantum efficiency at 420 nm, is mounted on the side and bottom of the water tank to detect Cherenkov photons produced by muons or other relativistic particles traversing the water.

The LSV is a 4 m-diameter stainless steel sphere filled with 30 tons of borated liquid scintillator. The scintillator consists of equal amounts of pseudocumene (PC) and trimethyl borate (TMB), with the wavelength shifter Diphenyloxazole (PPO) at a concentration of 3 g/L. The sphere is lined with Lumirror reflecting foils. An array of 110 Hamamatsu R5912 8″ PMTs, with low-radioactivity glass bulbs and high-quantum-efficiency photocathodes (37% average QE at 408 nm) is mounted on the inside surface of the sphere to detect scintillation photons. The neutron-capture reaction ^{10}B(n, α)^{7}Li makes the borated scintillator a very effective veto of neutron background [8]. A detailed description of the LAr TPC can be found in [9]. Additional information on the detectors employed in the DarkSide project can be found in [9–13].

3. Data acquisition and event reconstruction

The data acquisition system of DS-50 consists of two main sub-systems, handling the TPC and the vetoes respectively. The ground-referenced anodic signal from each of the thirty-eight TPC photomultiplier tube is first amplified by a cryogenic head-amplifier located in

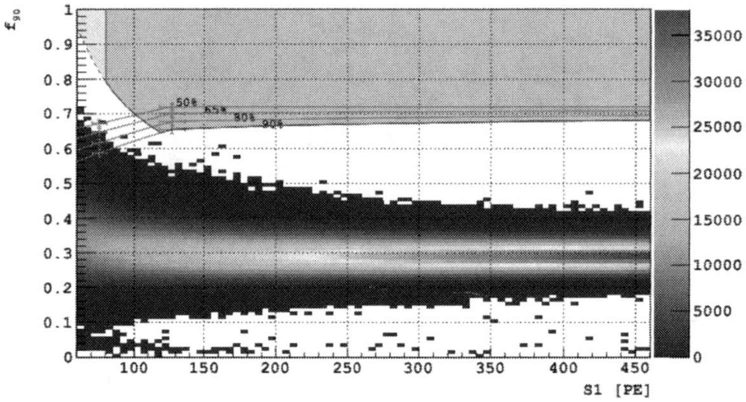

Figure 2. Distribution of the events in the scatter plot of S1 vs. f_{90} after all quality and physics cuts. Shaded Blue: dark matter search box in the f_{90} vs. S1 plane. Percentages label the f_{90} acceptance contours for nuclear recoils drawn connecting points (shown with error bars) determined from the corresponding SCENE measurements [14]. This picture has been taken from the DS-50 paper [9].

liquid argon on the phototube base divider. This first stage allows us to operate the PMTs at lower gain (typically 4×10^5), reducing the occurrence of flashers. The signal undergoes then a second stage that duplicates it into multiple branches. Ultimately, the signal is sent to a set of 12 bit, 250 MSamples/second, digitizers (CAEN 1720). In a similar way, the anodic signals from LSV and WCD phototubes undergo amplification and duplication by means of a custom front-end board. A $\times 10$ amplified signal for each PMT is digitised with NI PXIe-5162 National Instruments modules, by sampling them at 1.25 GSamples/second with 10 bit resolution.

An offline reconstruction code is used to analyze the stored waveforms from the TPC and the vetoes photomultiplier tubes. As far as the TPC is concerned, signals from each channel scaled by the corresponding single photoelectron mean are added to form a sum waveform that is used for identifying the pulses in the trigger gate and their start time. Due to the use of DAQ-level zero-suppression, reconstruction of LSV and WCD signals is different from the TPC reconstruction. Pulses are naturally defined as the non-zero portion of each raw waveform for each channel. The DAQ records 20 ns before and after the waveform rises above and drops below, respectively, the zero-suppression threshold. Each channel is then scaled by the corresponding single photoelectron mean and the channels in each veto detector are summed together. Additional information on the data acquisition, event reconstruction and detector calibrations can be found in [9, 13].

4. Data analysis and WIMP search

A dark matter search has been performed with data collected with atmospheric argon (AAr) target. This initial AAr data set contains a sample of ^{39}Ar equal to that expected in a campaign of 23 years using UAr at the upper limit of ^{39}Ar activity. The initial dark matter search consists of the data set acquired between November 2013 and May, 2014. We excluded from the data set any runs where one of more of the detectors where not running and where the DAQ, still in a developmental phase, showed signs of instability. We applied a set of data quality cuts, both for the LAr TPC and the vetoes, to exclude backgrounds and misconstructed events. All the details can be found in [9]. We performed a non-blind physics analysis on the 53.4 days of WIMP search data. The fiducial mass considered in this analysis (39.6 ± 0.6) kg. The accurate description of all cuts, their order of application, their effect on livetime, acceptance and fiducial volume, and their systematics and statistical uncertainties can be found in [9]. The total exposure remaining after all cuts prior to the WIMP search box is (1423 ± 67) kg d. The distribution of the remaining events in the scatter plot of f_{90} vs. S1 after all quality and physics cuts is shown in Fig. 2. There are (15×10^6) events in this plot, dominated by ^{39}Ar decays. Nuclear recoil acceptance curves in the f_{90} vs. S1 plane were derived from SCENE [14] data and translated to DS-50. f_{90} median values from SCENE, linearly interpolated and assumed to be constant above the highest SCENE NR energy, are translated from the SCENE nuclear recoil energies to DS-50 S1 values. This gives the 50% contour for DS-50. The other contours and associated errors depend also on the width of the DS-50 f_{90} distributions at each S1, which is established using the same analytical model in use to described the f_{90} spread for ^{39}Ar events described above. These curves are shown in Fig. 2. We observe 4 events passing all TPC cuts and with nuclear-recoil-like f_{90}, but with energy depositions in the LSV above our veto cut threshold. In coincidence with one of these 4 neutron candidates, we recorded signals near saturation in both the LSV and the WCD, and therefore we classify that event as a cosmogenic neutron, leaving 3 radiogenic neutron candidates. This number is in agreement with the neutron-induced events passing all TPC cuts expected from radiogenic neutrons from the PMTs based on Monte Carlo studies.

5. Conclusions

The DarkSide-50 direct dark matter detection system has completed his first physics data taking at LNGS. An innovative closed-loop argon circulation system with external purification and cooling allows the LAr TPC to achieve an electron drift lifetime of > 5 ms. Photoelectron yield of (7.9±0.4) PE/keV at null field is achieved for detection of the primary argon scintillation, giving the photoelectron statistics necessary for high performance pulse shape discrimination. Figure 2 covers the range of energies from 8.6 to 65.6 keV for ^{39}Ar, and a total of 15×10^6 ^{39}Ar β-decays were recorded over that energy range. Event selection based on the TPC cuts is shown to completely suppress ^{39}Ar background events in the present (1423 ± 67) kg d exposure. This exposure contains at least as many ^{39}Ar decays as 215000 kg d of running with underground argon, proving that DS-50 could run for two decades with underground argon and be free of ^{39}Ar background. A WIMP search with the present data-set gives a limit as low as 6.1×10^{-44} cm^2, the best result achieved to date with an argon target [9]. The DarkSide program is supported by the NSF and the DOE and by INFN.

References

[1] P. Benetti et al. (WARP Collaboration), Astopar. Phys. **28**, 495 (2008)
[2] M.G. Boulay and A. Hime, Astropart. Phys. **25**, 179 (2006)
[3] H.O. Back et al., arxiv:1204.6024
[4] H.O. Back et al., arxiv:1204.6061
[5] H.H. Loosli, Earth Plan. Sci. Lett. **63**, 51 (1983)
[6] P. Benetti et al. (WARP Collaboration), Nucl. Inst. Meth. A **574**, 83 (2007)
[7] J. Xu et al., arxiv:1204.6011
[8] A. Wright et al., Nucl. Instr. Meth. A **644**, 18 (2011)
[9] T. Alexander et al. (DarkSide Collaboration), arXiv:1410.0653
[10] S. Davini, Nucl. Instrum. Meth. A **742** (2014) 183

[11] T. Alexander et al. (DarkSide Collaboration), JINST **8**, C11021 (2013)
[12] C. Aalseth et al., The DarkSide multi-ton detector for the direct dark matter search, Advances in High Energy Physics, Accepted June 2014 (`http://www.hindawi.com/4journals/ahep/raa/541362`)
[13] P. Agnes et al. (DarkSide Collaboration), arXiv:1412.2969
[14] T. Alexander et al. (SCENE Collaboration), arxiv:1406.4825

EPJ Web of Conferences **121**, 06011 (2016) DOI: 10.1051/epjconf/201612106011
RICAP-2014

Quark nuggets search using gravitational waves aluminum bar detectors

Francesco Ronga[a] (ROG Collaboration)[b]

INFN Laboratori Nazionali di Frascati, Frascati, Italy

Abstract. Up to now there is no evidence of supersymmetric WIMPS dark matter. This may suggests to look for more exotic possibilities, for example compact ultra-dense quark nuggets. Nuclearites are an example of compact objects that could be constituent of the dark matter. After a short discussion on nuclearites, the result of a nuclearite search with the gravitational wave bar detectors NAUTILUS and EXPLORER is reported.

1. Introduction

During the last decade a very large experimental and theoretical effort has been devoted to understand the problem of dark matter (DM). Composite objects consisting of light quarks in a color super-conducting phase have been suggested as super-heavy quark nuggets DM; in addition, super-heavy DM anti-quark nuggets could exist and could perhaps solve the matter-antimatter asymmetry [1]; the detection of such anti-quark nuggets by cosmic ray experiments is discussed in [2]. The energy loss predicted for super-heavy DM particles varies in different models, but it is likely that such particles could be confused with meteors, since the velocity, 270 km/sec, is in the high-end tail of the meteor velocity distribution. Here, we will focus our attention only on one possible kind of very massive particle called "nuclearite". It consists of neutral matter including a strange quark among its constituents. We make this choice because nuclearites are an example of particles already searched for by other experiments.

Nuggets of Strange Quark Matter (SQM), composed of approximately the same numbers of up, down and strange quarks could be the true ground state of quantum chromodynamics [3].

According to [4] nuclearites are considered to be large strange quark nuggets, with overall neutrality ensured by an electron cloud which surrounds the nuclearite core, forming a sort of atom. Nuclearites with galactic velocities are protected by their surrounding electrons against coulomb interactions with the atoms they might hit.

As a consequence, the principal energy-loss mechanism for a nuclearite passing through matter is atomic collision. For a massive nuclearite the energy-loss rate is:

$$\frac{dE}{dx} = -A\rho v^2 \qquad (1)$$

[a] e-mail: francesco.ronga@lnf.infn.it

[b] ROG ollaboration: M. Bassan, E. Coccia, S. D'Antonio, V. Fafone, G. Giordano, A. Marini, Y. Minenkov, I. Modena, G.V. Pallottino, G. Pizzella, A. Rocchi, F. Ronga, M. Visco.

where ρ is the density of the traversed medium, v the nuclearite velocity and A is its effective cross-sectional area. The effective area can be obtained by the nuclearite density ρ_N. For a small nuclearite of mass less than 1.5 ng, the cross-section area A is controlled by its electronic atmosphere which is never smaller than 10^{-8} cm: $A = \pi 10^{-16}$ for $M < 1.5$ ng otherwise $A = \pi(3M/(4\pi\rho_N))^{2/3}$ where $\rho_N = 3.5 \cdot 10^{14}$ g/cm^3 is the nuclearite density and M its mass. According to Eq. (1), nuclearites having galactic velocity and mass heavier than 10^{-14} g penetrate the atmosphere, while those heavier than 0.1 g pass freely though an Earth diameter. Equation (1) has been used by [4] to compute the amount of visible light emitted in the atmosphere, assuming that the light is emitted as a black-body radiation from an expanding cylindrical thermal shock wave and to compute therefore the apparent magnitude as defined for meteors. It turns out that there are two important differences that can help to discriminate between nuclearites and meteors. The first one is that the amount of light emitted by nuclearites is proportional to the atmospheric density and therefore most of the light is emitted near the ground level; the second difference is that a nuclearite of mass bigger than 0.1 g can move upward and this is extremely unlikely for a meteor.

The nuclearite search could be an interesting by-product of cosmic ray space experiments looking from space to the light produced in the atmosphere. The possibility to study meteors and to search nuclearites and other quark nuggets in the planned JEM-EUSO mission has been studied in Ref. [5].

Nuclearites and similar particles have been searched using different approaches. Many techniques have been used to detect nuclearites: acoustic emission due to the thermal shock in aluminum gravitational wave cylindrical detectors, damages in plastic materials like CR39, Makrofol or Lexan, light emission in oil or sea water, seismic waves induced by big nuclearites. Due to the uncertainties in the conversion of the energy losses in a measurable signal it is important to have different techniques to detect such exotic particles. Figure 1 shows upper limits of a representative experiment of each technique. It is not aimed at being a full list of the experiments done so far to search nuclearites, but it is a reasonable summary of the state of the art in this field. The lower line represents the JEM-EUSO possible upper limit in 24 h.

2. The NAUTILUS and EXPLORER gravitational wave bar detectors used as particle detector

The gravitational wave (gw) detector NAUTILUS [10] is located in Frascati (Italy) National Laboratories of INFN. NAUTILUS started operations around 1998. The current run started in 2003. The detector EXPLORER, similar to NAUTILUS, was located in CERN (Geneva-CH). The EXPLORER run ended in June 2010.

Both detectors use the same principles of operation. EXPLORER and NAUTILUS consist of a large aluminum alloy cylinder (3 m long, 0.6 m diameter) suspended in vacuum by a cable around its central section to reduce the seismic and acoustic noise and cooled to about 2 K by means of a superfluid helium bath. The (gw) excites the odd longitudinal modes of the cylindrical bar. To record the vibrations of the bar first longitudinal mode, an auxiliary mechanical resonator tuned to the same frequency is bolted on one bar end face. This resonator is part of a capacitive electro-mechanical transducer that produces an electrical a.c. current that is proportional to the displacement between the secondary resonator and the bar end face. Such current is then amplified by means of a dcSQUID superconductive device and recorded on disk with an ADC sampled at 5 kHz.

Both detectors are equipped with cosmic ray telescopes, to veto excitations due to large showers. The two telescopes rely on different technologies (scintillators for Explorer,

 DOI: 10.1051/epjconf/201612106011

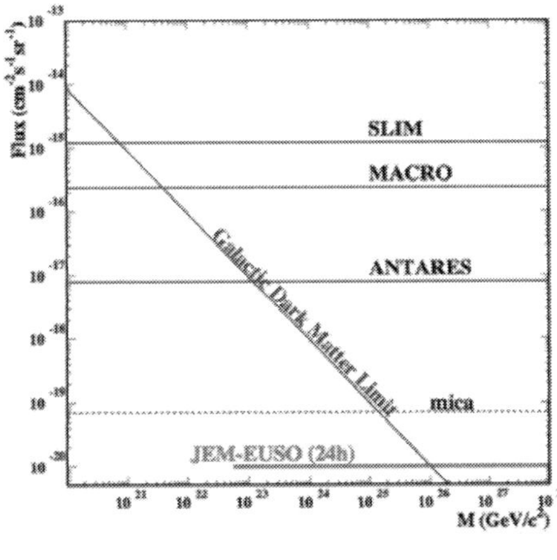

Figure 1. The 90% confidence level upper limit on the flux of nuclearites of experiments with different techniques [7–9, 12]. The "old" mica limits [12] are dependent from several additional assumptions, respect to the other experiments. The lower line shows the possible limit in 24h of the JEM-EUSO cosmic ray experiment planned for the international space station [5]. Note that most of the proposed quark nuggets have energy losses bigger than the neutral nuclearites. Therefore limits applies to many other candidates.

streamer tubes for NAUTILUS) but both provide a monitor of comparable performances and a continuous check of the antenna sensitivity. The data are processed off-line, applying adaptive frequency domain filters optimized for delta-like signals. The noise characteristics estimate is updated averaging the output over 10 minutes periods. Traditionally the noise is expressed as energy in Kelvin units.

The interaction of energetic charged particles with a normal mode of an extended elastic cylinder has been extensively studied over the years, both on the theoretical and on the experimental side [11].

According the so called "thermo acoustic model" a particle crossing the bar produces longitudinal vibrations originated from the local thermal expansion due to the warming up due to the energy loss. In particular, the vibration amplitude is directly proportional to the ratio of two thermo-physical parameters of the material, namely the thermal expansion coefficient and the specific heat at constant volume. The ratio of these two quantities appears in the definition of the Grüneisen parameter γ.

The vibrational energy E of the fundamental vibrational mode due the energy loss of a particle crossing an aluminum cylindrical bar is [11]:

$$E = \frac{4}{9\pi} \frac{\gamma^2}{\rho L v^2} \left(\frac{dW}{dx} \right)^2 \times \left[sin\left(\frac{\pi z_o}{L} \right) \frac{sin[(\pi l_o cos(\theta_o)/2L)]}{\pi R cos(\theta_o)/L} \right]^2 \quad (2)$$

where L is the bar length, R the bar radius, l_o the length of the particle track inside the bar, z_o the distance of the track mid point from one end of the bar, θ_o the angle between the particle track and the axis of the bar, $\frac{dW}{dx}$ the energy loss of the particle in the bar, ρ the density, v the longitudinal sound velocity in the material. This relation is valid for the normal-conducting (n) state material and some authors (see the references in [11]) have extended the model to a super-conducting (s) resonator, according to a scenario in which the vibration amplitude is due to two pressure sources, one due to $s - n$ transitions in small regions centered around the interacting particle tracks and the other due to thermal effects in these regions now in the n state.

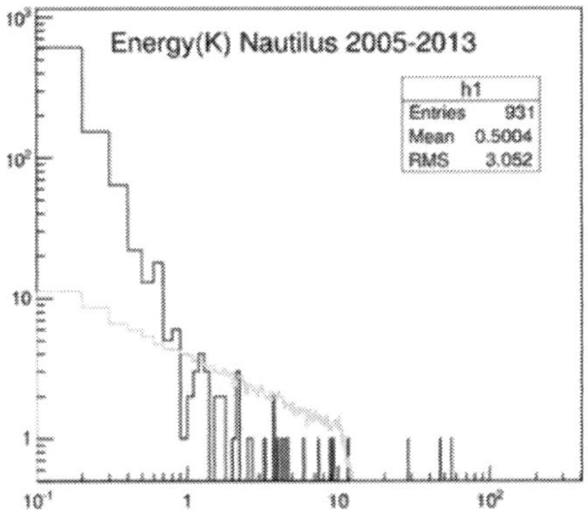

Figure 2. Vibrational energy distribution of the NAUTILUS events, after the cuts discussed in the text. The lower plot (in green) is the result of a Montecarlo simulation for $\beta\theta(M) = 0.004$ and is normalized to data.

It is important to note, at this point, that a *gw* bar antenna, used as particle detector, has characteristics very different from the usual particle detectors which are sensitive to ionization losses: indeed an acoustic resonator can be seen as a zero threshold calorimeter, sensitive to a vast range of energy loss processes. The usual *gw* software filter works well up to a time scale of the order of the order of 5 msec, corresponding to a $\beta = 4 \times 10^{-6}$ for a 60 cm particle track. So the antenna is sensitive to very slow tracks: this is another very important difference with to the usual particle detectors.

3. Nuclearite search in NAUTILUS and EXPLORER

The vibrational produced by a nuclearite crossing a bar can be evaluated inserting Eq. (1) in Eq. (2). Considering a nuclearite of mass M and velocity $c\beta$, intersecting orthogonally the center of the NAUTILUS (or EXPLORER) bar we have:

$$\Delta E[Kelvin] = 10.7 \left(\frac{\beta\theta(M)}{10^{-3}}\right)^4 \tag{3}$$

where the factor 10.7 is computed using the thermo acoustic parameters measured at T = 2K [11], ΔE is the energy variation of the bar fundamental mode measured in Kelvin and $\theta(M) = (M/1.5\,ng)^{1/3}$ if M> 1.5 ng. Otherwise $\theta(M) = 1$. The maximum geometrical acceptance for a nuclearite isotropic distribution is given by $2\pi S_{tot} = 19.54\,\mathrm{m^2 sr^{-1}}$, where S_{tot} is the bar surface.

In order to select clean events we have applied several cuts to the data. The most important cuts are based on: the gain of the electronic chain, the SQUID locking working point, the seismic monitors,the event shape. We have removed also periods with an high event rates and runs having live-time less than 10h. This procedure has been applied at the moment only for the NAUTILUS data in the years 2005–2013. The noise cut for NAUTILUS is $T_{eff} < 2.5$ mK. The total live-time for this data set is 1846 days. The 2003 and 2004 NAUTILUS data are not included because of the high noise. The efficiency of those cuts is verified continuously using the extensive air showers detected by the cosmic ray detector. The energy distribution of the 931 NAUTILUS events surviving the cuts and having E > 0.1 K is shown in Fig. 2, together with the result of a Montecarlo simulation for $\beta\theta(M) = 0.004$.

For EXPLORER the selection of clean events is in progress and at the moment only a fraction of the previous cuts is applied. The EXPLORER data are used in combination to the

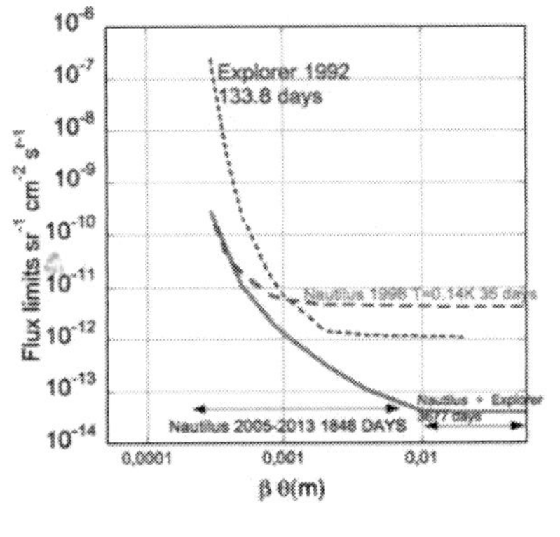

Figure 3. 90% C.L. isotropic flux upper limits compared to previous results (Explorer) [6]. The bottom curve, in red, shows the results of this analysis. The EXPLORER data are used in combination to the NAUTILUS data only for $\beta\theta(M) > 0.01$. Limits from a short run with NAUTILUS at $T = 0.14$ K, in green, may be interesting because of the different detection mechanism in the superconducting state. The limits are computed using the so called optimum interval method [13], for finding an upper limit of a one-dimensionally distributed signal in the presence of an unknown background. The signal is assumed to have the distribution computed by Montecarlo using Eqs. (2) and (3).

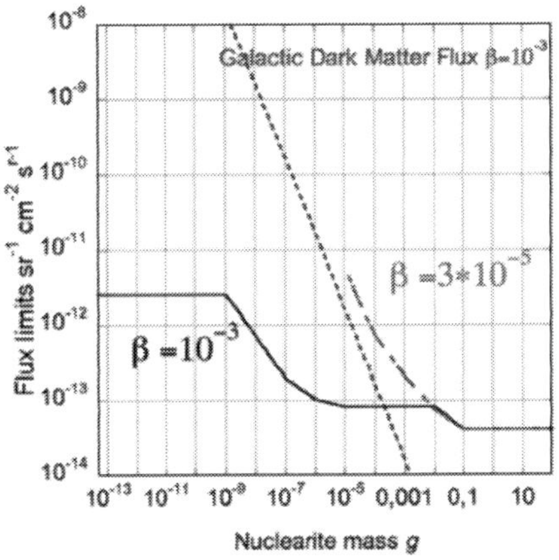

Figure 4. Flux upper limits for $\beta = 10^{-3}$ and $\beta = 3 \cdot 10^{-5}$ (earth escape velocity) vs mass. The limits are derived from Fig. 3, computing the appropriate $\beta\theta(M)$. For some combination of masses and β nuclearites cannot cross the earth, in this case a factor 2 is applied to the limits of Fig. 3. For mass between $5 \cdot 10^{-14}$ g (threshold due to the atmosphere) and 10^{-4}g this limit is significantly smaller than the flux of galactic dark matter for $\beta = 10^{-3}$.

NAUTILUS data only for $\beta\theta(M) > 0.01$. The noise cut for EXPLORER is $T_{eff} < 5$ mK and the live time of the years 2003–2010 is 1831.4 days.

The nuclearite flux upper limits have been computed starting from the energy distributions and is shown in Fig. 3. This figure also shows the limits from a short run with NAUTILUS at $T = 0.14$ K, with live-time = 35.1 days. This result may be interesting because of the different detection mechanism in the superconducting state. For $\beta\theta(M) > 0.01$, where the background is negligible, the flux upper limit is dominated only by the live-time. Note that in this search the events in coincidence with the cosmic ray detector are removed from the NAUTILUS data. Fast nuclearites could produce light in the EXPLORER scintillators (due to black body emission [4]) and could be confused with a cosmic ray event. Therefore cosmic rays are not removed from the EXPLORER data. Finally Fig. 4 shows the upper limits vs the nuclearite mass. For mass between $5 \cdot 10^{-14}$ g (threshold due to the atmosphere) and 10^{-4} g this limit is significantly smaller than the flux of galactic dark matter for $\beta = 10^{-3}$.

As shown in Fig. 1 other experiments above sea level using track etch detectors have obtained lower limits, but the detection mechanism is more complicated than the "calibrated" calorimetric technique used in this search.

References

[1] Gorham, P., W., Phys. Rev. D **86** (2012) 123005/1-8
[2] Lawson K., Phys. Rev. D **83** (2011) 103520/1-9
[3] Witten E., Phys. Rev. D **30** (1984) 272-285
[4] De Rujula, A. and Glashow, S.L., Nature **312** (1984) 734-737
[5] M. Bertaina, A. Cellino and F. Ronga Exp Astron. (2014) doi: 10.1007/ s10686-014-9375
[6] Astone P., et al., Phys. Rev. D **47** (1993) 4770-4773
[7] Cecchini, S., et al. (SLIM Coll.), Eur. Phys. J. C **57** (2008) 525-533
[8] Ambrosio, M., et al. (MACRO Coll.), Eur. Phys. J. C **13** (2000) 453-458
[9] Pavalas, G. E., et al., (ANTARES Coll.), Proc. 23rd ECRS, Moscow, 543 (2012)
[10] P. Astone et al., Astropart. Phys. **7** (1997) 231
[11] M. Bassan et al., Nucl. Instrum. Meth. A **659** (2011) 289
[12] Price, P. B., Phys. Rev. D **38** (1988) 3813-3814
[13] S. Yellin, arXiv:0709.2701 [physics.data-an]

AUTHOR INDEX

AUTHOR INDEX

AUTHOR INDEX

AUTHOR INDEX

AUTHOR INDEX